Intelligent Systems Reference Library

Volume 102

About this Series

The aim of this series is to publish a Reference Library, including novel advances and developments in all aspects of Intelligent Systems in an easily accessible and well structured form. The series includes reference works, handbooks, compendia, textbooks, well-structured monographs, dictionaries, and encyclopedias. It contains well integrated knowledge and current information in the field of Intelligent Systems. The series covers the theory, applications, and design methods of Intelligent Systems. Virtually all disciplines such as engineering, computer science, avionics, business, e-commerce, environment, healthcare, physics and life science are included.

More information about this series at http://www.springer.com/series/8578

James F. Peters

Computational Proximity

Excursions in the Topology of Digital Images

James F. Peters
University of Manitoba
Winnipeg, MB
Canada

ISSN 1868-4394 ISSN 1868-4408 (electronic)
Intelligent Systems Reference Library
ISBN 978-3-319-30260-7 ISBN 978-3-319-30262-1 (eBook)
DOI 10.1007/978-3-319-30262-1

Library of Congress Control Number: 2016933237

Printed on acid-free paper

This Springer imprint is published by Springer Nature
The registered company is Springer International Publishing AG Switzerland

This book is dedicated to
Somashekhar Naimpally, 1931–2014 and
two Annas, Amma, and Sheela

Preface

This book introduces computational proximity. Basically, **computational proximity** (CP) is an algorithmic approach to finding nonempty sets of points that are either close to each other or far apart. The basic notion of computational proximity draws its inspiration from the Preface written in 2009 by S.A. Naimpally in [1, pp. 23–28] and the Foreword in [2].

In CP, two types of near sets are considered, namely, spatially near sets and descriptively near sets. Spatially near sets contain points identified by their location and have at least one point in common. Descriptively near sets contain non-abstract points that have both locations and measurable features such as colour and gradient orientation. Connectedness, boundedness, mesh nerves, convexity, shapes and shape theory are principal topics in the study of nearness and separation of physical as well as abstract sets. CP has a hefty visual content. Applications of CP include computer vision, multimedia, brain activity, biology, social networks and cosmology.

CP leads to the study of structures in various forms in such things as sets of picture points in digital images or sets of nodes in location-based social networks. Typically in computational proximity, one starts with some form of proximity space (topological space equipped with a proximity relation) that has an inherent geometry.

A **topological space** is a nonempty set together with a collection of open sets that satisfy certain properties (for the details, see Appendix F.2). A **proximity space** is a topological space equipped with a proximity relation. Various forms of proximity relations are introduced in Sect. 1.4.

Using various algorithmic methods combined with instances of proximity spaces and computational geometry, it is possible to discover hidden objects and patterns in the selected mathematical structures. After that, one then highlights the presence of such things as connectedness, boundedness, bornologies, bornological nerves, mesh bornological nerves, convexity and shapes contained in such proximal structures, which are the focus of current research on proximity [3]. To see the importance of geometry in the study of connectedness in CP, recall what Stephen Willard succinctly observed: *The topological study of connected spaces is heavily geometric (or visual)* [4, §26, p. 191].

The region-based approach in CP represents a shift in focus from points to regions in the study of near sets. Both near (proximal) sets and classical, Zermelo–Fraenkel (ZF) sets are axiomatized. For various forms of proximity axioms, see Sect. 1.4 and Chap. 2. The ZF set theory axioms are given in Appendix E.

From a point-based perspective, sets of physical objects such as picture points are near sets, provided the sets contain points with matching descriptions. From a region-based perspective, sets of either abstract regions such as polygons in the Euclidean plane or picture regions are near sets, provided the regions have matching descriptions. This means that both purely geometric regions as well as regions occupied by sets of physical entities are susceptible to the computational proximity approach. Unlike points that have only location, regions in both types of sets of objects have features such as area, diameter, convex, concave and tessellated. Basically, region-based descriptions are helpful in detecting, characterizing, analyzing and classifying patterns in sets of objects that can be either purely geometric or physical.

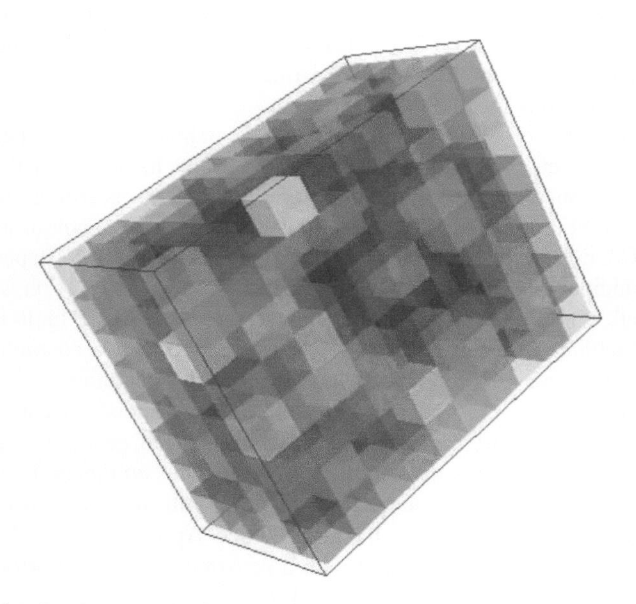

Fig. 1 $10 \times 8 \times 8 \times 3$ Array of cubes

Ultimately, computational proximity leads to various applications of topology in the study of set patterns in proximity spaces [5–7], new forms of topological spaces [6] and new forms of algebraic structures such as near groups [8–11] as well as leading to a new form of topology of digital images [12] useful in solving pattern recognition, analysis and classification problems.

CP utilizes the very rich collection of proximities such as Čech [13], Efremovič [14], Lodato [15], Wallman [16], and the more recent descriptive proximity [17–19]. Descriptive proximity grew out of a study of extensions of uniform topologies on metric spaces and descriptive forms of the traditional separation axioms. It is well known that an Efremovič proximity space [14] naturally defines a topology in the space [20]. A Leader uniform topology [21] is defined by finding all sets near each given set in a proximity space.

Computational proximity is important for the following reasons. Instantiation of proximities viewed in the context of computational geometry leads to the discovery of new forms of

(i) **Connectedness**. One need only look at a Delaunay triangulation on an annulus to find a connected space as well as collections of subsets that are strongly connected (See, e.g., Sect. 1.16). Connectedness leads to a study of adjacent as well as strongly connected mesh nerves (See, e.g. Sect. 1.17).

(ii) **Boundedness, Bornology and Bornological Nerves**. These structures are fundamental building blocks in CP (see, e.g. Sect. 1.18). A **bornology** is a boundedness $\mathscr{B}$ on a nonempty set X that it is a cover of X, i.e. $X \subseteq \mathscr{B}$. Examples of bornologies can be found collections of cubes in Euclidean 3D space such as the ones shown in Fig. 1.

(iii) **Proximal Boundedness and Proximal Boundedness Nerves**. In CP, these structures have inherent affinities that are fundamental in the detection and analysis of patterns in a topology of digital images as well as a topology of other sets of physical entities such as brain cells (see, e.g. Sect. 1.19).

(iv) **Mesh bornological nerves**. These structures tend to surround digital image as well as other forms of objects (see, e.g. Sects. 1.18, 1.19).

(v) **Local proximity spaces**. These structures provides a means of comparing, analyzing and classifying such things as mesh bornological nerves endowed with a proximity, which are not only local proximity spaces but also are a good source of tools in solving object recognition problems (see, e.g. Sect. 1.20).

(vi) **Convexity Structures**. A family of convex sets has the **convexity property**, provided the intersection of any number of sets in the family belongs to the family. In effect, a family of sets with the convexity property has a strong proximity. For more about this, see Sect. 1.7. An interest in convexity leads to the study of digital image tessellations in CP (see, e.g. Sect. 1.11).

(vii) **Homotopic Maps, Shapes and Borsuk–Ulam Theorem**. A pair of continuous maps f, g on a set X are homotopic, provided $f(X)$ continuously deforms into $g(X)$. It was K. Borsuk who first associated the geometric notion of shape and homotopies. For more about this, see Sect. 5.3.

(viii) **Manifolds**. A **manifold** is a topological space with added properties. Manifolds are very useful structures, especially if there is a need to solve dimensionality reduction problems. Digital image manifolds are a natural by-product of the excursions in the topology of digital images in this book. For more about this, see Sect. 8.1. ■

This book has been derived from my lectures in a graduate course on the topology of digital images taught over the past several years. Many of my students have provided important insights and valuable suggestions concerning topics in this book. I especially want to thank Binglin Li, Colin Gaudreau, Schail Younas, Clara Guadagni, Braden Cross, Dominic Villar, Doungrat Chitcharoen, Chido Uchime, Randima Hettiarachchi and Enoch A-iyeh for their comments and helpful observations. I especially want to thank Binglin Li for her work on Voronoï regions with a maximum number of neighbouring regions in a Dirichlet tessellation.

I am very grateful for the discussions I have had with Anna Di Concilio, Som Naimpally, Clara Guadagni, Zdzisław Pawlak, Andrzej Skowron, Mehmet Ali Öztürk, Mustafa Uçkun, Ebubekir İnan, Arturo Tozzi and Irakli Dochviri concerning a number of topics in this book.

The topics in this monograph introduce many forms of proximities with a computational flavour (especially, what has become known as the strong contact relation), many nuances of topological spaces, and point-free geometry that are a direct result of my discussions with Anna Di Concilio. Som Naimpally also introduced me to many aspects of the foundations of proximities and topology over a number of years. It was Som Naimpally who first suggested the importance of the proximity relation between digital images and original scenes represented in digital picture form. Both of these researchers have provided a great number of insights concerning these topics that underlie the foundations of computational proximity. My discussions with Clara Guadagni and Anna Di Concilio led to the introduction and deeper understanding of the various forms of strong proximity [Di Concilio strong contact] (see, e.g. [22–25]). I extend my thanks to Clara Guadagni for her many suggestions and corrections for the text.

Computational Proximity also has its roots in the closely related mathematics of rough sets and nearness approximation spaces. I have had many discussions concerning these topics with Zdzisław Pawlak and Andrzej Skowron as well with many others such as Jarosław Stepaniuk, Zbigniew Suraj, Ewa Orłowska, Lech Polkowski, Marcin Wolski, Piotr Wasilewski, L. Puzio, Jerzy W. Grzymała-Busse, Maria do C. Nicoletti, Roman Slowiński, Sankar K. Pal, Yiyu Yao, A. Czyżewski, B. Kostek, A. Gomolińska and G. Cattaneo. It was Zdzisław Pawlak who introduced me to rough sets and granular computing during FUZZ IEEE 1996 in New Orleans and it was Andrzej Skowron who introduced me to many of the nuances of rough set theory and Som Naimpally's and B.D. Warrack's 1970 book on proximity spaces [26] during one of my visits to the Warsaw University. A series of articles on nearness approximation spaces grew out of my discussions with Andrzej Skowron and Jarosław Stepaniuk (see, e.g. [27–28]), leading to [29–31].

Another offshoot of the original work on near sets[1] is the work by Tiwari [34], Wasilewski [35], Wolski [36, 37], Gomolińska [29], Fashandi [38], Henry [39], Ramanna [40] and Meghdadi [41, 42].

Eventually, the work on nearness approximation spaces led to a number of papers on near algebraic structures such as groups and semigroups in nearness approximation spaces by E. İnan, M. Uçkun and M.A. Öztürk (see, e.g. [8–10, 43]).

[1]For an overview of near set theory, see https://en.wikipedia.org/wiki/Near_sets and [32, 33].

Many insights concerning the algebra component of CP stem from my discussions with Mehmet Ali Öztürk, Mustafa Uçkun and Ebubekir İnan. This has led to a number of publications (see, e.g. [44–47]). I extend my thanks to Ebubekir İnan for his many suggestions and corrections for the text.

CP has benefited hugely from my discussions with Arturo Tozzi about the various applications of the Borsuk–Ulam Theorem (BUT). It was Arturo Tozzi who pointed out the many important applications of BUT in the study of brain activity [48], quantum entanglement [49], fractals and power laws [50] and multisensory neurons [51].

It was Irakli Dochviri who suggested the study of topological sorts of near sets in proximity spaces [52, 53]. It is not enough to identify sets that are near each other either spatially or descriptively. For near set theory and computational proximity to be useful in many applications, it is necessary to determine to what extent one set A is a near a particular set B relative to other sets that are near the same set B. This problem is directly related to the work by A. Gomolińska and M. Wolski on graded nearness [29]. A very interesting counterpart of the near sets sorting problem is the far sets sorting problem: *To what extent are various sets far (remote, distant) from a particular set*? The solution of this problem leads to the identification of separated patterns found in collections of sets that are far from each other.

At various times, a number of researchers have provided important insights concerning topics in this book. Among these researchers, I extend *many thanks* (*Grazi mille, Çok Teşekküler*) to

Brazil Maria C. Nicoletti, J.H. Saito.
Canada Sheela Ramanna, Chris Henry, Çenker Sangoz, Homa Fashandi, Dan Lockery, Amir H. Meghdadi and S. Shafar, Witold Pedrycz, Witold Kinsner, M. Pawlak and Nick Friesen.
Tbilisi, Georgia Irakli Dochviri.
Italy A. Di Concilio, G. Gerla, G. Di Maio, Clara Guadagni and Arturo Tozzi.
India Sankar K. Pal and S. Tiwari.
Poland Zdzisław Pawlak, Andrzej Skowron, J. Stepaniuk, Zbigniew Suraj, Ewa Orłowska, Lech Polkowski, Marcin Wolski, Piotr Wasilewski, L. Puzio and Piotr Artiemjew.
Russia Iskander A. Taimanov.
Thailand Doungrat Chitcharoen.
Turkey Mehmet Ali Öztürk, Mustafa Uçkun, Ebubekir İnan and Özlem Umdu.
U.S. Gerald Beer, Leon Schilmoeler, Bob Dumonceaux, Maciej Borkowski, Bill Hankley, Dave Schmidt, Jack Lange, Irving Sussman and Jerzy W. Grzymała-Busse.

Many thanks for the many helpful suggestions and corrections from A. Di Concilio, G. Di Maio, C. Guadagni, E. İnan, A. Tozzi, I. Dochviri and S. Ramanna.

This book contains a generous selection of implementations of CP algorithms in Mathematica scripts and a few implementations in Matlab scripts. For the most part, these scripts are given in Appendix A.

Chapter problems have been classified. Those problems that begin with 🚲 are the kind you can run with, and probably will not take much time to solve. Problems that begin with ☕ are the kind you can probably solve in about the time it takes to drink a cup of tea or coffee. The remaining problems will need varying lengths of time to solve.

The research leading to this book has been supported by grants from the University of Salerno, Tübitak, and the Natural Sciences and Engineering Research Council of Canada (NSERC).

References

[1] Di Maio, G., Naimpally, S.A. (eds.): Theory and Applications of Proximity, Nearness and Uniformity, 264 pp. Seconda Università di Napoli, Napoli, Italy (2009). MR1269778

[2] Naimpally, S., Peters, J., Wolski, M.: Foreword [near set theory and applications]. Math. Comput. Sci. **7**(1), 12 (2013)

[3] Guadagni, C.: Bornological convergences on local proximity spaces and ω_μ-metric spaces. Ph.D. thesis, Università degli Studi di Salerno, Salerno, Italy (2015). Supervisor: A. Di Concilio, 79 pp

[4] Willard, S.: General Topology. Dover Pub., Inc., Mineola, NY (1970). Xii + 369 pp, ISBN: 0-486-43479-6 54-02, MR0264581

[5] Peters, J.: Local near sets: Pattern discovery in proximity spaces. Math. Comp. Sci. **7**(1), 87106 (2013). doi:10.1007/s11786-013-0143-z, MR3043920

[6] Peters, J., Hettiarachchi, R.: Visual motif patterns in separation spaces. Theory Appl. Math. Comp. Sci. **3**(2), 36–58 (2013)

[7] Naimpally, S., Peters, J.: Topology with Applications. Topological Spaces via Near and Far. World Scientific, Singapore (2013). Xv + 277 pp, Amer. Math. Soc. MR3075111

[8] İnan, E., Öztürk, M.: Near semigroups on nearness approximation spaces. Annals Fuzzy Math. Inform. **10**(2), 287–297 (2015)

[9] Öztürk, M., Uçkun, M., İnan, E.: Near group of weak cosets on nearness approximation spaces. Fund. Inform. **133**, 433–448 (2014)

[10] İnan, E., Öztürk, M.: Near groups on nearness approximation spaces. Hacet. J. Math. Stat. **41** (4), 545–558 (2012)

[11] İnan, E., Öztürk, M.: Erratum and notes for near groups on nearness approximation spaces. Hacet. J. Math. Stat. **43**(2), 279–281 (2014)

[12] Peters, J.: Topology of Digital ImagesVisual Pattern Discovery in Proximity Spaces. Intelligent Systems Reference Library, vol. 63, Xv + 411 pp. Springer, Berlin (2014). Zentralblatt MATH Zbl 1295 68010

[13] Čech, E.: Topological Spaces. John Wiley & Sons Ltd., London (1966). Fr seminar, Brno, 1936-1939; rev. ed. Z. Frolik, M. Katětov

[14] Efremovič, V.: The geometry of proximity I (in Russian). Mat. Sb. (N.S.) **31(73)**(1), 189–200 (1952)

[15] Lodato, M.: On topologically induced generalized proximity relations, Ph.D. thesis Rutgers University (1962). Supervisor: S. Leader

[16] Wallman, H.: Lattices and topological spaces. Annals of Math. **39**(1), 112–126 (1938)

[17] Peters, J.: Near sets. Special theory about nearness of objects. Fundamenta Informaticae **75**, 407–433 (2007). MR2293708

[18] Peters, J.: Near sets. General theory about nearness of sets. Appl. Math. Sci. **1**(53), 2609–2629 (2007)

[19] Peters, J.: Near sets: An introduction. Math. Comp. Sci. **7**(1), 39 (2013). doi:10.1007/s11786-013-0149-6, MR3043914
[20] Smirnov, J.M.: On proximity spaces. Math. Sb. (N.S.) **31**(73), 543–574 (1952). English translation: Amer. Math. Soc. Trans. Ser. **2**(38), 535 (1964)
[21] Leader, S.: On clusters in proximity spaces. Fundamenta Mathematicae **47**, 205–213 (1959)
[22] Peters, J., Guadagni, C.: Strongly near proximity and hyperspace topology. arXiv: **1502** (05913), 16 (2015)
[23] Peters, J., Guadagni, C.: Strongly proximal continuity & strong connectedness. arXiv **1504** (02740), 111 (2015)
[24] Peters, J., Guadagni, C.: Strong proximities on smooth manifolds and Voronoï diagrams. Adv. Math. Sci. J. **4**(2), 91–107 (2015)
[25] Peters, J.: Visibility in proximal Delaunay meshes and strongly near Wallman proximity. Adv. Math. Sci. J. **4**(1), 41–47 (2015)
[26] Naimpally, S., Warrack, B.: Proximity Spaces. Cambridge Tract in Mathematics No. 59. Cambridge University Press, Cambridge (1970). X+128 pp.,Paperback (2008)
[27] Peters, J., Skowron, A., Stepaniuk, J.: Nearness in approximation spaces. In: Proceedings of the Concurrency, Specification and Programming (CS&P 2006), pp. 435–445. Humboldt Universität (2006)
[28] Peters, J., Skowron, A., Stepaniuk, J.: Nearness of objects: Extension of approximation space model. Fundamenta Informaticae **79**(34), 497–512 (2007). MR2346263
[29] Gomolińska, A., Wolski, M.: On graded nearness of sets. Fundamenta Informaticae **119**(34), 301–317 (2012). MR3050536
[30] Wang, L., Liu, X., Qiu, W.: Nearness approximation space based on axiomatic fuzzy sets. Int. J. Approx. Reason **52**(2), 200–211 (2012). MR2872908
[31] Peters, J., Skowron, A., Stepaniuk, J.: Nearness of visual objects. Application of rough sets in proximity spaces. Fundamenta Informaticae **128**(12), 159–176 (2013). MR3154898
[32] Peters, J., Naimpally, S.: Applications of near sets. Notices of the Amer. Math. Soc. **59**(4), 536–542 (2012). doi:10.1090/noti817. MR2951956
[33] Kiermeier, K.: Review of J.F. Peters, S.A. Naimpally, Applications of near sets, Notices Am. Math. Soc. **59**(4), 536–542 (2012). ISSN 0002-9920; 1088-9477. Zentralblatt MATH **an: 1251.68301**
[34] Tiwari, S.: Ultrafilter completeness in ?-approach nearness spaces. Math. Comput. Sci. **7**(1), 107–111 (2013). MR3043921
[35] Peters, J., Wasilewski, P.: Foundations of near sets. Inform. Sci. **179**(18), 3091–3109 (2009). MR2588809
[36] Wolski, M.: Toward foundations of near sets: (pre-)sheaf theoretic approach. Math. Comput. Sci. **7**(1), 125–136 (2013). MR3043923
[37] Wolski, M.: Perception and classification. A note on near sets and rough sets. Fundamenta Informaticae **101**(12), 143–155 (2010). MR2732874
[38] Fashandi, H.: Nearness of covering uniformities: theory and application in image analysis. Math. Comput. Sci. **7**(1), 43–50 (2013). MR3043917
[39] Henry, C.: Metric free nearness measure using description-based neighbourhoods. Math. Comput. Sci. **7**(1), 51–69 (2013). MR3043918
[40] Henry, C., Ramanna, S.: Signature-based perceptual nearness: Application of near sets to image retrieval. Math. Comput. Sci. **7**(1), 71–85 (2013). MR3043919
[41] Ramanna, S., Meghdadi, A.: Measuring resemblances between swarm behaviours: A perceptual tolerance near set approach. Fundamenta Informaticae **95**(4), 533–552 (2009). MR2582188
[42] Ramanna, S., Meghdadi, A., Peters, J.: Nature-inspired framework for measuring visual image resemblance: A near rough set approach. Theoret. Comput. Sci. **412**(42), 5926–5938 (2011). MR2866035
[43] Öztürk, M., İnan, E.: Soft nearness approximation spaces. Fund. Inform. **124**(1), 231–250 (2013)

[44] Peters, J., Öztürk, M., Uçkun, M.: Klee-Phelps convex groupoids. arXiv: **1411**(0934), 15 (2014). Mathematica Slovaca 2015, *accepted*
[45] Peters, J., Öztürk, M.A., Uçkun, M.: Exactness of homomorphisms on proximal groupoids. Fen Bilimleri Dergisi **X**(X), 114 (2014)
[46] Peters, J., İnan M.A. Öztürk, E.: Spatial and descriptive isometries in proximity spaces. Gen. Math. Notes **21**(2), 125–134 (2014)
[47] Peters, J., İnan, M.A, Öztürk, E.: Monoids in proximal Banach spaces. Int. J. of Algebra **8** (18), 869–872 (2014)
[48] Tozzi, A., Peters, J.: Our thoughts follow a donut-like trajectory in the brain. ResesarchGate Preprint, pp. 113 (2015). doi:10.13140/RG.2.1.3305.8008
[49] Peters, J., Tozzi, A.: The Borsuk-Ulam theorem explains quantum entanglement. ResesarchGate Preprint, pp. 17 (2015). doi:10.13140/RG.2.1.3860.1685
[50] Tozzi, A., Peters, J.: A topological concept explains fractals and power laws. ResesarchGate Preprint, pp. 19 (2015). doi:10.13140/RG.2.1.3657.9287
[51] Tozzi, A., Peters, J.: A topological approach explains multisensory neurons. J. Neurosci. (2015) *to appear*
[52] Dochviri, I., Peters, J.: Topological sorting of finitely many near sets. Math. Comput. Sci. 16 (2015), *communicated*
[53] Dochviri, I., Peters, J.: Near sets in bitopological spaces. Indian J. Math. 16 (2015), *communicated*

Contents

1 Computational Proximity 1
- 1.1 Computational Proximity Framework 5
- 1.2 Points, Regions, Connectedness and Point-Free Geometry 6
- 1.3 Choice of Probe 10
- 1.4 Proximities 13
- 1.5 Di Concilio Strong Contact 19
- 1.6 Strongly Far Proximity 21
- 1.7 Convexity Structures 24
- 1.8 Descriptive Proximity 26
- 1.9 Descriptive Strong Proximity 28
- 1.10 Nerves 30
- 1.11 Voronoï Diagrams and Mesh Nerves 32
- 1.12 Origins, Variations and Applications of Voronoï Diagrams 33
- 1.13 Mesh Nerves on a Digital Image 37
- 1.14 Singer Needle Tip Image Mesh: A Step Toward Object Recognition 39
- 1.15 Topological Spaces: Setting for Computational Proximity 42
- 1.16 Connectedness and Strongly Near Connectedness 43
- 1.17 Connected and Strongly Connected Mesh Nerves 47
- 1.18 Boundedness, Bornology and Bornological Nerves 49
- 1.19 Proximal Boundedness, Strong Proximal Boundedness and Proximal Boundedness Nerves 51
- 1.20 Local Proximity Spaces and Local Strong Proximity Spaces 55
- References 59

2 Proximities Revisited 63
- 2.1 Cech Proximity 65
- 2.2 Cech Closure of a Set 70
- 2.3 Near Edge Sets 71
- 2.4 Lodato Proximity 72

2.5 Descriptive Lodato Proximity . . . 73
2.6 Delaunay Triangulation . . . 74
2.7 Voronoï Diagrams Revisited . . . 79
2.7.1 Sites . . . 79
2.8 Some Results for Voronoï Regions . . . 80
2.9 Dirichlet Tessellation Quality and Digital Image Quality . . . 83
2.10 Tessellation Region Centroids . . . 88
2.11 Centroid-Based Voronoï Mesh on an Image . . . 89
References . . . 94

3 Distance and Proximally Continuous . . . 97
3.1 Metrics and Metric Topology . . . 98
3.2 Distances: Euclidean and Taxicab Metrics . . . 99
3.3 Metric Proximity . . . 101
3.4 Closeness Metric . . . 102
3.5 Some Recent History of Near Sets . . . 102
3.6 Descriptive Similarity Distance . . . 104
3.7 Descriptively Near Sets Can Be Spatially Disjoint Sets . . . 105
3.8 Neighbourhoods of Points in Euclidean Space . . . 106
3.9 2D Picture Descriptive Neighbourhood of a Point . . . 107
3.9.1 Unbounded Descriptive Neighbourhood of a Picture Point . . . 108
3.9.2 Bounded Descriptive Neighbourhood of a Picture Point . . . 110
3.9.3 Bounded Indistinguishable Descriptive Neighbourhood of a Picture Point . . . 113
3.9.4 Unbounded Indistinguishable Descriptive Neighbourhood of a Picture Point . . . 115
References . . . 116

4 Image Geometry and Nearness Expressions for Image and Scene Analysis . . . 119
4.1 Image Geometry . . . 122
4.1.1 Delaunay Triangulation . . . 123
4.1.2 Voronoï Diagrams . . . 124
4.2 Nearness Expressions: Basic Notions . . . 125
4.3 Near and Strongly Near Sets Revisited . . . 128
4.3.1 Lodato Proximity Revisited . . . 129
4.3.2 Descriptive Lodato Proximity Revisited . . . 129
4.3.3 Strong Proximity Revisited . . . 131
4.3.4 Descriptive Strong Proximity Revisited . . . 132
4.4 Di Concilio–Gerla Overlap Revisited . . . 133
4.5 Wallman Proximity Revisited . . . 134
4.6 Far and Strongly Far Sets Revisited . . . 137
References . . . 138

5 Homotopic Maps, Shapes and Borsuk–Ulam Theorem 141
5.1 Antipodal Points and Hyperspheres 142
5.2 Borsuk–Ulam Theorem 144
5.3 Homotopic Maps and Deformation Retracts 146
5.4 Object Description, Structured Sets and Shape Features 150
5.4.1 Euclidean Feature Space 150
5.4.2 Feature Vectors 152
5.5 Strongly Proximal Object and Feature Spaces 155
5.6 Strong Proximal Borsuk–Ulam Theorem 160
5.7 Strong Proximal Region-Based Borsuk–Ulam Theorem 162
5.8 Complexes, Nuclei and Applied Homotopy 164
5.9 Strong Proximal Homotopy 166
5.10 Borsuk–Ulam Theorem Extended to Hyperbolic Surfaces by A. Tozzi 169
References 171

6 Visibility, Hausdorffness, Algebra and Separation Spaces 175
6.1 Visibility 176
6.2 Separation Axioms 178
6.3 R0 Symmetric Space 179
6.4 Descriptive R0 Space 179
6.4.1 Near Sets in L-Proximity Spaces 179
6.4.2 Descriptive L-Proximity 181
6.4.3 Descriptive Proximity Space Revisited 183
6.5 T0 Anti-Symmetric Space 183
6.6 T1 Space 185
6.7 Hausdorff T2 Space 187
6.8 Hausdorffness and Visibility 189
6.9 Partial Descriptive Groupoids in Hausdorff Spaces 194
6.10 Set-Based Probes and Proximal Delaunay Groupoids 195
6.11 Voronoï Groupoids 201
6.12 T3 Space 203
6.13 T4 Space 204
6.14 Visual Set Patterns in Descriptive Separation Spaces 206
6.14.1 Visual Patterns in Descriptive T1 Spaces 207
6.14.2 Visual Patterns in Descriptive T2 Spaces 208
6.14.3 Set Pattern Generators 210
References 212

7 Strongly Near Sets and Overlapping Dirichlet Tessellation Regions 215
7.1 Known Mesh Seed or Generating Points 217
7.2 Julia Set Image Example 218

7.3 Strongly Near Regions and Mesh Cells . . . 218
7.4 Back to Earth . . . 222
References . . . 236

8 Proximal Manifolds . . . 237
8.1 Manifolds: Basic Notions . . . 239
8.2 Charts and Atlases . . . 241
8.3 Proximal Voronoï Manifolds, Atlases and Charts . . . 245
8.4 Mean Shift Manifold . . . 247
8.5 Digital Image Manifolds . . . 249
8.6 Mean-Shift Image Manifolds . . . 251
8.7 Image Saliency Manifolds . . . 253
8.8 Selfie Voronoï Manifold . . . 254
8.9 Manifolds with a Proximity . . . 255
References . . . 257

9 Watershed, Smirnov Measure, Fuzzy Proximity and Sorted Near Sets . . . 259
9.1 See-Through MM Segments Via Dirichlet Tessellation on an Image . . . 261
9.2 Strongly Near Voronoï Region Interiors on MM Segmentations . . . 266
9.3 Extension of Smirnov Proximity Measure . . . 267
9.4 Fuzzy Lodato Proximity and Fuzzy Lodato–Smirnov Membership Function . . . 268
9.5 Descriptive Smirnov Proximity . . . 274
9.6 Application: Sorting Near Sets Using Smirnov Proximity Measures . . . 275
9.7 Fuzzy Descriptive Lodato Proximity Space . . . 279
9.8 Image Dilation . . . 281
9.9 MorphologicalBinarize Operation with Lower and Upper Bounds . . . 282
9.10 Dilation Method . . . 285
9.11 Dilation and Watershed Segmentation . . . 287
9.12 Gradient Filtering Watershed Components . . . 288
References . . . 290

10 Strong Connectedness Revisited . . . 291
10.1 Connectedness: Basic Concepts . . . 292
10.2 Strongly Connected Subsets in a Voronoï Mesh . . . 298
References . . . 303

11 Helly's Theorem and Strongly Proximal Helly Theorem . . . 305
11.1 Helly's Theorem . . . 308
11.2 Strongly Near Version of Helly's Theorem . . . 310

11.3 Descriptive Helly's Theorem 311
11.4 Polyforms 313
References 316

12 Nerves and Strongly Near Nerves 317
12.1 Polyform Nerves 318
References 323

13 Connnectedness Patterns 325
13.1 Connectedness in Voronoï Meshes 328
13.2 Nearness of Collections of Strongly Connected Sets 329
13.3 Picture Mesh Patterns 331
13.4 Pattern Axioms and Object Signature Axioms 333
13.5 Keypoints Mesh Sites 337
13.6 Keypoints Mesh Patterns 339
References 342

14 Nerve Patterns 343
14.1 Voronoï Mesh Nerves 345
14.2 Nerves with a Nucleus 346
14.3 Partial Ordering of Nuclei 347
14.4 Voronoï Mesh Nerve Patterns 352
References 353

Appendix A: Mathematica and Matlab Scripts 355

Appendix B: Kuratowski Closure Axioms 391

Appendix C: Sets. A Topological Perspective 393

Appendix D: Basics of Proximities 395

Appendix E: Set Theory Axioms, Operations and Symbols 397

Appendix F: Topology of Digital Images 401

Author Index 411

Subject Index 417

List of Figures

Figure 1.1 Strongly near torus mesh segments. 1
Figure 1.2 CP components 2
Figure 1.3 Dirichlet tessellation 5
Figure 1.4 2D convex region and 3D Wolfram Stanford Bunny centroids 7
Figure 1.5 UNISA coin and UNISA coin centroid • 8
Figure 1.6 Dirichlet tessellations 11
Figure 1.7 Proximity (nextness) of adjacent Feudi di San Gregorio wine casks 14
Figure 1.8 Descriptively close, spatially remote torus polygons 18
Figure 1.9 Near sets: $A = \{(x, 0) : x > 0\}$, $B = \{(x, y) : y = 1/x\}$ 20
Figure 1.10 Near sets: $A = \{(x, 0) : 0.1 \leq x \leq 1\}$, $B = \{(x, \sin(5/x)) : 0.1 \leq x \leq 1\}$ 21
Figure 1.11 $E \overset{\wedge\wedge}{\delta} B, A \overset{\wedge\wedge}{\delta} B$ but $E \underset{\mathbb{W}}{\not\delta} C, A \underset{\mathbb{W}}{\not\delta} C$ 22
Figure 1.12 Strongly far proximity $\underset{\mathbb{W}}{\not\delta}$ 23
Figure 1.13 Family of all subsets in $2^X, X = \{1, 2, 3, 4\}$ 25
Figure 1.14 $A \not\delta_{\Phi} B \Rightarrow A \not\delta_{\Phi} C$ and $B \not\delta_{\Phi} \mathbf{C^c}$ for some $C \in 2^X$ 28
Figure 1.15 Descriptive strongly near sets: $A_i \overset{\wedge\wedge}{\delta}_{\Phi} B, i \in \{1, 2, 3, 4, 5, 6, 7\}$ 29
Figure 1.16 Sample torus closure nerve 31
Figure 1.17 $V_p =$ intersection of closed half-planes 32
Figure 1.18 Voronoï mesh 33
Figure 1.19 Partial Möbius diagram [54, Sect. 2.4.1] 35
Figure 1.20 Voronoï mesh nerve 36
Figure 1.21 Drawing of a microscopic image of a needle point by Robert Hooke 37
Figure 1.22 Corners in Hooke needle image 38
Figure 1.23 Hooke needle point corner-based Voronoï mesh. 38
Figure 1.24 Hooke needle point Voronoï mesh on image 39

Figure 1.25 Zeiss microscope image showing ≈5 mm Singer needle point tip . . . 40
Figure 1.26 Singer needle. . . . 40
Figure 1.27 Singer needle point tip corners. . . . 40
Figure 1.28 Singer needle point tip corners Voronoï mesh . . . 41
Figure 1.29 Singer needle point tip corners Voronoï mesh . . . 41
Figure 1.30 Connected sets. . . . 43
Figure 1.31 Strongly connected set $\overset{\wedge\!\!\wedge}{\text{conn}}\, H$ on Hooke needle point . . . 45
Figure 1.32 Connected *Brown* polygons on torus surface . . . 45
Figure 1.33 Strongly connected sets . . . 46
Figure 1.34 Strongly connected mesh nerves on Hooke needle point . . . 47
Figure 1.35 Boundedness in $\mathbb{R}^3$ . . . 50
Figure 1.36 Proximal boundedness in $(\mathbb{R}^3, \delta, \mathscr{B})$ . . . 52
Figure 1.37 Strong proximal boundedness in $(\mathbb{R}^3, \overset{\wedge\!\!\wedge}{\delta}, \mathscr{B})$ . . . 53
Figure 1.38 Annulus mesh proximal boundedness in $(\mathbb{R}^2, \overset{\wedge\!\!\wedge}{\delta}, \mathscr{B})$. . . . 53
Figure 1.39 Annulus mesh nerves in proximal boundedness in $(\mathbb{R}^2, \overset{\wedge\!\!\wedge}{\delta}, \mathscr{B})$. . . . 54
Figure 1.40 Mesh nerve that is a local proximity space . . . 56
Figure 1.41 Mesh bornological nerve surrounding an object . . . 57
Figure 1.42 Sierpinski triangles, containing $N_n = 3^n, 1, 3, 9, 27, \ldots$ *Black* triangles . . . 58
Figure 1.43 $8 \times 5 \times 5$ array of cubes. . . . 58
Figure 2.1 Som Naimpally . . . 63
Figure 2.2 $A\ \delta_\Phi\ B$ (descriptively near sets) . . . 64
Figure 2.3 $x\ \delta_\Phi\ B$ . . . 68
Figure 2.4 $A\ \overset{\wedge\!\!\wedge}{\delta}\ B$ . . . 70
Figure 2.5 Near edge sets. . . . 71
Figure 2.6 Delaunay triangle $\Delta(\boldsymbol{pqr})$ . . . 72
Figure 2.7 Proximal cameraman elbow sets . . . 73
Figure 2.8 Descriptive Lodato near apple sets . . . 73
Figure 2.9 Strongly near Delaunay triangles $\triangle(pqr)$ and $\triangle(qrt)$ . . . 74
Figure 2.10 Corner-based Delaunay mesh and corresponding image corners . . . 75
Figure 2.11 Corners in a hand image. . . . 75
Figure 2.12 Corner-based Delauny mesh on a hand image . . . 76
Figure 2.13 Corner-based Voronoï mesh and corresponding image mesh . . . 77
Figure 2.14 Combined corner-based Voronoï and Delaunay meshes on a hand image . . . 77
Figure 2.15 Hand image Voronoï mesh nerve. . . . 78
Figure 2.16 Voronoï region = convex polygon. . . . 79

Figure 2.17	Cameraman edges detected	81
Figure 2.18	Dirichlet tessellation of the cameraman image	82
Figure 2.19	Corner and edge centroid tessellations and quality histograms [39, Sect. 6]	85
Figure 2.20	Sample tessellation derived from randomly selected sites	90
Figure 2.21	Region centroids detected	91
Figure 2.22	Centroidal Voronoï mesh	91
Figure 2.23	Centroidal Voronoï mesh on image	92
Figure 2.24	Centroidal Voronoï mesh on a Nikon® camera image	92
Figure 3.1	J. Hadamard	97
Figure 3.2	Distances in the Euclidean plane	98
Figure 3.3	Four different norms of a vector	100
Figure 3.4	$\mathrm{cl}A \cap \mathrm{cl}B \neq \emptyset$ implies $A\ \delta\ B$	102
Figure 3.5	Several houses	105
Figure 3.6	Neighbourhood of point p with coordinates $(120, 114), \varepsilon = 30$	106
Figure 3.7	Unbounded descriptive neighbourhood of point at $(120, 114), \varepsilon_c = 30$	108
Figure 3.8	Coordinates of Points in Sparrow Image	109
Figure 3.9	Unbounded descriptive Nbd of p(300, 190) in Sparrow Image	110
Figure 3.10	Tree Nymph Butterfly	110
Figure 3.11	Unbounded descriptive Nbd of p(343, 324) in Tree Nymph Image	111
Figure 3.12	Unbounded descriptive Nbd of p(306, 279) in tree Nymph image	112
Figure 3.13	Bounded indistinguishable descriptive Nbd of p(302, 275) in tree Nymph image	114
Figure 4.1	Dirichlet tessellation of fisherman scene	119
Figure 4.2	Hunting grounds for scene information: corner-based Delaunay and Voronoï meshes	122
Figure 4.3	Visible points	123
Figure 4.4	$\Delta(\boldsymbol{pqr})$ = Delaunay triangle	123
Figure 4.5	$V(p), p \in S$ = Intersection of closed half-planes	124
Figure 4.6	Voronoï regions $V_{a_i}, i \in \{1, 2, 3, 4, 5, 6, 7, 8\}$	125
Figure 4.7	Strongly separated points	126
Figure 4.8	Hunting ground for near sets: Delaunay on Voronoï mesh	127
Figure 4.9	Image geometry = Source of Lodato proximity space	128
Figure 4.10	$\mathrm{int}\boldsymbol{A}\ \overset{\wedge\!\!\wedge}{\boldsymbol{\delta}}\ \mathrm{int}\boldsymbol{B},\ \mathrm{int}\boldsymbol{D}\ \overset{\wedge\!\!\wedge}{\boldsymbol{\delta}}\ \boldsymbol{E}\cdot$	132
Figure 4.11	$C\ \overset{\wedge\!\!\wedge}{\delta}_{\Phi}\ H$ and $C\ \overset{\wedge\!\!\wedge}{\delta}_{\Phi}\ B\cdot$	133

Figure 4.12 $D \stackrel{\Phi}{\hat{\delta}} E$ 134
Figure 4.13 $A \stackrel{\Phi}{\hat{\delta}} B$ 136
Figure 5.1 Karol Borsuk. 141
Figure 5.2 Antipodal 5gon on circle. 142
Figure 5.3 $f : S^2 \longrightarrow \mathbb{R}^2$ 143
Figure 5.4 Two pairs of antipodal points on an S^2 circle perimeter 144
Figure 5.5 Concentric concrete 2-spaces with shades of *green*. 146
Figure 5.6 A bit more, Punch, 1845. 149
Figure 5.7 50 knight corners. 151
Figure 5.8 50 knight corners. 153
Figure 5.9 Knight edge pixel gradients. 155
Figure 5.10 Fisherman object space 156
Figure 5.11 Close voronoï regions. 159
Figure 5.12 Proximal half planes: $A \ \delta \ B$. 159
Figure 5.13 80° rotations 161
Figure 5.14 Family of adjacent triangles. 161
Figure 5.15 $C \ \underline{\delta}\!\!\!/ \ H$ and $C \, \delta_{\Phi} \, H$ 161
Figure 5.16 $f : 2^{S^2} \longrightarrow 2^{\mathbb{R}^2}$ 163
Figure 5.17 Shape J deformed into shape? 165
Figure 5.18 Mathematica widget that deforms j into ?. 166
Figure 5.19 Equivalent shapes with holes. 166
Figure 5.20 $S^n \mapsto M^n$ 170
Figure 6.1 Activity separation pattern, 1870 Punch 175
Figure 6.2 Visible sets in the plane 176
Figure 6.3 Strongly visible sets $A \stackrel{\wedge}{v} B$. 177
Figure 6.4 Delaunay mesh 177
Figure 6.5 Voronoï mesh 178
Figure 6.6 $\mathbf{A} \ \delta \ \mathbf{B}$ and $\mathbf{B} \ll \mathbf{C}$ implies $\mathbf{A} \ \delta \ \mathbf{C}$. 180
Figure 6.7 clA δ b and $b \ \delta$ clC for some $b \in$ clB implies clA δ clC 180
Figure 6.8 $A \ \delta \ E$ for every $E \in \mathcal{P}(X)$ or $B \ \delta \ X \setminus E \implies A \ \delta \ B$ 180
Figure 6.9 $\mathbf{A} \ \delta_{\Phi} \ \mathbf{B}$ and $\mathbf{B} \ll \mathbf{C}$ implies $\mathbf{A} \ \delta_{\Phi} \ \mathbf{C}$. 181
Figure 6.10 $A \ \delta_{\Phi} \ E$ for every $E \in \mathcal{P}(X)$ or $B \ \delta_{\Phi} \ X \setminus E$ implies $A \ \delta_{\Phi} \ B$ 182
Figure 6.11 Sample disjoint neighbourhoods. 182
Figure 6.12 Sample visual space 185
Figure 6.13 Sample T_1^{Φ} visual space 186
Figure 6.14 Manitoba dragonfly 188
Figure 6.15 Manitoba dragonfly 189
Figure 6.16 Descriptively distinct points $\mathbf{p}, \mathbf{p}'$ in $\mathbf{N_p}, \mathbf{N_{p'}}$, resp. 190
Figure 6.17 Voronoï regions: $V_{a_5} \stackrel{\Phi}{\hat{\delta}} V_{a_8}$ and $V_{a_5} \ \underline{\delta}\!\!\!/_{\Phi} \ V_{a_8}$. 191
Figure 6.18 Unbounded descriptive neighbourhood of pixel edge of righthand side of dish. 193

Figure 6.19	Unbounded descriptive neighbourhood of pixel closest to the Nhbd in Fig. 6.18	193
Figure 6.20	$\Phi(A)$ descriptively strongly near $\Phi(C)$	197
Figure 6.21	$X = \{$Spatially near Delaunay triangles$\}$	200
Figure 6.22	Sample T_4^{Φ} visual space	205
Figure 6.23	Descriptive separation space: dragonfly edges	206
Figure 6.24	Sample T_1^{Φ} visual patterns	207
Figure 6.25	Sample T_2^{Φ} descriptive neighbourhoods	209
Figure 7.1	Keypoint-based Dirichlet tessellation versus segmentation	215
Figure 7.2	Quadratic julia set image (see mathematica script 9)	218
Figure 7.3	$\boldsymbol{A} \stackrel{\wedge\wedge}{\delta} \boldsymbol{B}, \boldsymbol{E} \stackrel{\wedge\wedge}{\delta} (\boldsymbol{C} \cup \boldsymbol{D})$	219
Figure 7.4	$\stackrel{\wedge\wedge}{\delta}$-near segmentation sets	220
Figure 7.5	Photographer taking a picture of the sea	221
Figure 7.6	Photographer hand keypoint-based Voronoï mesh cells and diagram	221
Figure 7.7	Labelled keypoint-based Voronoï mesh cells and diagram	221
Figure 7.8	Labelled image polygon	223
Figure 7.9	Julia set image key points	223
Figure 7.10	Julia set key points-generated mesh	223
Figure 7.11	Alhambra floorplan	225
Figure 7.12	Edges in the Alhambra floorplan	226
Figure 7.13	Edge-key-points in the Alhambra floorplan	226
Figure 7.14	Edge-key-points Voronoï mesh on the Alhambra floorplan	227
Figure 7.15	Strongly near polygons in the Alhambra floorplan	228
Figure 7.16	Rainbow on a plant	229
Figure 7.17	Edges in the rainbow plant floorplan	229
Figure 7.18	Edge-key-points in the rainbow plant	230
Figure 7.19	Edge-key-points Delaunay mesh on the rainbow plant.	230
Figure 7.20	Strongly Delaunay mesh near triangles in the rainbow plant.	231
Figure 7.21	High Delaunay mesh common vertices in the rainbow plant.	231
Figure 7.22	Sparrow	232
Figure 7.23	Edges in the sparrow	233
Figure 7.24	Edge-key-points in the sparrow	233
Figure 7.25	Edge-Keypoint Voronoï mesh	234
Figure 7.26	Edge-Keypoint Delaunay mesh	234
Figure 7.27	Edge-Key-points clusters in the sparrow	235
Figure 7.28	Edge-Keypoint Voronoï mesh on sparrow	235
Figure 7.29	Edge-keypoint Delaunay mesh on sparrows	236

Figure 8.1 Colour image manifold $\mathbb{R}^5$ mapped to greyscale image $\mathbb{R}^3$ 237
Figure 8.2 Euclidean topology on a tessellation of $\mathbb{R}^2$ 240
Figure 8.3 $A_{P_{n-1}}$ strongly near A_{P_n} for each $n \geq 2$. 240
Figure 8.4 Voronoï Regions $V_{a_i}, i \in \{1,2,3,4,5,6,7,8\}$ 243
Figure 8.5 Smallest open set containing Voronï region of a4. 244
Figure 8.6 Voronoï diagram with respect to Theorem 8.17 245
Figure 8.7 Result of mean-shifting image r-neighbourhoods 249
Figure 8.8 Rainbow-on-a-shoe colour image manifold in $\mathbb{R}^3$ 251
Figure 8.9 Alhambra satellite image manifold 252
Figure 8.10 Mean-shift filtered Alhambra satellite image 253
Figure 8.11 Shoe saliency manifold map 254
Figure 8.12 Selfie Voronoï manifold 255
Figure 8.13 $A \overset{\wedge\wedge}{\delta} B$ 256
Figure 9.1 Watershed segments 259
Figure 9.2 Watershed salient boxed segments 260
Figure 9.3 Segment centroids 262
Figure 9.4 See-thru mesh via centroids on MM segments 262
Figure 9.5 Centroidal- versus keypoints mesh on MM segments 264
Figure 9.6 Centroidal- versus keypoints mesh segments on an image 265
Figure 9.7 Strongly near Voronoï regions covering MM segments 266
Figure 9.8 Fuzzy Lodato proximity space on a Voronoï mesh 269
Figure 9.9 Fuzzy Lodato proximity space on a Voronoï mesh 270
Figure 9.10 Decomposition of 2D bounded set into 3 parts 272
Figure 9.11 Overlapping MM segments in adjacent Voronoï regions 276
Figure 9.12 Morphed edges in a DragonFly image 283
Figure 9.13 White foreground (object) pixels (1), black background pixels (0) 283
Figure 9.14 Default binarization of the rainbow-on-a-plant image 283
Figure 9.15 Bounded binarization of the rainbow-on-a-plant image 284
Figure 9.16 Optimal bounded binarization of the rainbow-on-a-plant image 284
Figure 9.17 Rainbow-on-a-plant image dilation with disk structuring element. 286
Figure 9.18 Rainbow-on-a-plant image edges after binarized image subtraction 286
Figure 9.19 Rainbow-on-a-plant image after dilation of watershed components 288
Figure 9.20 Rainbow-on-a-plant image without dilation of watershed components 288

Figure 9.21 Rainbow-on-a-plant with Gradient Filtering $\mapsto$ Watershed Components . . . 289
Figure 10.1 Portrait exhibiting strong connectedness . . . 291
Figure 10.2 Descriptive strong connectedness subsets in a portrait . . . 292
Figure 10.3 Connected set defined by the Dirichlet L-function in $\mathbb{R}^3$ . . . 293
Figure 10.4 Sine curve . . . 294
Figure 10.5 Set defined by the Dirichlet L-function in $\mathbb{R}^3$ . . . 295
Figure 10.6 Connected subset $\boldsymbol{E}$ but not $\overset{\wedge\wedge}{\delta}$-connected . . . 296
Figure 10.7 Strongly connected subset $D := A \cup B$ in a Delaunay mesh . . . 297
Figure 10.8 Strongly connected subset $D := A \cup B$ in a Voronoï mesh . . . 299
Figure 10.9 1870Punch image . . . 300
Figure 10.10 Connected Voronoï regions . . . 301
Figure 10.11 conn$X = \{A_1, A_2, A_3\}$ . . . 303
Figure 11.1 Helly family of convex bodies in an apple mesh nerve . . . 305
Figure 11.2 $\mathcal{F} = \{V_p, V_{a_1}, V_{a_2}, V_{a_3}, V_{a_4}, V_{a_6}, V_{a_7}\}$ . . . 308
Figure 11.3 $\mathcal{F} = \{V_p, V_q, V_{a_1}, V_{a_2}, V_{a_4}, V_{a_8}\}$ . . . 312
Figure 11.4 3D Delaunay triangulation . . . 314
Figure 11.5 8×8 grid . . . 315
Figure 12.1 0-, 1-, 2-, 3-simplical complexes . . . 317
Figure 12.2 Multiple polydiamond nerves . . . 318
Figure 12.3 Poly2sim nerve . . . 319
Figure 12.4 Multiple polyhex nerves . . . 321
Figure 12.5 Multiple polypentagon nerves . . . 321
Figure 12.6 Multiple polyheptagon nerves . . . 321
Figure 12.7 Multiple polydecahedron nerves . . . 322
Figure 13.1 Voronoï mesh on archway image . . . 325
Figure 13.2 Shadows . . . 326
Figure 13.3 Connectedness patterns on a Voronoï mesh . . . 327
Figure 13.4 Strongly Connected Voronoï Regions whose Union is not Convex . . . 328
Figure 13.5 Connectedness on a Voronoï mesh . . . 330
Figure 13.6 Connectedness shape pattern on a Voronoï mesh . . . 331
Figure 13.7 Sample picture mesh patterns . . . 335
Figure 13.8 Increasing number of corner sites . . . 337
Figure 13.9 Increasing number of keypoint sites . . . 338
Figure 13.10 Increasing number of keypoint sites on snowdrops image . . . 339
Figure 13.11 Connectedness on a snowdrops Voronoï mesh . . . 340
Figure 13.12 Keypoints Voronoï meshes on snowdrops image . . . 340

Figure 14.1 Nucleus nerve patterns on a Voronoï mesh 343
Figure 14.2 Source of nucleus nerve patterns in Fig. 14.1 344
Figure 14.3 Mesh nerve 346
Figure 14.4 Lodato nerve with a nucleus A 347
Figure 14.5 Nucleus nerve in a Voronoï mesh 349
Figure 14.6 Nuclei complete poset on Voronoï mesh 349
Figure 14.7 Chipmunk 350
Figure 14.8 Sample keypoints on picture edges 351
Figure 14.9 Chipmunk edge-keypoints versus keypoints mesh 351
Figure 14.10 Chipmunk edge-keypoints versus keypoints-only picture mesh 352

Chapter 1
Computational Proximity

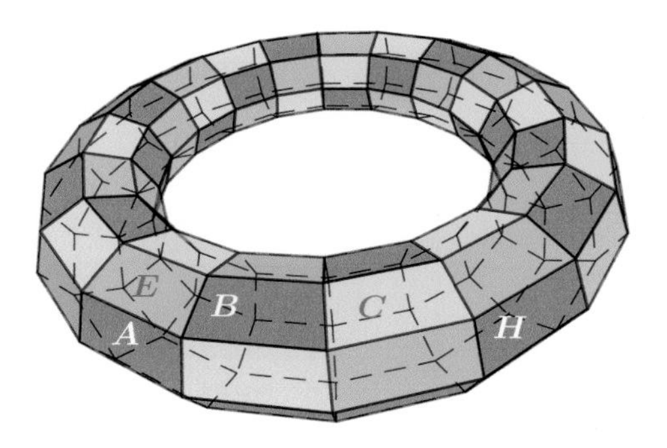

Fig. 1.1 Strongly near torus mesh segments

This chapter introduces computational proximity. Basically, *computational proximity* (CP) is an algorithmic approach to finding nonempty sets of points that are either close to each other or far apart. The methods used by CP to find either near sets or remote sets result from the study of structures called proximity spaces. The basic notion of computational proximity draws its inspiration from the Preface written in 2009 by S.A. Naimpally in [1, pp. 23–28] and the Foreword in [2].

In CP, two types of near sets are considered, namely, spatially near sets and descriptively near sets. Spatially near sets contain points identified by their location and have at least one point in common. Descriptively near sets contain non-abstract points that have both locations and measurable features such as colour and gradient orientation. Connectedness, boundedness, mesh nerves, convexity, shapes and shape theory are principal topics in the study of nearness and separation of physical as well as abstract sets. CP has a hefty visual content. Applications of CP include computer vision, multimedia, brain activity, biology, social networks, and cosmology.

J.F. Peters, *Computational Proximity*, Intelligent Systems Reference Library 102, DOI 10.1007/978-3-319-30262-1_1

Fig. 1.2 CP components

Computational Proximity Framework:

τ: **Finite topology** equipped with proximities in relator $\mathscr{R}_\delta$ [3].
τ-**basis**: Each subset of τ equals the union or intersection of basis elements [4, Example 3.8].
Axioms, Theorems: Sources, e.g., [5, Sect. 4.15], [6, Sect.2], [7, 8, 9, 10, 11].
Proximities:
e.g., δ[12], $\overset{\wedge\!\!\!\wedge}{\delta}$[6], δ_Φ[13].
Relator: e.g., $\mathscr{R}_\delta = \left\{\delta, \overset{\wedge\!\!\!\wedge}{\delta}, \delta_\Phi, \overset{\wedge\!\!\!\wedge}{\delta}_\Phi\right\}$.
Pseudometrics: e.g., point-to-manifold distance [14, Sect. 3.2, PMD], similarity distance [5, Sect.12.3.2].
Probes: Feature value extraction functions, e.g., [5, Sect.1.16.2].
Methods: Proximity-based methods, e.g., [14, Sect. 3.3], [15, p. 6].
Algebra: Proximal algebraic structures, e.g., [11, 16, 17, 18], [19, Sect. IV.1].
Geometry: Proximities in Tesselated subsets of Euclidean space, e.g., [6, 20, 21, 22, 23], [19, §III.3], [24] and Fig. 1.3.
Rules: Proximal decision rules, e.g., clustering rule [25, Sect.4].
Graph Properties: e.g., [25, Sect.4].

P. Pták and W.G. Kropatsch [26] were the first to connect nearness in digital images and proximity spaces. S.A. Naimpally [27] was the first to suggest a connection between proximity space theory and the nearness of digital images to the original scene displayed by an image and the nearness of musical sounds. *A digital image put into a computer consists of finitely many points which try to give a faithful representation of a continuous original* [28]. Naimpally also observes that transformations of digital images must preserve connectedness.

In CP, connectedness, boundedness, mesh nerves, convexity, shapes and shape theory are principal topics in the study nearness and the separation of structures that includes digital images and musical sounds as well as nearness and separation in the composition of mineral and fossil deposits, distributions of fauna and flora, interactions between animals, social networking, environmental events such as earth tremors, nearness and remoteness of stellar systems, nearness and apartness in such things as architectural designs of buildings and cities and in the design of machines and the design of structures such as bridges. CP has a strong theoretical component that gets its strength from the foundations of proximity spaces that have their roots in antiquity and in more recent times.

Philosophy. The notion of nearness of structures crops up very early in philosophy, e.g., Plato's notion of an Ideal Form is defined in terms of the qualities and structure that the quality and structure physical world objects resemble (approximate) and the Ideal Form of Number that individual numbers resemble. Platonism in the Philosophy of Mathematics postulates that there are abstract mathematical objects such as numbers and sets, independent of us. This can be seen in the dialogue between Glaucon and Socrates in Plato's Republic. In that dialogue, Socrates introduces the allegory of the cave in which he contrasts the shadows (appearances) on a cave wall cast by the glow of a fire and the real things that cast the shadows [29]. The question to ask is *How closely do the shadows on the wall resemble the objects that cast the shadows?* Basically, Platonism reduces to a consideration of the nearness of familiar objects to ideal forms. The controversy over the existence of abstract mathematical objects carries over into the work of G. Frege during the early 1950s [30]. And with introduction by A.N. Whitehead [31, 32] of *region* as a primitive in geometry instead of the usual *point* in Euclidean geometry, an entirely new, refreshing view of nearness of regions and what are known as *featured points* are ushered in by A. Di Concilio [22, Sect. 5, p. 42], providing a solid geometric foundation for computational proximity.

Microscopy. An interest in proximities (nearnesses) can be found in Robert Hooke's search for the points of physical objects that resemble the points in geometric drawings, i.e., Robert Hooke's 1665 Micrographia [33] (little pictures, where, for instance, tips of sharp small needles, hairs, bristles, claws of insects, and hairs of leaves are more or less visually similar to points in drawings containing intersections of very fine lines).

Computer Vision and Object Recognition. An interest in the closeness of geometric models and physical objects in pictures crops up repeatedly in computer vision, e.g., [34, Sect. 7.92]. Evidence can be seen in the search for object recognition methods. The basic approach is to formulate geometric models that can be used to recognize and predict the presence of objects. The use of geometry in object recognition leads to shape matching [35, Sect. 2.5.5]. The closer a geometric representation is to a corresponding physical object such as a facial expression or camouflaged shapes or obstructions in the path of a robot, the more likely that we can find applications of computational proximity.

Mathematics. Topics such as proximity, connectedness, boundedness, bornology and local proximity spaces [11] as well as simplical complexes, convex sets systems, Voronoï diagrams and Delaunay complexes in computational geometric topology [19, Sect. III], basic equivalence relations called homotopies (a homotopy is an equivalence relation on a set of continuous functions on a space X to a space Y) [36] and yet another view of geometry and topology via the fundamental group, homology, cohomology and homotopy theory provided by algebraic topology [37] are the focus of current research on proximity [11]. Homotopy theory gets its impetus from J.H.C. Whitehead's work on simplicial spaces and nuclei [38]. Any set of $n+1, n \geq -1$ vertices is called an open n-simplex (briefly, **simplex**). A **symbolic complex** is a closed set of simplexes. The adjective *symbolic* is usually omitted. Homology theory focuses on detecting holes in topological spaces. The common part of a pair of complexes is called the **nucleus**. Complexes that have a nucleus are strongly near, i.e., the complexes in some sense overlap, provided one complex can be deformed into another one. For example, a bounded manifold and its interior have the same Euclidean nucleus (see [38, Theorem 40, p. 326]). The empty simplex has dimension -1 and is analogous to zero. The homology of a space X is based on maps of Euclidean simplices into X. A very good introduction to homology theory is given by S. Krantz [39, Sect. 3.2].

Interest in nearness in mathematics is age-old. Evidence of this can be found, for instance, in the work of Euclid in the study of parallel lines or lines that are close but non-intersecting or circles that cut one another in the construction of equilateral triangles, e.g., [40, vol. 1, p. 241].

Examples of many sets of points with high proximity can be found among the adjacent quadrilaterals along the surface of the torus in Fig. 1.1. Recent work in algebra has called attention to the nearness of convex groupoids [18].

Again, more recently, cellular picture elements (pixels) have been introduced by C. Ronse in 1990. Digital pixels are elements of $\mathbb{Z}^2$ (integer coordinates space). The *cellular representation* of $\mathbb{Z}^2$ associates to each pixel $(i, j) \in \mathbb{Z}^2$ a subset $C(i, j) \in \mathbb{R}^2$ (called a *cellular pixel*) so that the union of all $C(i, j)$ for $(i, j) \in \mathbb{Z}^2$ covers the whole of $\mathbb{R}^2$ [41, Sect. 4.1]. From a mathematical perspective, CP, with an interest in the closeness of sets of points, has a number of practical outcomes in mathematics, science and engineering (see, e.g., [5, 42]).

Computational Geometry. A central thrust in CP is the application of computational geometry [19, 43] in the study of digital images, leading to the recognition of image objects. The basic approach is to detect the geometry of objects in digital images by covering each image with mesh polygons surrounding (in the vicinity of) image objects. Image mesh polygons tend to reveal the shapes and identity of objects. For a sample Dirichlet tessellation of a digital image, see Fig. 1.3. A **Dirichlet tessellation** of a plane surface is a tiling of the surface. For more about this, see Sect. 1.11.

An important fringe benefit of the CP approach in digital image processing is the discovery of image patterns. There are many forms of image patterns revealed by computational proximity methods that provide a basis for classifying images. Examples of geometric image patterns are connected sets of points that form what are known as mesh nerves. The notion of nerve, mesh nerve, proximal nerves, and nervous systems in digital images are introduced in this chapter.

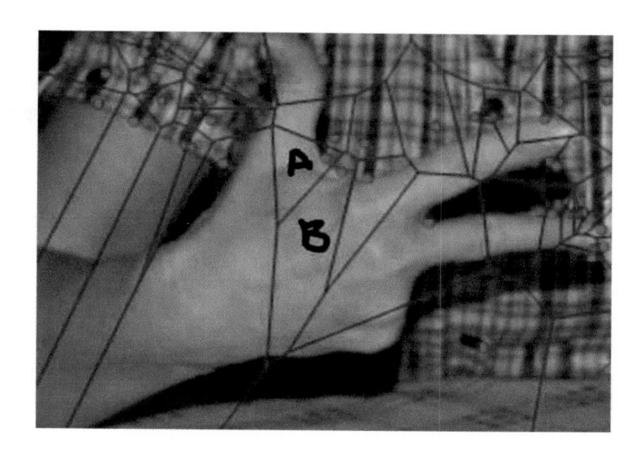

Fig. 1.3 Dirichlet tessellation

1.1 Computational Proximity Framework

A unified computational proximity framework typically includes the components shown in Fig. 1.2. Various incarnations of CP demonstrate the utility of computational proximity. CP components provide a basis for the design of systems useful in comparing, characterizing, analyzing, and classifying sets of objects such as pixels in digital images or videos and sets of nodes in social networks (see, e.g., [44]). These components are either implicitly or explicitly part of the fabric of computational systems useful in computer science, natural sciences, medical sciences and engineering applications.

Recent progress in this research provides stepping stones in the foundations and applications of computational proximity. From a foundation perspective, two types of near sets are considered, namely, spatially near and descriptively near sets. Spatially near sets were introduced by F. Riesz in 1908 [45] and axiomatized by E. Čech [46] and V.A. Efremovich in [47] during the 1930s with the axiomatizations published, starting in 1952 with Efremovich's papers and with the publication of Čech's 1930s far-reaching Brno seminar notes that first appeared in 1959 and later revised and completed in 1966. Descriptively near sets were introduced in [48] and completely axiomatized in [5, Sect. 4.15, p. 151]. Nonempty sets are Wallman near, provided the intersection of the closures of the sets is nonempty. However, near sets may not have points in common (see, e.g., [7, Sect. 1.3, p. 4]). By contrast, nonempty sets are strongly near, provided the sets have at least one point in common (introduced in [20], axiomatized in [49] and applied in [6]).

The nearness of sets in proximity spaces usually has visual (often geometric) representations. For example, let the set of points in the digital image in Fig. 1.3 be a finite topological space endowed with the Wallman proximity δ [50] and strong proximity $\overset{\wedge\wedge}{\delta}$. And let the sets of points A and B be represented by the filled polygons in the Dirichlet tessellation of the digital image in Fig. 1.3. A and B are strongly near (denoted by $A \overset{\wedge\!\!\wedge}{\delta} B$), since A and B have a common edge. The same sets are also Wallman near (denoted by $A \ \delta \ B$), since $\text{cl}A \cap \text{cl}B \neq \emptyset$.

1.2 Points, Regions, Connectedness and Point-Free Geometry

Points, regions, connectedness and point-free geometry are principal components in a computational proximity framework. A **point** is a 0-dimensional mathematical object in an n-dimensional space that has only location, specified by an *n-tuple* $(x_1, \ldots, x_n)$ containing n coordinates. For spaces with dimension $n \geq 2$, points are synonymous with vectors and are called *n-vectors*.[1] From Euclid [51, Book I, definitions], a *point* is that which has no part. The extremities of a line are points and the center of a circle is a point. Points in geometry are called **ideal** (or *abstract*) points.

In CP, the primitives are concrete points, regions and connectedness relations between such points and regions. A **concrete** point is that part of a physical surface that has both location and measurable features. A **region** is a collection of nonempty sets that can be either contain abstract points or physical objects such as picture points. The **simplest region** is a single set of points in a surface shape. A **probe** maps a geometric object to a feature value that is a real number.

There are two basic types of *object features*, namely, *object characteristic* and *object location*. For example, an object characteristic feature of a picture point is colour. And important object location of a single-set region is the geometric centroid, which is the center of mass of the region. Let X be a set of points in a $n \times m$ rectangular 2D region containing points with coordinates (x_i, y_i) , $i = 1, \ldots, n$ in the Euclidean plane.

Then, for example, the coordinates x_c, y_c of the centroid of a 2D region in the Euclidean space $\mathbb{R}^2$ are

$$x_c = \frac{1}{n}\sum_{i=1}^{n} x_i, y_c = \frac{1}{m}\sum_{i=1}^{m} y_i.$$

The coordinates x_c, y_c, z_c of the centroid of a 3D region in Euclidean space $\mathbb{R}^3$ are

$$x_c = \frac{1}{n}\sum_{i=1}^{n} x_i, y_c = \frac{1}{m}\sum_{i=1}^{m} y_i, z_c = \frac{1}{h}\sum_{i=1}^{h} z_i.$$

---

[1] For more about this, see C. Stover and E.W. Weisstein, Point, http://mathworld.wolfram.com/Point.html.

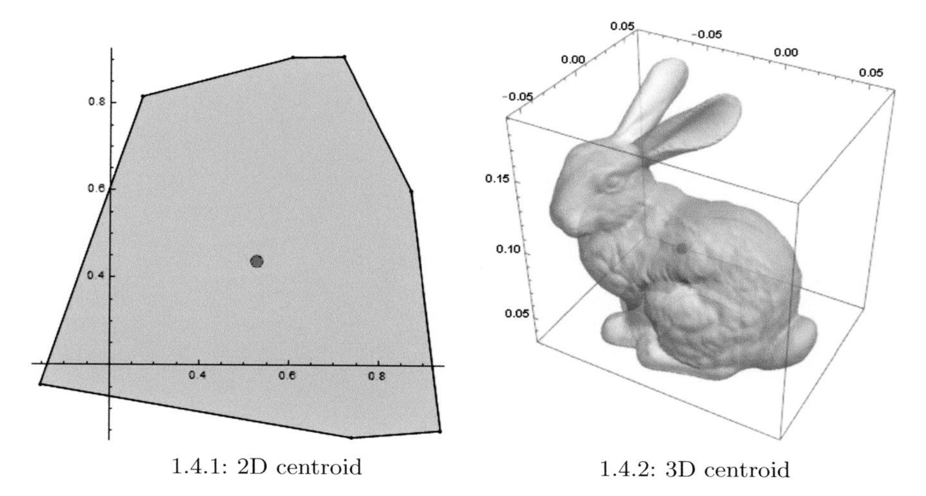

1.4.1: 2D centroid 1.4.2: 3D centroid

Fig. 1.4 2D convex region and 3D Wolfram Stanford Bunny centroids

Example 1.1 **2D and 3D Region Centroids.**
In Fig. 1.4, the red dot ● indicates the location of a region centroid. Two examples are shown, namely, centroid ● in a 2D convex region in Fig. 1.4.1 and centroid ● in a 3D region occupied with the Wolfram Stanford Bunny in Fig. 1.4.2. To experiment with finding other region centroids, see MScripts 16 and 17 in Appendix A. ■

A region **weighted centroid** is derived by multiplying each location coordinate by a quantity that represents the mass of the location. For example, in a digital image, the mass associated with each pixel is the pixel intensity. For a 2D region X of a greyscale digital image, let each pixel $p_i = (x_i, y_i)$ have an intensity g_i. Then the coordinates of the weighted centroid of region X are

$$x_c = \frac{\sum_{i=1}^{n} g_i x_i}{\sum_{i=1}^{n} g_i}, \; y_c = \frac{\sum_{i=1}^{m} g_i y_i}{\sum_{i=1}^{m} g_i}.$$

Example 1.2 **Digital Image Centroid.**
In Fig. 1.5, the black dot ● indicates the location of a digital image centroid. In Fig. 1.5.1, the digital image shown a UNISA coin from a 1982 football tournament in Salerno, Italy. In Fig. 1.5.2, the location of the centroid of the UNISA coin is identified with ●. To experiment with finding the centroids of other digital images, see MScript 18 in Appendix A. ■

Problem 1.3
🚲 Modify MScript 18 in Appendix A to find the centroid of a digital image based on the location of each picture element weighted with its intensity. ■

1.5.1: UNISA coin

1.5.2: Coin centroid at ●

Fig. 1.5 UNISA coin and UNISA coin centroid ●

Each object characteristic feature of a concrete point or region has a real value that is extracted by a probe ϕ which is a mapping $\phi : X \longrightarrow \mathbb{R}$. Let 2^X be the family of subsets of X in the Euclidean plane $\mathbb{R}^2$, A a plane region in 2^X and let (x, y) be the coordinates of the center of mass (centroid) of A. Also, let p be a point with coordinates (x_1, y_1). Each **object location** extracted by a location probe ϕ_L is a mapping

$$\begin{aligned}
\phi_L :& 2^X \longrightarrow \mathbb{R} \times \mathbb{R} \text{ (Location of a region centroid)},\\
\text{e.g., } \phi_L(A) &= (x, y) \text{ (Region } A \text{ centroid Coordinates)}.\\
\phi_L :& X \longrightarrow \mathbb{R} \times \mathbb{R} \text{ (Location of a point)},\\
\text{e.g., } \phi_L(p) &= (x_1, y_1) \text{ (Coordinates of point } p).
\end{aligned}$$

This means that each planar point or single-set region with n-features has a description defined by a feature vector in an $n + 2$-dimensional feature space. Let $\Phi(A), \Phi(p)$ denote feature vectors for a single-set region A and point p, respectively, in a space X. Then

$$\begin{aligned}
\Phi(A) &= (\phi_L(A), \phi_1(A), \dots, \phi_n(A)) \text{ (Region feature vector with location)},\\
\text{e.g., } \Phi(A) &= ((x_M, y_M), \phi_1(A), \dots, \phi_n(A)) \text{ (Region } A \text{ feature vector)},\\
&\text{with the centroid of } A \text{ at location } ((x_M, y_M)).\\
\Phi(p) &= (\phi_L(p), \phi_1(p), \dots, \phi_n(p)) \text{ (Region feature vector that includes location)},\\
\text{e.g., } \Phi(p) &= ((x, y), \phi_1(p), \dots, \phi_n(p)) \text{ (Point } p \text{ feature vector)},\\
&\text{with the centroid of } p \text{ at location } ((x, y)).
\end{aligned}$$

Remark 1.4 **Importance of location in a feature vector: Pullback.**
By including the location for either a region centroid or a point in a feature vector, it is then possible to have a pullback operation.[2] For example, let A be a single-set region in the Euclidean plane $\mathbb{R}^2$ and let a feature vector W containing $\phi_L(A)$ (location of the centroid of A and region A feature values $\phi_1(A), \ldots, \phi_n(A)$). Then the feature vector $\Phi(A)$ for A has the following form.

$$\Phi(A) = \left(\overbrace{\mathbb{R}, \mathbb{R}}^{\phi_L(A)}, \overbrace{\mathbb{R}, \mathbb{R}, \ldots, \mathbb{R}, \mathbb{R}}^{\phi_1(A),\ldots,\phi_n(A)} \right) = W$$

Let $2^{\mathbb{R}^2}$ be the family of all subsets of the Euclidean plane with subset $A \in 2^{\mathbb{R}^2}$. The mapping $\Phi : 2^{\mathbb{R}^2} \longrightarrow \mathbb{R}^{n+2}$ is defined by

$$\Phi(A) = (\phi_L(A), \phi_1(A), \phi_2(A), \ldots, \phi_{n-1}(A), \phi_n(A)) .$$

Then the **pullback mapping** $\Phi^* : \mathbb{R}^{n+2} \longrightarrow \mathbb{R}^2$ is defined by

$$\Phi^*(W) = \phi_L(A) = (x_M, y_M) \text{ (coordinates of centroid of } A).$$

The availability of a pullback for each location-based feature vector mapping $\Phi(A)$ means that each $\Phi(A)$ has an pseudo-inverse $\Phi^*(A)$. The pseudo-inverse $\Phi^*(A)$ maps to the location of the centroid of A but not back to the entire set A. To have an inverse mapping $\Phi^{-1}(A)$ to the entire set A, we can expand the definition of the mapping Φ to include the location of every point in A as part of the feature vector for the region A. This is important in the study of image manifolds in particular and topological manifolds in general. For more about this, see Sect. 8.1. ■

In a point-free geometry, the primitives are regions and a connection relation [21, 22], instead of the usual primitives, namely, *point* and *connectedness relation* in a digital geometry [52, Sect. 1.2.5]. In a computational point-free geometry, one can also speak in terms of concrete regions (instead of concrete points) and a connectedness relation. A **concrete region** is that part of a physical surface that has a non-zero diameter and measurable features in an n-dimensional space. In effect, a concrete region is a nonempty collection of sets of concrete points, which we can either ignore in the point-free geometry case or take into account in the point-based geometry case. A **spatial region** (also called an **ideal region**) is a nonempty collection of sets in a point-free geometry. A *small* region approximates an ideal point. A region-based geometry provides an alternative to point-based Euclidean geometry [22]. The intuition underlying point-free geometry is that we have a better knowledge of small

[2] Many thanks to Anna Di Concilio for pointing this out.

regions rather than points and the natural intuition that we have about the nearness of regions, either small or large.

Let X be a finite nonempty set, $A \in 2^X$ (a region *Re* in the family of subsets 2^X). In a point-free geometry, a **set** is a spatial region. Notice that the parts (sets) in both ideal (abstract) regions and concrete regions have measurable features. Each feature of a region has a real value that is extracted by a region-based probe. Briefly, a **region-part probe** or **set-based probe** is a mapping $\phi : 2^X \longrightarrow \mathbb{R}$. Diameter and boundedness are examples of features of regions and region parts, either concrete or ideal [21, Sect. 2, Definition 2.4]. This means that each region and each part of a region with n-features has a description defined by a feature vector in an n-dimensional feature space. A **connection relation** is a relation between regions that either overlap or have at least a common boundary point.

In CP, it makes sense to rise above the level of point-based geometry and consider the connectedness between regions in a point-free geometry instead of the connectedness between points in Euclidean geometry. This is the case, since region-based probes can be used to describe sets of points such as neighbourhoods of points in the usual Hausdorff space. This means that the nearness of abstract sets of points with n features can be defined descriptively by means of n-vectors in an n-dimensional feature space. And region-based probes can also be used to describe concrete regions that are nonempty sets on physical surfaces such as digital images or paintings or sets of physical objects such as nodes in a social network.

1.3 Choice of Probe

The choice of probes varies considerably, depending on the application.

Nonempty sets A, B in a topological space endowed with a descriptive Efremovich proximity are descriptively near, provided there are points $a \in A, b \in B$ such that the description of a matches the description of b. For example, the description of a picture point is defined by a feature vector containing feature values that are real numbers (values obtained using probes, e.g., colour brightness, gradient orientation). A **probe** ϕ maps a member of a set to a value in $\mathbb{R}$ (reals). There are two basic types of probes, namely, point-based probes and set-based probes. A **point-based probe** maps a point (singleton in a nonempty set) to a feature value of the point that is a real number. A **set-based probe** maps a nonempty set (non-singleton set of points) to a feature value of the set that is a real number. For more about this, see Sect. 7.1. Probe function values define feature vectors useful in comparing, clustering and classifying members of a set (see, e.g., [15]).

This notion of descriptive nearness is analogous to the similarity between art gallery paintings such as the closeness of descriptions of mountain winterscapes.

From an application of computational proximity perspective, CP carries forward extensive work in various forms of topology and proximity begun by F. Riesz [45] in 1908 and later axiomatized by V.A. Efremovich [47] and descriptively near sets introduced in [48] and later completely axiomatized in [5, Sect. 4.15, p. 151].

Computational geometry [19, 43, 52–54] is an important part of Computational Proximity inasmuch as it serves as a stepping stone in the analysis of large data sets. The application of finite topology in image analysis is not new (see, e.g., [5], [19, Sect. IX], [55]) and for many years there has been speculation about the applicability of proximity space theory in image and music analysis (see, e.g., [2]) as well as in visual merchandising, camouflage detection, microscopy and forgery detection [7]. By contrast, CP offers a thorough-going algorithmic approach that is both novel and highly useful in many applications.

Seed points (also called sites or generators) provide a basis for generating Dirichlet tessellations (also called Voronoï diagrams) and Delaunay triangulations of sets of points, providing a basis for the construction of meshes that cover a set with clusters of polygonal shapes. For an overview of the different types of mesh seeds or generating points, see Sect. 7.1. In general, a **tessellation** of a plane surface is a tiling of the surface. A **plane tiling** is a plane-filling arrangement of plane figures that cover the plane without gaps or overlaps [56]. The plane figures are closed sets. Taken by themselves or in combination, pixel intensity, corner, edge, centroid, salient, critical and key points are examples of seed points with many variations.

Example 1.5 **Sample Mesh Seed (Generating) Points**.
Three sample tessellation-hand hand segments are shown in Fig. 1.6. Initially, the image segments are polygons derived with seed (generating) points. Consider, for example, seed points that are image intensities in Fig. 1.6.1, corners in Fig. 1.6.2, and keypoints in Fig. 1.6.3. In this instance, the keypoint tessellation in Fig. 1.6.3 is the most promising in solving many image analysis problems.

There is no one best seed point. Rather the choice of seed points depends the characteristics of the members of a data set. Choices of seed points targeted for particular data sets provide a fundamentally important component in setting up a computational

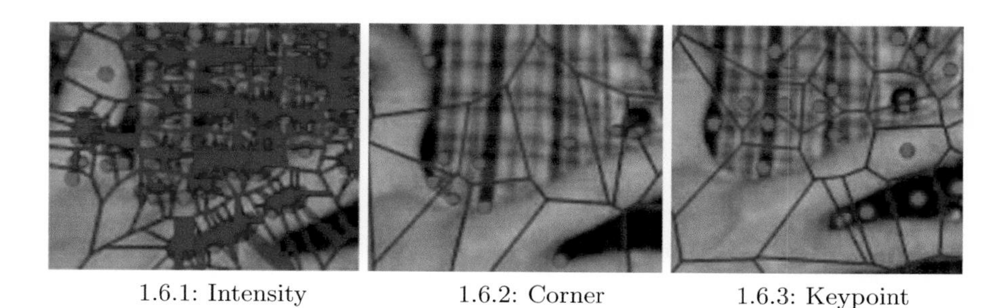

1.6.1: Intensity 1.6.2: Corner 1.6.3: Keypoint

Fig. 1.6 Dirichlet tessellations

proximity framework. The Mathematica script MScript 19 in Appendix A is used to construct the tessellated hand images in Fig. 1.6. ■

Remark 1.6 **Types of Mesh Seed (Generating) Points.**
For an overview of the different types of mesh seed or generating points, see Sect. 7.1. From an image segmentation perspective, observe that the choice of keypoint seeds yields a better distribution of mesh polygons in Fig. 1.6 than either corner or intensity seed points. However, if the aim is to use CP to construct new clusters by merging clusters of points extracted from neighbouring mesh polygons, then the centroid seed point is a better choice. It has been observed that *the resemblance between two clusters equals the resemblance between their centroids* [57].

> Resemblance between either points or regions in CP is defined in terms of the feature vectors that describe either the points or the regions. ■

Let ϕ be a probe function that maps a point to its feature value. And let $\phi_1, \ldots, \phi_n$ be n probes. For a point x, $\Phi(x)$ denotes the feature vector of x, where

$$\Phi(x) = (\phi_1(x), \ldots, \phi_i(x), \ldots, \phi_n(x)), i \in \{1, \ldots, n\}.$$

In terms of a CP approach to merging mesh clusters based on region centroids c_1, c_2, a comparison needs to be made between the feature vector $\Phi(c_1)$ and the feature vector $\Phi(c_2)$, e.g.,

$$\begin{aligned} &\varepsilon > 0, \\ \phi_j(x) = &\text{ probe } \phi_j \text{ maps } x \text{ to feature value of } x, j \in \{1, \ldots, n\} \\ \Phi(c_i) = &\text{ feature vector } (\phi_1(c_1), \ldots, \phi_j(x), \ldots, \phi_n(c_n)), i \in \{1, 2\} \\ \text{if } |\Phi(c_1) - \Phi(c_2)| < \varepsilon, &\text{ then merge regions containing } c_1, c_2. \end{aligned}$$

In effect, if a pair of centroids are descriptively close, their feature vectors will be close. This approach has proved to be effective in merging region clusters in [15]. ■

Proximity. **n.**, from **Fr.** *proximité*.

Nearness in space, time, or relationship.

Etymologically, *state of being next to*.

Nextness.

–Oxford English Dictionary, 1933.

The effect of the proximity was strong adhesion of the bodies.

–G. Adams, Nat. & Exp. Philos. III, xxv, 67, 1794.

Strong. **adj**.., cf. **Gk.** $\sigma\tau\rho\alpha\gamma\gamma\acute{o}s$.

Tightly twisted.

–An Etymological Dictionary of the English Language, 1983.

Where a strong clayey soil is covered with a healthy vegetation.
–London Suburban Hort., 1842.

Strong Proximity. **Expr.**.

1. *Tightly twisted nearness in space, time, or relationship.*
2. *Tightly twisted overlapping nextness*, e.g., shapes A, B:

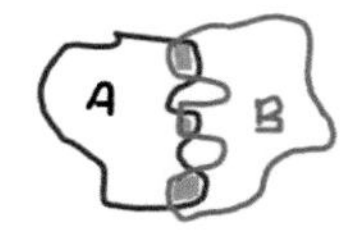

1.4 Proximities

This section introduces two basic types of proximities, namely, traditional *spatial proximity* and the more recent *descriptive proximity*. Nonempty sets that have **spatial proximity** are close to each other, either asymptotically or with common points. Nonempty sets that have **descriptive proximity** are close, provided the sets contain one or more elements that have matching descriptions. A commonplace example of descriptive proximity is a pair of paintings that have matching parts such as matching facial characteristics, matching eye, hair, skin colour, or matching nose, mouth, ear shape. Each of these proximities has a strong form. A **strong proximity** embodies a special form of tightly twisted nextness of nonempty sets. In simple terms, this means sets that share elements, have strong proximity. For example, the shaded parts , ,  in

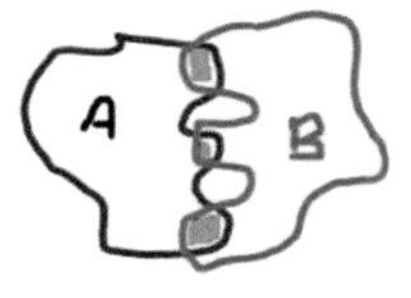

contain points in the Euclidean plane that are shared by the tightly twisted (overlapping) shapes A and B.

For an example of proximal but not strongly proximal physical objects, see the adjacent Feudi di San Gregorio wine casks in Fig. 1.7. The adjacent wine casks touch each other but do not have have common points. This illustrates one of the great subtleties of traditional spatial proximities. That is, proximal non-empty sets can be as close as possible (touch each other) and yet not share any points (e.g., there is no point on the surface of one adjacent wine cask that is also a point on the surface of the other wine cask).

Proximities are nearness relations. In other words, a *proximity* between nonempty sets is a mathematical expression that specifies the closeness of the sets. A **proximity**

Fig. 1.7 Proximity (nextness) of adjacent Feudi di San Gregorio wine casks

space results from endowing a nonempty set with one or more proximities. Typically, a proximity space is endowed with a common proximity such as the proximities from Čech [46], Efremovič [47], Lodato [10], and Wallman [50], or the more recent descriptive proximity [48, 58, 59].

A pair of nonempty sets in a proximity space are *near* (*close to each other*), provided the sets have one or more points in common or each set contains one or more points that are sufficiently close to each other. Let X be a nonempty set, $A, B, C \subset X$. E. Čech [46] introduced axioms for the simplest form of proximity δ_C, which satisfies

Čech Proximity Axioms [46, Sect. 2.5, p. 439]

(P1) $\varnothing \not\delta A, \forall A \subset X$.
(P2) $A \; \delta \; B \Leftrightarrow B\delta A$.
(P3) $A \cap B \neq \varnothing \Rightarrow A\delta B$.
(P4) $A \; \delta \; (B \cup C) \Leftrightarrow A \; \delta \; B$ or $A \; \delta \; C$. ■

A fifth axiom (called the separation or EF axiom) was introduced by V.A. Efremovič [47]. $A \not\delta B$ reads A is not close to (far from) B. A proximity δ_E is EF, provided it satisfies Čech's axioms and the following axiom.

Efremovič Proximity Axiom [47]

(P5$_E$) If, for any $A, B \subset X, A \not\delta B$, there exists $C, D \subset X$, $C \cup D = X$ such that $A \not\delta C$ and $B \not\delta D$. ■

M.W. Lodato [10] swapped out the EF axiom for the following axiom and introduced the proximity δ_L. The proximity δ_L satisfies the Čech proximity axioms and

Lodato Proximity Axiom [10]

(P5$_L$) $A\ \delta_L\ B$ and $\{b\}\ \delta_L\ C$ for each $b \in B\ \Rightarrow A\ \delta_L\ C$. ■

Recall that the closure of a nonempty set A (denoted cl$A$) is defined by

Closure [46, Sect. 2.5, p. 439]

$$\text{cl}A = \{x \in X : x\ \delta\ A\}. \quad ■$$

The closure of a nonempty set A satisfies the Kuratowski axioms [60, Sect. 4.III] [61, Chap. X, Sect. 1]. Let X be a topological space, $A, B \subset X$. The closure mapping $\text{cl} : 2^X \longrightarrow 2^X$ satisfies the following axioms.

cl.1 $\text{cl}A \cup \text{cl}B = \text{cl}(A \cup B)$.
cl.2 $A \subset \text{cl}A$.
cl.3 $\text{cl}\varnothing = \varnothing$.
cl.4 $\text{cl}(\text{cl}A) = \text{cl}A$. ■

For more about the closure of a set, see Appendix B.

A Čech, Efremovič or Lodato proximity becomes a Wallman proximity δ_0 (introduced by H. Wallman [50]), provided the following axiom is satisfied.

Wallman Proximity Axiom [50]

(P$_W$) $A\ \delta_0\ B$ if and only if $\text{cl}A \cap \text{cl}B \neq \varnothing$. ■

Descriptive Intersection [7, Sect. 4.3, p. 84].

($\boldsymbol{\Phi}$) $\Phi(A) = \{\Phi(x) \in \mathbb{R}^n : x \in A\}$, set of feature vectors.
($\underset{\Phi}{\cap}$) $A \underset{\Phi}{\cap} B = \{x \in A \cup B : \Phi(x) \in \Phi(A) \& \Phi(x) \in \Phi(B)\}$. ■

The descriptive proximity δ_Φ was introduced in [48, 58, 59]. Let $\Phi(x)$ be a feature vector for $x \in X$, a nonempty set of non-abstract points such as picture points. $A\ \delta_\Phi\ B$ reads A is descriptively near B, provided $\Phi(x) = \Phi(y)$ for at least one pair of points, $x \in A, y \in B$. The proximity δ in the Čech, Efremovič, and Wallman proximities is replaced by δ_Φ. Then swapping out δ with δ_Φ in each of the Lodato axioms defines a descriptive Lodato proximity that satisfies the following axioms.

Descriptive Lodato Axioms [5, Sect. 4.15.2]

(dP0) $\varnothing \not\delta_{\Phi} A, \forall A \subset X$.
(dP1) $A \ \delta_{\Phi} \ B \Leftrightarrow B \ \delta_{\Phi} \ A$.
(dP2) $A \underset{\Phi}{\cap} B \neq \varnothing \Rightarrow A \ \delta_{\Phi} \ B$.
(dP3) $A \ \delta_{\Phi} \ (B \cup C) \Leftrightarrow A \ \delta_{\Phi} \ B$ or $A \ \delta_{\Phi} \ C$.
(dP4) $A \ \delta_{\Phi} \ B$ and $\{b\} \ \delta_{\Phi} \ C$ for each $b \in B \ \Rightarrow A \ \delta_{\Phi} \ C$. ■

Further δ_{Φ} is *descriptively separated*, if

(dP5) $\{x\} \ \delta_{\Phi} \ \{y\} \Rightarrow \Phi(x) = \Phi(y)$ (x and y have matching descriptions). ■

A descriptive form of the closure of a set results from replacing the usual proximity δ with δ_{Φ}. This leads to a descriptive form of the Wallman proximity axiom.

Descriptive Closure [5, Sect. 1.21.2]

$$\mathrm{cl}_{\Phi}A = \{x \in X : x \ \delta_{\Phi} \ A\} . \quad ■$$

With descriptive proximity, the Čech and Lodato proximity axioms are rewritten, replacing δ with δ_{Φ}. The descriptive proximity δ_{Φ} is Wallman, provided the following axiom is satisfied.

Descriptive Wallman Proximity [5, Sect. 4.15.2, Theorem 4.3]

$(\delta_{\Phi_{cl}})$ $A\delta_{\Phi}B$ if and only if $\mathrm{cl}_{\Phi}A \ \cap \ \mathrm{cl}_{\Phi}B \neq \varnothing$. ■

Strong Proximity [Di Concilio Strong Contact] [20, Sect. 2].

The relation $\overset{\wedge\!\!\wedge}{\delta}$ is a **strong proximity** or *strong nearness* (briefly, **sn**), provided it satisfies the following axioms.

(snN0) $\varnothing \overset{\not\delta}{\vee\!\!\vee} A, \forall A \subset X$, and $X \overset{\wedge\!\!\wedge}{\delta} A, \forall A \subset X$.
(snN1) $A \overset{\wedge\!\!\wedge}{\delta} B \Leftrightarrow B \overset{\wedge\!\!\wedge}{\delta} A$.
(snN2) $A \overset{\wedge\!\!\wedge}{\delta} B$ implies $A \cap B \neq \varnothing$.
(snN3) If $\{B_i\}_{i \in I}$ is an arbitrary family of subsets of X and $A \overset{\wedge\!\!\wedge}{\delta} B_{i^*}$ for some $i^* \in I$ such that $\mathrm{int}(B_{i^*}) \neq \varnothing$, then $A \overset{\wedge\!\!\wedge}{\delta} (\bigcup_{i \in I} B_i)$
(snN4) $\mathrm{int}A \ \cap \ \mathrm{int}B \neq \varnothing \Rightarrow A \overset{\wedge\!\!\wedge}{\delta} B$. For the details, see Sect. 1.5. ■

When we write $A \overset{\wedge\!\!\wedge}{\delta} B$, we read A is *strongly near* B. For each *strong proximity*, we assume the following relations:

(snN5) $x \in \text{int}(A) \Rightarrow x \overset{\wedge\!\!\!\wedge}{\delta} A$

(snN6) $\{x\} \overset{\wedge\!\!\!\wedge}{\delta} \{y\} \Leftrightarrow x = y$ ■

Descriptive Strong Lodato Proximity [49].

To obtain a **descriptive strong Lodato proximity** (denoted by **dsn**), we swap out δ_{Φ} in each of the descriptive Lodato axioms with the descriptive strong proximity $\overset{\wedge\!\!\!\wedge}{\delta}_{\Phi}$, which satisfies the following axioms.

(dsnP0) $\varnothing \not{\overset{\wedge\!\!\!\wedge}{\delta}}_{\Phi} A, \forall A \subset X$, and $X \overset{\wedge\!\!\!\wedge}{\delta}_{\Phi} A, \forall A \subset X$.

(dsnP1) $A \overset{\wedge\!\!\!\wedge}{\delta}_{\Phi} B \Leftrightarrow B \overset{\wedge\!\!\!\wedge}{\delta}_{\Phi} A$.

(dsnP2) $A \overset{\wedge\!\!\!\wedge}{\delta}_{\Phi} B$ implies $A \underset{\Phi}{\cap} B \neq \varnothing$.

(dsnP4) $\text{int}A \underset{\Phi}{\cap} \text{int}B \neq \varnothing \Rightarrow A \overset{\wedge\!\!\!\wedge}{\delta}_{\Phi} B$. For the details, see Sect. 1.9. ■

When we write $A \overset{\wedge\!\!\!\wedge}{\delta}_{\Phi} B$, we read A is *descriptively strongly near* B. For each *descriptive strong proximity*, we assume the following relations:

(dsnP5) $\Phi(x) \in \Phi(\text{int}(A)) \Rightarrow x \overset{\wedge\!\!\!\wedge}{\delta}_{\Phi} A$.

(dsnP6) $\{x\} \overset{\wedge\!\!\!\wedge}{\delta}_{\Phi} \{y\} \Leftrightarrow \Phi(x) = \Phi(y)$. ■

When it is clear from the context what is meant, a plain δ denotes a particular form of proximity. ■

Among the different types of proximities in this section, Wallman proximity is fundamentally important, since it paves the way for a consideration of the nearness of closed sets, either with or without nonempty interiors. When it is clear from the context what is meant, we usually write $A\ \delta\ B$ to indicate the Wallman closeness of A and B.

Example 1.7 **Nearness of Torus Segments.**
For example, let X be the set of points on the surface mesh of the torus in Fig. 1.1. The rectangular shaped segments covering the torus surface are examples of nonempty sets in a proximity space. That is, let the points in the torus X be a topological space equipped with the Wallman proximity δ, $A, B \subset X$. Each pair of neighbouring torus segments shown in Fig. 1.1 are proximally close. The closure of the torus segments A and B in Fig. 1.1, for instance, have a vertex in common. Hence, $A\ \delta\ B$. The descriptive proximity case, let $\Phi(A)$, $\Phi(B)$ be descriptions of A and B. For example, each $x \in \Phi(A)$ is a feature vector in an n-dimensional Euclidean space $\mathbb{R}^n$. Again, also assume X is equipped with the Wallman descriptive proximity δ_{Φ}, so that X

is a descriptive Wallman proximity space. In that case, the torus segments that are the same colour form a chain of descriptively near sets wrapping around the torus surface. ■

Example 1.8 **Descriptive Nearness of Torus Segments.**
Let the points in the torus X be a topological space equipped with the Wallman proximity δ and Wallman descriptive proximity $\delta_\Phi, A, B \subset X$. If we take into account the colour of the torus segments in Fig. 1.1, then, for example, $A\ \delta_\Phi\ B$, i.e., torus surface polygon A is descriptively close to polygon B. In fact, all of the torus surface polygons with the same colour are descriptively close, independent of whether the polygons have points in common or not. That is,

$$\begin{aligned}
&A\ \delta_\Phi\ B \text{ and } A\ \delta\ B,\\
&\qquad \text{(Wallman descriptively close and Wallman close), i.e.}\\
&\qquad \mathrm{cl}_\Phi A\ \cap\ \mathrm{cl}_\Phi B \neq \emptyset \text{ and } \mathrm{cl}A \cap \mathrm{cl}B \neq \emptyset.\\
&A\ \delta_\Phi\ H \text{ and } A\ \not\delta H,\\
&\qquad \text{(Wallman descriptively close and Wallman far), i.e.}\\
&\qquad \mathrm{cl}_\Phi A\ \cap\ \mathrm{cl}_\Phi H \neq \emptyset \text{ and } \mathrm{cl}A \cap \mathrm{cl}H = \emptyset.\\
&B\delta_\Phi H \text{ and } B\ \not\delta H,\\
&\qquad \text{(Wallman descriptively close and Wallman far), i.e.}\\
&\qquad \mathrm{cl}_\Phi B\ \cap\ \mathrm{cl}_\Phi H \neq \emptyset \text{ and } \mathrm{cl}B \cap \mathrm{cl}H = \emptyset.
\end{aligned}$$

In other words, nonempty sets of non-abstract points can be either spatially and descriptively close or spatially far and descriptively close. This is the case with polygons A and B in Fig. 1.1.

For example, polygons A and B in Fig. 1.8 are both spatially close (A and B have a common edge) and descriptively close (A and B have the same colour). By contrast, consider the torus surface polygons A and H in Fig. 1.8. In this case, polygons A and H are descriptively close (both polygons have the same colour) and yet these polygons are spatially far apart, i.e., these polygons have no points in common. ■

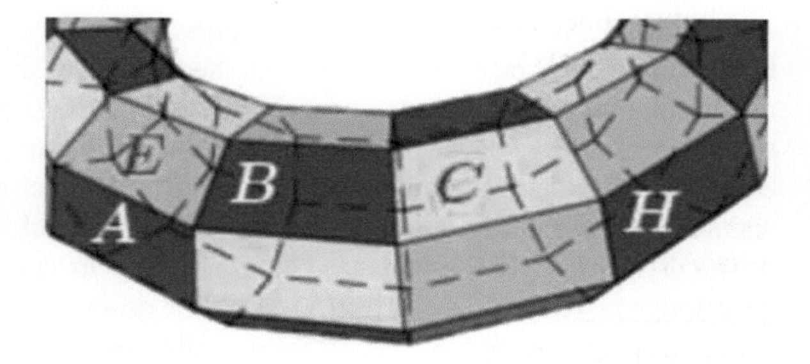

Fig. 1.8 Descriptively close, spatially remote torus polygons

1.5 Di Concilio Strong Contact

Nonempty sets A, B in a topological space X equipped with the relation $\overset{\wedge\wedge}{\delta}$, are **strongly near [strongly contacted]** (denoted $A \overset{\wedge\wedge}{\delta} B$), provided the sets have at least one point in common. The strong contact relation $\overset{\wedge\wedge}{\delta}$ was introduced in [20, Sect. 2] and axiomatized in [11, Sect. 6 Appendix], [49]. Strong Lodato proximities satisfy a number of axioms given in Definition 1.9.

Definition 1.9 Strong Proximity [49].

Let X be a topological space, $A, B, C \subset X$ and $x \in X$. The relation $\overset{\wedge\wedge}{\delta}$ on the family of subsets 2^X is a **strong proximity [Di Concilio strong contact]**, provided it satisfies the following axioms.

(snN0) $\emptyset \overset{\wedge\wedge}{\not\delta} A, \forall A \subset X$, and $X \overset{\wedge\wedge}{\delta} A, \forall A \subset X$

(snN1) $A \overset{\wedge\wedge}{\delta} B \Leftrightarrow B \overset{\wedge\wedge}{\delta} A$

(snN2) $A \overset{\wedge\wedge}{\delta} B \Rightarrow A \cap B \neq \emptyset$

(snN3) If $\{B_i\}_{i \in I}$ is an arbitrary family of subsets of X and $A \overset{\wedge\wedge}{\delta} B_{i^*}$ for some $i^* \in I$ such that $\text{int}(B_{i^*}) \neq \emptyset$, then $A \overset{\wedge\wedge}{\delta} (\bigcup_{i \in I} B_i)$

(snN4) $\text{int}A \cap \text{int}B \neq \emptyset \Rightarrow A \overset{\wedge\wedge}{\delta} B$ ■

When we write $A \overset{\wedge\wedge}{\delta} B$, we read A is **strongly near** B (A **strongly contacts** B). The notation $A \overset{\wedge\wedge}{\not\delta} B$ reads A is not strongly near B (A does not **strongly contact** B). For each *strong proximity* (*strong contact*), we assume the following relations:

(snN5) $x \in \text{int}(A) \Rightarrow x \overset{\wedge\wedge}{\delta} A$

(snN6) $\{x\} \overset{\wedge\wedge}{\delta} \{y\} \Leftrightarrow x = y$ ■

For strong proximity of the nonempty intersection of interiors, we have that $A \overset{\wedge\wedge}{\delta} B \Leftrightarrow \text{int}A \cap \text{int}B \neq \emptyset$ or either A or B is equal to X, provided A and B are not singletons; if $A = \{x\}$, then $x \in \text{int}(B)$, and if B too is a singleton, then $x = y$. It turns out that if $A \subset X$ is an open set, then each point that belongs to A is strongly near A. The bottom line is that strongly near sets always share points, which is another way of saying that sets with strong contact have nonempty intersection.

Remark 1.10 **Naming $\overset{\wedge\wedge}{\delta}$.**

If we consider the axioms that a proximity relation satisfies, the relation $\overset{\wedge\wedge}{\delta}$ is more appropriately called **overlap** or **strong contact**, since Axiom **(N2)** differs from the usual Axiom **(P3)** for each of the traditional forms of proximity.[3] This axiom indicates

---

[3]Many thanks to A. Di Concilio for pointing this out and suggesting *overlap* or *strong contact* as a more appropriate name for $\overset{\wedge\wedge}{\delta}$. For an in-depth view of *overlap*, see [21], especially [22, Sect. 3].

that for each instance of $A \overset{\wedge\!\!\!\wedge}{\delta} B$, A and B have one or more points in common. This is markedly different from traditional proximity, since near sets may not have any points in common. If we consider the significance of the adjective *strong* in naming $\overset{\wedge\!\!\!\wedge}{\delta}$, then *strong contact* aptly describes this form of nearness. That is, $\overset{\wedge\!\!\!\wedge}{\delta}$ is **strong**, since the *strong closeness* (**strong contactedness**) of nonempty sets means sets that share members.

However, if we consider the metric view of proximity, then *strong proximity* correctly names the relation $\overset{\wedge\!\!\!\wedge}{\delta}$. It is easy to verify that $D(A, B) = 0$ for $A \overset{\wedge\!\!\!\wedge}{\delta} B$, provided $\overset{\wedge\!\!\!\wedge}{\delta}$ is the strong metric proximity. By definition [9], whenever $D(A, B) = 0$, A and B are near sets. ■

Theorem 1.11 $A \overset{\wedge\!\!\!\wedge}{\delta} B \Rightarrow A\ \delta\ B$.

Proof Immediate from Axiom (snN2) and Čech Axiom (P3). □

Basically, strongly near sets have *at least* one point in common. This strong requirement contrasts with the usual view of the proximity of sets. This is the case, since, for each of the usual forms of proximities such as Lodato and Efremovič proximities, nonempty sets can be near and not have any points in common.

Example 1.12 **Near Sets with No Points in Common**
Let X be a topological space endowed with a proximity δ, $A = \{(x, 0) : x > 0\}$, $B = \{(x, y) : y = 1/x\}$ (see Fig. 1.9) and X is a metric space with the metric d, i.e., if $x, y \in X$, then $d(x, y)$ is the distance between x and y. Then, for nonempty sets $A, B \subset X$, the Čech distance $D(A, B)$ is defined by

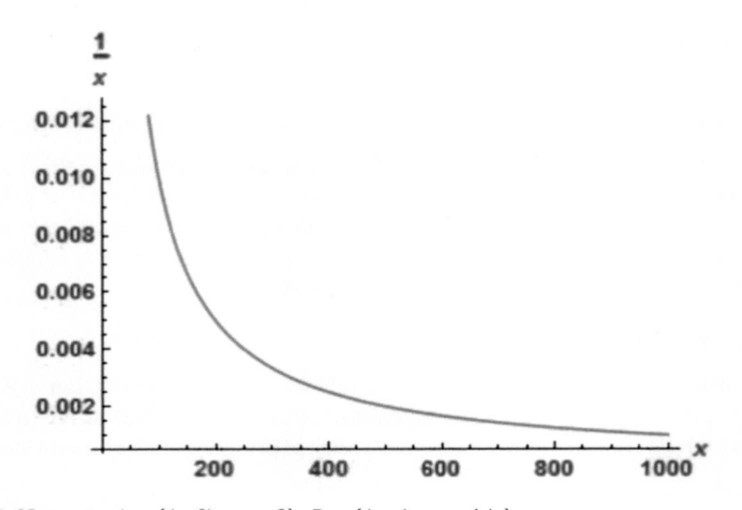

Fig. 1.9 Near sets: $A = \{(x, 0) : x > 0\}$, $B = \{(x, y) : y = 1/x\}$

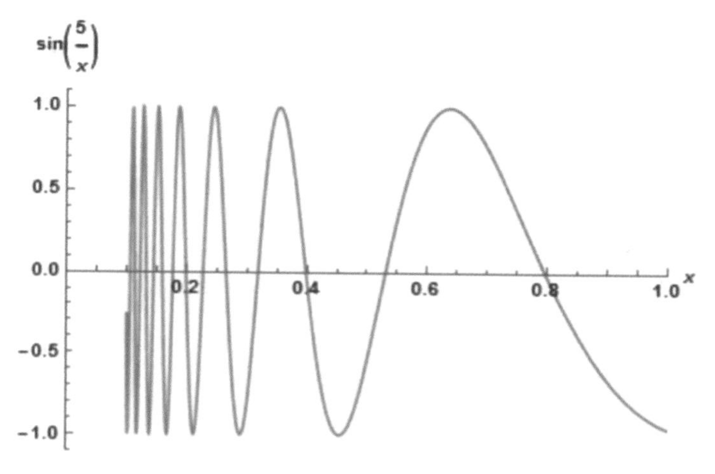

Fig. 1.10 Near sets: $A = \{(x, 0) : 0.1 \leq x \leq 1\}, B = \{(x, \sin(5/x)) : 0.1 \leq x \leq 1\}$

$$\begin{aligned} D(A, B) &= \inf\{d(a, b) : a \in A, b \in B\}, \\ &= \text{greatest lower bound of the distances } d(a, b). \end{aligned}$$

Recall that $A \ \delta \ B$, if and only if $D(A, B) = 0$, provided δ is the metric proximity [62, Sect. 1, p. 7]. Hence, A, B represented by Fig. 1.9 are near sets but have no points in common. For more about this, experiment with MScript 20 in Appendix A.1. ■

Example 1.13 **Strongly Near Sets with Points in Common**

Let X be a topological space endowed with the strong proximity $\overset{\wedge\!\!\wedge}{\delta}$, $\{(x, 0) : 0.1 \leq x \leq 1\}$, $B = \{(x, \sin(5/x)) : 0.1 \leq x \leq 1\}$. In this case, A, B represented by Fig. 1.10 are strongly near sets with many points in common. For more about this, see MScript 21 in Chap. A, Appendix A.A.1. ■

1.6 Strongly Far Proximity

The strongly far proximity (denoted $\underset{\vee\!\!\vee}{\not\delta}$) [63] is introduced in this section.

Definition 1.14 Strongly Far Proximity.
Let X be a nonempty set and δ be the *Lodato proximity* on the family of subsets 2^X.

We say that A and B are δ-**strongly far** and we write $\underset{\vee\!\!\vee}{\not\delta}$, if and only if $A \not\delta B$ and there exists a subset C of X such that $A \not\delta X \setminus C$ and $C \not\delta B$, i.e., the Efremovič property holds on A and B. ■

Notice that $A \not\delta B$ does not imply $A \overset{\not\delta}{\Vee} B$. In fact, this is the case when the proximity δ is not an *EF-proximity*.

Theorem 1.15 *The relation* $\overset{\not\delta}{\Vee}$ *is a basic proximity.*

Proof Immediate from the properties of δ. □

Example 1.16 In Fig. 1.12, let X be a nonempty set endowed with the euclidean metric proximity δ_e, $C, E \subset X, A \subset C, B \subset E$. Clearly, $A \overset{\not\delta_e}{\Vee} B$ (A is strongly far from B), since $A \not\delta_e B$ so that $A \not\delta_e X \setminus C$ and $C \not\delta_e B$. Also observe that the Efremovič property holds on A and B. ■

Example 1.17 **Strongly Near and Strongly Far Polygons.**
Let the points in the torus X be a topological space equipped with the strong proximity $\overset{\Wedge}{\delta}$ and each type of standard proximity $\delta, A, E, B, C \subset X$. In Fig. 1.11, $E \overset{\Wedge}{\delta} B$, since the quadrilaterals E and B have an edge in common. Similarly, $A \overset{\Wedge}{\delta} B$, since the quadrilaterals A and B have a vertex in common. However, $E \overset{\not\delta}{\Vee} C, A \overset{\not\delta}{\Vee} C$ in Fig. 1.11, since the quadrilaterals E, C and A, C have no points in common. In addition, $E \not\delta C, A \not\delta C$ for each type of proximity. ■

Remark 1.18 **Strongly Far Sets.**
In general, nonempty sets that have no points in common are remote, far from each other. The strongly far proximity $\overset{\not\delta}{\Vee}$ becomes useful with sets that not only have no points in common but also are, in some sense, very remote. For example, the sets of points in the torus solid surface polygons E and C in Fig. 1.11 are considered very remote, since they are separated from each other by the polygon B. Carrying this a step further, let $\mathcal{G} \setminus C$ be a collection of torus surface polygons that includes C, H and E is not a member of $\mathcal{G}$ in Fig. 1.11. In this case, $C \in \mathcal{G}$ is a proper subset in $\mathcal{G}$ (this is an example of what is known as *strong inclusion* of one set in another set). In looking for strongly far sets E and C, check whether C is strongly included in another set $\mathcal{G}$ that does not include E. In that case, $E \overset{\not\delta}{\Vee} C$. ■

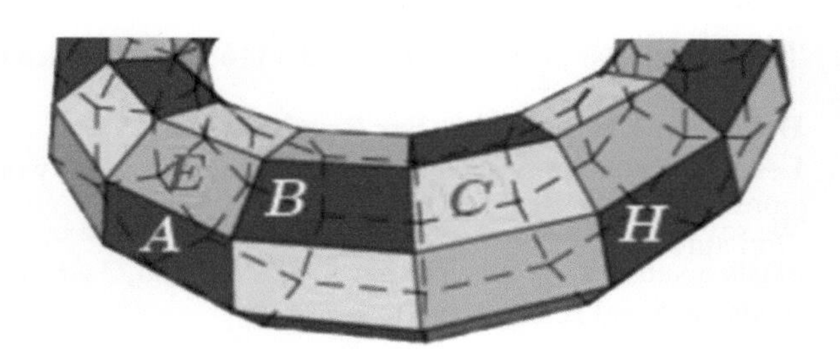

Fig. 1.11 $E \overset{\Wedge}{\delta} B, A \overset{\Wedge}{\delta} B$ but $E \overset{\not\delta}{\Vee} C, A \overset{\not\delta}{\Vee} C$

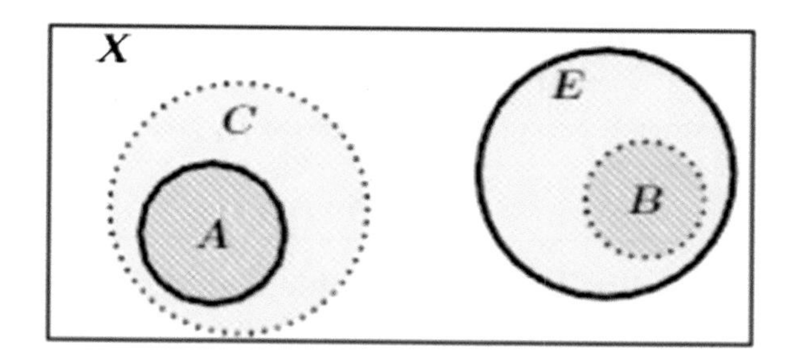

Fig. 1.12 Strongly far proximity $\underset{\doublevee}{\not\delta}$

Example 1.19 **Strongly Near Torus Surface Polygons.**
The torus surface rectangular segments B and C in Fig. 1.1 are examples of strongly near sets, since this pair of polygons has an edge in common. In that case, we write $B \stackrel{\doublewedge}{\delta} C$. None of the torus surface polygons with the same colour such as are strongly near, since none of the same colour torus polygons have more than one point in common. Even so, every pair of torus surface polygons are near, e.g., $A\ \delta\ B$ expresses this closeness of A and B. The counterpart of near and strongly near sets are those nonempty sets that are either *far* (not close) (denoted by $\not\delta$) or *strongly far* (denoted by $\underset{\doublevee}{\not\delta}$). The interior of polygon B is strongly far from polygon C in Fig. 1.1, i.e., $A \underset{\doublevee}{\not\delta} C$. *Far sets* have no points in common. For example, in Fig. 1.1, the set of points in the interior of A (denoted int$A$) are far from the set of points in int$B$, i.e., $\text{int}A\ \not\delta\ \text{int}B$. ■

Notice that although the torus segments A and B in Fig. 1.1 are neighbourly (in fact, strongly near), the interiors of these neighbourly sets are far, not close to each other. In the case where sets with nonempty interiors are not neighbourly, then the interiors of such sets are *strongly far* from each other (denoted $A \underset{\doublevee}{\not\delta} B$).

Example 1.20 **Strongly Far Torus Segments.**
For example, in Fig. 1.1, the interior of A is strongly far from the interior of C, i.e., $A \underset{\doublevee}{\not\delta} C$. In addition, the interior of A is strongly far from cl$C$, i.e., $\text{int}A \underset{\doublevee}{\not\delta} \text{cl}C$. ■

Remark 1.21 **Strongly Near and Strongly Far Sets.**
Strongly near sets [49] and strongly far [63, 64] are principal parts of computational proximity. Such sets are important in discovering closely linked parts of patterns found in collections of sets. By contrast, strongly far sets offer a means finding clearcut separation between sets in a family of sets. Both types of proximity expressions have many practical applications, especially in the study of patterns among sets of picture points in digital images or in sets of nodes in social networks. ■

1.7 Convexity Structures

The strongly near proximity can be used to give us a new formulation of the definition of convexity.

Zelins'kyi-Kay-Womble Convexity Structure.
V.P. Solan [65] defines a convexity structure in the following way: *A family of convex sets has the* **convexity property**, *provided the intersection of any number of sets in the family belongs to the family*. Y.B. Zelins'kyi [66] observes that Solan's view of convexity means that the set of all subsets of a set is convex. The notion of axiomatic convexity from Solan and Zelins'kyi has its origins in the 1971 paper by D.C. Kay and E.W. Womble [67], elaborated by V.V. Tuz [68]. Solan's view of axiomatic convexity leads to strong contact convexity spaces in terms of the family of all subsets of a set in which all subfamilies of subsets that have strong contact. ■

Definition 1.22 Zelins'kyi-Kay-Womble Convexity Structure.
Let $\mathscr{F} = 2^X$ be the family of all subsets of a nonempty set X and let subfamilies $\mathscr{A}, \mathscr{B} \in \mathscr{F}$. The family $\mathscr{F}$ on X is called a **Zelins'kyi-Kay-Womble convexity structure**, provided it satisfies the following axioms.

(nC0) $\varnothing$ and X belong to $\mathscr{F}$.
(nC1) $\mathscr{A} \cap \mathscr{B} \in \mathscr{F}$ for all subfamilies $\mathscr{A}, \mathscr{B} \in \mathscr{F}$. ■

The pair $(X, \mathscr{F})$ is a **Zelins'kyi-Kay-Womble convexity space**.

Theorem 1.23 *The family of all subsets $\mathscr{F} = 2^X$ of a nonempty set X is a Zelins'kyi-Kay-Womble convexity structure.*

Proof Let $\mathcal{A} \in \mathscr{F}$. X and $\varnothing$ are in $\mathscr{F}$. In addition, $\bigcap\limits_{A \in \mathcal{A}} A \in \mathscr{F}$. Hence, $\mathscr{F}$ is a Zelins'kyi-Kay-Womble convexity structure. □

Example 1.24 **Family of Subsets Convexity Structure.**
Let $\mathscr{F} := 2^X, X = \{1, 2, 3, 4\}$ (see Fig. 1.13). Observe that $X, \varnothing \in \mathscr{F}$ and

$$\bigcap_{\forall \mathcal{A} \in \mathscr{F}} \mathcal{A} \in \mathscr{F}.$$

For example, let $\mathcal{A} = \{\{1, 3\}, \{3, 4\}\}$ and $\{1, 3\} \cap \{3, 4\} = \{3\} \in \mathscr{F}$. Hence, $\mathcal{A}$ is a Zelins'kyi-Kay-Womble convexity structure. ■

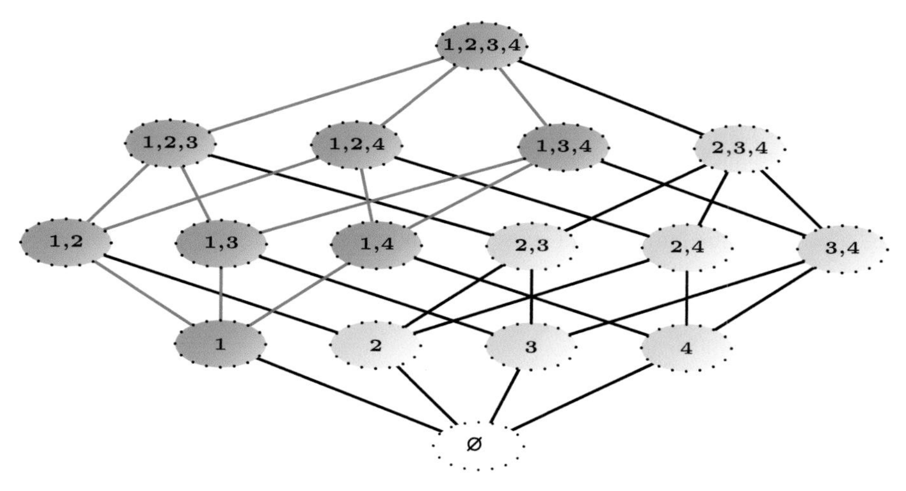

Fig. 1.13 Family of all subsets in 2^X, $X = \{1, 2, 3, 4\}$

A nonempty set A is a **convex set**, provided, for any pair of points $x, y \in A$, the line segment $\overline{xy}$ is also in A. The empty set $\emptyset$ and a one-element set $\{x\}$ are convex by definition. Let $\mathscr{F}$ be a family of convex sets. From the fact that the intersection of any two convex sets is convex, it follows that

$$\bigcap_{A \in \mathscr{F}} A \text{ is a convex set.}$$

This gives us the following result.

Theorem 1.25 *The family of all convex sets $\mathscr{F}$ of the convex set X is a Zelins'kyi-Kay-Womble convexity structure.*

Proof By definition, every subset of $\mathscr{F}$ is convex and the intersection of the members of any subfamily of $\mathscr{F}$ is convex. Hence, from Theorem 1.23, $\mathscr{F}$ is a Zelins'kyi-Kay-Womble convexity structure. □

Definition 1.26 Proximal Zelins'kyi-Kay-Womble Convexity Space.
Let $\mathscr{F} = 2^X$ be the family of all subsets of a nonempty set X equipped with the Wallman proximity δ and let subfamilies $\mathscr{A}, \mathscr{B} \in \mathscr{F}$. The family $\mathscr{F}$ on X is called a **proximal Zelins'kyi-Kay-Womble convexity space**, provided it satisfies the following axioms.

(nC0) $\emptyset$ and X belong to $\mathscr{F}$.
(nC1) $\mathscr{A}\ \delta\ \mathscr{B} \in \mathscr{F}$ for all subfamilies $\mathscr{A}, \mathscr{B} \in \mathscr{F}$. In other words, $\text{cl}\mathscr{A} \cap \text{cl}\mathscr{B} \neq \emptyset \Leftrightarrow \mathscr{A}\ \delta\ \mathscr{B}$. ■

The tuple $(X, \mathscr{F}, \delta)$ is a **proximal Zelins'kyi-Kay-Womble convexity space**.

Theorem 1.27 *Let X be a nonempty convex set equipped with the Wallman proximity δ. The family of all convex sets $\mathscr{F} := 2^X$ is a Zelins'kyi-Kay-Womble convexity structure.*

Proof Immediate from Theorem 1.25. □

Definition 1.28 Strong Contact Zelins'kyi-Kay-Womble Convexity Space. Let $\mathscr{F} = 2^X$ be the family of all subsets of a nonempty set X equipped with the Di Concilio strong contact $\overset{\wedge\!\!\wedge}{\delta}$ and let subfamilies $\mathscr{A}, \mathscr{B} \in \mathscr{F}$. The family $\mathscr{F}$ on X is called a **proximal Zelins'kyi-Kay-Womble convexity space**, provided it satisfies the following axioms.

(snC0) $\varnothing$ and X belong to $\mathscr{F}$.

(snC1) $\mathscr{A} \overset{\wedge\!\!\wedge}{\delta} \mathscr{B} \in \mathscr{F}$ for all subfamilies $\mathscr{A}, \mathscr{B} \in \mathscr{F}$. In other words, $\mathscr{A} \overset{\wedge\!\!\wedge}{\delta} \mathscr{B} \Rightarrow \mathscr{A} \cap \mathscr{B} \neq \varnothing$, i.e., all subfamilies of $\mathscr{F}$ have strong contact. ■

The tuple $\left(X, \mathscr{F}, \overset{\wedge\!\!\wedge}{\delta}\right)$ is a **strong contact Zelinskii-Kay-Womble convexity space**.

Theorem 1.29 *Let X be a nonempty convex set equipped with the strong proximity $\overset{\wedge\!\!\wedge}{\delta}$. The family of all convex sets $\mathscr{F} := 2^X$ is a Zelins'kyi-Kay-Womble convexity structure.*

Proof Immediate from Theorem 1.25. □

1.8 Descriptive Proximity

Let X be a topological space endowed with a descriptive proximity [13] δ_Φ, $x \in X$, $A, B \in \mathcal{P}(X)$, and let $\Phi = \{\phi_1, \ldots, \phi_i, \ldots, \phi_n\}$, a set of probe functions $\phi_i : X \to \mathbb{R}$ that represent features of each x, where $\phi_i(x)$ equals a feature value of x. Let $\Phi(x)$ denote a feature vector for the concrete object x, i.e., a vector of feature values that describe x, where

$$\Phi(x) = (\phi_1(x), \ldots, \phi_i(x), \ldots, \phi_n(x)).$$

A **concrete object** is an object in the physical world. Picture elements (pixels) in digital images, drawings, paintings and nodes in a social network are examples of concrete objects. Every concrete object has a description. In this work, a **description** is a feature vector in Euclidean space $\mathbb{R}^n$. In other words, the description of an object is a point in $\mathbb{R}^n$. This leads to an important economy in the notation. That is, if it is understood that the points in $\mathbb{R}^n$ describe physical objects, then reasoning about the descriptive nearness of sets of physical objects translates to sets in $\mathbb{R}^n$ that share points.

A feature vector provides a description of a concrete object such as a picture element (pixel) in a digital image or node in a social network. Let $A, B \in 2^X$, the collection of all subsets of X. $\Phi(A)$ is the set of feature vectors for all points in A, i.e.,

$$\Phi(A) = \{\Phi(a) : a \in A\}.$$

The expression $A\ \delta_\Phi\ B$ reads A *is descriptively near* B. The descriptive proximity of A and B is defined by

$$A\ \delta_\Phi\ B \Leftrightarrow \Phi(A) \cap \Phi(B) \neq \varnothing.$$

The expression $A\ \not\delta_\Phi\ B$ reads A *is descriptively far (remote) from* B (**Descriptive remoteness** of A and B) is defined by

$$A\ \not\delta_\Phi\ B \Leftrightarrow \Phi(A) \cap \Phi(B) = \varnothing.$$

Early informal work on the descriptive intersection of disjoint sets based on the shapes and colours of objects in the disjoint sets is given by N. Rocchi [69, p.159]. The *descriptive intersection* $\underset{\Phi}{\cap}$ of A and B is defined by

$$A \underset{\Phi}{\cap} B = \{x \in A \cup B : \Phi(x) \in \Phi(A) \text{ and } \Phi(x) \in \Phi(B)\}.$$

Definition 1.30 Let X be a topological space, $A, B, C \subset X$, C^c (complement of C) and $x \in X$. The relation δ_Φ on $\mathscr{P}(X)$ is a **descriptive EF proximity**, provided it satisfies the following axioms.

(***dEF***.1) $A\ \delta_\Phi\ B$ implies $A \neq \varnothing, B \neq \varnothing$.
(***dEF***.2) $A \cap B \neq \varnothing$ and $A \underset{\Phi}{\cap} B \neq \varnothing$ implies $A\ \delta_\Phi\ B$.
(***dEF***.3) $A\ \delta_\Phi\ B$ implies $B\ \delta_\Phi\ A$ (descriptive symmetry).
(***dEF***.4) $A\ \delta_\Phi\ (B \cup C)$, if and only if, $A\ \delta_\Phi\ B$ or $A\ \delta_\Phi\ C$.
(***dEF***.5) Descriptive Efremovič axiom:

$$A\ \not\delta_\Phi\ B \text{ implies } A\ \not\delta_\Phi\ C \text{ and } B\ \not\delta_\Phi\ C^c \text{ for some } C \in 2^X. \quad \blacksquare$$

The pair (X, δ_Φ) is called a **descriptive EF space**.

Example 1.31 **Descriptive EF Space.**
Let the finite topological space X represented in Fig. 1.14 be endowed with a descriptive proximity relation δ_Φ, where $A, B, C \in 2^X$. It can be observed that δ_Φ satisfies EF space axioms (***dEF***.1)–(***dEF***.4). Let $A \subset C^c$ and observe that

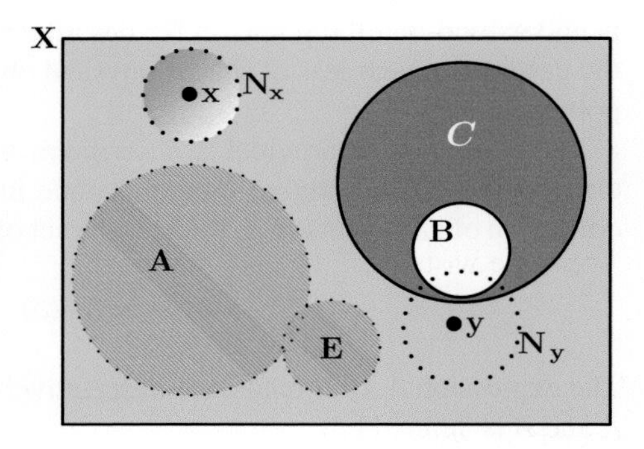

Fig. 1.14 $A\ \not\delta_{\Phi}\ B \implies A\ \not\delta_{\Phi}\ C$ and $B\ \not\delta_{\Phi}\ \mathbf{C^c}$ for some $C \in 2^X$

$$
\begin{aligned}
&B \subset C,\\
&C\ \not\delta_{\Phi}\ C^c\\
&A \subset\ C^c \text{ and}\\
&A\ \not\delta_{\Phi}\ B, \text{ hence}\\
&B\ \not\delta_{\Phi}\ C^c, \text{ and}\\
&A\ \not\delta_{\Phi}\ C.
\end{aligned}
$$

Consequently, δ_{Φ} satisfies axiom (***dEF***.5). Hence, (X, δ_{Φ}) is a descriptive EF-space. ■

1.9 Descriptive Strong Proximity

Nonempty sets A, B in a proximity space X are *strongly near* (denoted $A\ \overset{\wedge\wedge}{\delta}\ B$), provided the sets share points. Strong proximity $\overset{\wedge\wedge}{\delta}$ was introduced in [20, Sect. 2] and completely axiomatized in [49] (see, also, [11, Sect. 6 Appendix]). Strong proximities satisfy a number of axioms given in Definition 1.9. The descriptive strong proximity $\overset{\wedge\wedge}{\delta}_{\Phi}$ is the descriptive counterpart of $\overset{\wedge\wedge}{\delta}$.

Definition 1.32 Let X be a topological space, $A, B, C \subset X$ and $x \in X$. The relation $\overset{\wedge\wedge}{\delta}_{\Phi}$ on the family of subsets 2^X is a *descriptive strong Lodato proximity*, provided it satisfies the following axioms.

(dsnN0) $\emptyset\ \overset{\not\delta}{\mathrm{w}}_{\Phi}\ A, \forall A \subset X$, and $X\ \overset{\wedge\wedge}{\delta}_{\Phi}\ A, \forall A \subset X$

(dsnN1) $A\ \overset{\wedge\wedge}{\delta}_{\Phi}\ B \Leftrightarrow B\ \overset{\wedge\wedge}{\delta}_{\Phi}\ A$

(dsnN2) $A\ \overset{\wedge\wedge}{\delta}_{\Phi}\ B \Rightarrow A \underset{\Phi}{\cap} B \neq \emptyset$

(dsnN3) If $\{B_i\}_{i \in I}$ is an arbitrary family of subsets of X and $A \stackrel{\wedge\!\!\wedge}{\delta}_{\Phi} B_{i^*}$ for some $i^* \in I$ such that $\mathrm{int}(B_{i^*}) \neq \emptyset$, then $A \stackrel{\wedge\!\!\wedge}{\delta}_{\Phi} (\bigcup_{i \in I} B_i)$

(dsnN4) $\mathrm{int}A \underset{\Phi}{\cap} \mathrm{int}B \neq \emptyset \Rightarrow A \stackrel{\wedge\!\!\wedge}{\delta}_{\Phi} B$ ■

When we write $A \stackrel{\wedge\!\!\wedge}{\delta}_{\Phi} B$, we read A is *descriptively strongly near* B. The notation $A \not\stackrel{\wedge\!\!\wedge}{\delta}_{\Phi} B$ reads A is not descriptively strongly near B. For each *descriptive strong proximity*, we assume the following relations:

(dsnN5) $\Phi(x) \in \Phi(\mathrm{int}(A)) \Rightarrow x \stackrel{\wedge\!\!\wedge}{\delta}_{\Phi} A$

(dsnN6) $\{x\} \stackrel{\wedge\!\!\wedge}{\delta}_{\Phi} \{y\} \Leftrightarrow \Phi(x) = \Phi(y)$ ■

So, for example, if we take the strong proximity related to non-empty intersection of interiors, we have that $A \stackrel{\wedge\!\!\wedge}{\delta}_{\Phi} B \Leftrightarrow \mathrm{int}A \underset{\Phi}{\cap} \mathrm{int}B \neq \emptyset$ or either A or B is equal to X, provided A and B are not singletons; if $A = \{x\}$, then $\Phi(x) \in \Phi(\mathrm{int}(B))$, and if B is also a singleton, then $\Phi(x) = \Phi(y)$.

Example 1.33 **Descriptive Strong Proximity**.
Let X be a space of picture points represented in Fig. 1.15 with red, green or blue colours and let $\Phi : X \to \mathbb{R}$ be a description of X representing the colour of a picture point, where 0 stands for red (r), 1 for green (g) and 2 for blue (b). Suppose the range is endowed with the topology given by $\tau = \{\emptyset, \{r, g\}, \{r, g, b\}\}$. In Fig. 1.14, $A, E, C, B \subset X, x \in X, \Phi(x) =$ feature vector containing the r, g, b colour intensities of x. Then $A \stackrel{\wedge\!\!\wedge}{\delta}_{\Phi} E$, since $\mathrm{int}A \cap \mathrm{int}E \neq \emptyset$. Similarly, $B \not\stackrel{\wedge\!\!\wedge}{\delta}_{\Phi} C$, since $\mathrm{int}B \underset{\Phi}{\cap} \mathrm{int}C \neq \emptyset$. ■

Fig. 1.15 Descriptive strongly near sets: $A_i \stackrel{\wedge\!\!\wedge}{\delta}_{\Phi} B, i \in \{1, 2, 3, 4, 5, 6, 7\}$

Example 1.34 **Descriptive Strong Proximity**.
Let X be a space of picture points with the topology defined in Example 1.33. The picture of a Salerno port scene is tessellated with a Voronoï mesh constructed from keypoints using the Mathematica script MScript 19 in Appendix A. In this mesh, $A_i \overset{\doublewedge}{\delta}_{\Phi} B, i \in \{1, 2, 3, 4, 5, 6, 7\}$, since each set A_i has a common edge with B. ■

Theorem 1.35 *Let X be a topological space endowed with* $\left\{\overset{\doublewedge}{\delta}, \overset{\doublewedge}{\delta}_{\Phi}, \delta_{\Phi}\right\}$ *(called a proximal relator),* $A, B \subset X$.

1° $A \overset{\doublewedge}{\delta} B \Rightarrow A\ \delta_{\Phi}\ B$.

2° $A \overset{\doublewedge}{\delta}_{\Phi} B \Rightarrow A\ \delta_{\Phi}\ B$.

Proof
1°: Immediate from the definition of δ_{Φ}.

2°: $A \overset{\doublewedge}{\delta}_{\Phi} B \Rightarrow A\ \delta_{\Phi}\ B$ from Axiom **(dN2)** and the definition of δ_{Φ}. □

Problem 1.36 Use Mathematica to cover a digital image X with a Voronoï diagram $V(S)$ constructed using a set of seed points $S \subset X$. Let polygons $A, B, C, D, E, F, G, H \in V(S)$. Highlight in yellow the following cases:

1° $A \not\delta B$ and $A\ \delta_{\Phi}\ B$.

2° $C \overset{\doublewedge}{\delta} D$ and $C \not\delta_{\Phi} D$.

3° $E \overset{\doublewedge}{\delta} F$ and $E \overset{\doublewedge}{\delta}_{\Phi} F$.

4° $G \not\delta H$ and $G \not\delta_{\Phi} H$. ■

1.10 Nerves

Let $\mathscr{F}$ denote a collection of nonempty sets. An Edelsbrunner-Harer *nerve* of $\mathscr{F}$ [19, Sect. III.2, p. 59] (denoted by Nrv$\mathscr{F}$) consists of all nonempty sub-collections whose sets have a common intersection and is defined by

$$\text{Nrv}\mathscr{F} = \left\{X \subseteq \mathscr{F} : \bigcap X \neq \emptyset\right\}.$$

A natural extension of the basic notion of a nerve arises when we consider adjacent polygons and the closure of a set. Let A, B be nonempty subset in a topological space X. The expression $A\ \delta\ B$ (A near B) holds true for a particular proximity that we choose, provided A and B have nonempty intersection, i.e., $A \cap B \neq \emptyset$. Every nonempty set has a set of points in its interior (denoted int$A$) and a set of boundary points (denoted bdy$A$). A nonempty set is *open*, provided its boundary set is empty, i.e., bdy$A \neq \emptyset$. Put another way, a set A is open, provided all points $y \in X$ sufficiently close to $x \in A$

belong to A [70, Sect. 1.2]. A nonempty set is *closed*, provided its boundary set is nonempty. Notice that a closed set can have an empty interior.

Example 1.37 Circle, triangles, and or any quadrilaterals are examples of closed sets with either empty or nonempty interiors. Disks can be either closed or open sets with nonempty interiors. ■

The closure of a nonempty set (denoted cl$A$) in a Lodato proximity space, is defined by

$$\mathrm{cl}A = \{x \in X : x\ \delta\ A\}.$$

An extension of the notion of a nerve in a collection of nonempty sets $\mathscr{F}$ is a **closure nerve** (denoted NCL$\mathscr{F}$), defined by

$$\mathrm{NCL}\mathscr{F} = \left\{X \in \mathscr{F} : \bigcap \mathrm{cl}X \neq \emptyset\right\}.$$

Closure nerves are commonly found in a Voronoï diagram (see Sect. 1.11 for an introduction to Voronoï diagrams and mesh nerves).

Example 1.38 **Torus Nerve**.
Let $\mathscr{F}$ be the collection of rectangular segments in the mesh along the surface of the torus in Fig. 1.1. An example of a closure nerve is shown in Fig. 1.16. ■

Lemma 1.39 *A nonempty set containing more than one disjoint closed subset has more than one nerve.*

Proof Let X be a nonempty set containing disjoint closed subsets A, B. Let $A \in \mathscr{F}_1, B \in \mathscr{F}_2$. Since $A \not\delta B$, $\mathrm{Nrv}\mathscr{F}_1 \neq \mathrm{Nrv}\mathscr{F}_2$. Hence, there are at least two nerves in X. □

Theorem 1.40 *A nonempty set containing more than one disjoint closed subset has more than one closure nerve.*

Proof Immediate from Lemma 1.39. □

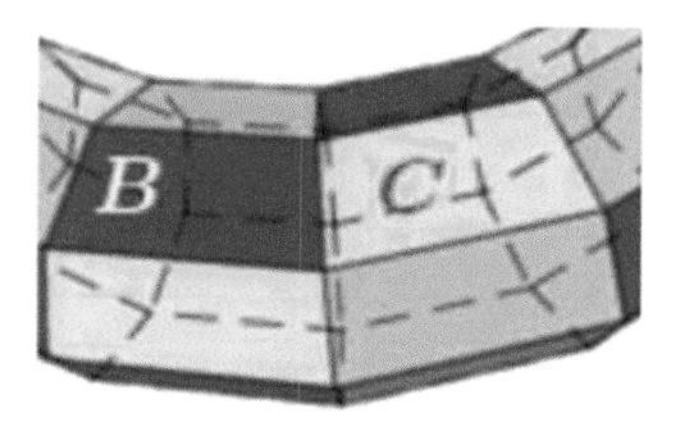

Fig. 1.16 Sample torus closure nerve

1.11 Voronoï Diagrams and Mesh Nerves

This section briefly introduces Voronoï diagrams (also called Dirichlet tessellations or meshes) and what are known as mesh nerves. The story starts with the notion of a mesh.

A **mesh** on a plane surface is a collection of nonempty subsets $\mathscr{M}$ such that near sets have only boundary points in common and the union of the sets equals $\mathscr{M}$. A **closed mesh** is a collection of nonempty closed subsets $\mathscr{M}$ on a plane surface such that near sets having only boundary points in common and the union of the closed subsets equals $\mathscr{M}$. Notice that the sets in a mesh can have empty interiors. The collection of rectangular polygons covering the surface of the torus in Fig. 1.1 is an example of a closed mesh.

Let S be a finite set in a n-dimensional Euclidean space. Elements of S are called sites to distinguish them from other points in E [71, Sect. 2.2, p. 10]. Sites (or seeds) are also called **generating points**, since sites are used to generate a mesh. Let $p \in S$. A *Voronoï region* of $p \in S$ (denoted V_p) is defined by

$$V_p = \left\{ x \in E : \|x - p\| \underset{\forall q \in S}{\leq} \|x - q\| \right\}.$$

Remark 1.41 **Voronoï regions, edges and vertices**.
A **Voronoï region of a site** $p \in S$ such as the one in Fig. 1.17 contains every point in the plane that is closer to p than to any other site in S [72, Sect. 1.1, p. 99]. A Voronoï region of a site is called a **cell** or Thiessen polytope or Voronoï polygon [73]. A **site** is a point used to construct or generate a Voronoï region. It is possible for a Voronoï region of a site to be unbounded in cases where the distance between the sites approaches infinity. A **Voronoï tessellation** of a surface is a tiling that covers the surface with convex polygonal shapes so that each polygon contains exactly one site. Let V_p, V_q be Voronoï polygons. If $V_p \cap V_q$ is a line, ray (half-line) or line segment, then it is called a *Voronoï edge*. If the intersection of three or more Voronoï regions is a point, that point is called a *Voronoï vertex*. ■

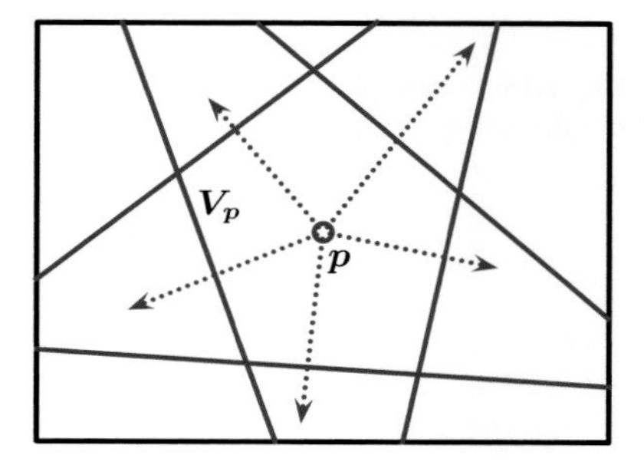

Fig. 1.17 V_p = intersection of closed half-planes

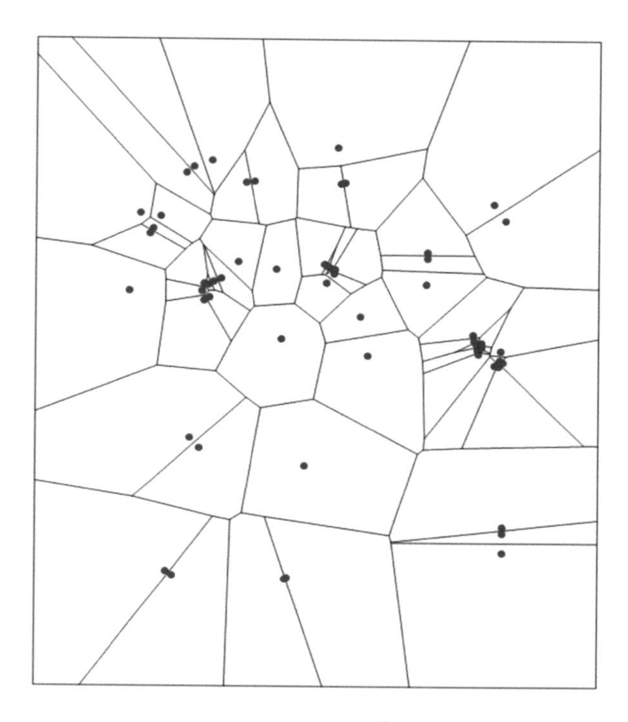

Fig. 1.18 Voronoï mesh

A Voronoï diagram of S (denoted by $V(S)$ or simply by $\mathbb{V}$) is the set of Voronoï regions, one for each site $p \in S$, defined by (Fig. 1.18)

$$\mathbb{V} = \bigcup_{p \in S} V_p.$$

1.12 Origins, Variations and Applications of Voronoï Diagrams

The origins of Voronoï diagrams are introduced by T.M. Liebling and L. Pourin [74, Sect. 2]. The Voronoï diagrams story starts with Kepler and Descartes in the study of a variety of phenomena ranging from snowflakes to galaxies. A detailed account of the origins of Voronoï diagrams from a computational perspective is given in [75].

This account includes an introduction to Poisson Voronoï Diagrams containing a countably infinite number of sites that generate infinite Voronoï diagrams. In such diagrams, the sites belong to a Poisson point process, which is a probabilistic generation of points in a space according to the Poisson probability distribution. A **point process** is a probabilistic model for random scatterings of points over some space, which is usually a subset of the Euclidean space $\mathbb{R}^n$ [76]. A *Poisson point process* satisfies the following axioms.

1° The number of changes in non-overlapping intervals are independent for all intervals.
2° Let ν be the probability of one change, n the number of trials. The probability of exactly one change occurring in a sufficiently small interval $h = \frac{1}{n}$ is $Pr[1 \text{ change in } h] = \frac{\nu}{n}$.
3° The probability of more than one change in a sufficiently small interval ν is approximately 0.

As the number of trials becomes large, the resulting distribution of the random scatterings of points over a space is called a Poisson distribution [77].

A good introduction to Voronoï diagrams is given by F. Aurenhammer [24]. Surface 3D Voronoï meshes are introduced P. Alliez and others in [78, starting on p. 54]. The statistical properties of planar Voronoï diagrams are given by H.J. Hilhorst [79]. For more about Voronoï diagrams, see, for example, [54, Sect. 2.1].

An important recent development in the tessellation of plane surfaces is the introduction of Möbius diagrams containing regions with curved edges This leads to curved Voronoï diagrams [54]. Let $p \in \mathbb{R}^2$, the Euclidean plane and let λ_i, μ_i be two real numbers, $i = 1, \ldots, n$.

For a point x in the Euclidean plane, the distance $\delta_i(x)$ from x to a Möbius site ω_i is defined by

$$\delta_i(x) = \lambda_i(x - p_i)^2 - \mu_i.$$

Then a Möbius region $M(\omega_i)$ of site ω_i is defined by

$$M(\omega_i) = \left\{x \in \mathbb{R}^2 : \delta_i(x) < \delta_j(x), 1 \leq j \leq n\right\} \text{ (\textbf{Möbius region})}.$$

A partial Möbius diagram is shown in Fig. 1.19. Planar Möbius diagrams are extendable to tessellation of 3D surfaces with lots of interesting applications.

An important recent development in the Dirichlet tessellations of plane surfaces is the Voronoï diagram of circles [80]. This approach to tessellating a plane surface uses the Poisson point process to generate the Johnson-Mehl tessellation [80, Sect. 5.1, starting on p.49]. Instead of the usual use of points as sites to generate Voronoï regions, sets of circles are used as a source of region generators so that regions have curved edges.

A recent application of Voronoï diagrams in a geometric approach to social networks is given in [81, 82] (see, also, J.F. Peters and S. Ramanna [25]).

Lemma 1.42 [71, Sect. 2.1, p. 9] *The intersection of convex sets is convex.*

Proof Let $A, B \subset \mathbb{R}^2$ be convex sets and let $K = A \cap B$. For every pair points $x, y \in K$, the line segment $\overline{xy}$ connecting x and y belongs to K, since this property holds for all points in A and B. Hence, K is convex. □

Lemma 1.43 [83] *A Voronoï region of a point is the intersection of closed half planes and each region is a convex polygon.*

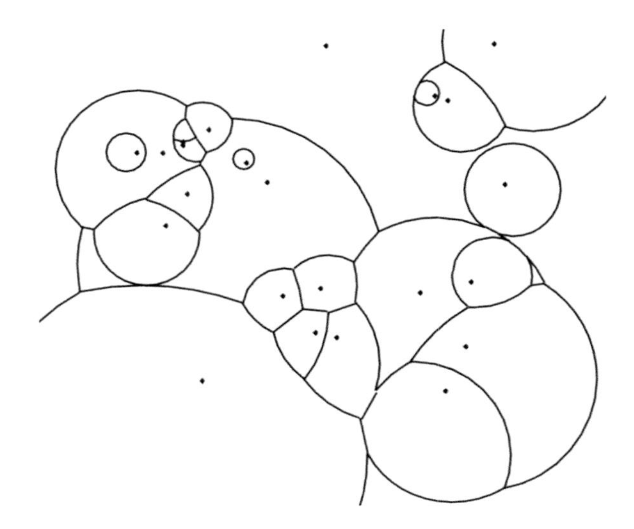

Fig. 1.19 Partial Möbius diagram [54, Sect. 2.4.1]

Proof From the definition of a closed half-plane

$$H_{pq} = \left\{ x \in R^2 : \|x - p\| \underset{q \in S}{\leq} \|x - q\| \right\},$$

V_p is the intersection of closed half-planes H_{pq}, for all $q \in S - \{p\}$ [43], forming a polygon. From Lemma 1.42, V_p is convex. □

Each mesh polygon is the centre of a **nervous system**, which is a collection of mesh polygons that are close to each other. Let $\mathscr{M}$ be a Voronoï mesh endowed with the $\overset{\wedge\wedge}{\delta}$ strongly near proximity, $V_p \subset \mathscr{M}$ (Voronoï region of a site (generating point) $p \in S$. A mesh nerve $NCL_{V_p}\mathscr{M}$ is defined by

$$NCL_{V_p}\mathscr{M} = \left\{ X \in \mathscr{M} : \mathrm{cl} V_p \overset{\wedge\wedge}{\delta} X \right\}. \text{ (Mesh nerve)}$$

In other words,

$$NCL_{V_p}\mathscr{M} = \left\{ X \in \mathscr{M} : \mathrm{cl} V_p \cap X \neq \varnothing \right\}. \text{ (Adjacent nerve polygons)}$$

An approach to constructing a collection of mesh nerves is given in Algorithm 1.

Remark 1.44 Algorithm 1 is used by Algorithm 2 to construct mesh nerves on a digital image. ■

Example 1.45 **Sample Mesh Nerve**.
A sample mesh nerve is shown in Fig. 1.20, i.e.,

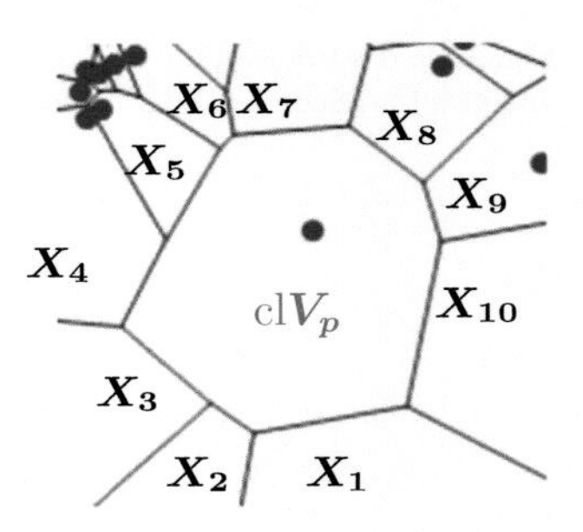

Fig. 1.20 Voronoï mesh nerve

Algorithm 1: Construct Collection of Mesh Nerves

Input : Read set of generators S, $p, q \in S$.
Output: Collection of mesh nerves NCL_{V_p}.

1 $S \longmapsto \mathscr{M}$ /* Construct mesh using sites in S */;
2 **while** ($\mathscr{M} \neq \varnothing$ *and* $V_p \in \mathscr{M}$) **do**
3   *Continue* $\leftarrow$ *True*;
4   *Select* V_p;
5   **while** ($\exists V_q \in \mathscr{M}$ & $V_q \neq V_p$ & *Continue*) **do**
6   *Select* $V_q \in \mathscr{M} \setminus V_p$;
7   **if** $V_p \cap V_q \neq \varnothing$ **then**
8          $V_q \in NCL_{V_p}$;
9          $\mathscr{M} \leftarrow \mathscr{M} \setminus V_q$;
10   /* V_q belongs to nerve NCL_{V_p} */
11   **else**
12   *Continue* $\leftarrow$ *False*;

$$NCL_{V_p}\mathscr{M} = \left\{\mathrm{cl}V_p, X_1, X_2, X_3, X_4, X_5, X_6, X_7, X_8, X_9, X_{10}\right\}. \quad \blacksquare$$

The closure of the Voronoï regions in a mesh nerve is strongly near each other, since, for each pair of nerve of Voronoï regions, the polygons share an edge and hence have more than one point in common. A method that can be used to construct a mesh nerve in given in Algorithm 1. In fact, Algorithm 1 finds all of the nerves in a mesh.

Lemma 1.46 *A mesh containing more than one disjoint subset has more than one nerve.*

Proof Immediate from Lemma 1.39. □

Theorem 1.47 *A mesh containing more than one disjoint closed subset has more than one closure nerve.*

Proof Immediate from Lemma 1.46. □

1.13 Mesh Nerves on a Digital Image

This section introduces an approach to constructing a collection of mesh nerves useful in solving object recognition and pattern problems.

Depending on the choice of mesh generating points, the nerves tend to surround (lay on top of) objects in images. In a binary image, an object is a white blob in the foreground (white areas of a binary image). For example, the white areas on the Hooke needle image in Fig. 1.21 are the ridges of visible grooves on the surface of the needle. Each ridge and its neighbouring ridges are objects of interest, providing a fingerprint characteristic of the needles from a particular manufacturer in the 17th century. Taking the identification of image objects a step further, the Mathematica script MScript 22 in Appendix A is used to find the corners in a either a binary or colour image.

Example 1.48 **Sample Foreground Object Corners.**
The corners found using MScript 22 are shown in Fig. 1.22. The corners tend to surround the places where there is an edge separating a foreground (white) region from a background (black) region. This is the first step in identifying image objects. ■

The next step in an approach to identifying image objects is to select a set of generating points to construct a Voronoï mesh on an image.

Example 1.49 **Sample Corner-Based Voronoï Mesh.**
The corners in Fig. 1.22 are used to construct a Voronoï mesh. The resulting mesh is shown in Fig. 1.23. This mesh can be viewed in a number of different ways. The tiny polygons along the center of the mesh reflect the groupings of the foreground ridges on the surface of the Hooke needle shown in Fig. 1.21. Collections of connected polygons define mesh nerves. And the collections of mesh nerves provide a basis for identifying large image objects such as the Hooke needle. ■

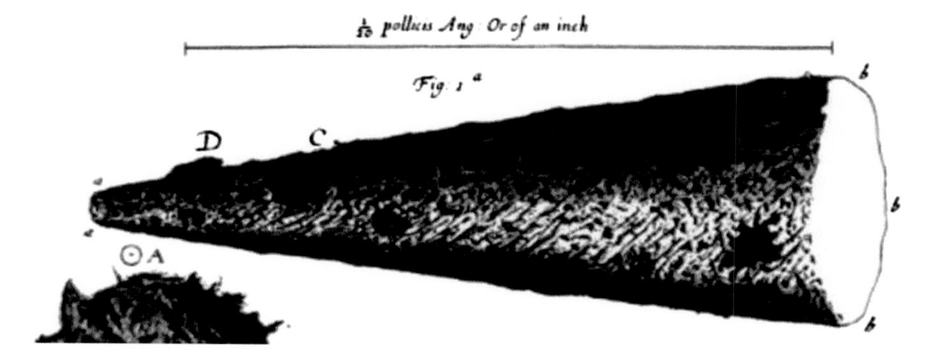

Fig. 1.21 Drawing of a microscopic image of a needle point by Robert Hooke

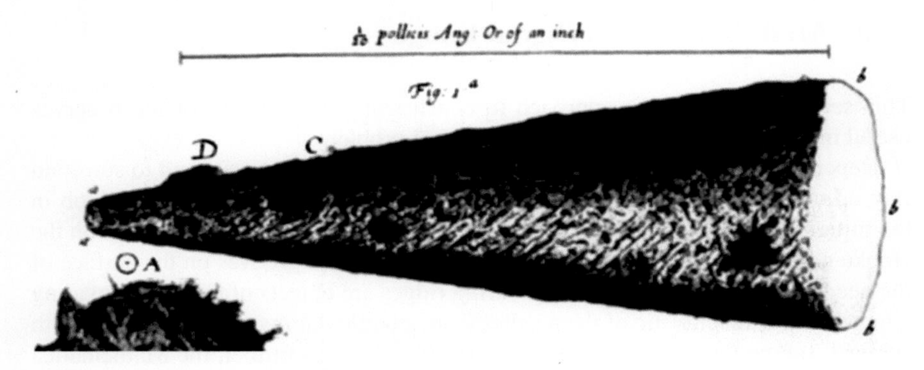

Fig. 1.22 Corners in Hooke needle image

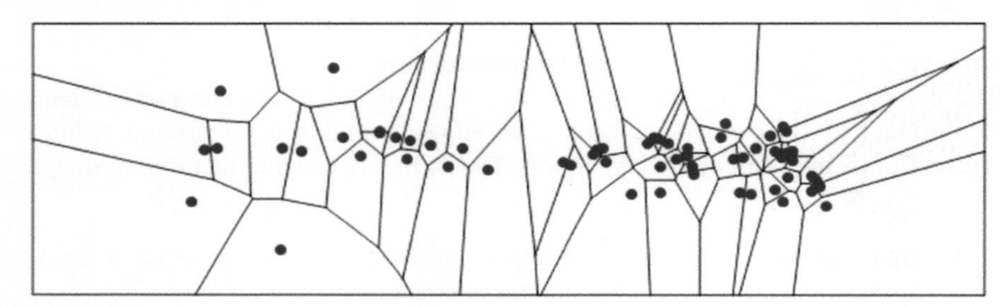

Fig. 1.23 Hooke needle point corner-based Voronoï mesh

Each given nerve is an image mesh belongs to a cluster of nerves, each with at least one common edge. Depending on the choice of a nerve, a cluster of nerves is defined. The nearness of clusters of nerves defines a pattern. In effect, the collection of clusters of nerves defines a Leader uniform topology on an image. An approach to constructing mesh nerves on an image Voronoï mesh is given in Algorithm 2.

Algorithm 2: Construct Mesh Nerves on a Digital Image

Input : Read digital image *img*.

Output: Collection of Image Mesh nerves NCL_{V_p}.

1 $img \longmapsto cornerCoordinates$;

2 $S \leftarrow cornerCoordinates$;

3 /* S contains corner coordinates used as mesh generating points (sites). */ ;

4 $S \longmapsto VoronoiMesh\ \mathscr{M}$;

5 $VoronoiMesh\ \mathscr{M} \longmapsto img$;

6 /* Use Algorithm 1 to construct image mesh nerves in $\mathscr{M}$. */ ;

An approach to superimposing a corner-based Voronoï mesh on a digital image is given in MScript 22 in Appendix A. From such a mesh, we can begin identifying image mesh nerves.

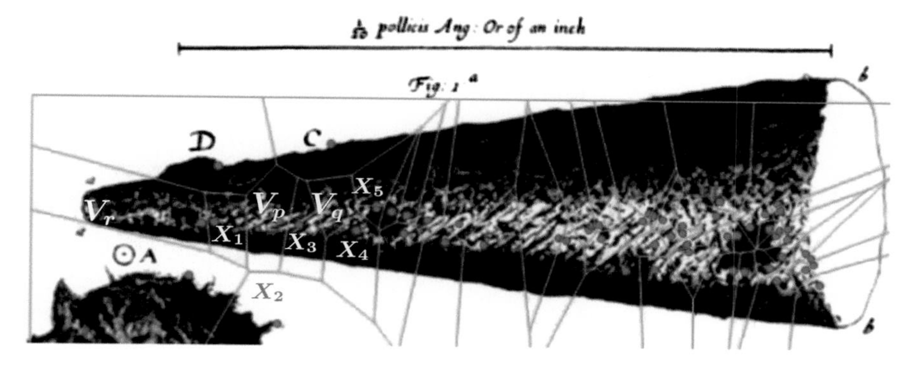

Fig. 1.24 Hooke needle point Voronoï mesh on image

Example 1.50 **Nerve on an Image Mesh.**
A corner-based Voronoï mesh $\mathscr{M}$ on the Hooke needle image is shown in Fig. 1.24. Then the following nerves can be identified.

$$
\begin{aligned}
NCL_{V_p} &= \left\{V_p, D, X_1, A, X_2, X_3, C\right\} : V_p \cap \{D, X_1, A, X_2, X_3, C\} \neq \varnothing.\\
NCL_{V_q} &= \left\{V_q, V_p, \ldots\right\} : V_q \cap \left\{V_p, \ldots\right\} \neq \varnothing.\\
NCL_{V_r} &= \{V_r, A, X_1, D\} : V_q \cap \{A, X_1, D\} \neq \varnothing.
\end{aligned}
$$

Observe that nerve NCL_{V_p} is $\overset{\wedge\!\!\wedge}{\delta}$-close to nerve NCL_{V_q} and NCL_{V_p} is $\overset{\wedge\!\!\wedge}{\delta}$-close to nerve NCL_{V_r}, since NCL_{V_p} has an edge in common with NCL_{V_q} and NCL_{V_r}. The collection of strongly connected nerves $\mathscr{N} = \left\{NCL_{V_p}, NCL_{V_q}, NCL_{V_r}\right\}$ characterize (identify) the Hooke needle tip. ■

1.14 Singer Needle Tip Image Mesh: A Step Toward Object Recognition

By way of comparison in a computational proximity approach to object recognition, the geometry of the mesh on the Hooke needle point tip is compared with a mesh on the microscope image of a Singer® needle point tip in this section. The results in this section stem from an application of MScript 22 in Appendix A.

The Singer needle point tip is about 4.3 cm long in Fig. 1.26.2. A Nikon AF-S Micro Nikkor 105 mm 1:2.28 G macro lens was used to capture the needle point image in Fig. 1.26.1, a closeup of roughly 2 cm of the length of the Singer needle. A Zeiss microscope was used to capture the image of the Singer needle point tip in

Fig. 1.25 Zeiss microscope image showing ≈5 mm Singer needle point tip

Fig. 1.25. This microscope image shows about 5 mm of the needle point tip, which is roughly equivalent to portion of the 17th needle point tip in Hooke's drawing shown in Fig. 1.21.

The study of the Singer needle point tip reported in this section parallels the study of the geometry of the Hooke needle point tip reported in Sect. 1.13. Notice that just

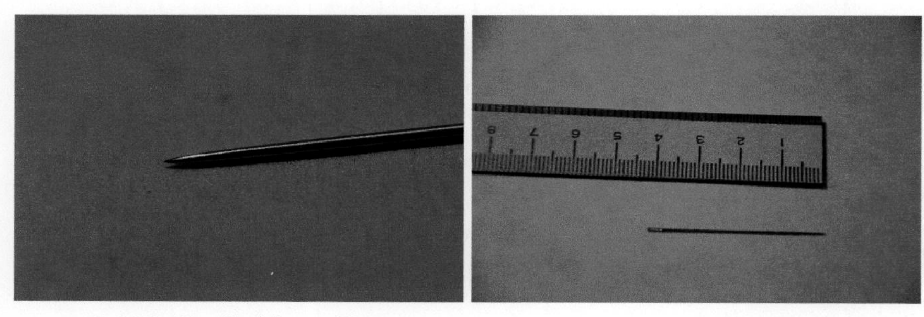

1.26.1: Needle Macro Image 1.26.2: Needle scale

Fig. 1.26 Singer needle

Fig. 1.27 Singer needle point tip corners

as Hooke had observed in looking at a needle point tip with a microscope, the Zeiss microscope image shown in Fig. 1.25 shows a surprisingly rough needle surface with a slightly blunt tip. It is the geometry of this microscope image that we consider next. The basic approach is to use the corners found in the microscope image as a set of sites or generating points in constructing a Voronoï mesh.

The needle point corners shown in Fig. 1.27 were found using MScript 22 in Appendix A. These corners tend to corroborate our conjecture that what are known as strongly connected mesh nerves are useful in object recognition (see Conjecture 1.74), since image corners and the corresponding mesh tend to surround the bumps on an image object. In this case, the corners are grouped around the bumps and the tip of the Singer needle point. These corners are then used to construct the Voronoï mesh shown in Fig. 1.28.

The next step is to superimpose the corner-based Voronoï mesh on the microscope needle point image in Fig. 1.25. This leads to the result shown in Fig. 1.29. The mesh polygons in this image cover the Singer needle point tip. This needle point mesh and the Hooke needle point mesh in Fig. 1.24 have striking similarity, suggesting the validity of the assumption that the basic approach to constructing Voronoï meshes on images is useful in solving object recognition problems. Corners are one among many options in the choice of mesh generating points. This choice depends on the type of image of interest. For more about this, see Sect. 7.1 on mesh generating points.

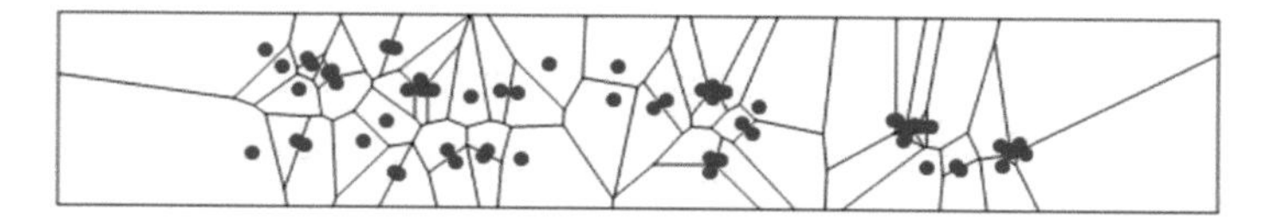

Fig. 1.28 Singer needle point tip corners Voronoï mesh

Fig. 1.29 Singer needle point tip corners Voronoï mesh

1.15 Topological Spaces: Setting for Computational Proximity

The notion of a topological structure (cf., [70]) provides a setting for Computational Proximity. Let X denote a nonempty open set. A set A in X is an **open set**, if and only if, for each $x \in A$, all points in X *sufficiently near* x belong to A [70, Sect. 2, p. 19].

Definition 1.51 **Topology** [7, Sect. 1.6, p. 11], [84, Sect. 12, p. 76], [85, Sect. 1.2, p. 1].
A collection of open sets τ on a nonempty open set X is a *topology* on X, provided

1° The empty set $\emptyset$ is open and $\emptyset$ is in τ.
2° The set X is open and X is in τ.
3° If $\mathcal{A}$ is a sub-collection of open sets in τ, then

$$\bigcup_{B \in \mathcal{A}} B \text{ is a open set in } \tau.$$

In other words, the union of open sets in τ is another open set in τ.
4° If $\mathcal{A}$ is a sub-collection open sets in τ, then

$$\bigcap_{B \in \mathcal{A}} B \text{ is a open set in } \tau.$$

In other words, the intersection of open sets in τ is another open set in τ. ■

The pair (X, τ) is called a **topological space**. In other words, a nonempty set X with a topology τ on it, is a *topological space*. Usually, X by itself is called a topological space, provided X has a topology on it. A topological space X endowed with a proximity δ is called a **proximity space**, denoted by (X, δ).

Remark 1.52 **Leader Uniform Topology**.
Let X be a nonempty open set endowed with a proximity δ. For each subset $A \subseteq X$, find all subsets in X that are near A. This constructive approach yields what S. Leader called a **uniform topology** [86]. ■

Theorem 1.53 *A Leader uniform topology can be constructed on any proximity space.*

Problem 1.54 ☕ Prove Theorem 1.53 for each type of proximity space, starting with the Čech proximity space. ■

Let X be an open set that is a digital image endowed with a proximity δ and let $\mathscr{B}$ be a digital basis that is a collection of digital open sets that are subimages in X. The topology τ with digital basis $\mathscr{B}$ is a **topology of digital images**. If X is endowed with a proximity δ, then τ is a **proximal topology of digital images**.

The computational proximity approach to digital images is inspired by a more general view of topological spaces (see, i.e., [1, 5, 70, 87–89]). For the details about digital open sets, digital basis and the topology of digital images, see Appendix F.

1.16 Connectedness and Strongly Near Connectedness

There is a natural strengthening of nerves in computational proximity that arises from the notion of connectedness of either open or closed sets in proximity spaces. Stephen Willard observed that *the topological study of connected spaces is heavily geometric (or visual)* [90, Sect. 26, p. 191].

A nonempty set X is **disconnected** if and only if there are disjoint nonempty open [closed] sets $A, B \subset X$ such that $X = A \cup B$. In that case, X is disconnected by A and B. The opposite situation leads us to the notion connectedness and connected spaces. The set X is **connected** (denoted by conn$X$), provided there are no disconnected sets in X. A nonempty set $E \subset X$ is a **connected set** (conn$E$), provided the set cannot be partitioned into two nonempty subsets that are open in X [91]. Put another way, a subset is a **connected subset**, provided the subset cannot be partitioned into two nonempty subsets so that each subset has no points in common with the closure of the other set. Let A, B be subsets in a set E. The subset E is **strongly connected** (denoted $\overset{\wedge\wedge}{\text{conn}}\, E$), provided $E = \text{cl}A \cup \text{cl}B$ and $\text{cl}A \cap \text{cl}B \neq \emptyset$ such that cl$A$ and cl$B$ have at least one point in common.

Example 1.55 **Connected Subset.**
Let X be a Euclidean topological space equipped with the Wallman proximity δ such that $X = X_1 \cup X_2$, where X_1, X_2 are connected and $X_1 \cap X_2 \neq \emptyset$ in Fig. 1.30 is connected (i.e., conn$X$), since $\text{cl}X_1 \cap \text{cl}X_2 \neq \emptyset$. See Theorem 1.62 in the sequel. Put another way, $X_1\ \delta_0\ X_2$ (Wallman proximity of X_1 and X_2). ■

Theorem 1.56 *Let X be a topological space, δ a Wallman proximity on X, and $\mathscr{F} := 2^X$ is the family of all subsets of X.*
1° The family $\mathscr{F} := 2^X$ is a convexity structure.
2° $\mathscr{F}$ is connected.

Proof
1°: From Theorem 1.25, the family $\mathscr{F} := 2^X$ is a convexity structure.
2°: Since

$$\mathscr{F} = \bigcup_{A \in \mathscr{F}} A \text{ and, for all } A \in \mathscr{F}, \bigcap_{A \in \mathscr{F}} A \neq \emptyset,$$

$\mathscr{F}$ is connected. □

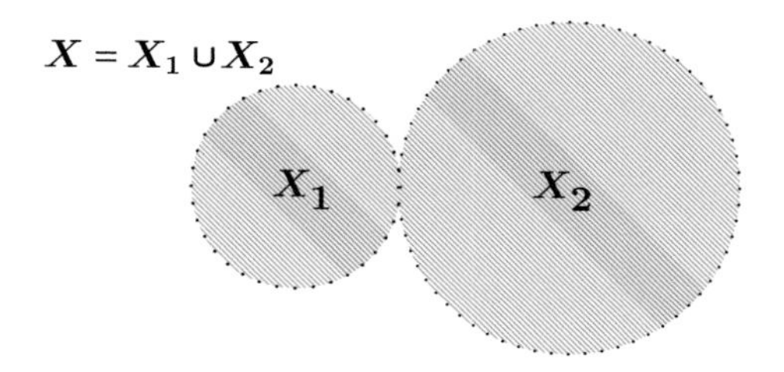

Fig. 1.30 Connected sets

From Theorem 1.56, every family of sets of a nonempty set is a connected convexity structure.

Theorem 1.57 *Let X be a topological space equipped with $\overset{\wedge\wedge}{\delta}$ strong proximity. Assume X is convex $\mathscr{F} := 2^X$ is the family of all subsets of X such that the members of each subfamily are strongly near.*

1° The family $\mathscr{F} := 2^X$ is a $\overset{\wedge\wedge}{\delta}$-convexity structure.

2° $\mathscr{F}$ is $\overset{\wedge\wedge}{\delta}$-connected.

Proof
1°: From Theorem 1.25, the family $\mathscr{F} := 2^X$ is a convexity structure. And, from Theorem 1.29, $\mathscr{F}$ is an $\overset{\wedge\wedge}{\delta}$-convexity structure.
2°: From 1° and the definition of strong connectedness, $\mathscr{F}$ is a $\overset{\wedge\wedge}{\mathrm{conn}}$-convexity structure. □

From Theorem 1.57, every family of $\overset{\wedge\wedge}{\delta}$-near sets of a nonempty set is an $\overset{\wedge\wedge}{\delta}$-connected convexity structure.

Example 1.58 **Strongly Connected Subset**.
Let H be a Euclidean topological space equipped with the strong Wallman proximity $\overset{\wedge\wedge}{\delta}$ such that

$$H = \bigcup X_i \in \{V_p, X_1, X_2\},$$

where V_p, X_1, X_2 are connected and

$$\bigcap X_i \in \{V_p, X_1, X_2\} \neq \emptyset.$$

in Fig. 1.24 is strongly connected (i.e., $\overset{\wedge\wedge}{\mathrm{conn}}\, H$), since $\mathrm{cl}X_1 \cap \mathrm{cl}X_2 \neq \emptyset$. In fact, $\mathrm{cl}X_1 \ \overset{\wedge\wedge}{\delta}\ \mathrm{cl}X_2$, since V_r and X_1 have at least one point in common, namely those points on the edge shared by V_p and X_1 (see Fig. 1.31). Similarly, $\mathrm{cl}V_p \ \overset{\wedge\wedge}{\delta}\ \mathrm{cl}X_1$ and $\mathrm{cl}V_p \ \overset{\wedge\wedge}{\delta}\ \mathrm{cl}X_2$. ■

Problem 1.59 The strong connectedness of H in Example 1.58 can be expanded by introducing other sets connected to V_p in the Hooke needle in Fig. 1.24. Indicate all sets in Fig. 1.24 that are strongly connected to V_p.

Example 1.60 Let $X = \{X_1, X_2, X_3\}$ be a set of pairwise near sets of single-colour such as the brown polygons in the mesh covering the surface of the torus in Fig. 1.32. Since

$$\bigcap_{i \in \{1,2,3\}} \mathrm{cl}X_i \neq \emptyset$$

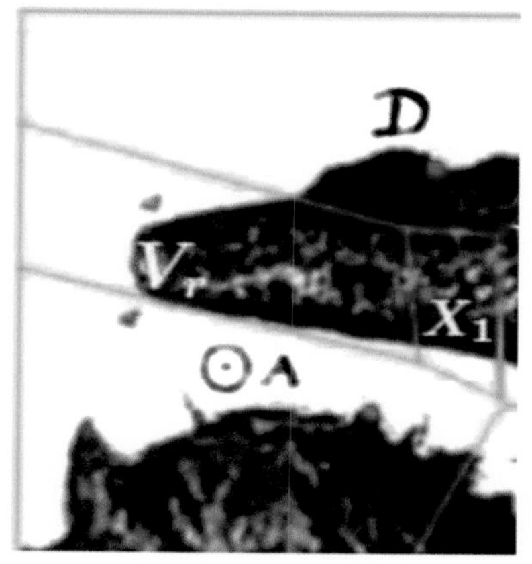

Fig. 1.31 Strongly connected set $\overset{\wedge\!\!\wedge}{\text{conn}}\, H$ on Hooke needle point

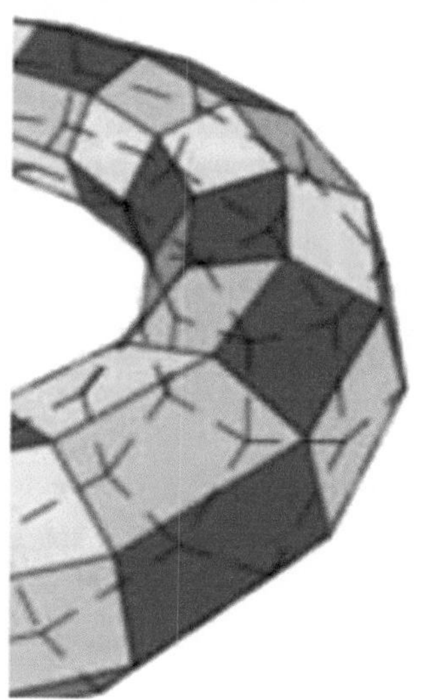

Fig. 1.32 Connected *Brown* polygons on torus surface

and

$$X = \bigcup_{i \in \{1,2,3\}} \text{cl} X_i$$

X is connected. For example, the set $Y = \{A, B\}$ in Fig. 1.1 is connected, since A and B are connected and $Y = A \cup B$. Again, for example, the set $Y = \{A, E\}$ in Fig. 1.1 is $\overset{\wedge\!\!\wedge}{\delta}$-connected, since A and E are $\overset{\wedge\!\!\wedge}{\delta}$-connected and $Y = A \cup E$. ■

Example 1.61 A connected space X is represented in Fig. 1.33, since $X_1 \cap X_2 \neq \emptyset$ and $X = X_1 \cup X_2$. ■

Theorem 1.62 [90, Theorem 26.7, 192–193].

(a) If $X = \bigcup X_n$, where each X_n is connected and $\bigcap X_n \neq \emptyset$, then X is connected.

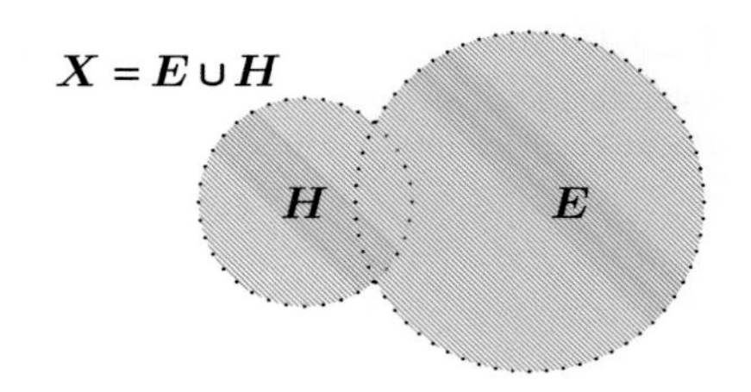

Fig. 1.33 Strongly connected sets

(*b*) *If each pair of points* x, y *in* X *lies in some connected subset* $E_{xy} \subset X$, *then* X *is connected.*
(*c*) *If* $X = \bigcup_{n=1}^{\infty} X_n$, *where each* X_n *is connected and* $X_{n-1} \cap X_n \neq \emptyset$ *for each* $n \geq 2$, *then* X *is connected.*

Theorem 1.63 [90, Theorem 26.8, p. 193].
If E *is a connected subset of* X *and* $E \subset A \subset clE$, *then* A *is connected.*

Example 1.64 Let int$E$ be a connected subset of the connected space X and int$E \subset A \subset$ cl$E$. Then A is connected by Theorem 1.63. In fact, cl$E$ is connected. ■

Problem 1.65 Let int$E$ be a connected subset of the connected space X and int$E \subset A \subset$ cl$E$. Prove cl$E$ is connected. ■

For a space endowed with the strong proximity $\stackrel{\wedge\wedge}{\delta}$, this naturally leads to strong connectedness [4].

Definition 1.66 $\stackrel{\wedge\wedge}{\delta}$**-Connected Topological Space**.
Let X be a topological space and $\stackrel{\wedge\wedge}{\delta}$ a strong proximity on X. We say that X is $\stackrel{\wedge\wedge}{\delta}$-connected if and only if $X = \bigcup_{i \in I} X_i$, where I is a countable subset of $\mathbb{N}$, X_i and int(X_i) or clX_i are connected for each $i \in I$, and $X_{i-1} \stackrel{\wedge\wedge}{\delta} X_i$ for each $i \geq 2$. ■

Example 1.67 **Strongly Connected Closures and Strongly Connected Interior of Sets**.
Let $X = E \cup H$ in Fig. 1.33. X is strongly connected (denoted by $\stackrel{\wedge\wedge}{\text{conn}}\, X$), since int$(E) \cap$ int$(H) \neq \emptyset$. ■

Theorem 1.68 [4] *Let* X *be a topological space and* $\stackrel{\wedge\wedge}{\delta}$ *a strong proximity on* X. *Then* $\stackrel{\wedge\wedge}{\delta}$*-connectedness implies connectedness.*

Proof This simply follows by Theorem 1.62 and axiom ($N2$) in the definition of strong proximities. □

Example 1.69 **Strongly Connected Polygons That Are Connected Polygons**.
Let E, B be the polygons on torus surface mesh in Fig. 1.1. $E \stackrel{\wedge\wedge}{\delta} B$, since E and B have a common edge. From Theorem 1.68, E and B are connected. ■

Example 1.70 **Strongly Connected Space.**
Let X equal the mesh nerve in Fig. 1.20. From Definition 1.66, X is strongly connected, $X = \bigcup_{i=1}^{11} \mathrm{cl}(X_i)$, where $X_{11} = \mathrm{cl}\,V_p$, each $\mathrm{cl}(X_i)$ is connected and $\bigcap \mathrm{cl}(X_i) \neq \emptyset$. Hence, from Theorem 1.68, X is connected. ■

1.17 Connected and Strongly Connected Mesh Nerves

This section highlights the presence of connectedness commonly found in collections of mesh nerves. The nerves V_p, V_q, V_r in Fig. 1.34 are examples of adjacent nerves. **Adjacent mesh nerves** are nerves that have at least one edge in common.

Example 1.71 **Sample Adjacent Mesh Nerves.**
In Fig. 1.34, mesh nerves NCL_{V_p} and NCL_{V_r} have an edges in common on polygons D, V_r, A and X_2. Similarly, mesh nerves NCL_{V_p} and NCL_{V_q} have edges in common on polygons C, V_q, X_4 and X_2. ■

From the observations in Examples 1.50, 1.71 and the closeup of three adjacent Voronoï mesh nerves shown in Fig. 1.34, we obtain the following results.

Theorem 1.72 *Let $\mathscr{M}$ be a Voronoï mesh on the topological space X endowed proximities δ, $\overset{\wedge\wedge}{\delta}$, adjacent nerves $A, B \in \mathscr{M}$, $X = A \cup B$. Then*

1° X is connected.

2° A and B strongly connected.

3° A and B connected.

4° $A \overset{\wedge\wedge}{\delta} B$, i.e., A and B are $\overset{\wedge\wedge}{\delta}$-close nerves.

5° $A\,\delta B$, i.e., A and B are close nerves.

Proof

1° The connectedness of X results from $X = A \cup B$ and $A \cap B \neq \emptyset$, since A and B are adjacent.

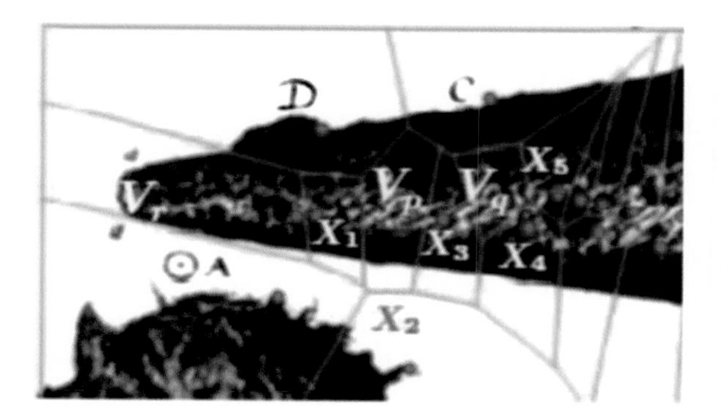

Fig. 1.34 Strongly connected mesh nerves on Hooke needle point

2° Since A and B are adjacent. Consequently, A and B have at least one common edge. Hence, A and B are $\overset{\wedge\!\!\wedge}{\delta}$-connected.

$2^o \Rightarrow 3^o$ Immediate from Theorem 1.68, i.e., A, B $\overset{\wedge\!\!\wedge}{\delta}$-connected $\Rightarrow A, B$ connected.

$2^o \Rightarrow 4^o$ From 2°, A, B are $\overset{\wedge\!\!\wedge}{\delta}$-connected. Hence, by definition, $\mathrm{cl}A \cap \mathrm{cl}B \neq \emptyset$. Moreover, $|A \cap B| > 1$, since A and B have at least one edge in common. Consequently, $A \overset{\wedge\!\!\wedge}{\delta} B$.

$4^o \Rightarrow 5^o$ $A \overset{\wedge\!\!\wedge}{\delta} B \Leftrightarrow A\ \delta\ B$.

□

Theorem 1.73 *Let $\mathscr{M}$ be an finite-dimensional mesh on the topological space X endowed proximities δ, $\overset{\wedge\!\!\wedge}{\delta}$, adjacent nerves $A, B \in \mathscr{M}$, $X = A \cup B$. Then*

1° X is connected.
2° A and B strongly connected.
3° A and B connected.
4° $A \overset{\wedge\!\!\wedge}{\delta} B$, i.e., A and B are $\overset{\wedge\!\!\wedge}{\delta}$-close nerves.
5° $A\ \delta B$, i.e., A and B are close nerves.

Proof Immediate from Theorem 1.72. □

For example, in a 3D Euclidean space, adjacent nerves in a mesh have at least one common surface. The strong connectedness of adjacent nerves is important in solving object recognition problems.

Conjecture 1.74 Object Recognition Via $\overset{\wedge\!\!\wedge}{\delta}$-Connected Nerves.
A $\overset{\wedge\!\!\wedge}{\delta}$-connected collection of image mesh nerves characterize image objects, provided the generating points used to construct an image mesh are not chosen randomly and the center of the initial nerve has a sufficiently high number of $\overset{\wedge\!\!\wedge}{\delta}$-polygons. ■

Conjecture 1.75 Object Recognition Via $\overset{\wedge\!\!\wedge}{\delta}$-Connected Voronoï Nerves.
A $\overset{\wedge\!\!\wedge}{\delta}$-connected collection of Voronoï mesh nerves characterize image objects, provided the generating points used to construct an image mesh are not chosen randomly and the initial nerve center has a maximal number of $\overset{\wedge\!\!\wedge}{\delta}$-polygons. ■

Problem 1.76 ☕ Verify Conjecture 1.74 for several different images using corners to generate an image mesh. ■

Problem 1.77 ☕ Verify Conjecture 1.75 for several different images using corners to generate an image mesh. ■

A repetition of similarly $\overset{\wedge\!\!\wedge}{\delta}$-connected mesh nerves signal the presence of an image pattern in a digital image. In fact, if there is a one-to-one mapping on an image mesh nerve into another mesh nerve in either the same image or in a different image, then an image pattern has been recognized.

Conjecture 1.78 Pattern Recognition Via $\overset{\wedge\!\!\wedge}{\delta}$-Connected Mesh Nerves.
Let $f : 2^X \longrightarrow 2^X$ be a 1-1 continuous mapping on a proximity space X into itself and define f to be

$$f(A) = NCL_A \text{ for nerve } A \in 2^X$$

Then $\overset{\wedge\!\!\wedge}{\delta}$-connected A close to $\overset{\wedge\!\!\wedge}{\delta}$-connected $f(A)$ defines a nerve pattern on X. ■

The approach in Conjecture 1.78 carries over in the search for patterns in Voronoï meshes on digital images.

Conjecture 1.79 Pattern Recognition Via $\overset{\wedge\!\!\wedge}{\delta}$-Connected Voronoï Mesh Nerves.
Let X be a proximity space and let $\mathscr{M}$ be a Voronoï mesh on X. Let $f : 2^X \longrightarrow 2^X$ be a 1-1 continuous mapping on a $\mathscr{M}$ into itself and define f to be

$$f(A) = NCL_A \text{ for nerve } A \in 2^X$$

Then $\overset{\wedge\!\!\wedge}{\delta}$-connected A close to $\overset{\wedge\!\!\wedge}{\delta}$-connected $f(A)$ defines a pattern on X. ■

Problem 1.80 Verify Conjecture 1.79 or give a counterexample for several different images using corners to generate an image mesh. ■

1.18 Boundedness, Bornology and Bornological Nerves

This section carries forward recent work on boundedness and $\overset{\wedge\!\!\wedge}{\delta}$-boundedness, which are fundamental structures useful in CP.

Definition 1.81 Boundedness.
Let X be a nonempty set and let 2^{2^X} be collections of subsets $\mathscr{C}$ in the family of sets 2^X. A nonempty collection $\mathscr{B} \in \mathscr{C}$ is a **boundedness** if and only if
1° $A \in \mathscr{B}$ and $B \subset A$ implies $B \in \mathscr{B}$ (boundedness is hereditary).
2° $A, B \in \mathscr{B}$ implies $A \cup B \in \mathscr{B}$, i.e., $\mathscr{B}$ is closed under finite unions. ■

The elements of a boundedness are called *bounded sets*.

Example 1.82 **Mesh Nerve Boundedness.**
Let NCL_{V_p} be the mesh nerve shown in Fig. 1.34, $A \in NCL_{V_p}$. For each subset $B \subset A$, we have $B \in NCL_{V_p}$. In addition, for $A, B \in NCL_{V_p}$, it also the case that $A \cup B \in NCL_{V_p}$. ■

A **bornology** on X is a boundedness on X that is also a cover of X. That is, a boundedness $\mathscr{B}$ is cover of X, provided $X \subseteq \mathscr{B}$.

Example 1.83 **Bornology in Euclidean 3D space**.
Let X be represented by the largest of the concentric cubes (call it A) in Fig. 1.35 and let A, B, C be the collection of concentric solid cubes (the space inside each box is filled) shown in Fig. 1.35. This collection of concentric cubes (denoted by $\mathscr{B}$) is a boundedness, since $C \subset B \subset A$ and if $G, H \in \mathscr{B}$, then $G \cup H \in \mathscr{B}$. Since the boundedness $\mathscr{B} \supseteq X$, this boundedness on X is a bornology. Let NCL_{V_p} be the mesh nerve shown in Fig. 1.34, $A \in NCL_{V_p}$. For each subset $B \subset A$, we have $B \in NCL_{V_p}$. In addition, for $A, B \in NCL_{V_p}$, it also the case that $A \cup B \in NCL_{V_p}$. ■

A **bornological nerve** A is a collection of subsets X in a boundedness $\mathscr{B}$ such that the closure of A is strongly near each X. A **proximal bornological nerve** is bornological nerve endowed with a proximity.

Example 1.84 **Bornological Nerves in Euclidean 3D space**.
From Example 1.83, the boundedness $\mathscr{B}$ on X is a bornology that consists of a set of solid (filled) concentric cubes. Let X be a proximity space in Euclidean space $\mathbb{R}^3$ and $\mathscr{M}$ be a 3D mesh so the mesh on the outer cube (call it A) in Fig. 1.35 penetrates that cube. As a result, the inner concentric cubes share the mesh on A. Observe that each of the concentric cubes in the boundedness $\mathscr{B}$ is $\overset{\wedge\!\!\wedge}{\delta}$-connected to each of the other cubes, since all of the concentric cubes share the same mesh. From this, we obtain

$$NCL_A(\mathscr{B}) = \left\{ X \in \mathscr{B} : \text{cl}A \overset{\wedge\!\!\wedge}{\delta} X \right\}.$$

That is, $NCL_A(\mathscr{B})$ is a bornological nerve. ■

Theorem 1.85 *A single mesh nerve is a boundedness.*

Proof Immediate from the definition of a mesh nerve. □

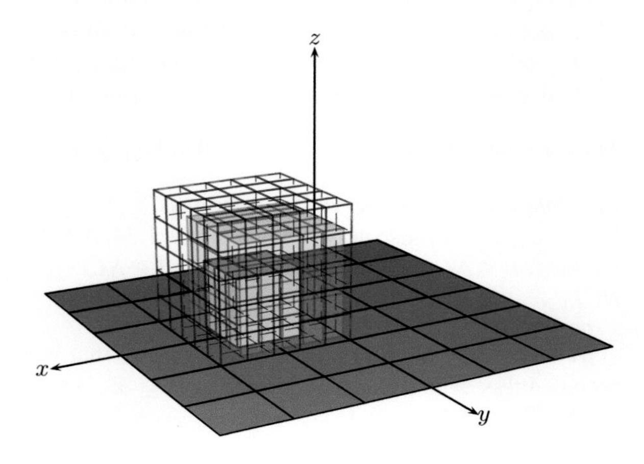

Fig. 1.35 Boundedness in $\mathbb{R}^3$

Theorem 1.86 *Let $\mathscr{M}$ be an finite-dimensional mesh on the topological space X endowed with proximities δ, $\overset{\wedge\wedge}{\delta}$. A collection of adjacent mesh nerves in $\mathscr{M}$ is a boundedness.*

Proof Immediate from Theorem 1.73. □

1.19 Proximal Boundedness, Strong Proximal Boundedness and Proximal Boundedness Nerves

Recall that a **bounded set** has a compact closure in a compact Hausdorff space X. A space X is **compact**, provided every cover of X has a finite subcover. A space X is **Hausdorff**, provided distinct points belong to disjoint neighbourhoods.

Definition 1.87 Filter.
A nonempty collection of subsets $\mathscr{F}$ on X is a **filter**, provided

$\mathscr{F}.1^o$ $\emptyset \notin \mathscr{F}$.
$\mathscr{F}.2^o$ $A, B \in \mathscr{F}$ implies $A \cap B \in \mathscr{F}$.
$\mathscr{F}.3^o$ $A \in \mathscr{F}$ and $A \subset B \subset X$ implies $B \in \mathscr{F}$. ■

Definition 1.88 Proximal Boundedness Structure.
A **proximal boundedness structure** $(X, \delta, \mathscr{B})$ [92] is a structure that has the following properties.

1° δ is a proximity relation.
2° $x\ \delta\ y$ implies $x = y$.
3° $A\ \delta\ B$ implies $A\ \delta\ E$, for some bounded subset of $E \subseteq B$.
4° Given $B \subset X$, A-bounded such that, for all $E \subset X$, $A\ \delta\ X - E$ or $B\ \delta\ E$, then $A\ \delta\ B$.
5° If, for every filter $\mathscr{F} \subset A$, there is some B-bounded $\delta\ E \in \mathscr{F}$, then A is bounded. ■

Example 1.89 **Proximal Boundedness Structure.**
From Example 1.83, $(X, \delta, \mathscr{B})$ is a proximal boundedness structure in 3D Euclidean space. To see this, let δ be the Lodato proximity and let $(X, \delta, \mathscr{B})$ be represented by the concentric solids in Fig. 1.36, assuming that $X \subseteq \mathscr{B}$, concentric solids in $\mathbb{R}^3$.

Proof

1°–2° Immediate from the definition of δ in $(X, \delta, \mathscr{B})$.
3°–4° Immediate from the geometry of the concentric solids on X.

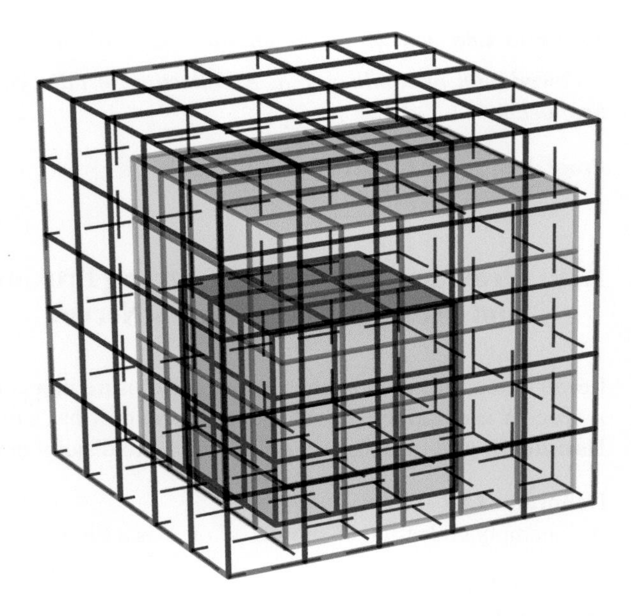

Fig. 1.36 Proximal boundedness in $(\mathbb{R}^3, \delta, \mathscr{B})$

5° This follows from the definition of a filter, since $\mathscr{B}$ is a filter. That is, $(\mathscr{F}.1^o)$ $\emptyset \notin \mathscr{B}$. $(\mathscr{F}.2^o)$ $A, B \in \mathscr{B}$ implies $A \cup B \in \mathscr{B}$. And, $(\mathscr{F}.3^o)$ if $A \in \mathscr{B}$ and $A \subset B \subset X$ implies $B \in \mathscr{B}$, since, by assumption, $X \subseteq \mathscr{B}$ (cover property).

□

Theorem 1.90 *Concentric solids form a proximal boundedness structure in 3D Euclidean space endowed with a proximity.*

A **strong proximal boundedness structure** $(X, \delta, \mathscr{B})$ is a proximal boundedness structure endowed with the strong proximity $\overset{\wedge\!\wedge}{\delta}$.

Example 1.91 **Proximal Boundedness Nerve.**
From Example 1.84, bornological nerve that is a cover of a space X endowed with the strong proximity $\overset{\wedge\!\wedge}{\delta}$ is a strong proximal boundedness structure. ■

Example 1.92 **Proximal Boundedness Geometric Structure.**
From Example 1.89, $\mathscr{B}$ (boundedness defined by the concentric solids in Fig. 1.36) is a cover of the space $X \subset \mathbb{R}^3$ endowed with the strong proximity $\overset{\wedge\!\wedge}{\delta}$ is a strong proximal boundedness structure. See, for example, the concentric solid cubes in Fig. 1.37, assuming that $\mathscr{B} \supseteq X$ (the boundedness represented by the concentric cubes is cover of X). ■

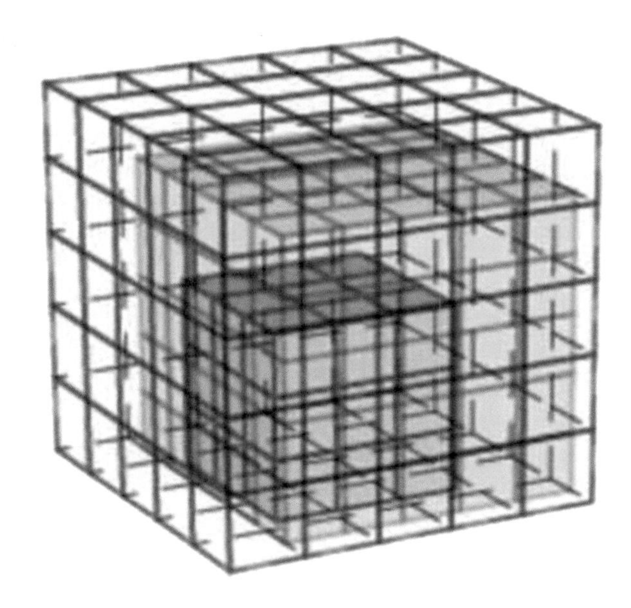

Fig. 1.37 Strong proximal boundedness in $(\mathbb{R}^3, \overset{\wedge\!\!\wedge}{\delta}, \mathscr{B})$

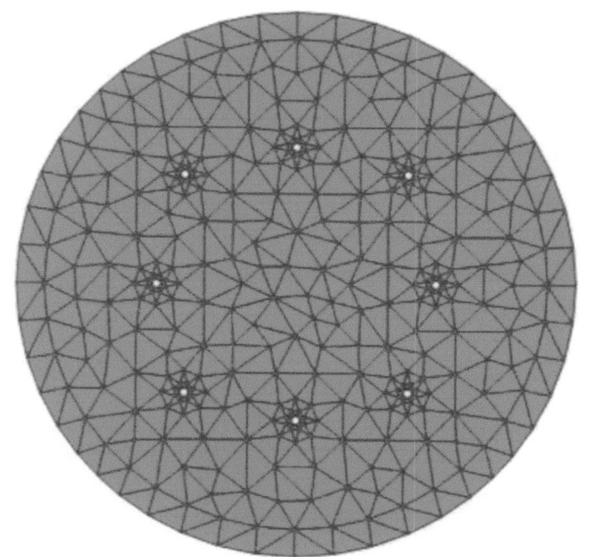

Fig. 1.38 Annulus mesh proximal boundedness in $(\mathbb{R}^2, \overset{\wedge\!\!\wedge}{\delta}, \mathscr{B})$

Example 1.93 **Strong Proximal Boundedness Structure in an Annulus Mesh**. Let boundedness $\mathscr{B}$ be represented by the annulus mesh in Fig. 1.38. This boundedness is in the strong proximal boundedness structure $(X, \overset{\wedge\!\!\wedge}{\delta}, \mathscr{B})$, where $X \subset \mathbb{R}^2$ (Euclidean plane). The assumption made here is that $X \subseteq \mathscr{B}$. Clearly, this structures satisfies the properties for proximal boundedness. The MScript 25 in Appendix A constructs a mesh on an annulus such as the one shown in Fig. 1.38. ■

Example 1.94 **Annulus Strong Proximal Boundedness Nerves.**
Let $\mathscr{B}$ be strong proximal boundedness in the annulus mesh in Example 1.93. The annulus mesh contains nerves with different shapes. Each of the inner circular regions inside the annulus mesh is an annulus nerve (see, e.g., Fig. 1.39.1). The polygons clusters centered on each tiny disk • is a nerve. ■

Problem 1.95 ☕ Prove that the circular and polygon clusters in Fig. 1.39 are mesh nerves. Give an example of a nervous system, which is a collection of adjacent annulus mesh nerves. ■

Problem 1.96 Use Mathematica to construct an example of a proximal boundedness structure in 3D Euclidean space endowed with a proximity δ. Prove that the example satisfies each of the properties of a proximal boundedness structure. ■

Problem 1.97 Use Mathematica to construct an example of a strong proximal boundedness structure in 3D Euclidean space endowed with a proximity δ. Prove that the example satisfies each of the properties of a strong proximal boundedness structure. ■

Problem 1.98 ☕ Let $\mathscr{B}$ be a collection of non-solid (non-filled) concentric cubes that do not have any points in common. Prove or disprove that such non-solid concentric cubes in 3D Euclidean space endowed with a proximity δ is not a proximal bornology. ■

Problem 1.99 Let $\mathscr{B}$ be a collection of non-solid concentric cubes that do not have any points in common. Prove or disprove that such non-solid concentric cubes in 3D Euclidean space endowed with the strong Lodato proximity $\overset{\wedge\!\!\wedge}{\delta}$ is not a strong proximal bornology. ■

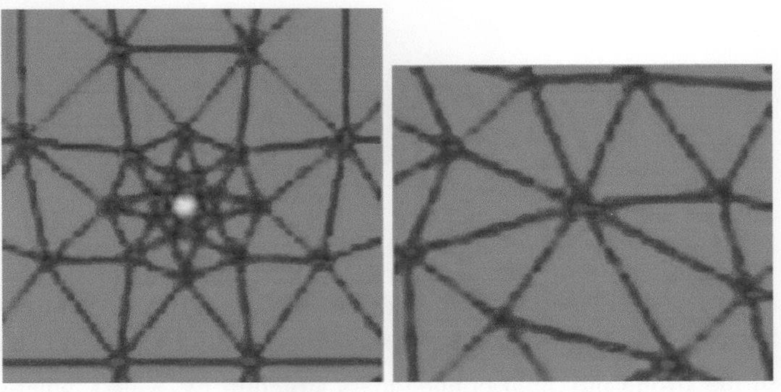

1.39.1: Circular nerve 1.39.2: Polygons nerve

Fig. 1.39 Annulus mesh nerves in proximal boundedness in $(\mathbb{R}^2, \overset{\wedge\!\!\wedge}{\delta}, \mathscr{B})$

Problem 1.100 Prove that non-solid concentric cubes in 3D Euclidean space endowed with a proximity δ is not a proximal boundedness structure. ■

Problem 1.101 Prove that non-solid concentric cubes in 3D Euclidean space endowed with the strong Lodato proximity $\overset{\wedge\!\!\wedge}{\delta}$ is not a strong proximal boundedness structure. ■

1.20 Local Proximity Spaces and Local Strong Proximity Spaces

This section carries forward the notion of local proximity spaces in the study of computational proximity. Local proximity spaces were introduced by S. Leader in 1967 [93].

Definition 1.102 Local Proximity Space.
A **local proximity space** $(X, \delta, \mathscr{B})$ consists of a nonempty set X, proximity δ on the collections of subsets in 2^{2^X} and a boundedness $\mathscr{B} \in 2^{2^X}$ that satisfies the following axioms.

P.1° δ is a proximity relation.
P.2° $\mathscr{B}$ is a boundedness.
P.3° $A \ \delta \ B$ implies $A \ \delta \ E$, for some bounded subset of $E \subseteq B$.
P.4° Given $B \subset X$, A-bounded such that, for all $E \subset X$, $A \ \delta \ X - E$ or $B \ \delta \ E$, then $A \ \delta \ B$.
P.5° $A \ \delta \ B$ implies $A \ \delta \ B \cup C$ for some $C \in \mathscr{B}$. ■

Example 1.103 Annulus Local Proximity Space.
Let X be the set of polygons on the annulus mesh in Fig. 1.38 endowed with the Lodato proximity δ. Each polygon $A \subset X$ is a nonempty set of points and $X = \bigcup A$ is therefore a nonempty set of points. It is a straightforward task to show that the required proximity space axioms for δ are satisfied. Hence, (X, δ) is a Lodato proximity space. Let $\mathscr{B} \in 2^{2^X}$ be a boundedness on X. It is also a straightforward task to show that the bounded subset axioms (P.3°), (P.4°) and axiom (P.5°) are satisfied the boundedness $\mathscr{B}$. Hence, $(X, \delta, \mathscr{B})$ is a local proximity space. ■

Problem 1.104 ☕ Prove that annulus space $(X, \delta, \mathscr{B})$ in Example 1.103 is a local proximity space. For this proof, let δ be the Lodato proximity that is also a Wallman proximity. That is, assume that the lodato proximity δ satisfies the Wallman proximity axiom. **Hint**: First prove that δ satisfies the Lodato axioms and the Wallman axiom. Then prove that axioms (P.3°), (P.4°) and axiom (P.5°) are satisfied. ■

From the local proximity space axioms, we get the following results.

Theorem 1.105 [93, Sect. 1.1, p. 276] *Every subset $A \subset X$ is a local proximity space with δ and boundedness $\mathscr{B}$ restricted to subsets of A.*

Proof Let the proximity space be $(A, \delta, \mathscr{B}), A \subset X$ endowed with proximity δ and $\mathscr{B}$ confined to subsets of A. The result is then immediate from the definition of a local proximity space. □

Example 1.106 Annulus Sub-Local Proximity Space.
Let

$$X = \{X_1, X_2, X_3, X_4, X_5, X_6, X_7, X_8\} \text{ (collection of annulus wedges)}$$

be the set of mesh wedges on the annulus sub-mesh in Fig. 1.40 endowed with the Lodato-Wallman proximity δ. Each wedge $\mathrm{cl}X_i$ is a closed set with a nonempty interior. In addition,

$$\mathscr{B} = X.$$

In fact, $X \subseteq \mathscr{B}$, i.e., X is covered by the boundedness $\mathscr{B}$. The collection of subsets $\mathscr{B} \in 2^{2^X}$ is an example of a mesh nerve boundedness, which is a bornology, since $X \subseteq \mathscr{B}$. That is, $\mathscr{B}$ is a bornological nerve. From Theorem 1.105, $(X, \delta, \mathscr{B})$ is a local proximity space. ■

Theorem 1.107 [93, Sect. 1.3, p. 276] *If* (X, δ) *is a proximity space and* $\mathscr{B} \in 2^{2^X}$, *then* $(X, \delta, \mathscr{B})$ *is a local proximity space.*

Proof Let (X, δ) be a proximity with boundedness $\mathscr{B} \in 2^{2^X}$ endowed with the same proximity δ. Then, from Theorem 1.105, $(X, \delta, \mathscr{B})$ is a local proximity space. □

Theorem 1.108 *A proximal bornological nerve is a local proximity space.*

Problem 1.109 Prove Theorem 1.108. ■

Conjecture 1.110 *Local proximity space structures in digital images lead to object recognition in images.* ■

Example 1.111 **Proximal Bornological Nerves in a Digital Image.**
Proximal bornological nerves in digital images tend to surround image objects. For

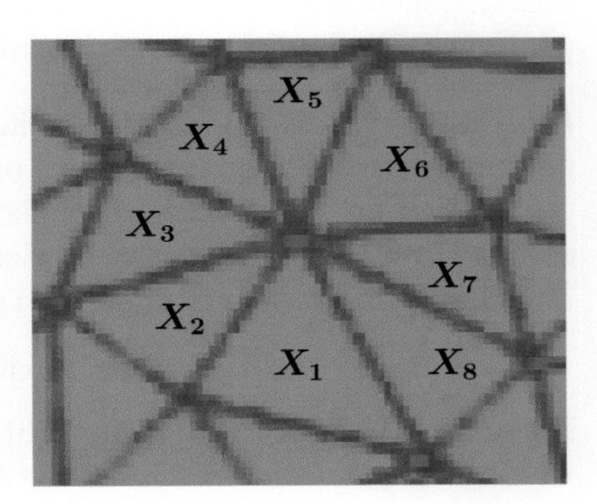

Fig. 1.40 Mesh nerve that is a local proximity space

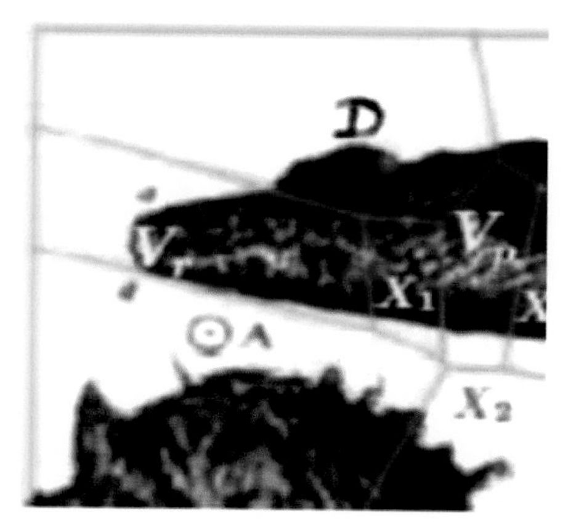

1.41.1: Bornological Nerve

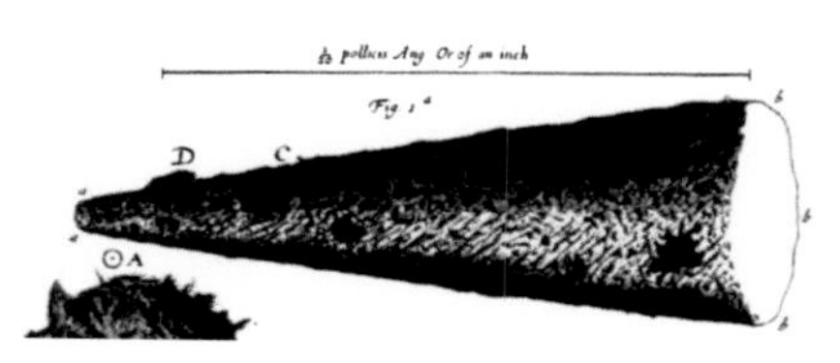

1.41.2: Hooke Needle Tip

Fig. 1.41 Mesh bornological nerve surrounding an object

example, the mesh bornological nerve $NCL_{X_1}(\mathscr{B})$ in Fig. 1.41.1 surrounds the needle point tip in the Hooke drawing in Fig. 1.41.2 with $NCL_{X_1}(\mathscr{B})$ defined by

$$NCL_{X_1}(\mathscr{B}) = \left\{X_1, V_p, D, V_r, A, X_2\right\}. \textbf{ (Bornological Nerve)}$$

In this case, the boundedness $\mathscr{B}$ consists of the subsets in the bornological nerve. Endowing the nerve with a proximity leads to the proximity structure $(NCL_{X_1}(\mathscr{B}), \delta, \mathscr{B})$. From Theorem 1.108, a mesh bornological nerve endowed with a proximity is not only a local proximity space, it is also a good tool in solving object recognition problems in digital images. In the search for objects in digital images using mesh bornological nerves, the key is to select an appropriate set of mesh generating points. ■

Problem 1.112 Use Mathematica to verify Conjecture 1.110 in a series of images containing proximal bornological nerves. ■

Conjecture 1.113 *A repetition of Local proximity space structures in digital images lead to pattern recognition in meshes constructed on the images.* ■

Problem 1.114 Use Mathematica to verify Conjecture 1.113 in a series of images containing proximal bornological nerves in meshes constructed on the images. ■

The Sierpiński triangle is a fractal introduced by Sierpiński in 1915 [94], appearing in Italian art in the 13th century [95]. The first 4 Sierpiński triangles are shown in Fig. 1.42. Let N_n, L_n, A_n be the number of black triangles after n iterations, length of the side of a triangle, and fractional area which is black after the nth iteration, respectively, where

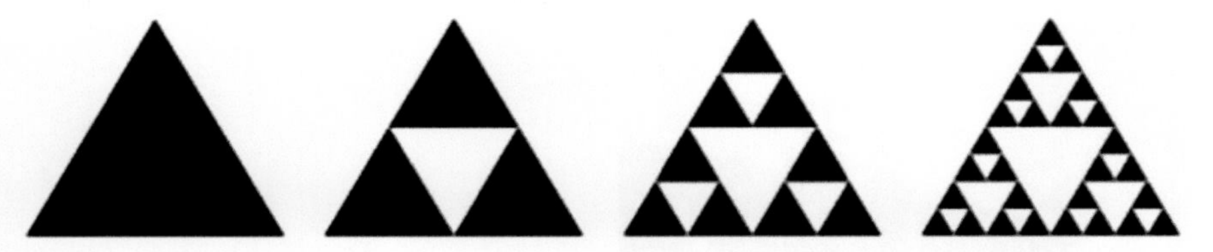

Fig. 1.42 Sierpinski triangles, containing $N_n = 3^n, 1, 3, 9, 27, \ldots$ *Black* triangles

$$\begin{aligned}
n \in \mathbb{N} &= 0, 1, 2, 3, \ldots, \\
N_n &= 3^n, \\
L_n &= \left(\frac{1}{2}\right)^n = 2^{-n}, \\
A_n &= L_n^2 N_n = \left(\frac{3}{4}\right)^n.
\end{aligned}$$

Problem 1.115 🚲 Use Mathematica to produce and display the first 5 Sierpiński triangles. ■

Problem 1.116 Prove that a collection of $N_n = 3^n$ black triangles in a Sierpiński triangle is a disconnectedness. Give an example of a subset of black triangles in a Sierpiński triangle that is connected. ■

Problem 1.117 Give an example of a subset of black triangles in a Sierpiński triangle that is strongly connected. State and prove a theorem that indicates when a collection of black triangles in a Sierpiński triangle is connected. ■

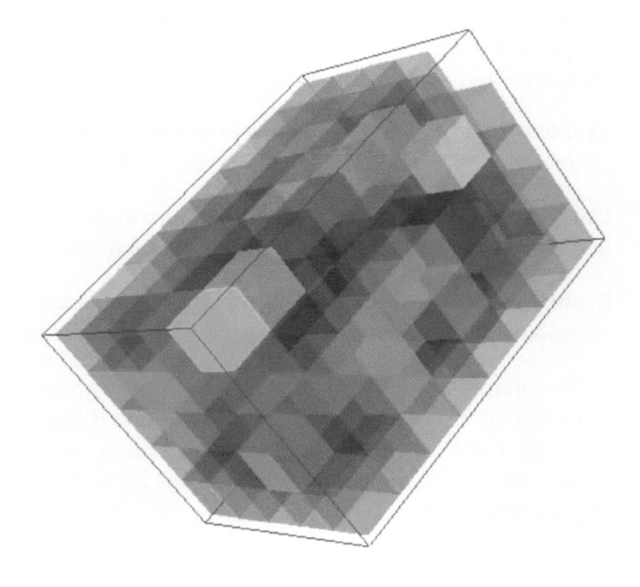

Fig. 1.43 $8 \times 5 \times 5$ array of cubes

Problem 1.118 MScript 26 in Appendix A produces the $8 \times 5 \times 5$ array of cubes in Fig. 1.43. Give an example of an example of a bornological nerve in Fig. 1.43. Prove that your choice of a bornological nerve endowed with a proximity δ is a local proximity space. To do this, show that each of the requirements for a local proximity space are satisfied. ■

Problem 1.119 Modify MScript 26 in Appendix A to produce an $10 \times 8 \times 8$ array of cubes. Give an example of an example of a bornological nerve in the array of cubes. Prove that your choice of a bornological nerve endowed with the strong proximity $\overset{\wedge\!\!\!\wedge}{\delta}$ is a local proximity space. To do this, show that each of the requirements for a local proximity space are satisfied. ■

References

1. Di Maio, G., Naimpally, S.A., Meccariello, E.: Theory and applications of proximity, nearness and uniformity. Seconda Università di Napoli, Napoli (2009), 264 pp., MR1269778
2. Naimpally, S., Peters, J., Wolski, M.: Foreword [near set theory and applications]. Math. Comput. Sci. **7**(1), 1–2 (2013)
3. Peters, J.: Proximal relator spaces. Filomat (2016). Accepted
4. Peters, J., Guadagni, C.: Strongly proximal continuity and strong connectedness, pp. 1–11 (2015). arXiv:1504.02740
5. Peters, J.: Topology of digital images - visual pattern discovery in proximity spaces, intelligent systems reference library, vol. 63. Springer, Berlin (2014). Xv + 411 pp. Zentralblatt MATH Zbl 1295 68010
6. Peters, J., Guadagni, C.: Strong proximities on smooth manifolds and Voronoï diagrams. Adv. Math.: Sci. J. **4**(2), 91–107 (2015)
7. Naimpally, S., Peters, J.: Topology with applications. Topological spaces via near and far. World Scientific, Singapore (2013). Xv + 277 pp. Am. Math. Soc. MR3075111
8. Dochviri, I., Peters, J.: Topological sorting of finitely many near sets. Math. Comput. Sci. 1–6 (2015). Communicated
9. Naimpally, S.: Proximity Spaces. Cambridge University Press, Cambridge (1970). X + 128 pp., ISBN 978-0-521-09183-1
10. Lodato, M.: On topologically induced generalized proximity relations. Ph.D. thesis. Rutgers University (1962). Supervisor: S. Leader
11. Guadagni, C.: Bornological convergences on local proximity spaces and ω_{μ}-metric spaces. Ph.D. thesis, Università degli Studi di Salerno, Salerno (2015). Supervisor: A. Di Concilio, 79 pp
12. Di Concilio, A.: Proximity: a powerful tool in extension theory, functions spaces, hyperspaces, boolean algebras and point-free geometry. In: Mynard, F., Pearl, E. (eds.) Beyond Topology. AMS Contemporary Mathematics, vol. 486, pp. 89–114. American Mathematical Society, Providence (2009)
13. Peters, J.: Local near sets: pattern discovery in proximity spaces. Math. Comput. Sci. **7**(1), 87–106 (2013). doi:10.1007/s11786-013-0143-z, MR3043920
14. Hettiarachchi, R., Peters, J.: Multi-manifold LLE learning in pattern recognition. Pattern Recognit. Elsevier **48**(9), 2947–2960 (2015)
15. Hettiarachchi, R., Peters, J.: Multi-manifold skin classifier for feature space voronoï region-based skin segmentation. Image Vis. Comput. (2015). Communicated
16. İnan, E., Öztürk, M.: Near groups on nearness approximation spaces. Hacet. J. Math. Stat. **41**(4), 545–558 (2012)

17. Peters, J., İnan, M.A., Öztürk, E.: Spatial and descriptive isometries in proximity spaces. Gen. Math. Notes **21**(2), 125–134 (2014)
18. Peters, J., Öztürk, M., Uçkun, M.: Klee-Phelps convex groupoids, pp. 1–5 (2014). arXiv:1411.0934, Mathematica Slovaca 2015. Accepted
19. Edelsbrunner, H., Harer, J.: Computational Topology. An Introduction. American Mathematical Society, Providence (2010). Xii + 110 pp., MR2572029
20. Peters, J.: Visibility in proximal Delaunay meshes and strongly near Wallman proximity. Adv. Math.: Sci. J. **4**(1), 41–47 (2015)
21. Di Concilio, A., Gerla, G.: Quasi-metric spaces and point-free geometry. Math. Struct. Comput. Sci. **16**(1), 115–137 (2006). MR2220893
22. Di Concilio, A.: Point-free geometries: proximities and quasi-metrics. Math. Comput. Sci. **7**(1), 31–42 (2013). MR3043916
23. Beer, G., Lucchetti, R.: Weak topologies for the closed subsets of a metrizable space. Trans. Am. Math. Soc. **335**(2), 805–822 (1993)
24. Aurenhammer, F.: Voronoi diagrams–a survey of a fundamental geometric data structure. ACM Comput. Surv. **23**(3), 345–405 (1991)
25. Peters, J., Ramanna, S.: Proximal three-way decisions: theory and applications in social networks. Knowl.-Based Syst. 1–12 (2014). http://dx.doi.org/10.1016/j.knosys.2015.07.021 (in press)
26. Pták, P., Kropatsch, W.: Nearness in digital images and proximity spaces. LNCS. In: Proceedings of the 9th International Conference on Discrete Geometry, vol. 1953, pp. 69–77 (2000)
27. Beer, G., Di Concilio, A., Di Maio, G., Naimpally, S., Pareek, C., Peters, J.: Somashekhar Naimpally, 1931–2014. Topol. Appl. **188**, 97–109 (2015). doi:10.1016/j.topol.2015.03.010, MR3339114
28. Di Maio, G., Naimpally, S.: Preface, theory and applications of proximity, nearness and uniformity, Quaderni di Matematica [Mathematics Series], vol. 22. Department of Mathematics, Seconda Universit di Napoli, Caserta (2008). viii + 364 pp., MR2760945
29. Plato: the allegory of the cave, in Republic, VII, 514a,2–517a, 7. Stanford University, Stanford (c300BC, 2015). Translated by T. Sheehan
30. Frege, G.: Foundations of arithmetic, Translated by J.L. Austin. Blackwell, Oxford (1953)
31. Whitehead, A.: An Inquiry into the Principles of Natural Knowledge. Cambridge University Press, Cambridge (1920)
32. Whitehead, A.: Process and Reality. Macmillan, London (1929)
33. Hooke, R.: Micrographia: or, some physiological descriptions of minute bodies made by magnifying glasses. With observations and inquiries thereupon. Martyn and J. Allestry, London (1665). 270 pp
34. Prince, S.: Computer Vision. Models, Learning, and Inference. Cambridge University Press, Cambridge (2012). Xvii + 580 pp
35. Favorskaya, M., Jain, L.C.: Computer Vision in Control Systems-1. Mathematical Theory. Springer, Berlin (2015). Xvii + 371 pp
36. Arkowitz, M.: Introduction to Homotopy Theory. Springer, New York (2011). Xiv + 344 pp. ISBN: 978-1-4419-7328-3, MR2814476
37. Hatcher, A.: Algebraic Topology. Cambridge University Press, Cambridge (2002). Xii + 544 pp. ISBN: 0-521-79160-X; 0-521-79540-0, MR1867354
38. Whitehead, J.: Simplicial spaces, nuclei and m-groups. Proc. Lond. Math. Soc. **45**, 243–327 (1939)
39. Krantz, S.: Essentials of topology with applications. CRC Press, Boca Raton (2010). Xvi + 404 pp. ISBN: 978-1-4200-8974-5, MR2554895
40. Euclid: The Thirteen Books of Euclid's Elements, 2nd ed. Dover Publications, New York (c300BC, 1956). Translated by T.L. Heath, from the text by Heiberg, xi + 432pp; i + 436pp; i + 546 pp, MR0075873
41. Ronse, C.: Regular open or closed sets. Philips Research Laboratory Series, Brussels WD59, pp. 1–8 (1990)

42. Peters, J., Naimpally, S.: Applications of near sets. Notices Am. Math. Soc. **59**(4), 536–542 (2012). http://dx.doi.org/10.1090/noti817, MR2951956
43. Edelsbrunner, H.: Geometry and Topology of Mesh Generation. Cambridge University Press, Cambridge (2001). 209 pp
44. Deritei, D., Lázár, Z., Papp, I., Járai-Szabó, F., Sumi, R., Varga, L., Regan, E., Ercsey-Ravasz, M.: Community detection by graph voronoi diagrams. New J. Phys. **16**, 1–17 (2014)
45. Riesz, F.: Stetigkeitsbegriff und abstrakte mengenlehre. Atti del IV Congresso Internazionale dei Matematici, vol. II, pp. 182–109 (1908)
46. Čech, E.: Topological Spaces. Wiley, London (1966). Fr seminar, Brno, 1936–1939; rev. ed. Z. Frolik, M. Katětov
47. Efremovič, V.: The geometry of proximity I (in Russian). Mat. Sb. (N.S.) 31(73)(1), 189–200 (1952)
48. Peters, J.: Near sets. Special theory about nearness of objects. Fundamenta Informaticae **75**, 407–433 (2007). MR2293708
49. Peters, J., Guadagni, C.: Strongly near proximity and hyperspace topology, pp. 1–6 (2015). arXiv:1502.05913
50. Wallman, H.: Lattices and topological spaces. Ann. Math. **39**(1), 112–126 (1938)
51. Euclid: Elements. Alexandria (300 B.C.). English translation by R. Fitzpatrick, from Euclidis Elementa Latin text by B.G. Teubneri, 1883–1885 and the Greek text by J.L. Heiberg, 1883–1885
52. Klette, R., Rosenfeld, A.: Digital Geometry. Geometric Methods for Digital Picture Analysis. Morgan-Kaufmann Publishers, Amsterdam (2004)
53. Du, Q., Faber, V., Gunzburger, M.: Centroidal voronoi tessellations: applications and algorithms. SIAM Rev. **41**(4), 637–676 (1999). MR
54. Boissonnat, J.D., Wormser, C., Yvinec, M.: Curved Voronoi diagrams. In: Boissonnat, J.D., Teillaud, E. (eds.) Effective Computational Geometry for Curves and Surfaces, pp. 67–116. Springer, New York (2006)
55. Kovalevsky, V.: Finite topology as applied to image analysis. Comput. Vis. Graph. Image Process. **46**, 141–161 (1989)
56. Grünbaum, B., Shephard, G.: Tilings and Patterns. W.H. Freeman and Co., New York (1987). Xii + 700 pp., MR0857454
57. Romesburg, H.: Cluster Analysis for Researchers. Lulu Press, North Carolina (2004). Xii + 334 pp., MR3155265
58. Peters, J.: Near sets. General theory about nearness of sets. Appl. Math. Sci. **1**(53), 2609–2629 (2007)
59. Peters, J.: Near sets: an introduction. Math. Comput. Sci. **7**(1), 3–9 (2013). doi:10.1007/s11786-013-0149-6, MR3043914
60. Kuratowski, C.: Topologie I. Panstwowe Wydawnictwo Naukowe, Warsaw (1958). XIII + 494 pp
61. Kuratowski, K.: Introduction to Set Theory and Topology, 2nd edn. Pergamon Press, Oxford (1962, 1972). 349 pp
62. Naimpally, S., Warrack, B.: Proximity Spaces. Cambridge Tract in Mathematics, vol. 59. Cambridge University Press, Cambridge (1970). X + 128 pp., Paperback (2008)
63. Peters, J., Guadagni, C.: Strongly far proximity and hyperspace topology, pp. 1–6 (2015). arXiv:1502.02771
64. Peters, J., Guadagni, C.: Strongly hit and far miss hypertopology and hit and strongly far miss hypertopology, pp. 1–8 (2015). arXiv:1503.02587
65. Solan, V.: Introduction to the axiomatic theory of convexity [Russian with English and French summaries]. Shtiintsa, Kishinev (1984). 224 pp., MR0779643
66. Zelins'kyi, Y.: Generalized convex envelopes of sets and the problem of shadow. J. Math. Sci. **211**(5), 710–717 (2015)
67. Kay, D., Womble, E.: Automatic convexity theory and relationships between the carathèodory, helly and radon numbers. Pac. J. Math. **38**(2), 471–485 (1971)

68. Tuz, V.: Axiomatic convexity theory [Russian]. Rossiĭskaya Akademiya Nauk. Matematicheskie Zametki [Math. Notes Math. Notes] **20**(5), 761–770 (1976)
69. Rocchi, N.: Parliamo Di Insiemi. Instituto Didattico Editoriale Felsineo. Bologna (1969). 316 pp
70. Bourbaki, N.: Elements of Mathematics. General Topology, Part 1. Hermann and Addison-Wesley, Paris and Reading (1966). I-vii, 437 pp
71. Edelsbrunner, H.: A Short Course in Computational Geometry and Topology. Springer, Berlin (2014). 110 pp
72. Frank, N., Hart, S.: A dynamical system using the Voronoi tessellation. Am. Math. Monthly **117**(2), 92–112 (2010)
73. Weisstein, E.: Voronoi diagram. Wolfram MathWorld (2015). http://mathworld.wolfram.com/VoronoiDiagram.html
74. Liebling, T., Pourin, L.: Voronoi diagrams and Delaunay triangulations: Ubiquitous siamese twins. Documenta Mathematica Extra volume: Optimization stories, 419–431 (2012). MR2991503
75. Okabe, A., Boots, B., Sugihara, K., Chiu, S.: Spatial Tessellations: Concepts and Applications of Voronoi Diagrams. Wiley, Chichester (2000). Xvi + 671 pp. ISBN: 0-471-98635-6, MR1770006
76. Stover, C.: Point Process. Wolfram MathWorld (2015). http://mathworld.wolfram.com/PointProcess.html
77. Weisstein, E.: Poissonprocess. Wolfram MathWorld (2015). http://mathworld.wolfram.com/PoissonProcess.html
78. Floriani, L.D., Spagnuolo, M.: Shape Analysis and Structuring. Springer, Berlin (2008). Xiv + 296 pp. ISBN 978-3-540-33264-0
79. Hilhorst, H.: Statistical properties of planar Voronoi tessellations. Eur. Phys. J. B **64**, 437–441 (2008)
80. Anton, F., Mioc, D., Gold, C.: The Voronoi diagram of circles and its application to the visualization of the growth of particles. In: Gavrilova, M., Tan, C.K. (eds.) Transactions on Computational Science III, pp. 20–54. Springer, Berlin (2009). MR2912541
81. Surendran, S., Chitraprasad, D., Kaimal, M.: Voronoi diagrams-based geometric approach to social network analysis. In: Krishnan, G.S.S., et al. (eds.) Computational Intelligence, Cyber Security and Computational Models, Advances in Intelligent Systems and Computing, vol. 246 pp. 359–369. Springer, India (2014)
82. Liu, H.: Dynamic concept cartography for social networks. Master's thesis, School of Information Technologies (2007)
83. Peters, J.: Proximal Voronoï regions, convex polygons, and leader uniform topology. Adv. Math.: Sci. J. **4**(1), 1–5 (2015)
84. Munkres, J.: Topology, 2nd edn. Prentice-Hall, Englewood Cliffs (2000). Xvi + 537 pp., 1st edn. in 1975, MR0464128
85. Krantz, S.: A Guide to Topology. The Mathematical Association of America, Washington (2009). Ix + 107 pp
86. Leader, S.: On clusters in proximity spaces. Fundamenta Mathematicae **47**, 205–213 (1959)
87. Naimpally, S.: Proximity Approach to Problems in Topology and Analysis. Oldenbourg Verlag, Munich (2009). 73 pp., ISBN 978-3-486-58917-7, MR2526304
88. Mozzochi, C., Gagrat, M., Naimpally, S.: Symmetric Generalized Topological Structures. Exposition Press, Hicksville (1976). I + 73 pp
89. Di Concilio, A., Naimpally, S.: Proximal convergence. Monatsh. Math. **103**, 93–102 (1987)
90. Willard, S.: General Topology. Dover Publications Inc, Mineola (1970). Xii + 369 pp, ISBN: 0-486-43479-6 54-02, MR0264581
91. Install, M., Weisstein, E.: Connected set. Mathworld. A Wolfram Web Resource (2015). http://mathworld.wolfram.com/ConnectedSet.html
92. Leader, S.: Extensions based on proximity and boundedness. Math. Z. **108**, 137–144 (1969)
93. Leader, S.: Local proximity spaces. Mathematische Annalen **169**, 275–281 (1967)
94. Sierpiński, W.: Sur une courbe dont tout point est un point de ramification. C.R.A.S. **160**, 302–305 (1915)
95. Wolfram, S.: A New Kind of Science. Wolfram Media Inc, Champaign (2002). Xiv + 1197 pp., MR1920418

Chapter 2
Proximities Revisited

Fig. 2.1 Som Naimpally

This chapter takes another look at the very rich proximity landscape. An overview of the proximity landscape is given in the life and work of S.A. Naimpally (Som). For a recent picture of Som, see Fig. 2.1 and for an overview of Som's research contributions, see [1]. This is a remarkable story of a mathematician who began studying proximity space theory after he completed his Ph.D. as a result of a chance meeting at the University of Michigan between Som and a visitor from Cambridge University Press, who invited him to write a monograph on proximity. This he did together with his graduate student B.D. Warrack, leading to a complete overview of proximity space theory until 1970 [2].

The study of the nearness of sets now spans more than 100 years, starting with the address by F. Riesz at the International Congress of Mathematicians in Rome in 1908 [3], recently commented on by S.A. Naimpally [4, 5] and A. Di Concilio [6–8]. One of the earliest introductions to nearness (proximity) relations was given by E. Čech during a 1936–1939 Brno seminar, published in 1966 [9, Sect. 25.A.1]. Čech used the symbol p to denote a proximity relation defined on a nonempty set X, which Čech axiomatized. Čech's work on proximity spaces started two years after V.A. Efremovič's work (in 1933), who introduced a widely considered axiomatization of proximity, which was not published until 1951 [10]. For a detailed

J.F. Peters, *Computational Proximity*, Intelligent Systems
Reference Library 102, DOI 10.1007/978-3-319-30262-1_2

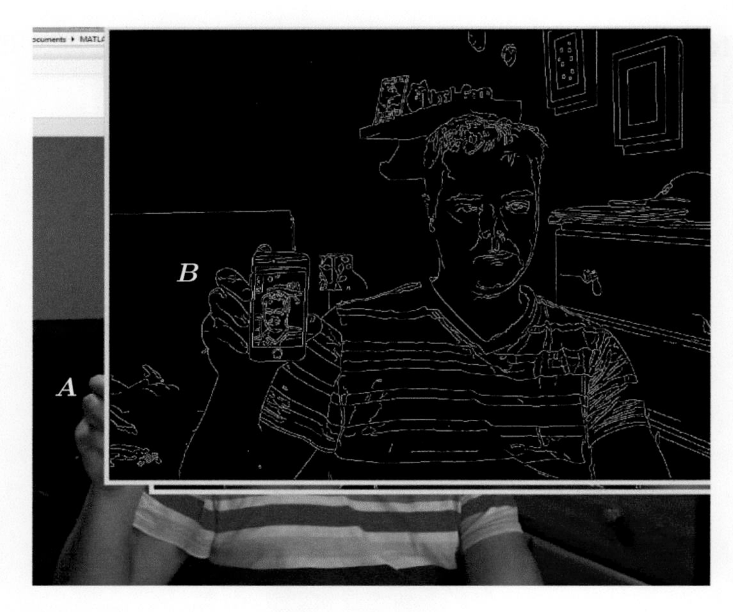

Fig. 2.2 $A\ \delta_{\Phi}\ B$ (descriptively near sets)

presentation of Efremovič's proximity axioms, see, e.g., [8, 11] and for applications, see, e.g., [12–16].

This chapter revisits a number of familiar proximities. It also introduces Delaunay triangulation and takes another look at Voronoï diagrams. The focus in this second look at proximities is on finite proximity spaces and the strong nearness of sets and points to sets. For example, let X be a finite topological space on the set of points in the digital image in Fig. 2.2 and let X be endowed with the descriptive Lodato proximity δ_{Φ}.

Let A be that part of Fig. 2.2 showing the hand and torso and let B be that part of the image showing detected edges.[1] Let a description $\Phi(A)$, $A \subset X$ be defined by the shape descriptor gradient orientation of the edge pixels in A. Clearly, $A\ \delta_{\Phi}\ B$, since, for instance, the gradient orientation of the edge pixels along the top edges of the hand in A are exactly the same as the gradient orientation of the edge pixelsalong the top edges of the hand in B. In fact, this is an example of strong descriptive nearness with $A\ \overset{\wedge\!\!\wedge}{\delta}_{\Phi}\ B$.

[1]Many thanks to Braden Cross for the webcam image in Fig. 2.2, captured using the Matlab Computer Vision System toolbox and Matlab implementation of the Canny edge detection algorithm.

2.1 Cech Proximity

A nonempty set X endowed with the Čech proximity relation δ is a proximity space, denoted (X, δ). *Distant* (*far*, *remote*) sets A, B are not near, denoted by $A \not\delta B$. In a metric topology endowed with a proximity, sets are remote, provided $D(A, B) > 0$. In a finite proximity space, remote sets can be defined without reference to a metric. That is, $A \not\delta B$, provided A and B have no points in common. Remote sets play an important role in applications of topology [14] and the topology of digital images [17]. That is, remote sets are disjoint, having no elements in common.

Remote sets are **separated sets** in any space. For example, separated sets of picture elements in the layered digital image in Fig. 2.2 correspond to the remote image regions such as the set of hand pixels A, which is separated from the set of hand edge pixels B. By identifying image regions remote from each other, we have a means of identifying separated image patterns, distinguishing one from the other by characteristics such as

1° smooth, continous, thick region edges such as those in the hand region A in Fig. 2.2.

2° non-smooth, fragmented, thin region edges such those in as in the hand region B in Fig. 2.2.

> Edgewise in Fig. 2.2, A and B are remote (edges are separated *spatially* and are different *descriptively*). Hence, we write $A \ \not\delta\ B$ (edges A are not close to edges B) and $A \ \not\delta_{\Phi}\ B$ (edges A are not descriptively the same as B). ■

3° skin-colour hand with solid shapes and with highlighted (chiaroscuro) parts in region A in Fig. 2.2.

4° white hand edge-enclosed shapes on a densely black background and with no highlighting (chiaroscuro) parts in region B in Fig. 2.2.

> Shape-wise in Fig. 2.2, A and B are remote (shapes are separated *spatially* and are different *descriptively*). Hence, again, we write $A \ \not\delta\ B$ (shape A is not close to shape B) and $A \ \not\delta_{\Phi}\ B$ (shape A is not descriptively the same as B). ■

The Čech proximity δ is the most elementary form of proximity relation. The relation δ (*close, near, proximal*) is a Čech proximity relation on the family of all subsets 2^X of X, provided it satisfies the axioms (**P1**)–(**P4**) given earlier in Sect. 1.4, introduced during the mid-1930s by E. Čech [9, Sect. 25, p. 439]. Let A, B be nonempty subsets in X. The expression $A\ \delta\ B$ reads A *close to* B and $A \ \not\delta\ B$ reads A *is not close to* B. Also recall that $\varnothing$ denotes the empty set. A set E is empty, provided E has no elements, members, points. The intersection of A and B (denoted by $A \cap B$) is the set of points that are common to (*in both*) A and B, i.e.

$$A \cap B = \{x \in A \cup B : x \in A \text{ and } x \in B\}.$$

Let X be a nonempty set, $x \in X$, $A, B \subset X$. The Čech proximity δ (or any of the other proximities in Sect. 1.4) can be used to define the closure of a set. Recall that

$$\mathrm{cl}B = \{x \in X : x\ \delta\ B\} \text{ (\textbf{Closure of a set})}.$$

The union of A and B (denoted by $A \cup B$) is the set of points that are in A or B. The Kuratowski [18] closure operator cl leads to refinements of each of the usual proximities.

Definition 2.1 Let X be a nonempty set, $A, B \subset X$. The *closure operator* is a self map on 2^X the power set of X (collection of all subsets of X) satisfies the following axioms.

K.1 $\mathrm{cl}\varnothing = \varnothing$.
K.2 $B \subset \mathrm{cl}B$.
K.3 $\mathrm{cl}(A \cup B) = \mathrm{cl}A \cup \mathrm{cl}B$.
K.4 $\mathrm{cl}(\mathrm{cl}B) = \mathrm{cl}B$.

The following axioms stem from finite sets.

K.5 If X is finite, then $\mathrm{cl}X = X$ (Kuratowski [18, Sect. 4, III]).
K.6 A subset $B \subset X$ is closed, if and only if $\mathrm{cl}B = B$.

For a brief overview of Kuratowski closure, see Appendix B. ■

From Axiom **K**.6, a subset B in X is **closed** if and only if B coincides with its closure, i.e., $\mathrm{cl}B = B$ [19].

Example 2.2 **Closed Sets**.
Let X be a finite subset in Euclidean space $\mathbb{R}^2$, $A, B, S \subset X$, $p, q, r \in S$. Further, let bdy$A$ denote the boundary of A, $\mathrm{int}A = A \setminus \mathrm{bdy}$ the interior of A. Here are some examples of closed sets.

1° Each region V_p in a Voronoï diagram $V(S)$ is a closed set, since $\mathrm{cl}V_p = V_p$ (from Axiom **K**.6). Every region V_p is a solid polygon with nonempty interior and includes its boundary.
2° Each triangle $\triangle(pqr)$ in a Delaunay triangulation $D(S)$ is a closed set with an empty interior.
3° In the psychology of human vision, it has been suggested that perceived objects always include their boundaries [20].
4° **Regular closed**. A set A is **regular closed**, provided $A = \mathrm{cl}(\mathrm{int}A)$. On the other hand, a set A is **regular open**, provided $B = \mathrm{int}(\mathrm{cl}B)$. That is, a nonempty set is *open*, if it coincides with its interior of its closure. Notice that $B \subseteq A$.
5° The description $\Phi(x)$ of a pixel x in a digital image A is a closed set, since each pixel can be completely described without reference to the description of its adjacent pixels, i.e., $\{\Phi(x)\} = \mathrm{cl}(\{\Phi(x)\})$. ■

Problem 2.3 🚲 Is a set $\{x\}$ containing a single pixel x in a digital image a regular open set? ■

If $A \subset X$ contains a single element $x \in X$, then, for simplicity, we write $x\ \delta\ B$ instead of $\{x\}\ \delta\ B$. That is, x is near B.

Then the closure axioms can be rewritten using the nearness relation between points and sets as in [19, Sect. 2]. This is done by defining $x\ \delta\ B \Leftrightarrow x \in \text{cl}B$. In that case, the point x is near B. From this, we obtain the following Lodato point-set proximity axioms.

Definition 2.4 Lodato Point-Set Proximity Revisited.
Let X be a proximity space endowed with a proximity δ, $x \in X$, nonempty sets $A, B, C \subset X$. δ is point-set proximity, provided it satisfies the following axioms.
T.1 $x\ \delta\ B \Rightarrow B \neq \emptyset$
T.2 $x \cap B \neq \emptyset \Rightarrow x\ \delta\ B$ (uniformity).
T.3 $x\ \delta\ (B \cup C) \Leftrightarrow x\ \delta\ B$ or $x\ \delta\ C$.
T.4 $x\ \delta\ B$ and $b\ \delta\ C$ for each $b \in B \Rightarrow x\ \delta C$.
In addition, if X is finite, we have
T.5 $x\ \delta\ B \Leftrightarrow \text{cl}x \cap \text{cl}B \neq \emptyset$ (Wallman). ■

A relation δ that satisfies Axiom **T**.2 is called a *discrete proximity* on the family of sets of X [8, Sect. 2.1, p. 93].

Further, if X is finite, we have

Theorem 2.5 *Let X be a finite Čech proximity space, $x, y \in X, x \neq \emptyset$, $B \subset X$. Then the following are equivalent.*
1° $x\ \delta\ B$.
2° $x \in B$ implies $x \cap B \neq \emptyset$.
3° $clx \subset clB \Leftrightarrow \{x\} \subset B$.
4° $cl(x \cap B) \subset (clx \cap clB) \subset x \cap B$

Proof
$1^o \Leftrightarrow 2^o$: $x\ \delta\ B \Leftrightarrow \text{cl}x \cap \text{cl}B \neq \emptyset$ (Axiom **T**.5) $\Leftrightarrow x \in B$ implies $x \cap B \neq \emptyset$.
$2^o \Leftrightarrow 3^o \Leftrightarrow 4^o$. □

Theorem 2.6 *Let X be a finite Čech proximity space, $x, y \in X, x \neq \emptyset$, $B \subset X$.*
1° Points x, y are remote, if and only if $clx \cap cly = \emptyset$.
2° A point x is separated from B, if and only if $clx \cap clB = \emptyset$.

Proof
1°: Let $B = \{x\}$ and the result follows from Axiom **T**.5.
2°: From Axiom **T**.5, $\text{cl}x \cap \text{cl}B = \emptyset \Leftrightarrow x\ \not\delta\ B$. □

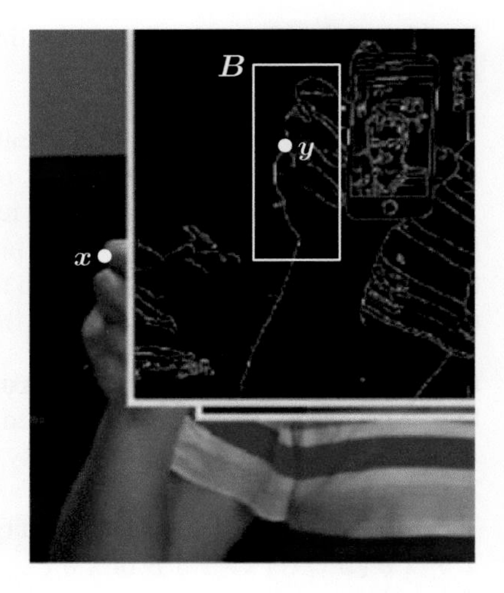

Fig. 2.3 $x\ \delta_{\Phi}\ B$

The **closure of a picture point** (digital image pixel or painting spot) equals the union of itself and its adjacent (boundary) points, i.e.,

$$
\begin{aligned}
X &= \text{ set of picture points.}\\
\|x - y\| &= 1, \text{ if and only if } x, y \text{ are } \textbf{adjacent}.\\
\text{bdy}(x) &= \{y \in X : y \text{ adjacent to } x\}.\\
\text{cl}x &= \text{bdy}(x) \cup \{x\} \text{ (Closure of picture point).}
\end{aligned}
$$

Remote picture points are those points that do not have common adjacent points.

Definition 2.7 Types of Remote Picture Points.
Let X be a finite set of picture points endowed with proximities δ, δ_{Φ}. Two basic types of remote picture points can be found in X.

1° **Spatially remote points**. *Spatially remote picture points* are those points that do not have common adjacent points.

2° **Descriptively remote points**. Descriptively remote picture points are those points that do not have matching descriptions. Notice that adjacent points can be descriptively remote. ■

Example 2.8 **Remote Picture Points.**
Let X be a finite set of picture points in Fig. 2.3 endowed with proximities δ, δ_{Φ}, $x, y \in X$. Then x is spatially remote from y, i.e., $x \ \not\delta\ y$, since $\text{cl}x \ \not\delta\ \text{cl}y$. Assuming that x and y have the different gradient orientations, x is descriptively remote from y, i.e., $x\ \not\delta_{\Phi}\ y$. ■

Example 2.9 **Descriptive Nearness of Points to Sets: Finite Case.**
Let X be a finite descriptive point-set proximity space (X, δ_{Φ}) that consists of the picture points shown in Fig. 2.3, B be the finite set of hand-edge picture points in Fig. 2.3, $x \in X$. The picture point x lies on the outer edge of the hand image and B is restricted to the set of edge picture points in Fig. 2.3. Assume that each picture point is described by its gradient orientation. In that case, $x\ \delta_{\Phi}\ B$, since the gradient orientation of x matches the gradient orientation of at least one of the picture points in B. ■

Definition 2.10 Types of Close Picture Points.
Let X be a finite set of picture points endowed with proximities δ, δ_{Φ}. Several basic types of close picture points can be found in X.

1° **Spatially close points.** *Spatially close picture points* are those points that are adjacent.

2° **Strongly close point to a set.** A picture point x **Strongly close** to a set $A \subset X$ is a member of A, i.e., $x \in A$ and $x\ \overset{\wedge\!\!\wedge}{\delta}\ A$.

3° **Descriptively close points.** *Descriptively close picture points* are those points that have matching descriptions, i.e., $\Phi(x) \in \Phi(A)$ (description of x belongs to the set of descriptions of the points in A. Notice that spatially remote points can be descriptively close). ■

Example 2.11 **Near and Strongly Near Picture Points.**
Let X be a finite set of picture points in Fig. 2.3 endowed with proximities δ, δ_{Φ}, $x, y \in X$. Then x is spatially close to all of is adjacent points. The point y is strongly close to B, i.e., $y\ \overset{\wedge\!\!\wedge}{\delta}\ B$, since $y \in B$. From Example 2.9, the point x is descriptively close to at least one edge point in B. ■

Example 2.12 **Near Polygons.**
Polygons A and B in Fig. 2.4 are examples of Voronoï regions in the mesh on hand image.[2] A and B have a common edge, $\mathrm{cl}A \cap \mathrm{cl}B \neq \emptyset$. Hence, from Axiom prox.2, $A\ \delta\ B$, i.e., A is close to B. In fact, $A\ \overset{\wedge\!\!\wedge}{\delta}\ B$, since $\mathrm{cl}A \cap \mathrm{cl}B$ contains more than one point.

Every point in the common edge shown in Fig. 2.4 is strongly near both A and B, since A and B overlap. All points y in A not on the common edge are spatially remote from B, i.e., $y\ \not\delta\ B$. The point x labeled with a red dot • in A is descriptively near the point y labeled with a red dot • in B, provided $\Phi(x) = \Phi(y)$. In that case, $x\ \overset{\wedge\!\!\wedge}{\delta}_{\Phi}\ B$, since $\Phi(x) \in \Phi(B)$, x is strongly close descriptively to B, since, by assumption, the description of x belongs to the set of descriptions of the points in B. ■

Remark 2.13 From this point forward, assume that δ is the Wallman proximity. ■

[2]Many thanks to Binglin Li for this hand image in Fig. 2.4.

Fig. 2.4 $A \stackrel{\wedge\!\!\wedge}{\delta} B$

Problem 2.14 Let X be the set of polygons in the Voronoï mesh on the hand image in Fig. 2.4. Let δ be a nearness (proximity) relation on X. Prove that each of the Čech axioms are satisfied by δ. In effect, prove that δ is the Čech proximity and that (X, δ) is an example of a Čech proximity space.
Hint: From Example 2.12, we know that δ satisfies Axiom prox.2. So it is only necessary to prove that the remaining three Čech axioms are satisfied. ■

2.2 Cech Closure of a Set

From the Čech proximity relation δ, we can derive the closure of a nonempty set [9]. Recall that the closure of a set A (denoted cl$A$) is the union of its interior points (denoted int$A$) and its boundary points (denoted bdy$A$). In fact, in a proximity space, the closure of a set A is the union of all points near A. Hence, the earlier formulation of the closure of nonempty set is rewritten in terms of the union of all of its near points. In a proximity space X, the *closure* of a subset $A \subset X$ is defined by

$$\mathrm{cl}A = \bigcup_{x \in X} \{x \ \delta \ A\} \text{ (closure of the set } A).$$

This formulation of the closure of a set is one of the properties of a proximity space[3] introduced by V.A. Efremovič [10]. The space (X, cl) is called a *closure space*. A point is *proximal* in (X, cl) if and only if $x \in \mathrm{cl}X$. The closure of a set X contains all points proximal to X. In a proximity space with the closure property, δ is called a *Wallman proximity*, named after H. Wallman [21].

[3]Pointed out by I. Dochviri.

Let subsets $A, B \subset X$, which is a proximity space. The negative of the Wallman proximity is very interesting, since it paves the way for the study of the separation of structures, i.e., those that have no points in common. That is,

$$A \not\delta B, \text{ if and only if } \mathrm{cl}A \cap \mathrm{cl}B = \emptyset.$$

Wallman proximity is particularly important in finding hidden patterns in digital images, especially if we are comparing and contrasting geometric structures such as edges, corners, line segments and other convex polygons in images.

2.3 Near Edge Sets

A set endowed with a proximity relation is called a proximity space. A *relator* is a nonvoid family of relations $\mathcal{R}$ on a nonempty set X. The pair $(X, \mathcal{R})$ (also denoted $X(\mathcal{R})$) is called a relator space. Sets A, B are descriptively near (have descriptive proximity), provided there are one or more pairs of points $a \in A, b \in B$ with matching descriptions.

Example 2.15 In Fig. 2.5, sets A_1, A_2, M' have EF-proximity (spatially near), since these sets have points in common. That is, $A_1\ \delta\ M', A_2\ \delta\ M', A_1\ \delta\ A_2$. Sets A_1, A_2, B_1, B_2, M' have descriptive proximity (descriptively near), since one can find pairs of points with matching descriptions. All of these sets contain black edge points. That is, $A_1\ \delta_\Phi\ M',\ A_2\ \delta_\Phi\ M', A_1\ \delta_\Phi\ A_2$ and $A_1\ \delta_\Phi\ B_1, A_2\ \delta_\Phi\ B_2, B_1\ \delta_\Phi\ B_2$ and so on. ■

Fig. 2.5 Near edge sets

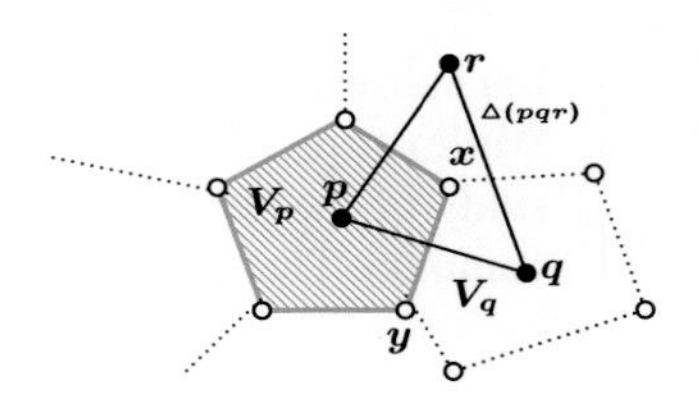

Fig. 2.6 Delaunay triangle $\Delta(pqr)$

2.4 Lodato Proximity

Recall how a *Lodato proximity* is defined [22–24] (see, also, [5, 11]). Let X be a nonempty set. A *Lodato proximity* δ is a relation on $\mathscr{P}(X)$ which satisfies the following axioms for all subsets A, B, C of X:

Lodato Proximity Axioms

(P0) $\emptyset \not\delta A, \forall A \subset X.$
(P1) $A \ \delta \ B \Leftrightarrow B \delta A.$
(P2) $A \ \cap \ B \neq \emptyset \Rightarrow A \delta B.$
(P3) $A \ \delta \ (B \cup C) \Leftrightarrow A \ \delta \ B$ or $A \ \delta \ C.$
(P4) $A \ \delta \ B$ and $\{b\} \ \delta \ C$ for each $b \in B \ \Rightarrow A \ \delta \ C.$ ■

Further δ is *separated* , if

(P5) $\{x\} \ \delta \ \{y\} \Rightarrow x = y.$ ■

We can associate a topology with the space (X, δ) by considering as closed sets the ones that coincide with their own closure, where for a subset A we have

$$\text{cl}A = \{x \in X : x \ \delta \ A\}.$$

This is possible because of the correspondence of the Lodato axioms with the well-known Kuratowski closure axioms (see Appendix B).

Example 2.16 **Lodato proximal Cameraman Elbow Sets**.
Let (X, δ) be a Lodato proximity space represented by set of points in the partial cameraman image in Fig. 2.7, $A_1, A_2, M', B_1, B_2 \subset X$. The Lodato axiom **(P4)** is satisfied by

$$M' \ \delta \ A_1, \ A_1 \ \delta \ A_2 \forall a \in A_1 \Leftrightarrow M' \ \delta \ A_2,$$
$$M' \ \delta \ B_1, \ B_1 \ \delta \ B_2 \forall b \in B_1 \Leftrightarrow M' \ \delta \ B_2. \quad ■$$

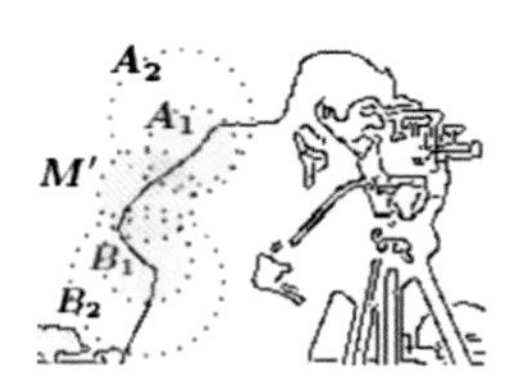

Fig. 2.7 Proximal cameraman elbow sets

2.5 Descriptive Lodato Proximity

Descriptive Lodato proximity was introduced in [17]. Let X be a nonempty set, $A, B \subset X, x \in X$. Recall that $\Phi(x)$ is a feature vector that describes x, $\Phi(A)$ is the set of feature vectors that describe points in A. The descriptive intersection between A and B (denoted by $A \underset{\Phi}{\cap} B$) is defined by

$$A \underset{\Phi}{\cap} B = \{x \in A \cup B : \Phi(x) \in \Phi(A) \text{ and } \Phi(x) \in \Phi(B)\}.$$

A *Descriptive Lodato proximity* δ_Φ is a relation on $\mathscr{P}(X)$ which satisfies the following axioms for all subsets A, B, C of X, which is a descriptive Lodato proximity space:

Descriptive Lodato Proximity Axioms Revisited

(dP0) $\emptyset \not\delta_\Phi A, \forall A \subset X$.
(dP1) $A \ \delta_\Phi \ B \Leftrightarrow B \ \delta_\Phi \ A$.
(dP2) $A \underset{\Phi}{\cap} B \neq \emptyset \Rightarrow A \ \delta_\Phi \ B$.
(dP3) $A \ \delta_\Phi \ (B \cup C) \Leftrightarrow A \ \delta_\Phi \ B$ or $A \ \delta_\Phi \ C$.
(dP4) $A \ \delta_\Phi \ B$ and $\{b\} \ \delta_\Phi \ C$ for each $b \in B \Rightarrow A \ \delta_\Phi \ C$. ■

Further, the Lodato space (X, δ_Φ) **descriptively separated**, if, for $x, y \in X$,

(dP5) $\{x\} \ \delta_\Phi \ \{y\} \Rightarrow \Phi(x) = \Phi(y)$ (x and y have matching descriptions). ■

Example 2.17 **Descriptive Lodato Proximity on Apple Sets.**
Let (X, δ) be a Lodato proximity space represented by set of points in the partial image of an apple in Fig. 2.8, with subsets $I, J, A, K, B \subset X$. The Lodato axiom **(P4)** is satisfied by

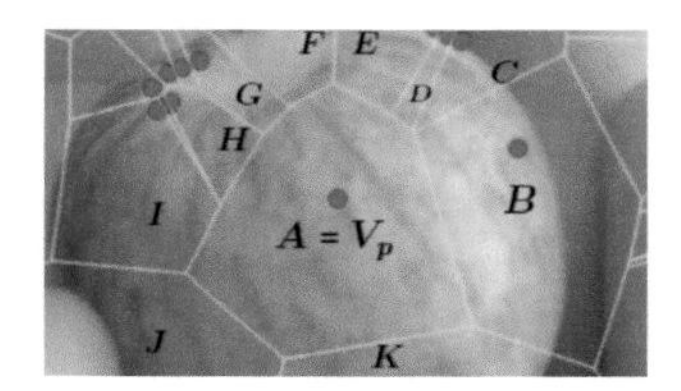

Fig. 2.8 Descriptive Lodato near apple sets

$$I \ \delta_{\Phi} \ A, \ A \ \delta_{\Phi} \ B \forall \ \Phi(a) \in \phi(A) \Leftrightarrow I \ \delta_{\Phi} \ B,$$
$$H \ \delta_{\Phi} \ A, \ A \ \delta_{\Phi} \ B \forall \ \Phi(a) \in \phi(A) \Leftrightarrow H \ \delta_{\Phi} \ B. \quad \blacksquare$$

2.6 Delaunay Triangulation

Delaunay triangulations, introduced by B.N. Delone [Delaunay] [25], represent pieces of a continuous space. This representation supports numerical algorithms used to compute properties such as the density of a space. A *triangulation* is a collection of triangles, including the edges and vertices of the triangles in the collection. A 2D *Delaunay triangulation* of a set of sites (generators) $S \subset \mathbb{R}^2$ is a triangulation of the points in S. Let $p, q \in S$. A straight edge connecting p and q is a *Delaunay edge* if and only if the Voronoï region of p [26, 27] and Voronoï region of q intersect along a common line segment [28, Sect. I.1, p. 3]. For example, in Fig. 2.6, $V_p \cap V_q = \overline{xy}$. Hence, $\overline{pq}$ is a Delaunay edge in Fig. 2.6.

A triangle with vertices $p, q, r \in S$ is a *Delaunay triangle* (denoted $\triangle(pqr)$ in Fig. 2.9), provided the edges in the triangle are Delaunay edges.

Example 2.18 **Near Sets in a Delaunay Triangulation.**
A *triangulation* is a collection of triangles, including the edges and vertices of the triangles in the collection. A 2D *Delaunay triangulation* of a set of sites (generators) $S \subset \mathbb{R}^2$ is a triangulation of the points in S, forming a collection of Delaunay triangles $\mathcal{D}$. Also let $\mathcal{D}$ be endowed with the Čech proximity δ. Let A, B be sets of points along the edges of the Delaunay triangles $\triangle(pqr)$ and $\triangle(qrt)$ in Fig. 2.9. $A \cap B \neq \emptyset$, since this pair of triangles have a common edge. Hence, $A \overset{\wedge\!\!\wedge}{\delta} B$. In addition, $(\mathcal{D}, \delta)$ is called a Delaunay proximity space [29]. $\blacksquare$

Example 2.19 **Image Corners.**
The hand image in Fig. 2.10.1 has dimension 552×685. This is reflected in the plot of the image corners in Fig. 2.11. Up to 50 corners were found in the hand image using MScript 27 in Appendix A.2. The corners are displayed as • dots in Fig. 2.10.1. These corners form a set of sites that can be used to generate a Delaunay mesh. By connecting each pair of nearest sites belonging to neighbouring Voronoï regions, the

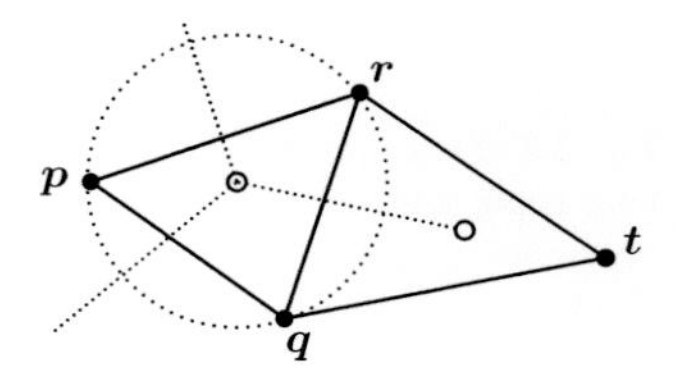

Fig. 2.9 Strongly near Delaunay triangles $\triangle(pqr)$ and $\triangle(qrt)$

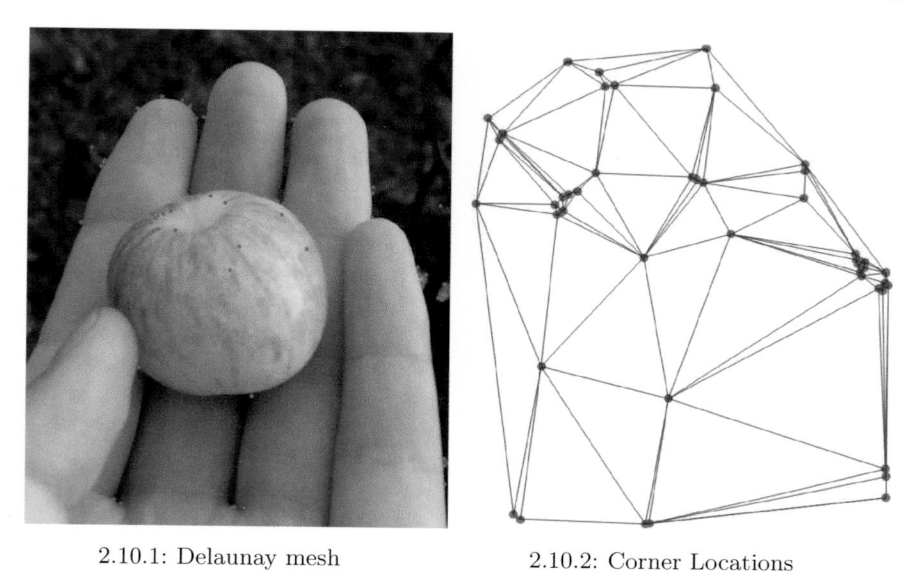

2.10.1: Delaunay mesh

2.10.2: Corner Locations

Fig. 2.10 Corner-based Delaunay mesh and corresponding image corners

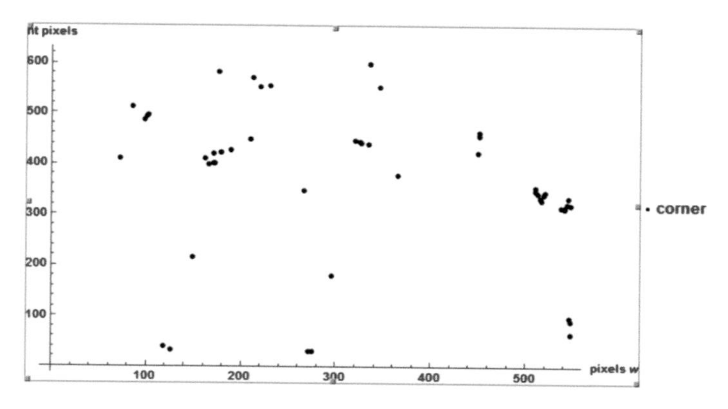

Fig. 2.11 Corners in a hand image

mesh in Fig. 2.10.2 is constructed. The locations of the Delaunay mesh corners are shown in the plot in Fig. 2.11. For the most part, the corners in the plot follow the contour of the hand (polygon A in Fig. 2.4 has the highest number of $\hat{\hat{\delta}}$-neighbouring polygons). ■

Fig. 2.12 Corner-based Delauny mesh on a hand image

Example 2.20 **Near and Strongly Near Delaunay Mesh Triangles**.
The corner-based Delaunay mesh in Fig. 2.10.2 is superimposed on the hand image in Fig. 2.12. This superimposition of a mesh on an image facilitates image analysis, since neighbouring triangles are then associated with image regions. In Fig. 2.12, let the set $X = \bigcup X_i, i \in \{1, 2, 3, 4\}$. Then the following things can be observed.

$$\begin{aligned}
&\text{cl}X_1 \ \delta_0 \ \text{cl}X_3, \ \text{Wallman proximity.}\\
&\text{cl}X_1 \ \delta_0 \ \text{cl}X_4, \ \text{Wallman proximity.}\\
&\text{cl}X_1 \ \overset{\wedge\!\!\wedge}{\delta} \ \text{cl}X_3, \ \text{strong proximity.}\\
&\text{cl}X_2 \ \overset{\wedge\!\!\wedge}{\delta} \ \text{cl}X_3, \ \text{strong proximity.}\\
&\text{cl}X_2 \ \overset{\wedge\!\!\wedge}{\delta} \ \text{cl}X_4, \ \text{strong proximity.}\\
&X \text{ is connected.}\\
&X \text{ is a mesh nerve.}
\end{aligned}$$

The fact that X is connected and a mesh nerve is of interest, since this nerve sits on top of that part of the underlying Voronoï mesh containing a polygon that has the greatest number of adjacent polygons. ■

MScript 27 in Appendix A.2 is also used to construct the corner-based Voronoï mesh in Fig. 2.13.1 and the corresponding mesh is superimposed on the hand image in Fig. 2.13.2. The Voronoï regions of corner points grouped around the contour of the and the apple held in the hand, lay underneath the Delaunay nerve that includes $X4$ in Fig. 2.12.

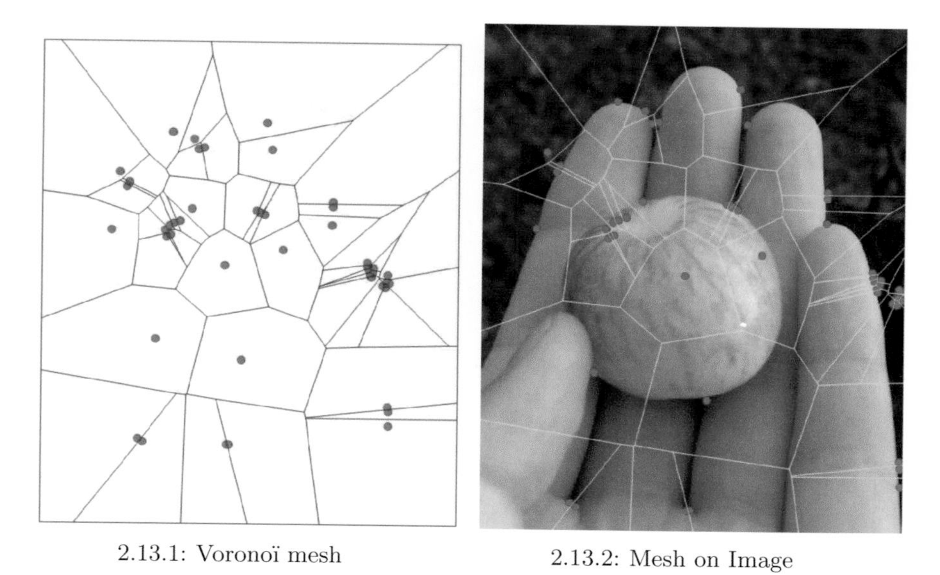

2.13.1: Voronoï mesh   2.13.2: Mesh on Image

Fig. 2.13 Corner-based Voronoï mesh and corresponding image mesh

Fig. 2.14 Combined corner-based Voronoï and Delaunay meshes on a hand image

Example 2.21 **Delaunay Triangle-Voronoï Mesh Connectedness.**
The combined corner-based Voronoï and Delaunay image meshes in Fig. 2.14 are connected but not strongly connected. Triangles X_4 and B in Fig. 2.14 correspond to Voronoï regions A and B in Fig. 2.4. ■

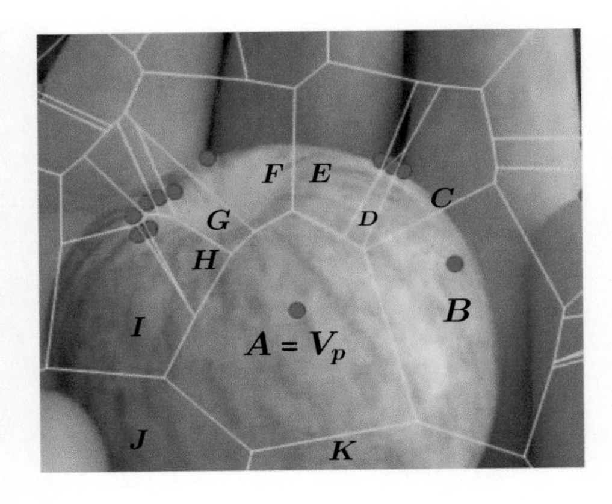

Fig. 2.15 Hand image Voronoï mesh nerve

Theorem 2.22 *A Delaunay triangle with a straight edge between a pair of Voronoï region sites is Wallman near but not* $\overset{\wedge\!\!\wedge}{\delta}$*-near to the pair of Voronoï regions.*

Problem 2.23 🚲 Prove Theorem 2.22. **Hint**: Use the fact that a Delaunay triangle has an empty interior, cuts each neigbouring Voronoï edge in only one point on the edge common to a pair of Voronoï regions. ■

Example 2.24 **Voronoï Mesh Closure Nerve**.
In Fig. 2.12, let X be the set of labelled Voronoï regions shown in Fig. 2.15, i.e.,

$$X = \bigcup X_i : X_i \in \{V_p, B, C, D, E, F, G, H, I, J, K\}. \text{(Set of adjacent Voronoï regions)}$$

Then the following things can be observed.

$$\mathrm{cl}A \ \delta_0 \ \mathrm{cl}X_i \forall X_i \in X, \text{ Wallman proximity.}$$
$$A \ \overset{\wedge\!\!\wedge}{\delta} \ X_i \forall X_i \in X, \text{ strong proximity.}$$
$$\bigcap \mathrm{cl}X \neq \emptyset.$$
$$X \text{ is a mesh closure nerve.}$$
$$X \text{ is not connected, since,e.g., } \mathrm{cl}B \cap \mathrm{cl}E = \emptyset.$$
$$X \text{ is not a bornology, since } X \text{ is not a boundedness.}$$

X is a mesh closure nerve that has nonempty intersection with a Delaunay mesh closure nerve. ■

Problem 2.25 Prove the set X in Example 2.24 is a mesh closure nerve. ■

Problem 2.26 Prove the set X in Example 2.24 is not connected. ■

Problem 2.27 Prove the set X in Example 2.24 is not a bornology. ■

2.7 Voronoï Diagrams Revisited

This section revisits Voronoï diagrams, introduced during the first decade of the 1900s by G. Voronoï [30–32]. A *simple convex set* is a closed half plane (all points on or on one side of a line in R^2).

Let $S \subset \mathbb{R}^2$ be a finite set of n points called sites, $p \in S$. The set S is called the *generating set* [33]. Let H_{pq} be the closed half plane of points at least as close to p as to $q \in S \setminus \{p\}$, defined by

$$H_{pq} = \left\{ x \in R^2 : \|x - p\| \underset{q \in S}{\leq} \|x - q\| \right\}.$$

A *convex polygon* is the intersection of finitely many half-planes [28, Sect. I.1, p. 2]. See, for example, Fig. 2.16.

Remark 2.28 The Voronoï region V_p depicted as the intersection of finitely many closed half planes in Fig. 2.16 is a variation of the representation of a Voronoï region in the monograph by H. Edelsbrunner [26, Sect. 2.1, p. 10], where each half plane is defined by its outward directed normal vector. The rays from p and perpendicular to the sides of V_p are comparable to the lines leading from the center of the convex polygon in G.L. Dirichlet's drawing [34, Sect. 3, p. 216]. ■

2.7.1 Sites

Let $S \subset E$, a finite-dimensional normed linear space. Elements of S are called sites (mesh generating points) to distinguish them from other points in E [26, Sect. 2.2, p. 10]. Let $p \in S$. A *Voronoï region* of $p \in S$ (denoted V_p) is defined by

$$V_p = \left\{ x \in E : \|x - p\| \underset{\forall q \in S}{\leq} \|x - q\| \right\}.$$

Remark 2.29 A Voronoï region of a site $p \in S$ contains every point in the plane that is closer to p than to any other site in S [33, Sect. 1.1, p. 99]. Let V_p, V_q be Voronoï

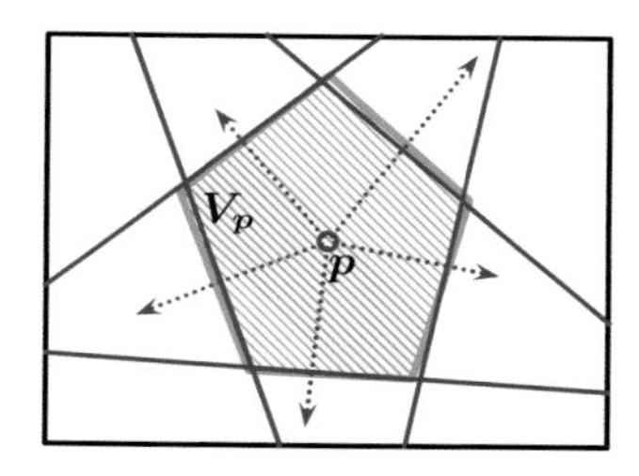

Fig. 2.16 Voronoï region = convex polygon

polygons. If $V_p \cap V_q$ is a line, ray or line segment, then it is called a *Voronoï edge*. If the intersection of three or more Voronoï regions is a point, that point is called a *Voronoï vertex*. ■

Lemma 2.30 *A Voronoï region of a point is the intersection of closed half planes and each region is a convex polygon.*

Proof From the definition of a closed half-plane

$$H_{pq} = \left\{ x \in R^2 : \|x - p\| \underset{q \in S}{\leq} \|x - q\| \right\},$$

V_p is the intersection of closed half-planes H_{pq}, for all $q \in S - \{p\}$ [28], forming a polygon. From Lemma 1.42, V_p is a convex. □

A Voronoï diagram of S (denoted by $\mathbb{V}$) is the set of Voronoï regions, one for each site $p \in S$, defined by

$$\mathbb{V} = \bigcup_{p \in S} V_p.$$

Example 2.31 **Centroids as Sites in an Image Tessellation**.
Let E be a segmentation of a digital image and let $S \subset E$ be a set of sites, where each site is the centroid of a segment in E. In a centroidal approach to the Voronoï tessellation of E, a Voronoï region V_p is defined by the intersection of closed half plains determined by centroid $p \in S$. Centroids can be found using **regionprops** available in Matlab®. The centroidal approach to Voronoï tessellation was introduced by Q. Du, V. Faber, M. Gunzburger [35]. ■

2.8 Some Results for Voronoï Regions

Let V_p, V_z be Voronoï regions of $p, z \in S$, a set of Voronoï sites in a finite-dimensional normed linear Space E that is topological, $\mathrm{cl} A$ the closure of a nonempty set A in E. V_p, V_z are *proximal* (denoted by $V_p \ \delta \ V_z$), provided $\mathbb{P} = \mathrm{cl} V_p \cap \mathrm{cl} V_z \neq \emptyset$ [8]. The set $\mathbb{P}$ is called a *proximal Voronoï region*.

Theorem 2.32 *Proximal Voronoï regions are convex polygons.*

Proof Let $\mathbb{P}$ be a proximal Voronoï region. By definition, $\mathbb{P}$ is the nonempty intersection of convex sets. From Lemma 1.42, $\mathbb{P}$ is convex. Consequently, $\mathbb{P}$ is the intersection of finitely many closed half planes. Hence, from Lemma 2.30, $\mathbb{P}$ is a Voronoï region of a point and is a convex polygon. □

Corollary 2.33 *The intersection of proximal Voronoï regions is either a Voronoï edge or Voronoï point.*

Any two adjacent Voronoï regions intersect along one of their boundaries and have at most one edge in common. Together, the complete set of Voronoï regions $\mathbb{V}$ cover the entire plane [28, Sect. 2.2, p. 10]. For a set of sites $S \subset E$, a Voronoï diagram $\mathbb{D}$ of S is the set of Voronoï regions, one for each site in S.

Corollary 2.34 *A Voronoï diagram* $\mathbb{D}$ *equals* $\mathbb{V}$.

The partition of a plane E with a finite set of n sites into n Voronoï polygons is known as a Dirichlet tessellation, named after G.L. Dirichlet [36] (see [34]). A *cover* (covering) of a space X is a collection $\mathcal{U}$ of subsets of X whose union contains X (i.e., $\mathcal{U} \supseteq X$) [37, Sect. 15], [14, Sect. 7.1].

Corollary 2.35 *A Dirichlet tessellation* $\mathbb{D}$ *of the Euclidean plane* E *is a covering of* E.

Using Mathematica script 1, we detect the edges in the cameraman image shown in Fig. 2.17.

Mscript 1 Detecting Image Edges.

$img =$ ;

2.17.1: Cameraman

2.17.2: Edges

Fig. 2.17 Cameraman edges detected

***edges = EdgeDetect*[*img*, 5]** ■

Using Mathematica script 2, we obtain a Dirichlet tessellation of the cameraman image shown in Fig. 2.18.

Mscript 2 Dirichlet Tessellation.

***imgBounds = Transpose*[{{0, 0}, *ImageDimensions*[*img*]}];**

***vm = VoronoiMesh*[*ImageValuePositions*[*edges*, *White*], *imgBounds*]**
***HighlightMesh*[*vm*, *Style*[2, *Opacity*[0.1], *Yellow*]]** ■

Example 2.36 **Sample Dirichlet Tessellation of an Image.**
A sample covering of an image with Voronoï regions is accomplished with Mathematica10® by first detecting the main edges in an image and then using points along image edges as the sites in a Dirichlet tessellation of a selected image. For example, Mscript 1 to find the edges in the cameraman image (see Fig. 2.17). Then Mscript 2 is used to tessellate the edges in the cameraman image (see Fig. 2.18). ■

Recall that the Euclidean space $E = R^2$ is a metric space. The topology in a metric space results from determining which points are close to each set in the space. A point $x \in E$ is close to $A \subset E$, provided the Hausdorff distance $d(x, A) = inf\{\|x - a\| : a \in A\} = 0$. Let X, Y be a pair of metric spaces, $f : X \longrightarrow Y$ is a function such that for each $x \in X$, there is a unique $f(x) \in Y$. A continuous function preserves the closeness (proximity) between points and sets, i.e., $f(x)$ is

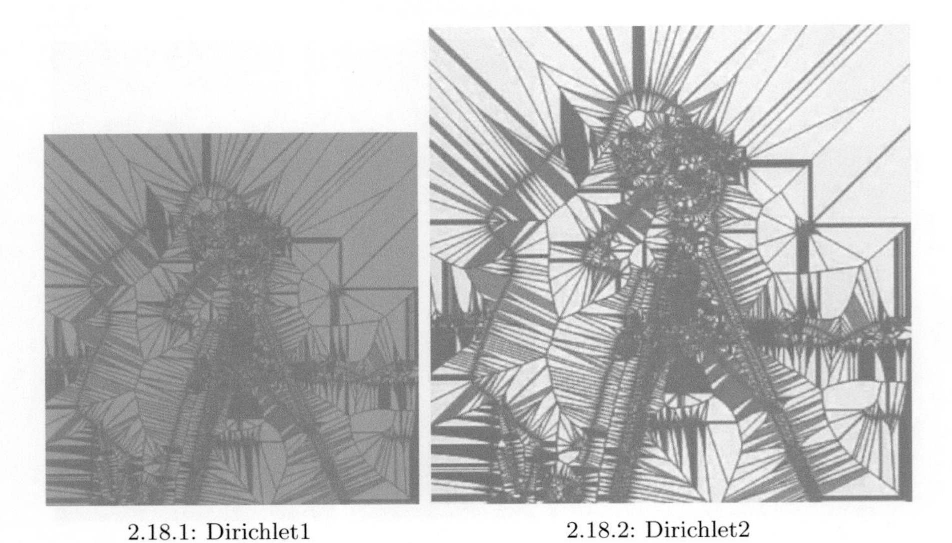

2.18.1: Dirichlet1 2.18.2: Dirichlet2

Fig. 2.18 Dirichlet tessellation of the cameraman image

close to $f(B)$ whenever x is close to B. In a proximity space, one set A is near another set B, provided $A\ \delta\ B$, i.e., the closure of A has at least one element in common with the closure of B. The set A is close to the set B, provided the Čech distance $D(A, B) = inf\{\|a - b\| : a \in A, b \in B\} = 0$. In that case, we write $A\ \delta\ B$ (A and B are proximal). A *uniformly continuous mapping* is a function that preserves proximity between sets, i.e., $f(A)\ \delta\ f(B)$ whenever $A\ \delta\ B$. A *Leader uniform topology* is determined by finding those points that are close to each given set in E.

Theorem 2.37 *Let S be a set of two or more sites, $p \in S$, $V_p \in \mathbb{D}$ in the Euclidean space R^2. Then*

1° V_p is near at least one other Voronoï region in $\mathbb{D}$.
2° Let p, y be sites in S. $\{y\}\ \delta\ \{p\} \Rightarrow \{y\}\ \delta\ V_p$.
3° V_p is close to Voronoï region V_y if and only if $d(x, V_y) = 0$ for at least one $x \in V_p$.
4° A mapping $f : V_p \longrightarrow V_y$ is uniformly continuous, provided $f(V_p)\ \delta\ f(V_y)$ whenever $V_p\ \delta\ V_y$.

Proof
1°: Assume S contains at least 2 sites. Let $p \in S$, $y \in S \setminus \{y\}$ such that V_p, V_y have at least one closed half plane in common. Then $V_p\ \delta\ V_y$.
2°: If $\{y\}\ \delta\ \{p\}$, then $\|y - p\| = 0$, since $y \in \{y\} \cap \{p\}$. Consequently, $\{y\} \cap \mathrm{cl}(V_p) \neq \emptyset$ Hence, $\{y\}\ \delta\ \mathrm{cl}(V_p)$.
3°: $V_p\ \delta\ V_y \Leftrightarrow$ exists $x \in \mathrm{cl}(V_p) \cap \mathrm{cl}(V_y) \Leftrightarrow d(x, V_y) = 0$.
4°: Let $f(V_p)\ \delta\ f(V_y)$ whenever $V_p\ \delta\ V_y$. Then, by definition, $f : V_p \longrightarrow V_y$ is uniformly continuous. □

Theorem 2.38 *Every collection of proximal Voronoï regions has a Leader uniform topology (application of* [38]*).*

Proof Assume $\mathbb{D}$ has more than one Voronoï region. For each $V_p \in \mathbb{D}$, find all $V_y \in \mathbb{D}$ that are close to V_p. For each V_p, this procedure determines a family of Voronoï regions that are near V_p. Let τ be a collection of families of proximal Voronoï regions. Let $A, B \in \tau$. $A \cap B \in \tau$, since either $A \cap B = \emptyset$ or, from Theorem 2.37.1^o, there is at least one Voronoï region $V_p \in A \cap B$, i.e., $V_p\ \delta\ A$ and $V_p\ \delta\ B$. Hence, $A \cap B \in \tau$. Similarly, $A \cup B \in \tau$, since $V_p\ \delta\ A$ or $V_p\ \delta\ B$ for each $V_p \in A \cup B$. Also, $\mathbb{D}$, $\emptyset$ are in τ. Then, τ is a Leader uniform topology in $\mathbb{D}$. □

2.9 Dirichlet Tessellation Quality and Digital Image Quality

The choice of sites influences the quality of the cells in a Dirichlet tessellations (Voronoï diagrams) [39]. Let X be a nonempty set of polygons in a Dirichlet tessellation, $x, y \in X$. A polygon $x \in X$ in a tessellation is called a **cell**. A number of cell quality measures are reported by J.R. Shewchuk in [40, Sect. 6.3]. A **fair** (**quality**) measure $\mathscr{Q} : X \longrightarrow \mathbb{R}$ satisfies the following axioms.

Q.1 $\mathscr{Q}(x) = 0$ for 2D cells with zero area.
Q.2 $\mathscr{Q}(x) = \mathscr{Q}(y)$ if and only if the 2D cells x and y are similar.
Q.3 $\mathscr{Q}(x)$ is finite.
Q.4 $\mathscr{Q}(x) \in [0, 1]$.

Let S be a set of tessellation cells, A the area of a tessellation containing a 3-sided polygon cell $s \in S, l_1, l_2, l_3$ the lengths of the sides of s with $Q(s)$ the quality of cell s. Then, for example, R.P. Bhatia and K.L. Lawrence [41], R.E. Bank and J. Xu [42] as well as D.A. Field [43] use the following smooth quality measure of a 3-sided cell.

$$Q_3(s) = 4\sqrt{3}\frac{A}{l_1^2 + l_2^2 +, l_3^2}.$$

Field observes, for triangles with vertices at (0,0), (1,0) and $(x, y), x \geq 0, y > 0$, we have

$$Q_3(s) = \frac{4\sqrt{3}y}{1 + (1-x)^2 + 2y^2 + x^2}.$$

Problem 2.39 ☕ What does a high quality Dirichlet tessellation of a digital image tell us about the image? **Hint**: See introduction to convex bodies and Helly's theorem, starting in Chap. 11. ■

Problem 2.40 ☕ Do the following:

1° ☕ Prove that $Q_3(s)$ satisfies the axioms for a fair measure of triangular tessellation cell quality.
2° Plot $Q_3(s)$ for triangles with vertices at (0,0), (1,0) and (x, y) for fixed $x \geq 0$ and varying $y > 0$. ■

Problem 2.41 Do the following:

1° Give a fair measure $Q_4(s)$ that satisfies the axioms for a fair measure of a 4-sided tessellation cell quality.
2° ☕ Prove that your $Q_4(s)$ satisfies the fair measure axioms.
3° For fixed 4gon lengths l_1, l_2, l_3, varying area A and varying length l_4, plot $Q_4(s)$ for 5 different values of A and l_4.
4° For fixed 4gon lengths l_1, l_2, varying area A and varying lengths l_3, l_4, plot $Q_4(s)$ for 5 different values of A and l_3, l_4. ■

Let $l_1, l_2, \ldots, l_n$ be the lengths of the edges of an n-sided tessellation polygon. It has been shown that mesh quality is maximum, provided $l_1 = l_2 = \cdots = l_n$ [39, Sect. 5, Theorem 5.1]. When a digital image is the source of sites for tessellation, then tessellation quality tells us about the quality of the image. This observation can be used to prove Theorem 2.42.

Theorem 2.42 Mesh Quality [39].
For any plane, there exists a set of sites for which the mesh quality is maximum.

Problem 2.43 Do the following:

1° Prove Theorem 2.42.
2° Select three digital images.
3° Select several different sets of sites on each of the images. Include centroidal and keypoint sites in your choices of sites.
4° Tessellate the selected with image with Voronoï diagrams using the selected sites.
5° Use Mathematica to measure the quality of the tessellated images.
6° Give the quality measurement for each of the tessellated images.
7° Comment on why one choice of sites leads to a higher quality tessellation than the other choices of sites. ■

Support for Image Object Geometry and Analysis.
When the Voronoï regions in a tessellated digital image have sides approaching equal length, then any line segment in any direction inside a Voronoï region can be used to identify and measure the geometry of image objects covered by the Voronoï region. ■

Example 2.44 **Digital Image Quality.**
For Fig. 2.19, let $\ulcorner$Centroids denotes sites that centroids that are also corners and let $\overline{\wedge}$Centroids denoted centroids that are also edge pixels. In this example, if the sites for a tessellation are image $\ulcorner$Centroids, then the quality of the image is a function of the number of centroid-corners, the positions of the centroid-corners and how evenly the

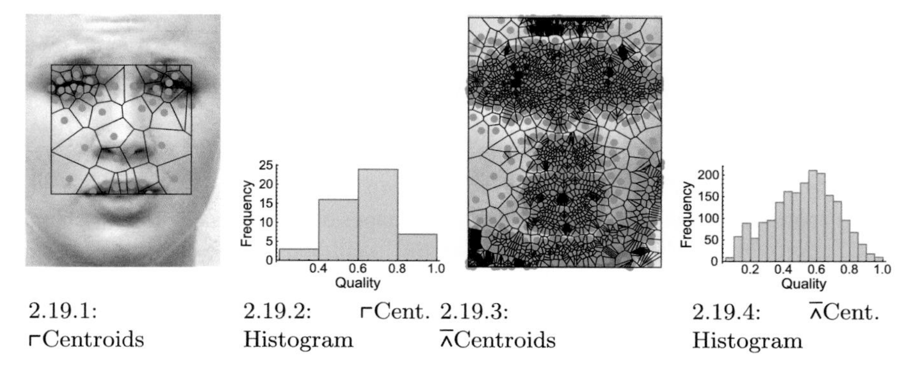

2.19.1: $\ulcorner$Centroids 2.19.2: $\ulcorner$Cent. Histogram 2.19.3: $\overline{\wedge}$Centroids 2.19.4: $\overline{\wedge}$Cent. Histogram

Fig. 2.19 Corner and edge centroid tessellations and quality histograms [39, Sect. 6]

centroid-corners are distributed in the image. The tessellation of the face in Fig. 2.19.1 is generated by sites that are image $\ulcorner$Centroids that are also corner pixels and the tessellation of the face in Fig. 2.19.3 is generated by sites that are image $\overline{\wedge}$Centroids that are also edge pixels. The poor quality of the $\overline{\wedge}$Centroids-based tessellation can be seen in the unevenness of the site quality distribution in Fig. 2.19.4. From this, it is apparent that the greater the number of high quality tessellation cells, the higher the quality of the image tessellation. ■

Conjecture 2.45 *A high quality digital image tessellation reflects the fact that high quality cells are more prevalent in the tessellated image.* ■

In practical terms, for the tessellation of a digital image, if each pair of sites is connected by a straight edge for an image $img1$, the resulting shape more readily approximates a corresponding shape derived from the edge-connected sites from an image $img2$ representing a class of images. Let $s_1, \ldots, s_n \in S$, a set of n sites. For each image, the edge-connected site form a path from s_1 to s_n.

The set of sites S is **connected**, provided, for all $p, q \in S$, there is a sequence $p_0, \ldots, p_m$ of sites in S so that $p = p_0$ and $q = p_n$ and site p_i is closest to site p_{i-1} for $1 \leq i \leq m$. The sequence $p_0, \ldots, p_m$ is called a **path**. What we want is a pathwise-connected set of sites. The set S is **pathwise-connected**, provided, for every pair of sites in X, there is a path connecting the sites. Pathwise-connected sites yield a **connected-sites shape** (denoted by ⚽).

Let f be a continuous mapping of a set of sites $S1$ for a test image to the Euclidean plane $\mathbb{R}^2$. Further, let $f(S1)$ be the ⚽ shape of the pathwise-connected sites on test image $img1$. Similarly, let g be a continuous mapping of a set of sites $S2$ for a test image to the Euclidean plane $\mathbb{R}^2$. In addition, let $g(S2)$ be the ⚽ shape of the pathwise-connected sites on $img2$ that represents a class of shapes. If the ⚽ shape $f(S1)$ is similar in structure to the ⚽ shape of $g(S2)$, then $f(S1)$ can deformed (mapped) to $f(S2)$. This means that all of the points in $f(S1)$ map to $f(S2)$. If ⚽ shape $f(S1)$ deforms into ⚽ shape $g(S2)$, then ⚽ shape $f(S1)$ belongs to the $g(S2)$ shapes class and $img1$ belongs to the $img2$ class of images.

The approximation of ⚽ shape $g(S2)$ with ⚽ shape $f(S1)$ enters into this, by computing the distance $D(f(S1), g(S2))$ defined by

$$D(f(S1), g(S2)) = inf\,\{\|f(x) - g(y)\| : x \in S_1, y \in S_2\}.$$

Let $\varepsilon > 0$ be a small real number. The ⚽ shape of $g(S2)$ is a good approximation of the ⚽ shape of $f(S1)$, provided $D(f(S1), g(S2)) < \varepsilon$. In that case, ⚽ shape $f(S1)$ approximates the members in the $g(S2)$ shapes class. The more closely shape $f(S1)$ approximates shape $g(S2)$, the higher the likelihood that $img1$ belongs to the $img2$ class of images. For more about the comparison of shapes, see Sect. 5.1.

Remark 2.46 **Image Quality and Geometric Reconstruction**.
Image geometry and the structure of objects embedded in an image are revealed by careful selection of mesh generating points, leading to an image cover with a high

quality mesh. A digital image often is a very complex object with its own inherent geometry, which is difficult to detect. The basic computational proximity approach in solving the image geometry detection problem is to

1° Reconstruct the geometry of an image, approximating what escapes the eye. This can be done using an approach such as the one introduced by A. Vacavant, D. Cocurjolly, and L. Tougne [44]. See, also, A. Kuba, L.G. Nyúl and K. Palágyi [45] on image geometry.

2° Tessellate part or all of an image with convex shapes.

3° Inspect the structural similaries of the sets of convex shapes in an image tessellation.

4° Carry out image quality assessment using a measure such as the structural similarity index measure (SSIM) introduced by Z. Wang, A.C. Bovik, H.R. Sheikh, and E.P. Simoncelli [46]:

$$SSIM(x, y) = \frac{(2\mu_x\mu_y + C_1)(2\sigma_{xy} + C_2)}{(\mu_x^2 + \mu_y^2 + C_1)(\sigma_x^2 + \sigma_y^2 + C_2)}.$$

The SSIM measures the statistical characteristics of signals x, y in the terms the mean value, variances, cross correlation between the standard deviations and constants C_1 and C_2. ■

Problem 2.47 Do the following:

1° Select a digital image img.

2° Select a set of sites (generating points) S.

3° 🚲 Tessellate the image img with a Voronoï diagram $V(S)$.

4° Compute the SSIM(S) for the sets of sites S, using values of C_1 and C_2 of your own choosing.

5° Let q_i be the quality of each polygon in the diagram $V(S)$ using Field's approach. Then, for N polygons in $V(S)$, compute q_{all} from [39]:

$$q_{all} = \frac{1}{N}\sum_{i=1}^{N} q_i$$

6° ☕ Compare the SSIM value with the q_{all} value.

7° Repeat the above steps for 10 different images and construct a table showing the comparison between the SSIM(S) and q_{all} values. What can you conclude for the values in your comparison? ■

Remark 2.48 **Line Detection and Fragmentation Quality.**
Another approach to measuring the quality of a tessellation and, indirectly, the quality of a digital image is to consider the fragmentation quality measure[4] introduced by L. Wenyin and D. Dori [47]. The CP approach in applying the Wenyin-Dori line segment fragmentation quality measure is summarized in the following steps.

---

[4]Many thanks to A. Vacavant for pointing this out.

1° Select a digital image img.
2° Select a set of sites (generating points) S.
3° Tessellate the image img with a Voronoï diagram $V(S)$.
4° Tessellate the image img with a Delaunay triangulation $De(S)$. Notice the edges of the Delaunay triangles and the Voronoï region polygons overlap. It is this overlap of each pair of line segments (one from a Delaunay triangle and the other from a Voronoï polygon) that provides a basis for the next step.
5° Compute the quality of overlappingline segments using $Q_b(k)$ (formula (23) in [47]). ■

Problem 2.49 Do the following:
1° Select a digital image img.
2° Select a set of sites (generating points) S.
3° 🚲 Tessellate the image img with a Voronoï diagram $V(S)$.
4° 🚲 Tessellate the image img with a Delaunay triangulation $De(S)$.
5° Compute the SSIM(S) for the sets of sites S, using values of C_1 and C_2 of your own choosing.
6° Let q_i be the quality of each polygon in the diagram $V(S)$ using Field's approach. Then, for N polygons in $V(S)$, compute

$$q_{all} = \frac{1}{N}\sum_{i=1}^{N} q_i$$

7° Compute the line fragmentation measure $Q_b(k)$ from Remark 2.48. Do this for each of pair of line segments from the overlapping line segments in diagram $V(S)$ and triangulation $De(S)$.
8° ☕ Compare the SSIM value with the q_{all} and $Q_b(k)$ values.
9° Repeat the above steps for 10 different images and construct a table showing the comparison between the SSIM(S), q_{all} and $Q_b(k)$ values. What can you conclude for the values in your comparison? ■

2.10 Tessellation Region Centroids

A *Voronoï tessellation* of a plane surface X is a collection of closed planar regions derived from a set of sites S (generating points). Let $s \in S$ be one of the sites. A Voronoï region V_s consists of all points in X that are closer to s than to any other site in S. Let $\rho : X \to (R)$ be a density function on X, $x \in X$. A *centroid* is a center of mass s^* of a region V. It corresponds to a measure of central location for a region, defined by

$$s^* = \frac{\int_V x\rho(x)dx}{\int_V \rho(x)dx}.$$

For more details about region centroids, see [35, p. 638]. Using Mathematica, generate Voronoï regions from a given set of sites.

Using Mathematica script 3 on a set of random numbers used as region sites, we generate the Voronoï regions shown in Fig. 2.20.

Mscript 3 Generating Voronoï Regions.

(*Find centroids in a Dirichlet tessellation using RegionCentroid.*)

pts = *RandomReal*[1, {100, 2}];

$\mathcal{R}$ = *VoronoiMesh*[*pts*, {{0, 1}, {0, 1}}] ∎

Next, using MScript 28 in Appendix A.2 on regions in a Voronoï tessellation, determine the region centroids shown in Fig. 2.21.

Example 2.50 **Sample Region Centroids.**
Sample collections of Voronoï regions in Fig. 2.20 are generated using Mathematica10 Mscript 3 using as sites random numbers. These sites are shown as a collection of black dots in Fig. 2.21. Next, Mscript 2 is used to find the centroid of each Voronoï region (see Fig. 2.21). ∎

2.11 Centroid-Based Voronoï Mesh on an Image

This section introduces the construction of centroid-based Voronoï meshes on digital images. Voronoï regions are a source of centroids used as a source of sites in the generate a Voronoï mesh $\mathscr{M}$. The steps in the construction of centroid-based Voronoï mesh are given in Algorithm 3.

Example 2.51 **Centroidal Voronoï Mesh Scripts.**
Algorithm 3 is implemented in two different ways in Appendix A.2. Matlab Listing A.1 in Appendix A.2 finds image segment centroids, which are used to construct a Voronoï mesh on an image. MScript 29 uses a different approach in implementing Algorithm 3. First, image corners are used to construct a Voronoï mesh (each Voronoï region is an image segment). Then the centroids in each corner-based Voronoï region are found. Those region centroids are a source of sites used to construct a centroid-based Voronoï mesh. ∎

Example 2.52 **Centroidal Voronoï Mesh on an Image.**
MScript 29 in Appendix A.2 is used on a camera image to find a set centroids that serve as sites in constructing a Voronoï mesh (shown in Fig. 2.22.1). The locations of centroids in the Voronoï regions are shown with • bullets in Fig. 2.22.2. The centroids on a Nikon® camera image are shown in Fig. 2.23.1 and centroidal Voronoï mesh superimposed on the camera image is shown in Fig. 2.23.2. Putting these results to together leads to the Voronoï mesh plus region centroids superimposed on the camera image in Fig. 2.24. ∎

Fig. 2.20 Sample tessellation derived from randomly selected sites

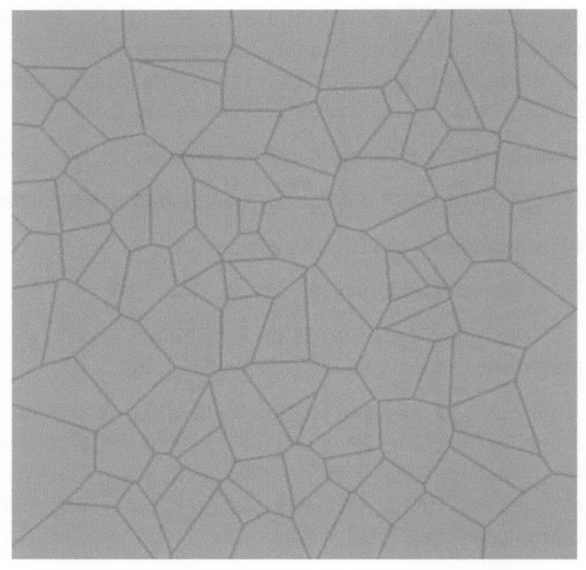

2.20.1: Centroid-Based Tessellation

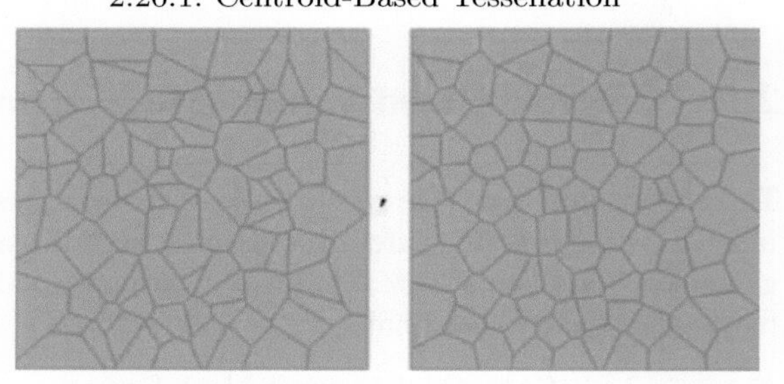

2.20.2: Tessellation Regions1

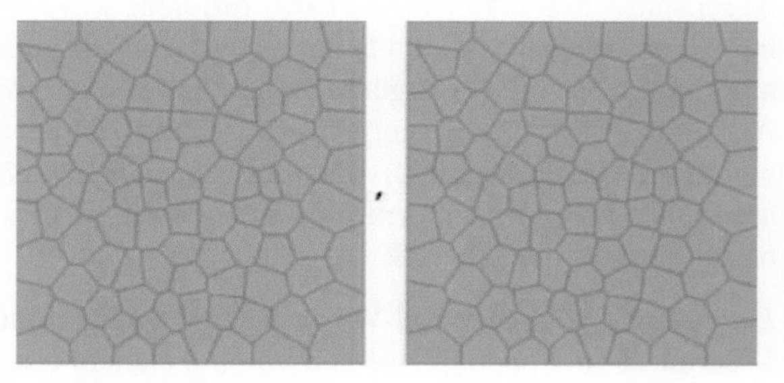

2.20.3: Tessellation Regions2

Problem 2.53 Use Mathematica to construct a centroid-based Voronoï mesh on 3 digital images of your choosing. In solving this problem, do the following:

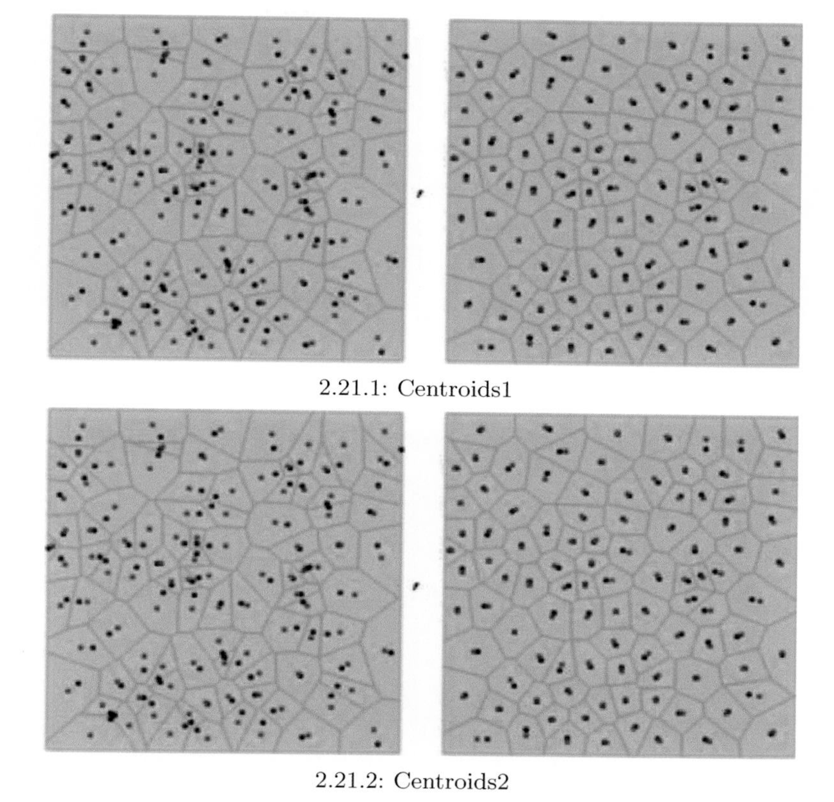

2.21.1: Centroids1

2.21.2: Centroids2

Fig. 2.21 Region centroids detected

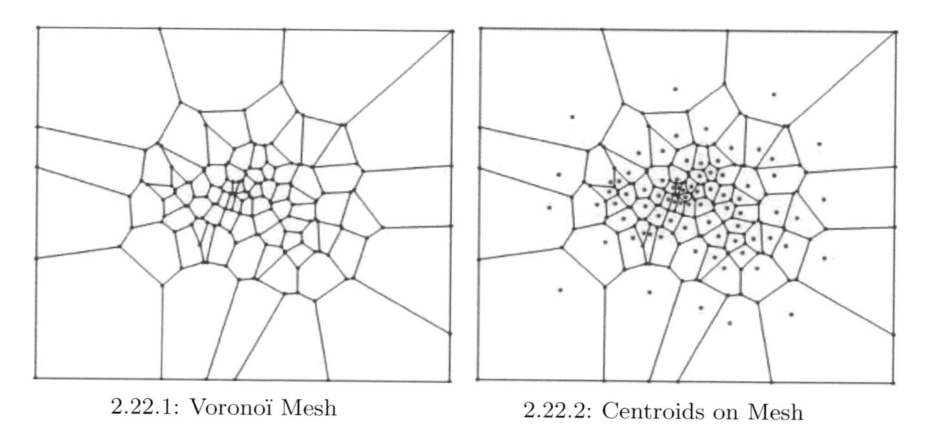

2.22.1: Voronoï Mesh

2.22.2: Centroids on Mesh

Fig. 2.22 Centroidal Voronoï mesh

2.23.1: Centroids on Mesh

2.23.2: Voronoï Mesh

Fig. 2.23 Centroidal Voronoï mesh on image

Fig. 2.24 Centroidal Voronoï mesh on a Nikon® camera image

1° Segment each image using the watershed method. **Hint**: See Sect. 9.11 on watershed segmentation.
2° Find the centroid of watershed segment.
3° Following steps 3 to 7 in Algorithm 3, superimpose a centroidal mesh on each image. ■

Problem 2.54 Use Mathematica to implement Step 8 in Algorithm 3 to find several mesh nerves in 3 digital images of your choosing. In each case, the mesh nerves are embedded in a centroidal Voronoï mesh. Highlight the Voronoï regions in each mesh nerve with an appropriate colour, using the opacity property on the parts of the underlying regions are visible after colouring the regions. ■

Algorithm 3: Construct Centroidal Voronoï Mesh on a Digital Image

Input : Read digital image img.
Output: Centroid-based Voronoï mesh $\mathscr{M}$

1 $img \longmapsto segmentedImg$;
2 $segmentedImg \longmapsto segmentCentroids$;
3 $S \leftarrow segmentCentroidsCoordinates$;
4 /* S contains segment centroid coordinates used as mesh generating points (sites). */ ;
5 $S \longmapsto\ img$;
6 $S \longmapsto\ VoronoiMesh\ \mathscr{M}$;
7 $VoronoiMesh\ \mathscr{M} \longmapsto\ img$;
8 /* Use Algorithm 1 to construct image mesh nerves in $\mathscr{M}$. */ ;

Problem 2.55 Use Mathematica to implement Step 8 in Algorithm 3 to find several mesh nerves so that the centers of mesh nerves are $\underset{\text{w}}{\not\delta}$-strongly far from each other in 3 digital images of your choosing. In each case, the mesh nerves are embedded in a centroidal Voronoï mesh. Highlight the Voronoï regions in each mesh nerve center with an appropriate colour, using the opacity property so that the parts of the underlying regions are visible after colouring the regions. ■

Problem 2.56 Use Mathematica to implement Step 8 in Algorithm 3 to find several mesh nerves so that the centers of mesh nerves are $\overset{\wedge\!\!\wedge}{\delta}_{\Phi}$-strongly far *descriptively* from each other in 3 digital images of your choosing. In each case, the mesh nerves are embedded in a centroidal Voronoï mesh. Highlight the Voronoï regions in each mesh nerve center with an appropriate colour, using the opacity property so that the parts of the underlying regions are visible after colouring the regions. ■

Problem 2.57 ☕ Give three examples of centers of mesh nerves that are $\underset{\text{w}}{\not\delta}$-strongly far from each other and $\overset{\wedge\!\!\wedge}{\delta}_{\Phi}$-strongly near each other *descriptively*. This means that strongly far centers of mesh nerves have a nonempty descriptive intersection. Let X a nonempty collection of image mesh nerves, $(X, \left\{\underset{\text{w}}{\not\delta}, \overset{\wedge\!\!\wedge}{\delta}_{\Phi}\right\})$ a proximity space endowed with the $\underset{\text{w}}{\not\delta}$ and $\overset{\wedge\!\!\wedge}{\delta}_{\Phi}$ proximities, and let $A, B \subset X$ be a pair of image mesh nerves. We know that $A \ \underset{\text{w}}{\not\delta}\ B$ is possible from the examples already given. Now give examples so that

$$A \ \underset{\text{w}}{\not\delta}\ B \text{ and } A \underset{\Phi}{\cap} B \neq \emptyset. \qquad ■$$

Problem 2.58 Use Mathematica to implement Step 8 in Algorithm 3 to find several mesh nerves so that the mesh nerves are $\overset{\wedge\!\!\wedge}{\delta}$-strongly near each other in 3 digital images of your choosing. In each case, the mesh nerves are embedded in a centroidal Voronoï mesh. Highlight the adjacent Voronoï regions in the strongly near mesh nerves with an appropriate colour, using the opacity property so that the parts of the underlying regions are visible after colouring the regions. **Hint**: Voronoï mesh nerves A and B are $\overset{\wedge\wedge}{\delta}$-strongly near, provided A has a Voronoï region edge in common with B. ■

Conjecture 2.59 *The centers of* $\overset{\wedge\wedge}{\delta}$*-strongly near image mesh nerves are* $\underset{\vee\vee}{\not\delta}$*-strongly far from each other.* ■

Problem 2.60 ☕ Prove Conjecture 2.59. ■

Conjecture 2.61 *The centers of all image mesh nerves are* $\underset{\vee\vee}{\not\delta}$*-strongly far from each other.* ■

Problem 2.62 ☕ Prove Conjecture 2.61. ■

References

1. Beer, G., Di Concilio, A., Di Maio, G., Naimpally, S., Pareek, C., Peters, J.: Somashekhar Naimpally, 1931–2014. Topol. Appl. **188**, 97–109 (2015). doi:10.1016/j.topol.2015.03.010, MR3339114
2. Naimpally, S., Warrack, B.: Proximity Spaces. Cambridge Tract in Mathematics, vol. 59. Cambridge University Press, Cambridge (1970). X + 128 pp., Paperback (2008)
3. Riesz, F.: Stetigkeitsbegriff und abstrakte mengenlehre. Atti del IV Congresso Internazionale dei Matematici, vol. II, pp. 182–109 (1908)
4. Naimpally, S.: Near and far. A centennial tribute to Frigyes Riesz. Siberian Electronic Mathematical Reports, vol. 2, pp. 144–153 (2009)
5. Naimpally, S.: Proximity Approach to Problems in Topology and Analysis. Oldenbourg Verlag, Munich (2009). 73 pp., ISBN 978-3-486-58917-7, MR2526304
6. Di Concilio, A.: Proximal set-open topologies on partial maps. Acta Math. Hung. **88**(3), 227–237 (2000). MR1767801
7. Di Concilio, A.: Topologizing homeomorphism groups of rim-compact spaces. Topol. Appl. **153**(11), 1867–1885 (2006)
8. Di Concilio, A.: Proximity: a powerful tool in extension theory, functions spaces, hyperspaces, boolean algebras and point-free geometry. In: Mynard, F., Pearl, E. (eds.) Beyond Topology, AMS Contemporary Mathematics, vol. 486, pp. 89–114. American Mathematical Society, Providence (2009)
9. Čech, E.: Topological Spaces. Wiley, London (1966). Fr seminar, Brno, 1936–1939; rev. ed. Z. Frolik, M. Katětov
10. Efremovič, V.: The geometry of proximity I (in Russian). Mat. Sb. (N.S.) 31(73)(1), 189–200 (1952)
11. Naimpally, S.: Proximity Spaces. Cambridge University Press, Cambridge (1970). X + 128 pp., ISBN 978-0-521-09183-1
12. Peters, J., Ramanna, S.: Pattern discovery with local near sets. In: Alarcón, R., Barceló, P. (eds.) Proceedings of the Jornadas Chilenas de Computación 2012 Workshop on Pattern Recognition, pp. 1–4. The Chilean Computing Society, Valparaiso (2012)
13. Peters, J., Naimpally, S.: Applications of near sets. Notices Am. Math. Soc. **59**(4), 536–542 (2012). http://dx.doi.org/10.1090/noti817, MR2951956

14. Naimpally, S., Peters, J.: Topology with applications. Topological spaces via near and far. World Scientific, Singapore (2013). Xv + 277 pp. Am. Math. Soc. MR3075111
15. Di Maio, G., Naimpally, S.A., Meccariello, E.: Theory and applications of proximity, nearness and uniformity. Seconda Università di Napoli, Napoli (2009), 264pp., MR1269778
16. Naimpally, S., Peters, J., Wolski, M.: Foreword [near set theory and applications]. Math. Comput. Sci. **7**(1), 1–2 (2013)
17. Peters, J.: Topology of digital images - visual pattern discovery in proximity spaces, intelligent systems reference library, vol. 63. Springer, Berlin (2014). Xv + 411 pp. Zentralblatt MATH Zbl 1295 68010
18. Kuratowski, C.: Topologie I. Panstwowe Wydawnictwo Naukowe, Warsaw (1958). XIII + 494 pp
19. Guadagni, C.: Bornological convergences on local proximity spaces and ω_μ-metric spaces. Ph.D. thesis, Università degli Studi di Salerno, Salerno (2015). Supervisor: A. Di Concilio, 79 pp
20. Ronse, C.: Regular open or closed sets. Philips Research Laboratory Series, Brussels WD59, pp. 1–8 (1990)
21. Wallman, H.: Lattices and topological spaces. Ann. Math. **39**(1), 112–126 (1938)
22. Lodato, M.: On topologically induced generalized proximity relations. Ph.D. thesis. Rutgers University (1962). Supervisor: S. Leader
23. Lodato, M.: On topologically induced generalized proximity relations I. Proc. Am. Math. Soc. **15**, 417–422 (1964)
24. Lodato, M.: On topologically induced generalized proximity relations II. Pac. J. Math. **17**, 131–135 (1966)
25. Delaunay, B.D.: Sur la sphère vide. Izvestia Akad. Nauk SSSR, Otdelenie Matematicheskii i Estestvennyka Nauk **7**, 793–800 (1934)
26. Edelsbrunner, H.: A Short Course in Computational Geometry and Topology. Springer, Berlin (2014). 110 pp
27. Peters, J.: Proximal Voronoï regions, convex polygons, and leader uniform topology. Adv. Math.: Sci. J. **4**(1), 1–5 (2015)
28. Edelsbrunner, H.: Geometry and Topology of Mesh Generation. Cambridge University Press, Cambridge (2001). 209 pp
29. Peters, J.: Proximal Delaunay triangulation regions, pp. 1–4 (2014). arXiv:1411.6260
30. Voronoï, G.: Sur un problème du calcul des fonctions asymptotiques. J. für die reine und angewandte Math. **126**, 241–282 (1903)
31. Voronoï, G.: Nouvelles applications des paramètres continus à la théorie des formes quadratiques. J. für die reine und angewandte Math. **133**, 97–178 (1907). JFM 38.0261.01
32. Voronoï, G.: Nouvelles applications des paramètres continus à la théorie des formes quadratiques. J. für die reine und angewandte Math. **134**, 198–287 (1908). JFM 39.0274.01
33. Frank, N., Hart, S.: A dynamical system using the Voronoi tessellation. Am. Math. Monthly **117**(2), 92–112 (2010)
34. Dirichlet, G.: Über die reduktion der positiven quadratischen formen mit drei unbestimmten ganzen zahlen. J. für die reine und angewandte **40**, 221–239 (1850). MR
35. Du, Q., Faber, V., Gunzburger, M.: Centroidal voronoi tessellations: applications and algorithms. SIAM Rev. **41**(4), 637–676 (1999). MR
36. Weisstein, E.: Voronoi diagram. Wolfram MathWorld (2014). http://mathworld.wolfram.com/VoronoiDiagram.html
37. Willard, S.: General Topology. Dover Publications Inc, Mineola (1970). Xii + 369 pp, ISBN: 0-486-43479-6 54-02, MR0264581
38. Leader, S.: On clusters in proximity spaces. Fundamenta Mathematicae **47**, 205–213 (1959)
39. A-iyeh, E., Peters, J.: Measure of tessellation quality of voronoï meshes. Theory Appl. Math. Comput. Sci. **5**(2), 158–185 (2015)
40. Shewchuk, J.: What is a good linear element? interpolation, conditioning, and quality measures. Technical report, University of California at Berkeley (2015). http://www.cs.berkeley.edu/jrs/papers/elem.pdf

41. Bhatia, R., Lawrence, K.: Two-dimensional finite element mesh generation based on stripwise automatic triangulation. Comput. Struct. **36**(2), 309–319 (1990)
42. Bank, R., Xu, J.: An algorithm for coarsening unstructured meshes. Numer. Math. **73**, 1–36 (1996)
43. Field, D.: Qualitative measures for initial meshes. Int. J. Numer. Methods Eng. **47**, 887–906 (2000)
44. Vacavant, A., Cocurjolly, D., Tougne, L.: Topological and geometrical reconstruction of complex objects on irregular isothetic grids. In: Kuba, A., Nyúl, L., Palágyi, K. (eds.) Discrete Geometry for Computer Imagery. Lecture Notes in Computer Science, vol. 4245, pp. 470–481. Springer, Heidelberg (2006)
45. Kuba, A., Nyúl, L., Palágyi, K.: Discrete Geometry for Computer Imagery. Lecture Notes in Computer Science, vol. 4245. Springer, Berlin (2006). Xiv + 688 pp. ISBN: 3-540-47651-2, MR2307269
46. Wang, Z., Bovik, A., Sheikh, H., Simoncelli, E.: Image quality assessment: from error visibility to structural similarity. IEEE Trans. Image Process. **13**(4), 600–612 (2004)
47. Wenyin, L., Dori, D.: A protocol for performance evaluation for line detection algorithms. Mach. Vis. Appl. **9**, 240–250 (1997)

Chapter 3
Distance and Proximally Continuous

Fig. 3.1 J. Hadamard

This chapter introduces distance functions called metrics and continuous functions. A **metric** is a distance function that maps each pair of points in a set to a real number. A nonempty set endowed with a distance function d is called a *metric space*, provided d satisfies a number of properties. Metric spaces were introduced by M. Fréchet,

J.F. Peters, *Computational Proximity*, Intelligent Systems
Reference Library 102, DOI 10.1007/978-3-319-30262-1_3

a student of J. Hadamard, in his Ph.D. thesis completed in 1906 [1]. Hadamard[1] was also supervisor of S. Mandelbrot, who introduced fractals, and A. Weil, who was the founding member of the Bourbaki group. Hadamard was a geometer [2], who won the Bordin prize from the French Academy of Sciences for his work on geodesics in the differential geometry of surfaces and introduced the psychology of invention in mathematics [3]. Fréchet's main interest was the introduction of what he called a functional calculus. Thanks to metrics, the notion of the distance between points was made precise. In the first edition of his book on set theory, it was F. Hausdorff [4, 5] who introduced metrics that measure the distance between a point and a set. It was E. Čech who introduced a metric for the distance between sets [6, Sect. 18 A.2, pp. 301]. Either metrics or metric-free proximities provide a means of defining continuous functions in a precise way.

3.1 Metrics and Metric Topology

A real-valued function $d : X \times X \longrightarrow \mathbb{R}$ is a **metric** on X, provided
M.1 $d(x, y) \geq 0$,
M.2 $d(x, y) = 0$, if and only if, $x = y$,
M.3 $d(x, y) = d(y, x)$ (symmetry),
M.4 $d(x, z) \leq d(x, y) + d(y, z)$ (triangle inequality).

A nonempty set X endowed with a metric d is called a **metric space**, denoted by (X, d) or simply by X when it is understood that X is endowed with a metric. If all

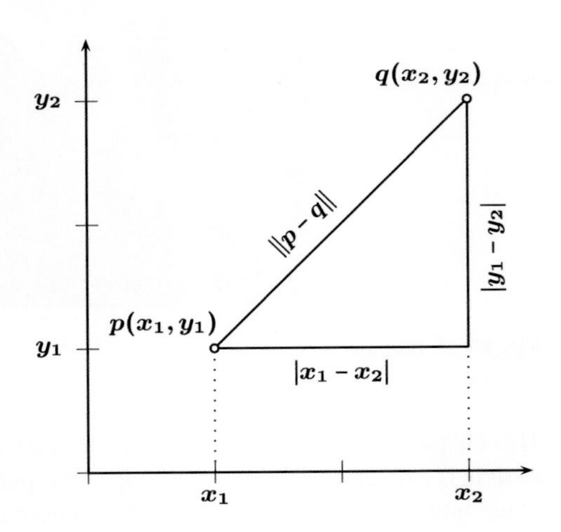

Fig. 3.2 Distances in the Euclidean plane

[1] See Fig. 1 for a picture of Hadamard, a public domain photo from the School of Mathematics and Statistics, University of St. Andrews, Scotland: http://www-history.mcs.standrews.ac.uk/Biographies/Hadamard.html.

of the axioms except M.2 are satisfied, then d is called a *pseudometric*. Functions that measure distance but have not been tested to see if the functions are metrics or pseudometrics, are simply called distance functions. An overview of distance functions is given in [7, 8].

Proximity spaces arise from underlying topological structures. A **topological structure** on X is a structure of a family of subsets τ of X, having the following properties.

(**O**.1) Every union of sets in τ is a set in τ.
(**O**.2) Every finite intersection of sets in τ is a set in τ.

A family of sets τ of a set X is a *topology* on X, provided τ has the union and intersection topological structures and one additional property, namely,

$$X \text{ and the empty set } \emptyset \text{ are in } \tau.$$

A topology endowed with a metric is called a *metric topology*.

3.2 Distances: Euclidean and Taxicab Metrics

This section briefly introduces two of the most commonly used means of measuring distance, namely, Euclidean distance metric and Manhattan distance metric. Let $\mathbb{R}^n$ denote the real Euclidean space. In Euclidean space in $\mathbb{R}^n$, a **vector** is also called a **point** (also called a *vector* with n coordinates). The *Euclidean line* (or real line) equals $\mathbb{R}^1$ for $n = 1$, usually written $\mathbb{R}$. A line segment $\overline{x_1 x_2}$ between points x_1, x_2 on the real line has length that is the absolute value $\overline{x_1 - x_2}$.

The *Euclidean plane* (or 2-space) $\mathbb{R}^2$ is the space of all points with 2 coordinates. The *Euclidean 3-space* $\mathbb{R}^3$ is the space of all points each with 3 coordinates. In general, the *Euclidean n-space* is the n-dimensional space $\mathbb{R}^n$. The elements of $\mathbb{R}^n$ are points (also called vectors), each with n coordinates.

For example, let points $x, y \in \mathbb{R}^n$ with n coordinates, then $x = (x_1, \ldots, x_n)$, $y = (y_1, \ldots, y_n)$. The norm of $x \in \mathbb{R}^n$ (denoted $\|x\|$) is

$$\|x\| = \sqrt{x_1^2 + x_2^2 + \ldots + x_n^2} \text{ (vector length from the origin).}$$

The distance between vectors x, y is the *norm* of $x - y$ (denoted by $\|x - y\|$). The *Euclidean norm* $\|x - y\|$ in the plane is computed with the *Euclidean metric* defined by

$$\|x - y\| = \sqrt{\sum_{i=1}^{n} \left(x_i^2 - y_i^2\right)} \text{ (Euclidean distance).}$$

Sometimes the Euclidean distance is written $\|x - y\|_2$ (see, e.g., [7, Sect. 5, p. 94]).

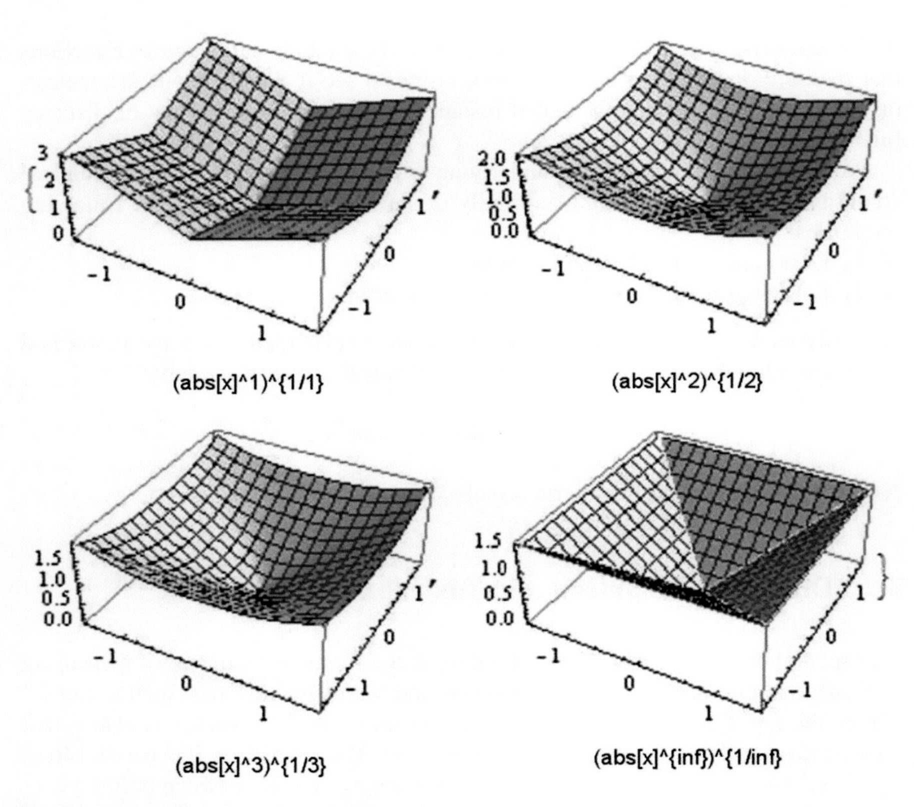

Fig. 3.3 Four different norms of a vector

Example 3.1 **Euclidean norm in the Plane**.
For the points p, q in Fig. 3.2, the Euclidean norm $\|p - q\|$ is the length of the hypotenuse in a right triangle. ■

In general, norm of a vector x has the form

$$\|x\|_p = \left\{|x|^p\right\}^{\frac{1}{p}}$$

Example 3.2 $\|x\|_p$ **Plots**.
Using Mathematica script 4, we obtain the 4 plots shown in Fig. 3.3.

Mscript 4 Norm of a Vector.

***Table*[*Plot3D*[*Norm*[{*x*, *y*}, *p*], {*x*, −1.5, 1.5}, {*y*, −1.5, 1.5}, *Mesh* → *True*,**

***Ticks* → *True*],**

{*p*, {1, 2, 3, *Infinity*}}] ■

The taxicab metric is computed using the absolute value of the differences between points along the vertical and horizontal axes of a plane grid . Let $|x_1 - x_2|$ equal the absolute value of the distance between x_1 and x_2 (along the horizontal axis of a digital image). The *taxicab metric* d_{taxi}, also called the Manhattan distance between points p at (x_1, y_1) and q at (x_2, y_2), is distance in the plane defined by

$$d_{taxi} = |x_1 - x_2| + |y_1 - y_2| \text{ (Taxicab distance between two vectors in } \mathbb{R}^2\text{)}.$$

In general, the taxicab distance between two points in $\mathbb{R}^n$ mimics the distance logged by a taxi moving down one street and up another street until the taxi reaches its destination. The taxicab distance between two points $x = (x_1, \ldots, x_n)$, $y = (y_1, \ldots, y_n)$ in n-dimensional Euclidean space $\mathbb{R}^n$ is defined by

$$d_{taxi} = \sum_{i=1}^{n} |x_i - y_i| \text{ (Taxicab distance in } \mathbb{R}^n\text{)}.$$

Example 3.3 **Taxicab Distance in the Plane.**
For the points p, q in Fig. 3.2, the taxicab distance is the sum of the lengths of the two sides in a right triangle. ■

Euclidean distance and the taxicab distance are two of commonest metrics used in measuring distances in digital images. For example, see the sample distances computed in Matlab listing 3.1.

```
% distance between pixels
clc, clear all, close all
im0 = imread('liftingbody.png'); % built-in greyscale image
image(im0), axis on, colormap (gray(256)); % display image
% select vector components
x1 = 100; y1 = 275; x2 = 325; y2 = 400;
im0 (x1, y1),im0 (x2, y2), % display pixel intensities
p = [100 275]; q = [325 400]; % vectors
norm (p), norm (q), % 2-norm values
norm (p-q), % norm(p-q)=Euclidean dist.
EuclideanDistance = sqrt((x1-x2)^2 + (y1-y2)^2),
ManhattanDistance = abs(x1-x2) + abs(y1-y2)
```

Listing 3.1 Use the Matlab code in `distance.m`to experiment with distances between pixels.

Example 3.4 **Distance Between Image Pixels.**
Try using Matlab listing 3.1 to experiment with distances between pixels. ■

Problem 3.5 🚲 In computing the Euclidean or Manhattan distance between image pixels, what is the unit of measurement? For example, given pixels p at (x_1, y_1) and q at (x_2, y_2), the distance $\|x - y\|$ is either dimensionless or has a unit measurement.
■

3.3 Metric Proximity

A proximity space results from endowing a metric topology with a proximity relation. A nearness relation between sets is called a *proximity*.

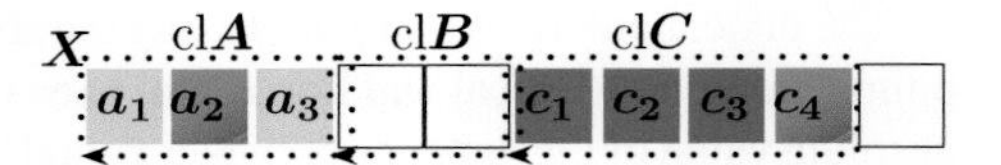

Fig. 3.4 $\mathrm{cl}A \cap \mathrm{cl}B \neq \emptyset$ implies $A\ \delta\ B$

3.4 Closeness Metric

This section introduces sets that are spatially near. In keeping with an interest in finding a basis for spatial nearness, we first consider the closure of a set. Briefly, the closure of a set A in a space X (denoted by $\mathrm{cl}A$) is the set of all points that are close to A.

Let $\mathcal{P}(X)$ denote the collection of all subsets of X. Subsets $A, B \in \mathcal{P}(X)$ are spatially near (denoted by $\mathrm{cl}A\ \delta\ \mathrm{cl}B$), provided the intersection of closure of A and the closure of B is nonempty, which implies $A\ \delta\ B$. That is, nonempty sets are spatially near, provided the sets have at least one point in common. This interpretation of spatial nearness is directly related to the connection algebra approach proposed by D. Vakarelov, G. Dimov, I. Düntsch and B. Bennett [9]. The connection between regions of a proximity space is formalized with the introduction of a connection relation $\mathscr{C}$. Given A, B in a proximity space X, the connection relation between closed regions is defined by

$$A\ \mathscr{C}\ B, \text{ if and only if, the closed regions } A \text{ and } B \text{ share a common point.}$$

In effect,

$$A\ \mathscr{C}\ B, \text{ if and only if } \mathrm{cl}A\ \cap\ \mathrm{cl}B \neq \emptyset.$$

3.5 Some Recent History of Near Sets

This section briefly introduces some of the recent history of near sets, starting the 1970 monograph on proximity spaces by S.A. Naimpally and B.D. Warrack [10]. Spatially near sets are defined in terms of the distance between sets. Let X be a metric normed topological space equipped with the Lodato proximity δ, nonempty subsets $A, B \subset X$, $\|a - b\|$ the distance between points $a \in A, b \in B$. The Čech distance $D(A, B)$ between A and B is defined by

$$D(A, B) = inf\,\{\|a - b\| : a \in A, b \in B\}\,.$$

Sets A, B are **spatially near** (denoted $A\ \delta\ B$), provided $D(A, B) = 0$. This is a traditional approach to defining the nearness of sets (see, e.g., [10, Chap. 1, p. 7]). However, in the sequel we consider the descriptive nearness of disjoint sets.

Disjoint sets are not spatially near but can be descriptively near, provided what is known as the descriptive intersection between the sets is not empty. That is, disjoint sets A, B are **descriptively near**, provided there is at least one pair $a \in A$, $b \in B$ such that the description of a matches the description of b. In other words, descriptive nearness between a pair of sets A, B is defined non-spatially in terms of a set of points $a \in A, b \in B$ with matching descriptions and belonging to the descriptive intersection of A and B. In other words, sets are descriptively near, provided $A \underset{\Phi}{\cap} B \neq \emptyset$, i.e., the descriptive intersection of A and B is nonempty.

Typically, a **probe** maps an object to a real number that is a characteristic feature value of the object. An *object* can either be physical such as a granule of rice, picture point or radar signal or abstract such as a shape in the Euclidean plane or topological manifold or a Voronoï region of a point. A **point-based probe** function $\varphi : X \longrightarrow \mathbb{R}, x \in A$ returns a feature value of x, which is a concrete point such as a picture point with features such as red, green, blue colours. The description of a point $x \in A$ is a feature vector (denoted by $\Phi(x)$)

$$(\varphi_1(x), \ldots, \varphi_i(x), \ldots, \varphi_n(x)), \text{ with } \varphi_i : X \longrightarrow \mathbb{R}.$$

Point-based probes were introduced in [11, Sect. 3, p. 415], closely related to the idea of a probe in M. Pavel [12, Sect. 2.3, p. 9].

The recent work on proximities and quasi-metrics in point-free geometry by A. Di Concilio [13] and quasi metrics in point-free geometry jointly by A. Di Concilio and G. Gerla [14] has led to the introduction of region-based probe functions.

For recent work on descriptively near sets, see C.J. Henry, who was the first one to introduce a metric for tolerance near sets (TNSs) [15]. For recent applications of near sets, see [16–19]. For recent work on near set theory and the distinction between near sets and rough sets, see M. Wolski [20–22].

Zdzisław Pawlak's seminal work on the classification of objects by means of attributes [23] and the joint work on nearness in approximation spaces by J.F. Peters, A. Skowron and J. Stepaniuk [24] led to the introduction of descriptively near sets [11].

Recent work on nearness spaces has been done by S. Tiwari [25]. An approach to solving matching (spatial nearness) problems in digital images with geometric and approximation space methods is given by M. Borkowski [26]. Maciej Borkowski's research on solving coarse, point, feature and dense matching in digital images was the forerunner of the introduction of spatially near and descriptively near sets in images [11], later axiomatized with descriptive nearness axioms for new forms of Efremovič and Lodato proximity spaces in a topology of digital images [27, Sect. 4.15.2]. Maciej Borkowski's work in the recent history of near sets is very important.

A perceptual information systems approach to mining the nearness of sets objects such as sets of picture points in digital images is given in [28]. In the near set approach, every perceptual granule is a set of objects that have their origin the physical world. Piotr Wasilewski's seminal work on similarity and tolerance relations [29] led to the study of the origins, theory and applications of tolerance spaces [30] and tolerance relations as another means of investigating the nearness of sets.

The work by Clara Guadagni [31] on proximity spaces in general, especially Lodato proximity, and on bornological convergences on local proximity spaces in particular, was the forerunner of the introduction of the $\overset{\wedge\!\wedge}{\delta}$ of strong proximity [Di Concilio strong contact]. This form of proximity marks a paradigm shift in the study of proximity, namely, a definiteness about near sets that strongly contact inasmuch as such sets share points.

An overview of applications of near sets from topological and category theory perspectives is given in [32] and a review of this recent work is given by K.D. Kiermeier [33]. In addition, 14 applications of near sets is given in [34].

3.6 Descriptive Similarity Distance

Since we are interested in recognizing objects and set patterns across disjoint regions of digital images that resemble each other, we introduce a descriptive similarity measure on pairs of collections of sets. Let X be a descriptive proximity space and $\mathcal{A}, \mathcal{B} \in 2^X$ be collections containing sets A, B, respectively. The descriptive distance D_Φ is a descriptive form of the distance between sets introduced by E. Čech [6, Sect. 18.A.2]. The distance D_Φ is used to define the **descriptive similarity distance** $\mathbb{D}_\Phi$ between collections of sets. The descriptive distance $\mathbb{D}_\Phi : \mathcal{P}^2(X) \times \mathcal{P}^2(X) \to \mathbb{R}$ between collections $\mathcal{A}, \mathcal{B}$ is defined by

$$\mathbb{D}_\Phi(\mathcal{A}, \mathcal{B}) = \inf\{D_\Phi(A, B) : A \in \mathcal{A}, B \in \mathcal{B}\}, \text{ where,}$$
$$D_\Phi(A, B) = \inf\{d(\Phi(a), \Phi(b)) : a \in A, b \in B\}.$$

The descriptive distance $\mathbb{D}_\Phi$ can be used to measure the distance between collections of nonempty sets that are descriptively near each other.

Definition 3.6 Measure of Descriptive Similarity Between Collections of Sets. Let X be a nonempty set endowed with the descriptive proximity δ_Φ, $\mathbb{D}_\Phi$ the **descriptive similarity distance** between collections of sets $\mathcal{A}, \mathcal{B} \in 2^X$, $\varepsilon > 0$. $\mathcal{A}, \mathcal{B}$ are descriptively similar (denoted by $\mathcal{A}\ \delta_\Phi\ \mathcal{B}$), if and only if

$$\mathbb{D}_\Phi(\mathcal{A}, \mathcal{B}) \leq \varepsilon, \quad \text{(Similarity distance).}$$

That is, the pair of collections $\mathcal{A}, \mathcal{B}$ are descriptively similar, provide they have close descriptions. ■

For more this, see [27, Sect. 1.22].

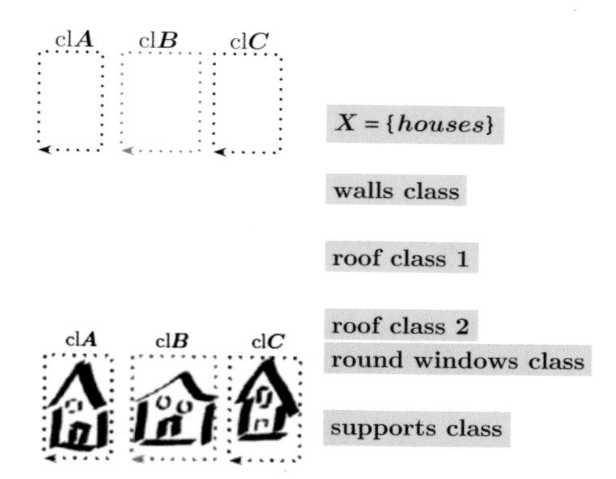

Fig. 3.5 Several houses

3.7 Descriptively Near Sets Can Be Spatially Disjoint Sets

Sets of objects can be spatially far apart and yet be descriptively close to each other.

Example 3.7 **Descriptively Near Disjoint Sets**
Again, choose Φ to be a set of probe functions representing weave cell colours. Let the set of cells X in Fig. 3.4 be endowed with the descriptive proximity relation δ_Φ. Again, observe that sets $A, C \in \mathcal{P}(X)$ are disjoint, i.e., $A \cap C = \emptyset$. Sets A and C contain cells with matching colours, namely, cells $a_2 \in A$ and $c_4 \in C$. Then $\mathrm{cl}A \underset{\Phi}{\cap} \mathrm{cl}B \neq \emptyset$. Hence, $A\ \delta_\Phi\ C$. ■

In classifying subsets of point samples (briefly, *points*) in, for example, a digital image or in a drawing or in the overlapping threads in a fabric weave, it is helpful to choose a set of probe functions that make it possible to compare shapes. In either a digital image or in a drawing or in a weave, a *point sample* is a number representing intensity of light (e.g., $0 =$ lowest intensity (black) and $255 =$ highest intensity (white)). For example, let $\phi \in \Phi$ be a probe function that represents the gradient orientation of points along an edge in a drawing.

Example 3.8 **Shape Near Sets**
Let X be a finite set of point samples for the drawing of house fronts[2] in Fig. 3.5 and let Φ be a set of probe functions that includes a gradient direction probe ϕ. Further, let $\mathrm{cl}A$ be the set of point samples in the leftmost house and $\mathrm{cl}C$ be the set of point samples in the rightmost house in Fig. 3.5. Since the points in the walls in $\mathrm{cl}A$ and $\mathrm{cl}C$ have almost the same gradient orientation (i.e., $\mathrm{cl}A \underset{\Phi}{\cap} \mathrm{cl}C \neq \emptyset$), then $A\ \delta_\Phi\ C$. In addition, for the same reason, $A\ \delta_\Phi\ B$. By collecting together in separate

[2]This drawing was made using Inkscape, a public domain, vector graphics system that makes it possible to make freehand drawings that can be saved as a LaTeX (.tex) file containing a pspicture environment for the drawing. Many thanks to Mario Liziér, Universidade Federal de São Carlos (UFSCar), for suggesting the use of Inkscape in this way.

sets containing points with the same gradient direction, we obtain point equivalence classes like those shown in Fig. 3.5. ■

3.8 Neighbourhoods of Points in Euclidean Space

In a metric space X, endowed with a distance function $d : X \times X \to \mathbb{R}$, $x \in X$, $\varepsilon > 0$, a neighbourhood of x (denoted by $N_{x,\varepsilon}$) is defined by

$$N_{x,\varepsilon} = \{y \in X : d(x, y) < \varepsilon\}. \text{ (Neighbourhood of a } x\text{)}$$

A neighbourhood $N_{x,\varepsilon}$ is also called a ε-disk about x [35, Sect. 2.4, p. 17]. For simplicity, $N_{x,\varepsilon}$ is denoted by N_x, when ε is understood.

Example 3.9 **Neighbourhood of a point in the Euclidean plane.**
In the Eucliean plane with $x \in \mathbb{R}^2$, the $N_{x,\varepsilon}$ is defined by

$$N_{x,\varepsilon} = \{y \in X : \|x - y\| < \varepsilon\}. \text{ (Neighbourhood of a } x\text{)} \quad ■$$

Example 3.10 **Neighbourhood of a point in a 2D digital image.**
In a 2D digital image (picture) Im (subset of the Euclidean plane) with vector $p \in Im \times Im$, the $N_{p,\varepsilon}$ is also defined by

$$N_{p,\varepsilon} = \{q \in Im \times Im : \|p - q\| < \varepsilon\}. \text{ (Neighbourhood of a } p\text{)} \quad ■$$

Fig. 3.6 Neighbourhood of point p with coordinates $(120, 114)$, $\varepsilon = 30$

For example, a picture neighbourhood $N_{p,\varepsilon}$ of a pixel p at location (120, 114), $\varepsilon = 30(pixels)$ is shown in Fig. 3.6.

In general, for $x \in \mathbb{R}^n$, $N_{x,\varepsilon}$ is the set of points inside an n-ball[3] with center x and radius $\varepsilon > 0$. For this reason, $N_{x,\varepsilon}$ is called a **spherical neighbourhood**.

3.9 2D Picture Descriptive Neighbourhood of a Point

This section introduces the a number of different 2D Picture descriptive neighbourhoods of points $N_{\Phi(x),\varepsilon}$ from [27, Sect. 1.16], namely,

Unbounded $N_{\Phi(x)}$ This is a descriptive neighbourhood in which there is no restriction on the location of a picture point in relation to the neighbourhood centre $\Phi(x)$. That is, if $N_{\Phi(x)}$ is an unbounded descriptive neighbourhood of a picture point in X, then every $y \in X$ with a description close to the description of x belongs to $N_{\Phi(x)}$.

Example 3.11 **Red Channel Unbounbed Descriptive Neighbourhood of a picture point.**
Let x be a pixel at location (120, 114) in the picture of Lena in Fig. 3.7 and let $\Phi(x)$ equal the intensity (brightness) of the red channel for pixel x, $\varepsilon_c = 30$ (red channel intensity). Informally, the unbounded descriptive neighbourhood $N_{\Phi(x)}$ is the set of all pixels with red intensity ε_c-close to the red channel intensity of x. This particular unbounded descriptive neighbourhood is shown in Fig. 3.7. ■

Bounded $N_{\Phi(x)}$ This is a descriptive neighbourhood in which there is a restriction on the location of a picture point in relation to the neighbourhood centre $\Phi(x)$. That is, if $N_{\Phi(x)}$ is a bounded descriptive neighbourhood of a picture point in X, then every $y \in X$ is within a fixed distance $\varepsilon > 0$ from x and with a description close to the description of x belongs to $N_{\Phi(x)}$.

Indistinguishable bounded $N_{\Phi(x)}$ This is a descriptive neighbourhood in which there is a restriction on the location of a picture point in relation to the neighbourhood centre $\Phi(x)$. That is, if $N_{\Phi(x)}$ is an indistinguishable bounded descriptive neighbourhood of a picture point in X, then every $y \in X$ within a fixed distance $\varepsilon > 0$ from x and with a description that matches the description of x belongs to $N_{\Phi(x)}$.

Indistinguishable unbounded $N_{\Phi(x)}$ This is a descriptive neighbourhood in which there is no restriction on the location of a picture point with a description that matches the description of the neighbourhood centre $\Phi(x)$. That is, if $N_{\Phi(x)}$ is an indistinguishable unbounded descriptive neighbourhood of a picture point in X, then every $y \in X$ with a description that matches the description of x belongs to $N_{\Phi(x)}$.

[3] M. Barile, E.W. Weisstein, Neighbourhood, MathWorld, http://mathworld.wolfram.com/Neighborhood.html.

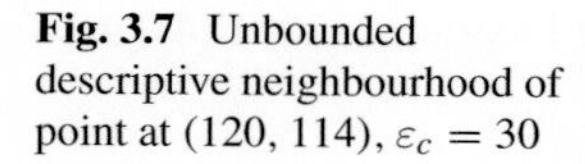

Fig. 3.7 Unbounded descriptive neighbourhood of point at (120, 114), $\varepsilon_c = 30$

3.9.1 Unbounded Descriptive Neighbourhood of a Picture Point

Let (img, δ_Φ) be a local descriptive proximity space, img, an $n \times m$ digital image, $p(x, y)$, a picture point at location (x, y), $\Phi(p(x, y)) = img(x, y)$, the description of p equal to the intensity of the picture point. In other words, $img(x, y)$ is a feature of the picture p at location (x, y) in the image img, which we call a picture. An unbounded descriptive neighbourhood of a picture point $N_{\Phi(x)}$ is defined by

Requirements 1 Requirements for an Unbounded Descriptive Neighbourhood of a Picture Point

$$img = \textit{digital image},\ p(x, y), q(x', y') \in img.$$
$$img(x, y), img(x', y') = \textit{image intensity at locations}\ (x, y), (x', y'), respectively.$$
$$\varepsilon_c > 0\ \textit{colour intensity difference threshold}.$$
$$p(x, y), q(x', y') \in im\ \textit{(picture points in image)}.$$
$$N_{\Phi(p)} = \left\{q \in im : \left|im(x, y) - im(x', y')\right| < \varepsilon_c\right\}.$$

Basically, an **unbounded descriptive neighbourhood of a point** p in an image img is the set of all points with intensities close to the intensity of p. Algorithm 4 gives the steps to find an unbounded descriptive nbd of a selected point.

Fig. 3.8 Coordinates of Points in Sparrow Image

Algorithm 4: Unbounded Descriptive Nbd of a Point on a Digital Image

---

Input : Read digital image img, selected point $p \in img$ at (x, y), $\varepsilon_c > 0$.
Output: $N_{\Phi(p)}$: unbounded descriptive neighbourhood of p

1 $p(x, y) \longmapsto img(x, y)$;
2 $img \longmapsto imgCopy$;
3 /* $img(x, y)$ equals the intensity of picture point $p(x, y)$ */;
4 **while** ($\exists q \in imgCopy : q(x', y') \neq 0$ *and* $q \neq p$) **do**
5   *Select* $q(x', y') > 0$;
6   **if** $|img(x, y) - img(x', y')| < \varepsilon_c$ **then**
7         $img(x', y', 1) \leftarrow 255$;
8         $imgCopy(x', y') \leftarrow 0$;
9   /* $img(x', y')$ intensity close to the intensity of $img(x, y)$ */

Example 3.12 **Unbounded Descriptive Neighbourhood of a Point.**
A 363×254 colour image is shown in Fig. 3.8. In this example, MScript 30 in Appendix A.3 implements Algorithm 4, where the selected point is located at (300, 190), centering on a dark region of the eye of the sparrow. In this example, the intensity threshold ε_c in Algorithm 4 is set equal to 0.002, which is actually fairly high. The basic idea then is to use a false colour with • for all pixels with a colour intensity close to 0.002. The resulting unbounded descriptive neighbourhood of p(300,190) is shown in Fig. 3.9. ■

Example 3.13 **Second Unbounded Descriptive Neighbourhood of a Point.**
A 1152×656 colour image of a tree nymph is shown in Fig. 3.10. In this example, the Matlab script A.2 in Appendix A.3 implements Algorithm 4, where the selected point is located at (343, 324), centering on the dark region of the left wing of the

Fig. 3.9 Unbounded descriptive Nbd of p(300, 190) in Sparrow Image

Fig. 3.10 Tree Nymph Butterfly

tree nymph at . In this implementation, the selected point (found by using a mouse to click on a particular point of interest) is false coloured with a • bullet. In this example, the intensity threshold ε_c in Algorithm 4 is again set equal to 0.002. Then all pictures points with a colour intensity close to 0.002 are false coloured with •. The resulting unbounded descriptive neighbourhood of p(343, 324) is shown in Fig. 3.11. ■

Problem 3.14 Modify MScript 30 in Appendix A.3 so that the following things are displayed:

1° A • bullet is displayed so that the selected point p of the unbounded neighbourhood $N_{\Phi(p)}$ is false-coloured.

2° The horizontal and vertical axes of a selected image are displayed. ■

3.9.2 Bounded Descriptive Neighbourhood of a Picture Point

An bounded descriptive neighbourhood of a picture point $N_{\Phi(x),\varepsilon_c,r}$ is defined by

Fig. 3.11 Unbounded descriptive Nbd of p(343, 324) in Tree Nymph Image

Requirements 2 Requirements for a Bounded Descriptive Neighbourhood of a Picture Point

$$\begin{aligned}
img &= digital\ image,\ p(x, y), q(x', y') \in img.\\
img(x, y), img(x', y') &= \ image\ intensity\ at\ locations\ (x, y), (x', y'), respectively.\\
\varepsilon_c &> 0\ colour\ intensity\ difference\ threshold.\\
r &> 0\ neighbourhood\ radius.\\
p(x, y), q(x', y') &\in im\ (picture\ points\ in\ image).\\
N_{\Phi(p)} &= \left\{q \in im : \left|im(x, y) - im(x', y')\right| < \varepsilon_c\ and\ \|p - q\| < r\right\}. \quad \blacksquare
\end{aligned}$$

Basically, a **bounded descriptive neighbourhood of a picture point** p is the set of all points with intensities close to the intensity of p with a fixed radius. Algorithm 5 gives the steps to find an unbounded descriptive nbd of a selected point.

Example 3.15 **Bounded Descriptive Neighbourhood of a Picture Point.**
A 1152 × 656 colour image of a tree nymph is shown in Fig. 3.10. In this example, the Matlab script A.2 in Appendix A.3 implements Algorithm 5, where the selected point is located at (306,279), again centering on a dark region of the left wing of the tree nymph at . In this implementation, the selected point (found by using a mouse to click on a particular point of interest) is false coloured with a • bullet. In this example, the intensity threshold ε_c in Algorithm 5 is again set equal to

Algorithm 5: Bounded Descriptive Nbd of a Point on a Digital Image

Input : Read digital image img, selected point $p \in img$ at (x, y), $\varepsilon_c > 0, r > 0$.
Output: $N_{\Phi(x),\varepsilon_c,r}$: bounded descriptive neighbourhood of p

```
1 p(x, y) ⟼ img(x, y);
2 img ⟼ imgCopy;
3 /* img(x, y) equals the intensity of point p(x, y) */;
4 while (∃q ∈ N_{p,r} : q(x', y') ≠ 0 and q ≠ p) do
5     Select q(x', y') > 0;
6     if |im(x, y) − im(x', y')| < ε_c then
7         img(x', y', 2) ← 255;
8         imgCopy(x', y') ← 0;
9         /* im(x', y') intensity close to the intensity of im(x, y) */
```

Fig. 3.12 Unbounded descriptive Nbd of p(306, 279) in tree Nymph image

0.002 and the radius $r = 80$. Then all pictures points with a colour intensity close to 0.002 are false coloured with •. The resulting bounded descriptive neighbourhood of p(306,279) is shown in Fig. 3.12. ■

Problem 3.16 Prove that Algorithm 5 satisfies Requirements 2. ■

3.9.3 *Bounded Indistinguishable Descriptive Neighbourhood of a Picture Point*

A bounded indistinguishable descriptive neighbourhood of a picture point $N_{\Phi(x),r}$ is defined by

Requirements 3 Requirements for a Bounded Indistinguishable Descriptive Neighbourhood of a Picture Point

$$img = digital\ image,\ p(x, y), q(x', y') \in img.$$
$$img(x, y), img(x', y') = \ image\ intensity\ at\ locations\ (x, y), (x', y'), respectively.$$
$$r > 0\ neighbourhood\ radius.$$
$$p(x, y), q(x', y') \in im\ (picture\ points\ in\ image).$$
$$N_{\Phi(p)} = \left\{q \in im : im(x, y) = im(x', y')\ and\ \|p - q\| < r\right\}.$$
■

Basically, a **bounded indistinguishable descriptive neighbourhood of a picture point** p is the set of all points with intensities equal to the intensity of p within a fixed radius. Algorithm 6 gives the steps to find a bounded indistinguishable descriptive nbd of a selected point.

Algorithm 6: Bounded Indistinguishable Descriptive Nbd of a Point on a Digital Image

---

Input : Read digital image img, selected point $p \in img$ at (x, y), $\varepsilon_c > 0, r > 0$.
Output: $N_{\Phi(x),r}$: bounded indistinguishable descriptive neighbourhood of p

1 $p(x, y) \longmapsto img(x, y)$;
2 $img \longmapsto imgCopy$;
3 /* $img(x, y)$ equals the intensity of point $p(x, y)$ */;
4 **while** ($\exists q \in N_{p,r} : q(x', y') \neq 0$ *and* $q \neq p$) **do**
5   *Select* $q(x', y') > 0$;
6   **if** $im(x, y) = im(x', y')$ **then**
7         $img(x', y', 1) \leftarrow 255$;
8         $imgCopy(x', y') \leftarrow 0$;
9   /* $im(x', y')$ equals the intensity of $im(x, y)$ */

Example 3.17 **Bounded Indistinguishable Descriptive Neighbourhood of a Picture Point**.
A 1152 × 656 colour image of a tree nymph is shown in Fig. 3.10. In this example, the Matlab script A.2 in Appendix A.3 implements Algorithm 6, where the selected point is located at (302, 275), again centering on a dark region of the left wing of the tree nymph at . In this implementation, the selected point (found by using a mouse to click on a particular point of interest) is false coloured with a • bullet. In this example, the radius $r = 120$. Then all pictures points with a colour intensity equal to the intensity of $p(302, 275)$ are false coloured with •. The result-

Fig. 3.13 Bounded indistinguishable descriptive Nbd of p(302, 275) in tree Nymph image

ing bounded indistinguishable descriptive neighbourhood of p(302, 275) is shown in Fig. 3.13. ■

Problem 3.18 Prove that Algorithm 6 satisfies Requirements 3. ■

3.9.4 Unbounded Indistinguishable Descriptive Neighbourhood of a Picture Point

An unbounded indistinguishable descriptive neighbourhood of a picture point $N_{\Phi(x),r}$ is defined by

Requirements 4 Requirements for an Unbounded Indistinguishable Descriptive Neighbourhood of a Picture Point

$$img = digital\ image,\ p(x, y), q(x', y') \in img.$$
$$img(x, y), img(x', y') = image\ intensity\ at\ locations\ (x, y), (x', y'), respectively.$$
$$p(x, y), q(x', y') \in im\ (picture\ points\ in\ image).$$
$$N_{\Phi(p)} = \left\{q \in im : im(x, y) = im(x', y')\right\}. \quad ■$$

Basically, an **unbounded indistinguishable descriptive neighbourhood of a picture point** p is the set of all points with intensities close to the intensity of p with a fixed radius.

Problem 3.19 Give an algorithm that implements an unbounded indistinguishable descriptive nbd of a selected point. ■

Problem 3.20 Use the Mathematica **Manipulate, PopupMenu, InputField** function in a script to make it possible to do the following:

1° Use a mouse to click on a selected picture point p.
2° Display the coordinates of the selected point p in a digital image.
3° Use **InputField** to select a neighbourhood radius r (this is ∞ for unbounded descriptive neighbourhoods).
4° Use **InputField** to select an intensity threshold ε_c (this is used for each type of descriptive neighbourhood of a picture point).
5° Use **PopupMenu** to selected one of the following types of neighbourhoods:

(a) $N_{p,\varepsilon}$: Spherical neigbourhood of p.
(b) Unbounded descriptive neighbourhood $N_{\Phi(p),\varepsilon_c}$.
(c) Bounded descriptive neighbourhood $N_{\Phi(p),\varepsilon_c,r}$.
(d) Unbounded indistinguishable descriptive neighbourhood $N_{\Phi(p),\varepsilon_c}$.
(e) Bounded indistinguishable descriptive neighbourhood $N_{\Phi(p),\varepsilon_c,r}$.

6° A ● bullet is displayed so that the selected point p of the unbounded neighbourhood $N_{\Phi(p)}$ is false-coloured.

7° The horizontal and vertical axes of a selected image are displayed.
8° Display the selected neighbourhood of the point p. ■

Problem 3.21 ☕ Modify the Mathematica script in Problem 3.20 so that more than one point p can be selected. Repeat the steps in Mathematica script in Problem 3.20 so that more than one neighbourhood can be displayed. Refresh the image (showing no neighbourhoods of points) each time a neighbourhood type is selected. ■

Problem 3.22 ☕ A Computable Document Format (CDF) file can be created from existing Mathematica notebooks. Use a CDF to create an interactive version of the Mathematica script in Problem 3.21

Problem 3.23 Use the CDF in Problem 3.22 to construct examples of image nervous systems centered on a spherical neighbourhood, doing the following:
1° Construct a spherical neighbourhood $N_{p,\varepsilon}$ of a selected picture point p.
2° Surround $N_{p,\varepsilon}$ with spherical neighbourhoods $N_{q,\varepsilon}$, $N_{r,\varepsilon}$ so that $N_{q,\varepsilon} \cap N_{r,\varepsilon} \neq \emptyset$, i.e., each pair of neighbourhoods have nonempty intersection in the nervous system. ■

Problem 3.24 Let (X, δ), (Y, δ') be proximity spaces. Let $B \subset X$. A function $f : X \longrightarrow Y$ is *continuous* if and only

$$\text{for all } x\ \delta\ B \text{ implies } f(x)\ \delta\ f(B) \text{ [31, Sect. 1.1, p. 2].}$$

A mapping $f : X \longrightarrow Y$ is a **proximal homeomorphism**, provided f is a 1-to-1 mapping of the proximity space X onto the proximity space Y such that f and its inverse $f^{-1} : Y \longrightarrow X$ are continuous [6, Sect. 25.A.7, p. 442]. In the sequel, proximal homeomorphic mappings lead to what are known as proximal manifolds.

🚲 Give an example of a proximal homeomorphism.
■

References

1. Fréchet, M.: Sur quelques points du calcul fonctionnel. Rend. Circ. Mat. Palermo **22**, 1–74 (1906)
2. Hadamard, J.: Leçons de Géométrie élémentaire. Armand Colin & Cie, Paris (1898)
3. Hadamard, J.: An Essay on the Psychology of Invention in the Mathematical Field. Dover Publications, NY (1954)
4. Hausdorff, F.: Grundzüge der Mengenlehre, Veit and Company, Leipzig (1914). Viii + 476 pp
5. Hausdorff, F.: Set Theory, Translated by J.R. Aumann, AMS Chelsea Publishing, Providence (1957). 352 pp
6. Čech, E.: Topological Spaces. Wiley, London (1966). Fr seminar, Brno, 1936–1939; rev. ed. Z. Frolik, M. Katětov
7. Deza, E., Deza, M.M.: Encyclopedia of Distances. Springer, Berlin (2009)

8. Deza, E., Deza, M.M.: Dictionary of Distances. Elsevier, Amsterdam (2006). ISBN 0-444-52087-2, 391 pp
9. Vakarelov, D., Dimov, G., Düntsch, I., Bennett, B.: A proximity approach to some region-based theories of space. J. Appl. Non-Class. Log. **12**(3–4), 527–559 (2012). doi:10.3166/jancl.12.527-559
10. Naimpally, S.: Proximity Spaces. pp. X + 128. Cambridge University Press, Cambridge (1970). ISBN: 978-0-521-09183-1
11. Peters, J.: Near sets. Special theory about nearness of objects. Fundamenta Informaticae **75**, 407–433 (2007). MR2293708
12. Pavel, M.: Fundamentals of Pattern Recognition, 2nd edn. Marcel Dekker Inc, N.Y (1993). Xii+254 pp. ISBN: 0-8247-8883-4, MR1206233
13. Di Concilio, A.: Point-free geometries: Proximities and quasi-metrics. Math. Comput. Sci. **7**(1), 31–42 (2013). MR3043916
14. Di Concilio, A., Gerla, G.: Quasi-metric spaces and point-free geometry. Math. Struct. Comput. Sci. **16**(1), 115–137 (2006). MR2220893
15. Henry, C.: Near sets: Theory and applications. Ph.D. thesis. Department of Electrical and Computer Engineering (2010). URL http://130.179.231.200/cilab/. Supervisor: J.F. Peters
16. Poli, G., Llapa, E., Cecatto, J., Saito, J., Peters, J., Ramanna, S., Nicoletti, M.: Solar flare detection system based on tolerance near sets in a gpu-cuda framework. Knowl.-based Syst. **70**(1), 345–360 (2014)
17. Henry, C., Ramanna, S.: Quantifying nearness in visual spaces. Cybern. Syst. **44**(1), 38–56 (2013)
18. Ramanna, S., Meghdadi, A.: Measuring resemblances between swarm behaviours: a perceptual tolerance near set approach. Fundamenta Informaticae **95**(4), 533–552 (2009). MR2582188
19. Ramanna, S., Chitcharoen, D.: Flowgraphs: analysis with near sets. Math. Comput. Sci. **7**(1), 11–29 (2013)
20. Wolski, M.: Toward foundations of near sets: (pre-)sheaf theoretic approach. Math. Comput. Sci. **7**(1), 125–136 (2013). MR3043923
21. Wolski, M.: Perception and classification. A note on near sets and rough sets. Fundamenta Informaticae **101**(1–2), 143–155 (2010). MR2732874
22. Wolski, M.: Granular computing: topological and categorical aspects of near and rough set approaches to granulation of knowledge. Transactions on Rough Sets. Lecture Notes in Computer Science **7736**(16), 34–52 (2013)
23. Pawlak, Z.: Classification of Objects by Means of Attributes. Polish Academy of Sciences PAS, Warsaw (1981)
24. Peters, J., Skowron, A., Stepaniuk, J.: Nearness of objects: extension of approximation space model. Fundamenta Informaticae **79**(3–4), 497–512 (2007). MR2346263
25. Tiwari, S.: Some aspects of general topology and applications. Approach merotopic structures and applications. Ph.D. thesis, Department of Mathematics, Allahabad (U.P.), India (2010). Supervisor: M. Khare
26. Borkowski, M.: 2d to 3d conversion with direct geometrical search and approximation spaces. Ph.D. thesis, Department of Electrical and Computer Engineering (2007). Supervisor: J.F. Peters
27. Peters, J.: Topology of Digital Images - Visual Pattern Discovery in Proximity Spaces, Intelligent Systems Reference Library, vol. 63. Springer (2014). Xv + 411 pp., Zentralblatt MATH Zbl 1295 68010
28. Peters, J., Wasilewski, P.: Foundations of near sets. Inform. Sci. **179**(18), 3091–3109 (2009). MR2588809
29. Wasilewski, P.: On selected similarity relations and their applications into cognitive science. Ph.D. thesis, (in Polish), Department of Logic, Cracow (2004)
30. Peters, J., Wasilewski, P.: Tolerance spaces: Origins, theoretical aspects and applications. Inf. Sci. **185**, 211–225 (2012). doi:10.1016/j.ins.2012.01.023, MR2904846

31. Guadagni, C.: Bornological convergences on local proximity spaces and ω_μ-metric spaces. Ph.D. thesis, Università degli Studi di Salerno, Salerno, Italy (2015). Supervisor: A. Di Concilio, 79 pp
32. Peters, J., Naimpally, S.: Applications of near sets. Notices Am. Math. Soc. **59**(4), 536–542 (2012). doi:10.1090/noti817, MR2951956
33. Kiermeier, K.: Review of J.F. Peters, S.A. Naimpally, Applications of near sets. Notices Am. Math. Soc. 59(4), 536–542 (2012). issn 0002–9920; issn 1088–9477. Zentralblatt MATH an 1251 68301
34. Naimpally, S., Peters, J.: Topology with Applications. Topological spaces via near and far. World Scientific, Singapore (2013). Xv + 277 pp., Am. Math. Soc. MR3075111
35. Willard, S.: General Topology. Dover Publications Inc, Mineola, NY (1970). Xii + 369 pp., ISBN: 0-486-43479-6 54-02, MR0264581

Chapter 4
Image Geometry and Nearness Expressions for Image and Scene Analysis

Fig. 4.1 Dirichlet tessellation of fisherman scene

This chapter suggests an approach to image and scene analysis based on Dirichlet (also called Voronoï) tessellations and Delaunay triangulation of selected seed (generating or site) points in a digital image. A **visual scene** is a collection of objects in a visual field that captures our attention. In human vision, a **visual field** is the total area in which objects can be seen. A normal visual field is about 60° from the vertical meridian of each eye and about 60° above and 75° below the horizontal meridian.

A **digital image scene** is a collection of visual field objects recorded by a camera. A sample Dirichlet tessellation of a 640 $\times$ 480 digital image containing fisherman scene is shown in Fig. 4.1. Here, the locations of up to 60 image key colour-feature

J.F. Peters, *Computational Proximity*, Intelligent Systems
Reference Library 102, DOI 10.1007/978-3-319-30262-1_4

values are the source of sites used to generate the fisherman diagram. The • indicates the location of a keypoint location. To experiment with this image for different choices of keypoints, try the Mathemamtica Script 32 in Appendix A.4.

The foundations for scene analysis are built on the pioneering work by A. Rosenfeld work on digital topology [1–6] (later called digital geometry [7]) and others [8–12]. The work on digital topology runs parallel with the introduction of computational geometry by M.I. Shamos [13] and F.P. Preparata [14, 15], building on the work on spatial tessellations by G. Voronoï [16, 17] and others [18–23]. In terms of topology, convexity and point-free geometry, important foundation work leading to scene analysis and scene understanding has been done by G. Beer, A. Di Concilio, G. Di Maio, S.A. Naimpally (an overview is given in [24]).

For recent work on mappings that are continuous (very important for scene shape analysis, especially in considering the implications of the Borsuk–Ulam Theorem and homotopy theory, introduced in Sect. 5.1), see [25]. A standard reference in the study of continuous functions on topological spaces to the the space of reals $\mathbb{R}$ is by R. McCoy [26], revisited in [27].

Source of Image Geometry Information

The important thing to notice is that a Voronoï region $V(p)$ is a set of all points nearer to a particular generating point $p \in S$ than to any other generating point in the set of generating points S. Hence, the proximity of each point in the interior of a Voronoï region $V(p)$ is a source of image geometry information about a particular generating point such as a corner or centroid in a digital image scene. ■

To analyze and understand image scenes, it is necessary to identify the objects in the scenes. Such objects can be viewed geometrically as collections of connected edges (e.g., skeletonizations or edges belonging to shapes or edges in polygons) or image regions viewed as sets of pixels that are in some sense near each other or set of points near a fixed point (e.g., all points near a site (also, seed or generating point) in a Voronoï region [28]). For this reason, it is highly advantageous to associate geometric structures in an image with mesh-generating points (sites) derived from the fabric of an image. Image edges, corners, centroids, critical points, intensities, and keypoints (image pixels viewed as feature vectors) or their combinations provide ideal sources of mesh generators as well as sources of information about image geometry.

Algorithm 7: Digital Image Geometry via Mesh Covering Image

Input : Read digital image img.
Output: Mesh $\mathscr{M}$ covering an image.

1 $MeshSite \leftarrow MeshGeneratingPointType$;
2 $img \longmapsto MeshSitePointCoordinates$;
3 $S \leftarrow MeshSitePointCoordinates$;
4 /* S contains MeshSitePointType coordinates used as mesh generating points (seeds or sites). */ ;
5 $MeshType \leftarrow MeshChoice$;
6 /* $MeshType$ identifies a chosen form of mesh, e.g., Voronoï, Delaunay, polynomial. */ ;
7 $S \longmapsto MeshType\ \mathscr{M}$;
8 $MeshType\ \mathscr{M} \longmapsto img$;
9 /* Use $\mathscr{M}$ to gain information about image geometry. */ ;

Remark 4.1 **Image Geometry Issues**.
Algorithm 7 leads to a mesh covering a digital image. Image meshes can vary considerably, depending on the type of image and the type mesh generating points that are chosen. Image geometry tends to be revealed, whenever the choice of generating points accurately reflects the image visual content and the structure of the objects in an image scene. For example, corners would be the logical choice for image scenes containing buildings or objects with sharply varying contours such as hands or facial profiles. A sample implementation of Algorithm 7 is given in the Mathematica script 33 in Appendix A.4.

To carry this idea of image geometry further, it is necessary to correlate image geometric regions with semantic features such as oblong, pointed, concave, perpendicular, circular. For an approach to dealing with the gap between image visual content and semantic features, see [29]. Notice, however, that we are focusing on the gap between image geometry (instead of low level visual content) and semantic features. ■

Example 4.2 **Meshes Covering a Salerno Poste Auto Scene**.
A corner-based Voronoï mesh covering an image scene containing a Poste auto parked outside the train station in Salerno, Italy is shown in Fig. 4.2.1. This Voronoï mesh is also called a *Dirichlet tessellation*. Using the same set of corner generating points, a Delaunay triangulation cover in the Poste auto scene is shown in Fig. 4.2.2. To experiment with tessellating other image scenes, see Mathematica script 33 in Appendix A.4. ■

4.2.1: Corner-Based Voronoï mesh

4.2.2: Corner-Based Delaunay mesh

Fig. 4.2 Hunting grounds for scene information: corner-based Delaunay and Voronoï meshes

4.1 Image Geometry

This section gives an overview of Delaunay triangulations and Voronoï diagrams useful in extracting geometric scene information.

4.1.1 Delaunay Triangulation

Delaunay triangulations, introduced by B.N Delone [Delaunay] [30], represent pieces of a continuous space. A **triangulation** is a collection of triangles, including the edges and vertices of the triangles in the collection.

A 2D *Delaunay triangulation* of a set of sites (generators) $S \subset \mathbb{R}^2$ is a triangulation of the points in S. The set of vertices (called sites) in a Delaunay triangulation define a **Delaunay mesh**. A Delaunay mesh endowed with a nonempty set of proximity relations is a *proximal Delaunay mesh*. A proximal Delaunay mesh is an example of a proximal relator space [31], which is an extension a Száz relator space [32–34].

Let $S \subset \mathbb{R}^2$ be a set of distinguished points called *sites*, $p, q \in S$, $\overline{pq}$ straight line segment in the Euclidean plane. A site p in a line is *visible* to another site q in the same straight line segment, provided there is no other site between p and q.

Example 4.3 **Visible Points.**
A pair of Delaunay triangles $\triangle(pqr), \triangle(rst)$ are shown in Fig. 4.3. Points r, q is visible from p but in the straight line segment $\overline{ps}$, s is not visible from p. Similarly, points p, r are visible from q but in the straight line segment $\overline{qs}$, t is not visible from q. From r, points p, q, s, t are visible. ■

A straight edge connecting p and q is a *Delaunay edge* if and only if the Voronoï region of p [35, 36] and Voronoï region of q intersect along a common line segment [37, Sect. I.1, p. 3]. For example, in Fig. 4.4, the intersection of Voronoï regions V_p, V_q is a triangle edge, i.e., $V_p \cap V_q = \overline{xy}$. Hence, $\overline{pq}$ is a Delaunay edge in Fig. 4.4. A triangle with vertices $p, q, r \in S$ is a *Delaunay triangle* (denoted $\triangle(pqr)$ in Fig. 4.4), provided the edges in the triangle are Delaunay edges. Proximal Delaunay triangulation regions are derived from the sites of Voronoï regions in [36].

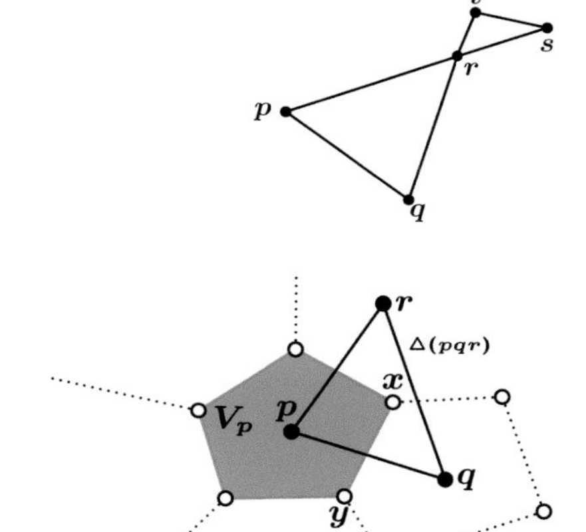

Fig. 4.3 Visible points

Fig. 4.4 $\triangle(pqr)$ = Delaunay triangle

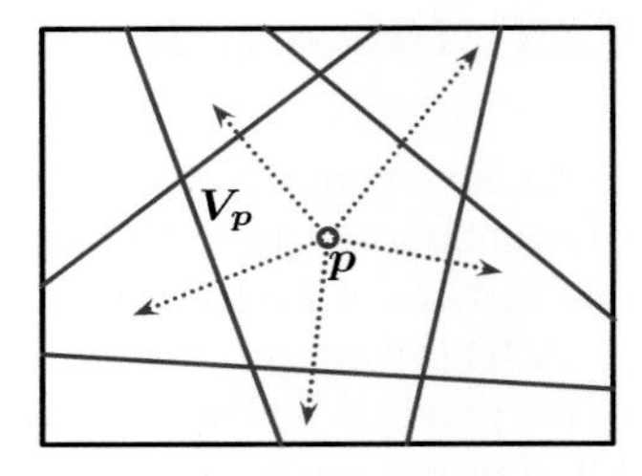

Fig. 4.5 $V(p), p \in S =$ Intersection of closed half-planes

4.1.2 Voronoï Diagrams

A *Voronoï diagram* represents a tessellation of the plane by convex polygons. It is generated by n site points and each polygon contains exactly one of these points. In each region there are points that are closer to its generating point than to any other. *Voronoï diagrams* were introduced by *René Descartes* (1667) looking at the influence regions of stars. They were studied also by Dirichlet (1850) and Voronoï (1907), who extended the study to higher dimensions.

To construct a Voronoï diagram, we have to start from a finite number of points. Consider a set S of n points in a finite-dimensional normed vector space $(X, \|\cdot\|)$. We call S the *generating set*. The Voronoï diagram based on S is constructed by taking for each point of S the intersection of suitable half planes (see, e.g., Fig. 4.5). Take $p \in S$ and let H_{pq} be the closed half plane of points at least as close to p as to $q \in S \setminus \{p\}$ given by

$$H_{pq} = \{x \in X : \|x - p\| \leq \|x - q\|\}.$$

The intersection of all the half planes for $q \in S \setminus \{p\}$ gives the *Voronoï region* V_p of p:

$$V_p = \bigcap_{q \in S \setminus \{p\}} H_{pq}.$$

Voronoï regions are named after Georgy Voronoï [16, 38, 39]. The simplifying notation $V(p)$ is sometimes used instead of V_p, when p is replaced by p_i for an indexed site.

Lemma 4.4 [35, Sect. 2.1, p. 9] *The intersection of convex sets is convex.*

Proof Let $A, B \subset \mathbb{R}^2$ be convex sets and let $K = A \cap B$. For every pair of points $x, y \in K$, the line segment $\overline{xy}$ connecting x and y belongs to K, since this property holds for all points in A and B. Hence, K is convex. □

Since a Voronoï region is the intersection of closed half planes, each *Voronoï region* is a closed convex polygon (see, e.g., Fig. 2.16).

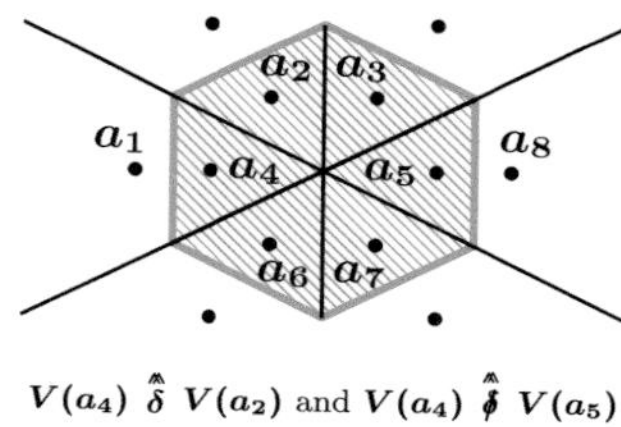

Fig. 4.6 Voronoï regions $V_{a_i}, i \in \{1, 2, 3, 4, 5, 6, 7, 8\}$

We want to define a strong proximity acting on Voronoï regions. We say that two Voronoï regions are strongly near, and we write $V_p \overset{\wedge\wedge}{\delta} V_q$, if and only if they share more than one point.

Theorem 4.5 [40, Sect. 6] *Let $(X, \|\cdot\|)$ be a finite-dimensional normed vector space and S a collection of points in X. The relation defined by saying $V_p \overset{\wedge\wedge}{\delta} V_q$ if and only if they share more than one point is a strong proximity on $\mathscr{V}(S)$, the class of Voronoï regions generated by S.*

Example 4.6 Let X be a space covered with a Voronoï diagram $\mathscr{V}(S)$, S, a set of sites. A partial view of $\mathscr{V}(S)$ is shown in Fig. 4.6, where

$$V_{a_i} \in \mathscr{V}(S), a_i \in S, i \in \{1, 2, 3, 4, 5, 6, 7, 8\}.$$

From Theorem 4.5, observe that

$$V_{a_4} \overset{\wedge\wedge}{\delta} V_{a_2},\ V_{a_4} \overset{\wedge\wedge}{\delta} V_{a_6} \text{ and } V_{a_4} \overset{\wedge\wedge}{\not\delta} V_{a_5},\ V_{a_2} \overset{\wedge\wedge}{\not\delta} V_{a_5},\ V_{a_6} \overset{\wedge\wedge}{\not\delta} V_{a_5},$$

since $\{V_{a_2}, V_{a_4}\}, \{V_{a_4}, V_{a_6}\}$ have a common edge. Further, $V_{a_2}, V_{a_4}, V_{a_6}$ are not strongly near V_{a_5}. $V_{a_2}, V_{a_4}, V_{a_6}$ share only one point with V_{a_5}. Similarly,

$$V_{a_5} \overset{\wedge\wedge}{\delta} V_{a_3}, V_{a_5} \overset{\wedge\wedge}{\delta} V_{a_7}, V_{a_5} \overset{\wedge\wedge}{\delta} V_{a_8},$$

since, taken pairwise, these Voronoï regions have a common edge. There are also Voronoï regions in Fig. 4.6 that are near but not strongly near, e.g., $V_{a_3} \overset{\wedge\wedge}{\not\delta} V_{a_6}, V_{a_7} \overset{\wedge\wedge}{\not\delta} V_{a_2}$. ■

4.2 Nearness Expressions: Basic Notions

Delaunay triangles are defined on a finite-dimensional normed linear space E that is topological. For simplicity, E is the Euclidean space $\mathbb{R}^2$. The *closure* of $A \subset E$ (denoted $\mathrm{cl}A$) is defined by

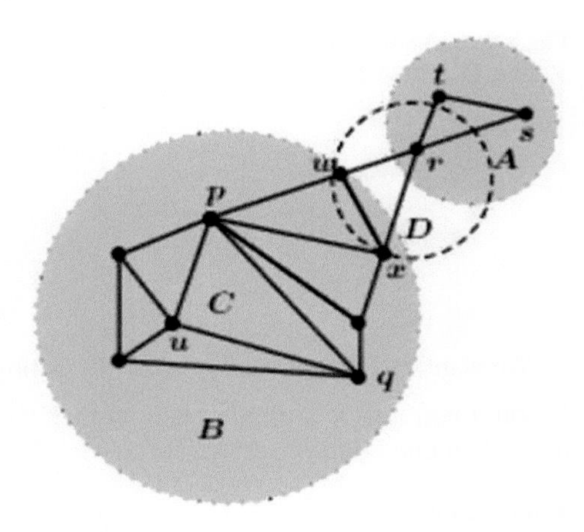

Fig. 4.7 Strongly separated points

$$\begin{aligned} \mathrm{cl}(A) &= \{x \in X : D(x, A) = 0\}, \text{ where} \\ D(x, A) &= inf\,\{\|x - a\| : a \in A\}, \end{aligned}$$

i.e., $\mathrm{cl}(A)$ is the set of all points x in X that are close to A ($D(x, A)$ is the Hausdorff distance [41, Sect. 22, p. 128] between x and the set A and $\|x, a\|$ is the Euclidean distance between x and a) (Fig. 4.7).

Let A^c denote the complement of A (all points of E not in A). The *boundary* of A (denoted $\mathrm{bdy}A$) is the set of all points that are near A and near A^c [42, Sect. 2.7, p. 62]. An important structure is the *interior* of A (denoted $\mathrm{int}A$), defined by $\mathrm{int}A = \mathrm{cl}A - \mathrm{bdy}A$. For example, the interior of a Delaunay edge $\overline{pq}$ are all of the points in the segment, except the endpoints p and q.

In general, a *relator* is a nonvoid family of relations $\mathcal{R}$ on a nonempty set X. The pair $(X, \mathcal{R})$ is called a relator space. Let E be endowed with the proximal relator

$$\mathcal{R}_\delta = \left\{\delta, \overset{\doublewedge}{\delta}, \underline{\delta}, \underset{\doublevee}{\delta}\right\} \text{ (Proximal Relator, cf. [31]).}$$

The Delaunay tessellated space E endowed with the proximal relator $\mathcal{R}_\delta$ (briefly, $\mathcal{R}$) is a *Delaunay proximal relator space*.

The proximity relations δ (near), $\overset{\doublewedge}{\delta}$ (strongly near) and their counterparts $\underline{\delta}$ (far) and $\underset{\doublevee}{\delta}$ (strongly far) facilitate the description of properties of Delaunay edges, triangles, triangulations and regions. Let $A, B \subset E$. The set A is near B (denoted $A\ \delta\ B$), provided $\mathrm{cl}A \cap \mathrm{cl}B \neq \emptyset$ [43]. The Wallman proximity δ (named after H. Wallman [44]) satisfies the four Čech proximity axioms [45, Sect. 2.5, p. 439] and is central in near set theory [42, 46]. Sets A, B are *far* apart (denoted $A\ \underline{\delta}\ B$), provided $\mathrm{cl}A \cap \mathrm{cl}B = \emptyset$. For example, Delaunay edges $\overline{pq}\ \delta\ \overline{qr}$ are near, since the edges have a common point, i.e., $q \in \overline{pq} \cap \overline{qr}$ (see, e.g., $\overline{pq}\ \delta\ \overline{qr}$ in Fig. 4.4). By contrast, edges $\overline{pr}, \overline{xy}$ have no points in common in Fig. 4.4, i.e., $\overline{pr}\ \underline{\delta}\ \overline{xy}$ (Fig. 4.8).

4.8.1: Poste auto

4.8.2: Poste Auto Delaunay-on-Voronoï image mesh covering

Fig. 4.8 Hunting ground for near sets: Delaunay on Voronoï mesh

4.3 Near and Strongly Near Sets Revisited

This section revisits a number of basic proximities. Let X be a finite topological space equipped with a proximal relator $\mathscr{R}_\delta = \left\{\delta, \delta_\Phi, \overset{\wedge\!\!\wedge}{\delta}, \overset{\wedge\!\!\wedge}{\delta}_\Phi, \not\delta, \underset{\vee\!\!\vee}{\not\delta}, \not\delta_\Phi, \underset{\vee\!\!\vee_\Phi}{\not\delta}\right\}$, $A, B \subset X$. Recall that a **relator** is a collection of relations [32–34]. An extension of the Szaz relator is in the form of a **proximal relator**, which is a collection of proximity relations [31]. This means that each of the proximities in the proximal relator $\mathscr{R}_\delta$ are defined on the space X. In other words, the space X is a realization of each of the relations in the proximal relator $\mathscr{R}_\delta$.

Lodato : Lodato proximity (denoted by $A\ \delta\ B$) [47–49].
$\boldsymbol{\delta_\Phi}$: **descriptive Lodato proximity** of nonempty sets (denoted by $A\ \delta_\Phi\ B$) [50, Sect. 4.15.2, p. 155] (see, also, Sect. 2.5).
$\overset{\wedge\!\!\wedge}{\delta}$: **strong proximity** of nonempty sets (denoted by $A\ \overset{\wedge\!\!\wedge}{\delta}\ B$) [51, 52].
$\overset{\wedge\!\!\wedge}{\delta}_\Phi$: **strong descriptive proximity** of nonempty sets (denoted by $A\ \overset{\wedge\!\!\wedge}{\delta}_\Phi\ B$).
$\not\delta$: **remoteness (farness)** of nonempty sets (denoted by $A\ \overset{\wedge\!\!\wedge}{\delta}_\Phi\ B$).
$\underset{\vee\!\!\vee}{\not\delta}$: **strong remoteness** of nonempty sets (denoted by $A\ \not\delta_\Phi\ B$).
$\not\delta_\Phi$: **descriptive remoteness** of nonempty sets (denoted by $A\ \not\delta_\Phi\ B$).
$\underset{\vee\!\!\vee_\Phi}{\not\delta}$: **strong descriptive remoteness** of nonempty sets (denoted by $A\ \not\delta_\Phi\ B$).

Fig. 4.9 Image geometry = Source of Lodato proximity space

4.3.1 Lodato Proximity Revisited

This section briefly revisits Lodato proximity δ.

Definition 4.7 Lodato Proximity.
Let X be a nonempty finite topological space. A *Lodato proximity* δ is a relation on X which satisfies the following properties for all subsets A, B, C of X:

(P0) $A\ \delta\ B \Rightarrow B\ \delta\ A$
(P1) $A\ \delta\ B \Rightarrow A \neq \emptyset$ and $B \neq \emptyset$
(P2) $A \cap B \neq \emptyset \Rightarrow A\ \delta\ B$
(P3) $A\ \delta\ (B \cup C) \Leftrightarrow A\ \delta\ B$ or $A\ \delta\ C$
(P4) $A\ \delta\ B$ and $\{b\}\ \delta\ C$ for each $b \in B\ \Rightarrow A\ \delta\ C$

Further δ is *separated*, if

(P5) $\{x\}\ \delta\ \{y\} \Rightarrow x = y$.

When we write $A\ \delta\ B$, we read "A is near to B", while when we write $A\ \not\delta\ B$ we read "A is far from B".

Example 4.8 **Lodato Proximity Space.**
Let Euclidean 2D-space be represented by $X = \{A, B, C, E, H\}$, the set of polygon regions in the mesh in Fig. 4.9. Assume that X is equipped with the Lodato proximity δ. It is easy to verify that proximity space (X, δ) satisfies the Lodato axioms. For Axiom (P4), Observe the following:

Axiom (P4) $A\ \delta\ B$ and $\{b\}\ \delta\ C$ for each $b \in B\ \Rightarrow A\ \delta\ C$, since $B \subset C$. ■

Conjecture 4.9 Ltd Lodato Proximity.
The relation δ is a Ltd Lodato proximity, if we change Axiom (P4) to Axiom (P4'):

(P4') $A\ \delta\ B$ and $\{b\}\ \delta\ C$ for some $b \in A \cap B\ \Rightarrow A\ \delta\ C$. ■

Problem 4.10 Prove or disprove Conjecture 4.9. If you think Conjecture 4.9 is false, then given a counterexample that demonstrates that at least one of the axioms (P0)–(P3) is not satisfied. If you think that Conjecture 4.9 holds, do two things:
1° Prove each of the axioms (P0)–(P3) is satisfied.
2° 🚲 Give a geometric example that illustrates Ltd Lodato Proximity. ■

4.3.2 Descriptive Lodato Proximity Revisited

This section briefly revisits descriptive Lodato proximity δ_Φ.

Definition 4.11 Descriptive Lodato Proximity [50, Sect. 4.15.2].
Let X be a nonempty finite topological space. A *Lodato proximity* δ is a relation on X which satisfies the following properties for all subsets A, B, C of X:

(dP0) $\emptyset \not\delta_\Phi A, \forall A \subset X$.
(dP1) $A \delta_\Phi B \Leftrightarrow B \delta_\Phi A$.
(dP2) $A \underset{\Phi}{\cap} B \neq \emptyset \Rightarrow A \delta_\Phi B$.
(dP3) $A \delta_\Phi (B \cup C) \Leftrightarrow A \delta_\Phi B$ or $A \delta_\Phi C$.
(dP4) $A \delta_\Phi B$ and $\{b\} \delta_\Phi C$ for each $b \in B \Rightarrow A \delta_\Phi C$. ■

Example 4.12 **Descriptive Lodato Proximity Space**.
Let Euclidean 2D-space be represented by $X = \{A, B, C, E, H\}$, the set of polygon regions in the mesh in Fig. 4.9. Assume that X is equipped with the Lodato proximity δ_Φ. Let the set of pixel feature Φ equal $\{\varphi\}$, where φ is a probe function that extracts colour intensity from an image pixel. It is easy to verify that proximity space (X, δ) satisfies the descriptive Lodato axioms. For Axioms (dP2) and (dP4), observe the following:

Axiom (dP2) $A \underset{\Phi}{\cap} E \neq \emptyset \Rightarrow A \delta_\Phi E$, since the description of some pixels $a \in A$ matches the description of at least one pixel in E.
Axiom (dP2) $C \underset{\Phi}{\cap} H \neq \emptyset \Rightarrow C \delta_\Phi H$, since the description of some pixels $c \in C$ matches the description of at least one pixel in H.
Axiom (dP4) $A \delta_\Phi E$ and $\{e\} \delta_\Phi B$ for each $e \in E \Rightarrow A \delta_\Phi B$, since the description of each pixel $e \in E$ matches the description of a pixel in B. ■

Conjecture 4.13 Etd Descriptive Lodato Proximity.
The relation δ_Φ is an extended (Etd) Lodato proximity, if we change Axiom (dP2) to Axiom (dP2'):

(dP2') $A \underset{\Phi}{\cap} B \neq \emptyset \Leftrightarrow A \delta_\Phi B$. ■

Problem 4.14 Prove or disprove Conjecture 4.13. If you think Conjecture 4.13 is false, then given a counterexample that demonstrates that at least one of the axioms (dP0)–(P4) is not satisfied. If you think that Conjecture 4.13 holds, do two things:
1° Prove each of the axioms (dP0)–(dP4) is satisfied.
2° Give a geometric example that illustrates Etd Descriptive Lodato Proximity. ■

Conjecture 4.15 Weak Descriptive Lodato Proximity.
The relation δ_Φ is weak Lodato proximity, if we change Axiom (dP4) to Axiom (dP4'):

(dP4') $A \delta_\Phi B$ and $\{b\} \delta_\Phi C$ for at least one $b \in B \Rightarrow A \delta_\Phi C$. ■

Problem 4.16 Prove or disprove Conjecture 4.15. If you think Conjecture 4.15 is false, then given a counterexample that demonstrates that at least one of the axioms (dP0)–(dP3) is not satisfied. If you think that Conjecture 4.15 holds, do two things:
1° Prove each of the axioms (dP0)–(dP4) is satisfied.
2° Give a geometric example that illustrates Weak Descriptive Lodato Proximity. ■

A *basic proximity* is one that satisfies $(P0)$–$(P3)$ introduced by E. Čech. *Lodato proximity* or *LO-proximity* is one of the simplest proximities.

4.3.3 Strong Proximity Revisited

Next, consider *strong proximity* [51] (see, also, [36, 46]) for subsets that at least have some points in common. This increased closeness of sets of points leads to strongly near proximity.

Definition 4.17 Let X be a topological space, $A, B, C \subset X$ and $x \in X$. We say that the relation $\overset{\wedge\wedge}{\delta}$ on $\mathcal{P}(X)$ is a *strong proximity*, provided that it satisfies the following axioms.

(N0) $\emptyset \not\overset{\wedge\wedge}{\delta} A, \forall A \subset X$, and $X \overset{\wedge\wedge}{\delta} A, \forall A \subset X$

(N1) $A \overset{\wedge\wedge}{\delta} B \Leftrightarrow B \overset{\wedge\wedge}{\delta} A$

(N2) $A \overset{\wedge\wedge}{\delta} B \Rightarrow A \cap B \neq \emptyset$

(N3) If $\text{int}(B)$ and $\text{int}(C)$ are not equal to the empty set, $A \overset{\wedge\wedge}{\delta} B$ or $A \overset{\wedge\wedge}{\delta} C \Rightarrow A \overset{\wedge\wedge}{\delta} (B \cup C)$

(N4) $\text{int}A \cap \text{int}B \neq \emptyset \Rightarrow A \overset{\wedge\wedge}{\delta} B$

When we write $A \overset{\wedge\wedge}{\delta} B$, we read A is *strongly near* to B. A **strong proximity space** is denoted by $(X, \overset{\wedge\wedge}{\delta})$ or simply by X, when it is understood that X is equipped with $\overset{\wedge\wedge}{\delta}$. If $A \subset X$ is an open set, then each point that belongs to A is strongly near A.

Example 4.18 A simple example of *strong proximity* that is also a *Lodato proximity* for nonempty sets A, B of a topological space X is $A \overset{\wedge\wedge}{\delta} B \Leftrightarrow A \cap B \neq \emptyset$. ■

Example 4.19 Another example of relation satisfying $(N0) - (N4)$ is given for nonempty sets A, B of a topological space X in Fig. 3.1, where $A \overset{\wedge\wedge}{\delta} B \Leftrightarrow \text{int}(A) \cap \text{int}(B) \neq \emptyset$ or either A or B is equal to X. If we want a strong proximity so that axioms $(N5)$, $(N6)$ also hold, we can take

$$A \overset{\wedge\wedge}{\delta} B \Leftrightarrow \text{int}A \cap \text{int}B \neq \emptyset \text{ or either } A \text{ or } B \text{ is equal to } X,$$

provided A and B are not singletons; if $A = \{x\}$, then $x \in \text{int}(B)$, and if B too is a singleton, then $x = y$. This is not a usual proximity. ■

Example 4.20 We can define another *strong proximity* for nonempty sets D, E of a topological space X in Fig. 4.10 in the following way: $D \overset{\wedge\wedge}{\delta} E \Leftrightarrow E \cap \text{int}(D) \neq \emptyset$ or $\text{int}(E) \cap D \neq \emptyset$, provided D and E are not singletons; if they are singletons, they are strongly near if and only if they coincide. ■

Example 4.21 Another example of *strong proximity* is given for nonempty sets $A, B \subset X$ in Fig. 4.10, where $A \overset{\wedge\wedge}{\delta} B \Leftrightarrow \text{int}(A) \cap \text{int}(B) \neq \emptyset$. ■

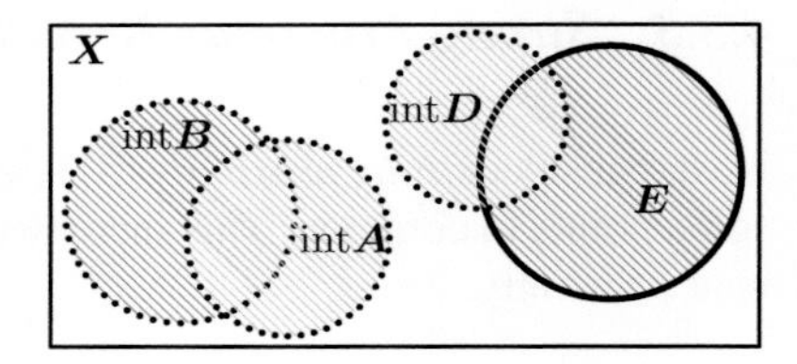

Fig. 4.10 $\mathrm{int}\boldsymbol{A}\ \overset{\wedge\!\!\wedge}{\delta}\ \mathrm{int}\boldsymbol{B},\ \mathrm{int}\boldsymbol{D}\ \overset{\wedge\!\!\wedge}{\delta}\ \boldsymbol{E}$

4.3.4 Descriptive Strong Proximity Revisited

In this section, we consider descriptive strong proximity in terms of image geometry. Let Φ denote a set of probes used to extract feature values for either a point or from a region. For a proximity space X with $A \subset X, a \in A$, we write $\Phi(a)$ to denote the feature vector of point a and $\Phi(A)$ to denote the set of feature vectors for all points in A. Recall the requirements for a descriptive strong proximity space.

Definition 4.22 Let X be a topological space, $A, B, C \subset X$ and $x \in X$. The relation $\overset{\wedge\!\!\wedge}{\delta}_{\Phi}$ on $\mathscr{P}(X)$ is a *descriptive strong proximity*, provided it satisfies the following axioms.

(dN0) $\varnothing \underset{\Phi}{\overset{\wedge\!\!\wedge}{\not\delta}} A, \forall A \subset X$, and $X \overset{\wedge\!\!\wedge}{\delta} A, \forall A \subset X$

(dN1) $A\ \overset{\wedge\!\!\wedge}{\delta}_{\Phi}\ B \Leftrightarrow B\ \overset{\wedge\!\!\wedge}{\delta}_{\Phi}\ A$

(dN2) $A\ \overset{\wedge\!\!\wedge}{\delta}_{\Phi}\ B \Rightarrow A \underset{\Phi}{\cap} B \neq \varnothing$

(dN3) If $\{B_i\}_{i\in I}$ is an arbitrary family of subsets of X and $A\ \overset{\wedge\!\!\wedge}{\delta}_{\Phi}\ B_{i^*}$ for some $i^* \in I$ such that $\mathrm{int}(B_{i^*}) \neq \varnothing$, then $A\ \overset{\wedge\!\!\wedge}{\delta}_{\Phi}\ (\bigcup_{i\in I} B_i)$

(dN4) $\mathrm{int}A \underset{\Phi}{\cap} \mathrm{int}B \neq \varnothing \Rightarrow A\ \overset{\wedge\!\!\wedge}{\delta}_{\Phi}\ B$ ■

When we write $A \overset{\wedge\!\!\wedge}{\delta} B$, we read A is *strongly near* B. The notation $A \overset{\wedge\!\!\wedge}{\not\delta} B$ reads A is not strongly near B. For each *descriptive strong proximity*, we assume the following relations:

(dN5) $\Phi(x) \in \Phi(\mathrm{int}(A)) \Rightarrow x\ \overset{\wedge\!\!\wedge}{\delta}_{\Phi}\ A$

(dN6) $\{x\}\ \overset{\wedge\!\!\wedge}{\delta}_{\Phi}\ \{y\} \Leftrightarrow \Phi(x) = \Phi(y)$ ■

Remark 4.23 **Descriptive Proximity versus Descriptive Strong Proximity**.
Put simply, nonempty sets have descriptive proximity, provided there is at least one point in each of the sets such that the points have matching descriptions. In other words, if we find point $a \in A$ and point $b \in B$ have matching descriptions, then $A\ \delta_{\Phi}\ B$. Recall that the feature vector for a point describes the point.

By contrast, nonempty sets have descriptive strong proximity, provided there is more than one point in each of the sets such that the points have matching descriptions.

Fig. 4.11 $C \overset{\wedge\!\!\wedge}{\delta}_{\Phi} H$ and $C \overset{\wedge\!\!\wedge}{\delta}_{\Phi} B$

In other words, if we find subset $X \subset A, |X| > 1$ and point $Y \subset B, |Y| > 1$ have matching descriptions, then $A \overset{\wedge\!\!\wedge}{\delta}_{\Phi} B$. That is, the feature vectors in $\Phi(X)$ match feature vectors in $\Phi(Y)$. ■

Example 4.24 **Wheels with Descriptive Strong Proximity**.
In Fig. 4.11 (extracted from the mesh in Fig. 4.9), the interior of the filled quadrilateral labelled C includes of part of rear wheel of the Poste auto. int$C$ has descriptive strong proximity with the interior filled quadrilateral H covering part of the front wheel of the Poste auto, since the dark areas as well as the silver areas have matching descriptions. Hence, int$C \overset{\wedge\!\!\wedge}{\delta}_{\Phi} H$. Similarly, notice that int$B \overset{\wedge\!\!\wedge}{\delta}_{\Phi} C$ and int$B \overset{\wedge\!\!\wedge}{\delta}_{\Phi} H$. Spatially, int$C$ is far from int$H$, i.e., int$C \not\delta H$. ■

4.4 Di Concilio–Gerla Overlap Revisited

This section briefly revisits Di Concilio–Gerla Overlap.

Definition 4.25 Di Concilio–Gerla Overlap Let A, B be nonempty subsets in a topological space X. The overlap relation $\mathscr{O}$ is defined by setting $A \mathscr{O} B$, provided that a region Z exists and is contained in both A and B [53]. ■

Remark 4.26 **Di Concilio–Gerla Overlap Issue Revisited**.
Earlier, it was pointed out that *overlap* is a good alternative descriptive of what we are calling strong proximity (see Remark 1.10 in Sect. 1.5). The overlap paradigm works well for spatial strong proximity, since spatially strongly proximal sets always share points, i.e., such sets are never disjoint. However, in the case of descriptive strong proximity between sets, the overlap paradigm fails to describe the situation that often arises with such sets. This is the case, since it is entirely possible that $\overset{\wedge\!\!\wedge}{\delta}_{\Phi}$-sets are both disjoint and yet have descriptive strong proximity. For example, in Fig. 4.11, C and H are disjoint and yet int$C$ and int$H$ are strongly near descriptively. ■

Conjecture 4.27 Di Concilio–Gerla Overlap Paradigm.
Under certain conditions, the Di Concilio overlap relation $\mathscr{O}$ (call it $\mathscr{O}^{+}$) describes (holds for) both $\overset{\wedge\!\!\wedge}{\delta}$ and $\overset{\wedge\!\!\wedge}{\delta}_{\Phi}$ proximities. ■

Problem 4.28 Do the following:

1° Give a set of axioms that the extended Di Concilio–Gerla Overlap relation $\mathcal{O}^+$ must satisfy so that Conjecture 4.27 holds. These axioms must cover both the spatial and descriptive cases, i.e., for non-disjoint and disjoint that overlap descriptively.

2° Prove $\mathcal{O}^+$ describes both $\overset{\wedge\wedge}{\delta}$ and $\overset{\wedge\wedge}{\delta}_\Phi$ proximities. ■

Problem 4.29 Do the following:

1° Select an appropriate set of seed points S and construct a Voronoï mesh on a digital image.

2° For the mesh in step 1, highlight adjacent $\overset{\wedge\wedge}{\delta}_\Phi$-regions in yellow and highlight in green disjoint $\overset{\wedge\wedge}{\delta}_\Phi$-regions.

3° Implement step 1 and step 2 in Mathematica.

4° Implement step 1 and step 2 in Matlab. ■

4.5 Wallman Proximity Revisited

Remark 4.30 **Nearness Expressions**.
The assumption made here is that δ is a Wallman proximity. This means that we require that close sets of points have at least one point in common. $A\ \delta\ B$ is an example of a nearness expression. Such an expression signals that A and B are close, i.e., A and B have at least one point in common. The strongly near proximity $\overset{\wedge\wedge}{\delta}$ indicates that set A has more than one point in common with the set B. If that is the case, then we write $A\ \overset{\wedge\wedge}{\delta}\ B$. See, for example, sets E and D in Fig. 4.12. ■

Voronoï regions are near, provided the regions have a common vertex. Voronoï regions V_p, V_q are *strongly near* (denoted $V_p\ \overset{\wedge\wedge}{\delta}\ V_q$) if and only if the regions have a common edge.

Example 4.31 **Strongly Near Voronoï regions**.
In Fig. 4.12, Voronoï regions D and E are near (denoted by $D\ \delta\ E$), since these

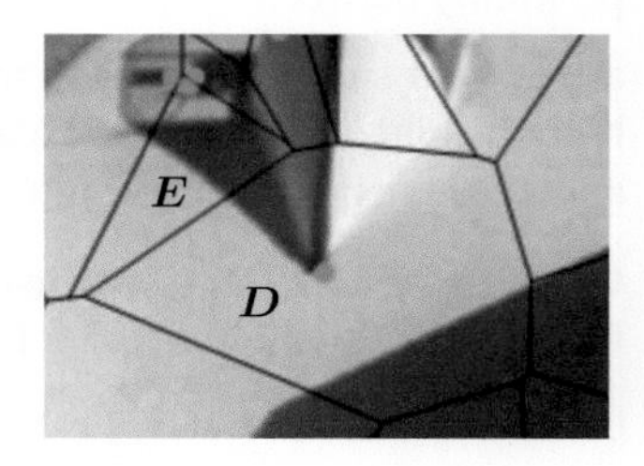

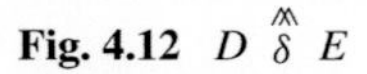
Fig. 4.12 $D\ \overset{\wedge\wedge}{\delta}\ E$

convex polygons have a vertex in common. In fact, the convex polygons D and E have an edge in common. Voronoï regions $D \stackrel{\wedge\!\wedge}{\delta} E$ in Fig. 4.12. ■

Theorem 4.32 *Near Voronoï regions are strongly near. That is, Voronoï regions that have a common vertex also have a common edge.*

Problem 4.33 ☕ Prove Theorem 4.32. If you think that Theorem 4.32 is not true, you must disprove it. To disprove Theorem 4.32, give a counterexample.
Hint: Each Voronoï region vertex lies on the border of a closed half plane halfway between a pair of sites used to construct an edge of the region. ■

Problem 4.34 Write a Mathematica script to do the following with your choice of three digital images:

1° Find the set of corners S in a digital image Im.
2° Construct a Voronoï mesh on Im.
3° Find the Voronoï region of a site $p \in S$ with the highest number of adjacent regions. A pair of Voronoï regions are *adjacent*, provided the regions have a common edge. There may be more than one region with a highest number of adjacent regions.
4° Use false-colouring to "paint" each region yellow that has the maximal number of adjacent regions. ■

N.B.: Identify the following things for each of the three digital images you choose: **source** (do not use images taken from the web. Instead, use digital images taken with your camera), type of camera used, image size.

Theorem 4.35 *Each Voronoï regions has a nonempty interior. That is, each Voronoï region inside its borders.*

Problem 4.36 ☕ Prove Theorem 4.35.
Hint: Consider the definition of a Voronoï region. Let E be the Euclidean plane. Let $p \in S$ (set of generators (also called sites). A *Voronoï region* of $p \in S$ (denoted V_p) is defined by

$$V_p = \left\{ x \in E : \|x - p\| \underset{\forall q \in S}{\leq} \|x - q\| \right\}.$$

■

Delaunay triangles are near (close), provided the triangles have a common vertex. Delaunay triangles are strongly near (very close), provided the triangles have a common edge.

Example 4.37 **Strongly Near Triangles**.
Delaunay triangles A and B are strongly near in Fig. 4.13, since A and B have a common edge. In that case, we write $A \stackrel{\wedge\!\wedge}{\delta} B$. ■

Theorem 4.38 *Delaunay triangles with a common edge are strongly near.*

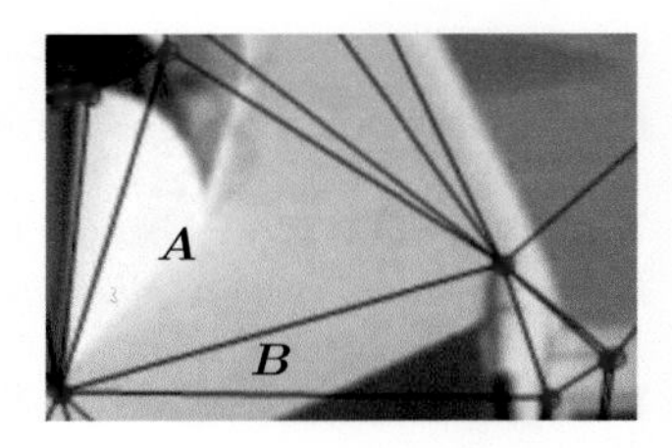

Fig. 4.13 $A \ \overset{\wedge\wedge}{\delta}\ B$

Problem 4.39 Prove Theorem 4.38.
Hint: Each Delaunay triangle vertex lies on a straight line segment between a pair of generating points (sites). The border of a closed half plane halfway between a pair of sites used to construct an edge of the region. ■

Theorem 4.40 *Every Delaunay triangle has an empty interior. That is, each Delaunay triangle has no points (belonging to the triangle) inside its borders.*

Problem 4.41 Prove Theorem 4.40.
Hint: Consider the definition of a Delaunay triangle. Let E be the Euclidean plane. Let $p, q \in S$ (set of generators (also called sites) used to construct Voronoï regions in E. The edge $\overline{pq}$ is a Delaunay triangle edge. ■

Problem 4.42 Write a Mathematica script to do the following with your choice of three digital images:

1° Find the set of corners S in a digital image Im.
2° Construct a corner-based Delaunay mesh on Im.
3° Find the Delaunay triangle with the highest number of adjacent regions. A pair of Delaunay triangles are *adjacent*, provided the regions have a common vertex. There may be more than one triangle with a maximal number of adjacent triangles.
4° Use false-colouring to "paint"each triangle yellow that has the highest number of adjacent triangles. ■

N.B.: Identify the following things for each of the three digital images you choose: **source** (do not use images taken from the web. Instead, use digital images taken with your camera), **image size**.

Problem 4.43 Write a Mathematica script to do the following with your choice of three digital images:

1° Find the set of corners S in a digital image Im.
2° Construct a corner-based Voronoï mesh on Im.
3° Find the Voronoï region with the highest number of adjacent regions. A pair of Voronoï regions are *adjacent*, provided the regions have a common edge. There may be more than one region with a maximal number of adjacent regions.
4° Find a digital image covered with a Voronoï diagram, where the diagram contains 2 or more Voronoï regions that (a) have the same number of adjacent regions and (b) the number of adjacent regions is maximal in each case.

5° Use false-colouring to "paint"each Voronoï region yellow that has the highest number of adjacent triangles. ■

N.B.: Identify the following things for each of the three digital images you choose: (i) subject, i.e., selfie, nature scene, machine, (ii) **source**: type of camera used (do not use images taken from the web. Instead, use digital images taken with your camera.), (iii) **image size**, (iv) date picture taken, (v) location that picture shows.

4.6 Far and Strongly Far Sets Revisited

Nonempty sets are far apart (not close), provided the sets have no points in common.

Nonempty sets A and C are strongly far apart (denoted $A \ \underset{\doublewedge}{\not\delta}\ C$), provided $C \subset \text{int}(\text{cl}B)$ and $A \ \underline{\delta}\ B$.

Example 4.44 **Far and Strongly Far Sets**.

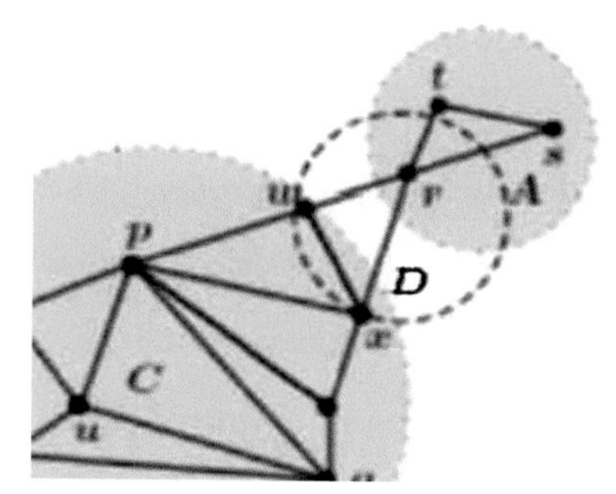

In the Delaunay mesh in Fig. 4.7, sets A and B have no points in common. Hence, $A \not\delta B$ (A is far from B). Also in Fig. 4.7, (see the image fragment at the beginning of this example) let $C = \{\triangle(pqu)\}$. Consequently, $C \subset \text{int}(\text{cl}B)$, such that triangle $\triangle(pqu)$ lies in the interior of the closure of B. Hence, $A \ \underset{\doublewedge}{\not\delta}\ C$. ■

Problem 4.45 Do the following:

(a) ☕ Give an example of two sets of Voronoï regions in a Voronoï mesh on a digital image of your own choosing that are strongly far apart.

(a) 🚲 Annotate the mesh by drawing a border around the sets to indicate the locations of the strongly far apart sets of Voronoï regions. Label the sets **A** and **B**. ■

Problem 4.46 Do the following:

(a) ☕ Give an example of two sets of Delaunay triangles in a Delaunay mesh on a digital image of your own choosing that are strongly far apart.

(a) 🚲 Annotate the mesh by drawing a border around the sets to indicate the locations of the strongly far apart sets of Delaunay triangles. Label the sets **D** and **E**. ■

References

1. Rosenfeld, A.: Distance functions on digital pictures. Pat. Rec. **1**(1), 33–61 (1968)
2. Rosenfeld, A.: Digital topology. Am. Math. Mon. **86**(8), 621–630 (1979). Am. Math. Soc. MR0546174
3. Rosenfeld, A.: Digital Picture Analysis. Springer, Berlin (1976). Xi + 351 pp
4. Rosenfeld, A., Kak, A.: Digital Picture Processing, vol. 1. Academic Press, New York (1976). Xii + 457 pp
5. Rosenfeld, A., Kak, A.: Digital Picture Processing, vol. 2. Academic Press, New York (1982). Xii + 349 pp
6. Kong, T., Roscoe, A., Rosenfeld, A.: Concepts of digital topology. Special issue on digital topology. Topol. Appl. **46**(3), 219262 (1992). Am. Math. Soc. MR1198732
7. Klette, R., Rosenfeld, A.: Digital Geometry Geometric Methods for Digital Picture Analysis. Morgan-Kaufmann Publisher, Amsterdam (2004)
8. Latecki, L.: Topological connectedness and 8-connectedness in digital pictures. Comput. Vis. Graph. Image Process. **57**, 261–262 (1993)
9. Latecki, L., Conrad, C., Gross, A.: Preserving topology by a digitization process. J. Math. Image Vis. **8**, 131–159 (1998)
10. Eckhardt, U., Latecki, L.J.: Topologies for the digital spaces $\mathbb{Z}^2$ and $\mathbb{Z}^3$. Comput. Vis. Image Underst. **90**(3), 295–312 (2003)
11. Kronheimer, E.: The topology of digital images. Special issue on digital topology. Topol. Appl. **46**(3), 279–303 (1992). MR1198735
12. Kong, T., Rosenfeld, A.: Topological Algorithms for Digital Image Processing. North-Holland, Amsterdam (1996)
13. Shamos, M.: Computational geometry. Ph.D. thesis, Yale University, New Haven, Conn., USA (1978). Supervisors: D. Dobkin, S. Eisenstat, M. Schultz
14. Preparata, F.: Steps into computational geometry. Coordinated science laboratory, University of Illinois, Technical report (1977)
15. Preparata, F.: Convex hulls of finite sets of points in two and three dimensions. Commun. Assoc. Comput. Mach. **2**(20), 87–93 (1977)
16. Voronoï, G.: Nouvelles applications des paramètres continus à la théorie des formes quadratiques. J. für die reine und angewandte Math. 134, 198–287 (1908). JFM 39.0274.01
17. Voronoi, G.: Sur une fonction transcendante et ses applications à la sommation de quelque séries. Ann. Sci. Ecole Norm. Sup. **21**(3) (1904)
18. M.L. Gavrilova, E.: Generalized Voronoi Diagrams: A Geometry-Based Approach to Computational Intelligence. Springer, Berlin (2008). Xv + 304 pp
19. Grünbaum, B., Shepherd, G.: Tilings with congruent tiles. Bull. Am. Math. Soc. **3**(3), 951–973 (1980). (New Series)
20. Gardner, M.: On tessellating the plane with convex polygon tiles. Sci. Am. 116–119 (1975)
21. Toussaint, G.: Computational geometry and morphology. In: Kato, Y., et al. (eds.) Proceedings of the First International Symposium for Science on Form, pp. 395–403. KTK Scientific Publishers, Tokyo (1986)
22. Chan, M.: Topical curves and metric graphs. Ph.D. thesis, University of California, Berkeley (2012). Supervisor: B. Sturmfels
23. Lai, R.: Computational differential geometry and intinsic surface processing. Ph.D. thesis, University of California, Los Angeles (2010). Supervisors: T.F. Chan, P. Thompson, M. Green, L. Vese
24. Beer, G., Di Concilio, A., Di Maio, G., Naimpally, S., Pareek, C., Peters, J.: (Somashekhar Naimpally, 1931–2014). Topology and its Applications **188**, 97–109 (2015). doi:http://dx.doi.org/10.1016/j.topol.2015.03.010, MR3339114
25. Naimpally, S., Peters, J.: Preservation of continuity. Scientiae Mathematicae Japonicae **76**(2), 305–311 (2013). MR3330078
26. McCoy, R.: Topological Properties of Spaces of Continuous Functions. Springer, Berlin (1988). iv+124 pp. ISBN: 3-540-19302-2, MR0953314

27. Di Maio, G., Holá, L., Holý, D., McCoy, R.: Topologies on the space of continuous functions. Topol. Appl. **86**, 105–122 (1998). MR1621396
28. Du, Q., Faber, V., Gunzburger, M.: Centroidal voronoi tessellations: applications and algorithms. SIAM Rev. **41**(4), 637–676 (1999). MR
29. Ma, H., Zhu, J., Lyu, M., King, I.: Bridging the semantic gap between image contents and tags. J. LaTeX Class Files **6**(1), 1–12 (2007)
30. [Delaunay], B.D.: Sur la sphère vide. Izvestia Akad. Nauk SSSR, Otdelenie Matematicheskii i Estestvennyka Nauk **7**, 793–800 (1934)
31. Peters, J.: Proximal Relator Spaces. Filomat (2016). accepted
32. Száz, A.: Basic tools and mild continuities in relator spaces. Acta Math. Hungar. **50**, 177–201 (1987)
33. Száz, A.: An extension of Kelley's closed relation theorem to relator spaces. FILOMAT **14**, 49–71 (2000)
34. Száz, A.: Applications of relations and relators in the extensions of stability theorems for homogeneous and additive functions. The Australian J. Math. Anal. Appl. **6**(1), 1–66 (2009)
35. Edelsbrunner, H.: A Short Course in Computational Geometry and Topology, p. 110. Springer, Berlin (2014)
36. Peters, J.: Proximal Voronoï regions, convex polygons, and Leader uniform topology. Adv. Math. Sci. J. **4**(1), 1–5 (2015)
37. Edelsbrunner, H.: Geometry and Topology of Mesh Generation. Cambridge University Press, Cambridge (2001)
38. Voronoï, G.: Sur un problème du calcul des fonctions asymptotiques. J. für die reine und angewandte Math. **126**, 241–282 (1903)
39. Voronoï, G.: Nouvelles applications des paramètres continus à la théorie des formes quadratiques. J. für die reine und angewandte Math. **133**, 97–178 (1907). JFM 38.0261.01
40. Peters, J., Guadagni, C.: Strong proximities on smooth manifolds and Voronoï diagrams. Adv. Math. Sci. J. **4**(2), 91–107 (2015)
41. Hausdorff, F.: Grundzüge der Mengenlehre. Veit and Company, Leipzig (1914). Viii + 476 pp
42. Naimpally, S., Peters, J.: Topology with Applications. Topological Spaces via Near and Far. World Scientific, Singapore (2013). Xv + 277 pp, Am. Math. Soc. MR3075111
43. Di Concilio, A.: Proximity: a powerful tool in extension theory, functions spaces, hyperspaces, boolean algebras and point-free geometry. In: F. Mynard, E. Pearl (eds.) Beyond Topology, AMS Contemporary Mathematics, vol. 486, pp. 89–114. American Mathematical Society, Providence (2009)
44. Wallman, H.: Lattices and topological spaces. Ann. Math. **39**(1), 112–126 (1938)
45. Čech, E.: Topological Spaces. Wiley, London (1966). Fr seminar, Brno, 1936–1939; rev. ed. Z. Frolik, M. Katětov
46. Peters, J., Naimpally, S.: Applications of near sets. Notices Am. Math. Soc. **59**(4), 536–542 (2012). doi:http://dx.doi.org/10.1090/noti817, MR2951956
47. Lodato, M.: On topologically induced generalized proximity relations, Ph.D. thesis. Rutgers University (1962). Supervisor: S. Leader
48. Lodato, M.: On topologically induced generalized proximity relations i. Proc. Am. Math. Soc. **15**, 417–422 (1964)
49. Lodato, M.: On topologically induced generalized proximity relations ii. Pac. J. Math. **17**, 131–135 (1966)
50. Peters, J.: Topology of Digital Images—Visual Pattern Discovery in Proximity Spaces, Intelligent Systems Reference Library, vol. 63. Springer, Heidelberg. Xv + 411 pp, Zentralblatt MATH Zbl **1295**, 68010 (2014)
51. Peters, J.: Visibility in Proximal Delaunay Meshes. pp. 1–5 (2015). arXiv:1501.02357v1 [Math.MG]
52. Peters, J., Guadagni, C.: Strongly Near Proximity and Hyperspace Topology, pp. 1–6 (2015). arXiv:1502.05913v3
53. Di Concilio, A., Gerla, G.: Quasi-metric spaces and point-free geometry. Math. Struct. Comput. Sci. **16**(1), 115137 (2006). MR2220893

Chapter 5
Homotopic Maps, Shapes and Borsuk–Ulam Theorem

Fig. 5.1 Karol Borsuk

This chapter introduces object spaces, where objects are located in a visual field, which are viewed in the context of the Borsuk–Ulam Theorem,[1] elementary homotopy theory and the equivalence of shapes. The objects we have in mind are those found in 2D digital images, viewed as regions containing concrete points with

[1]Many thanks to Andrzej Skowron, who contributed the picture of Karol Bursuk in Fig. 5.1.

J.F. Peters, *Computational Proximity*, Intelligent Systems
Reference Library 102, DOI 10.1007/978-3-319-30262-1_5

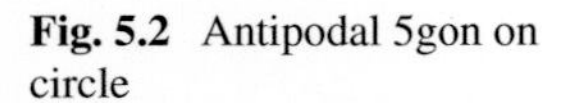
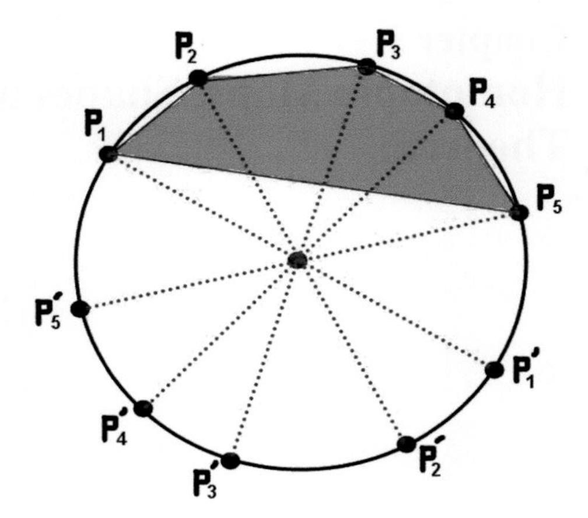

Fig. 5.2 Antipodal 5gon on circle

measurable features such as colour, gradient direction, gradient magnitude in the x-direction, and so on. The regions themselves can be probed to extract measurable region features such density, area, as well as colour and gradient magnitude. It also introduces feature spaces, containing descriptions of objects in Euclidean space $\mathbb{R}^n$. Let $\mathscr{O}$ denote a planar object space, respectively. Then $\mathcal{O} \mapsto \mathbb{R}^n$, i.e., the object space maps to an n-dimensional feature space.

5.1 Antipodal Points and Hyperspheres

Continuous mappings from object spaces to feature spaces lead to various incarnations of the Borsuk–Ulam Theorem, which is a remarkable finding about continuous mappings from antipodal points on an n-sphere to n-dimensional Euclidean space, found by Karol Borsuk [1].

Given a point p on circle, draw a line passing through p and the center o of the circle. The **antipode** p' of p is on the circle through which the line $\overline{po}$ passes through o [2]. The antipode p' is said to **mirror** the point p.

Let S be a set of $2n$ points on a circle such that, for each $p \in S$, the **antipodal** $p' \in S$ of p is mirrored with respect to the center of the circle. An **antipodal polygon** on S is a convex polygon having as vertices precisely one point from each pair of antipodal points (p, p') of S [3]. The points p_1, p_2, p_3, p_4, p_5 are vertices in a antipodal 5gon on the circle shown in Fig. 5.2.

For example, the pairs of vertices $(p_1, p_1^{'}), (p_2, p_2^{'}), (p_3, p_3^{'}), (p_4, p_4^{'}), (p_5, p_5^{'})$ of the convex 5gon in Fig. 5.2 are **antipodal points** which are the endpoints of line segments passing through the center of a circle or on what is known as a 2-sphere.

Problem 5.1 🚲 Give a sketch of an antipodal 5gon on the surface of 3D sphere. ■

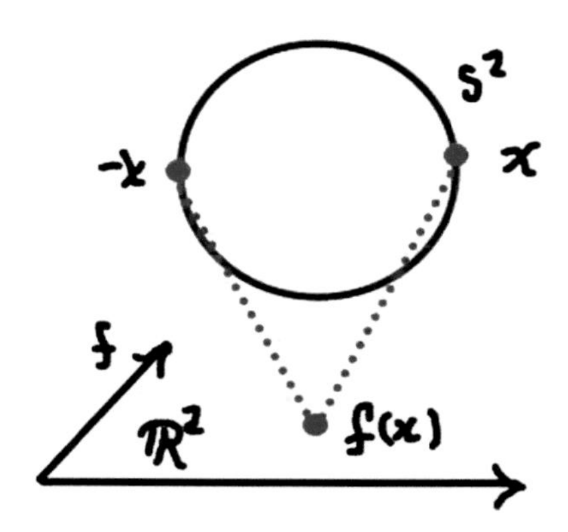

Fig. 5.3 $f : S^2 \longrightarrow \mathbb{R}^2$

An n-sphere (also called an ***n*-hypersphere**) is a generalization of the circle, which geometers call a 2-sphere. The usual sphere in 3-dimensional Euclidean space R^3 is called a **3-sphere** [4]. An n-sphere S^n with radius r is defined by

$$S^n = \left\{(x_1, \ldots, x_n) \in \mathbb{R}^n : x_1^2 + \cdots + x_n^2 = r^2\right\} \text{ (n-sphere in Euclidean space $\mathbb{R}^n$).}$$

A good introduction to S^1, S^2 and S^n spheres from a topologist's perspective is given by J.R. Munkres [5, pp. 106, 139, 156]. There are natural ties between Borsuk's result for antipodes and mappings called homotopies. The early work on n-spheres and antipodal points eventually led Borsuk to the study of retraction mappings and homotopic mappings [6–8].

The hypersphere S^n is a generalization of the circle. From a geometer's perspective, an ***n*-sphere** with radius R is a set of n-tuples of points $(x_1, \ldots, x_n)$, with

$$\begin{aligned}
\text{2-sphere} : x_1^2 + x_2^2 &= R^2, \text{ hypersphere } S^2, x \in S^2 \text{ is vector } (x_1, x_2)\\
&\quad \text{on a circle perimeter,}\\
\text{3-sphere} : x_1^2 + x_2^2 + x_3^2 &= R^2, \text{ hypersphere } S^3, x \in S^3 \text{ is a vector } (x_1, x_2, x_3)\\
&\quad \text{on the surface of the 3-sphere } S^3,\\
\text{4-sphere} : x_1^2 + x_2^2 + x_3^2 + x_4^2 &= R^2 \text{ hypersphere } S^4, x \in S^4 \text{ is a}\\
&\quad \text{a surface vector } (x_1, x_2, x_3, x_4) \text{ on } S^4,\\
&\quad \vdots\\
n\text{-sphere} : x_1^2 + \cdots + x_n^2 &= R^2 \text{ hypersphere } S^n, x \in S^n \text{ is a vector } (x_1, \ldots, x_n)\\
&\quad \text{on the surface of the n-sphere } S^n.
\end{aligned}$$

For $n \geq 4$, an n-sphere is a **hypersphere**. S^4 is the smallest hypersphere. Points are **antipodal**, provided the points are diametrically opposite.[2] Examples are the

[2]E.W. Weisstein, antipodal points, http://mathworld.wolfram.com/AntipodalPoints.html.

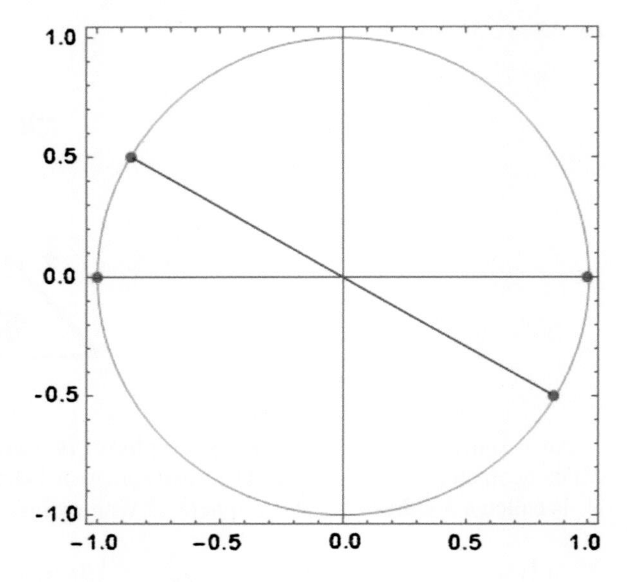

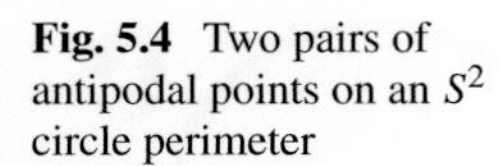
Fig. 5.4 Two pairs of antipodal points on an S^2 circle perimeter

endpoints of a line segment or opposite points along the circumference of a circle or poles of a sphere.

Example 5.2 **Antipodal Points**.
Two pairs of antipodal points are shown with red • dots in Fig. 5.4, namely,

Antipodes I: $p1 = (-0.86, 0.5)$; $p2 = (0.86, -0.5)$.
Antipodes II: $p5 = (-1.0, 0.0)$; $p8 = (1.0, 0.0)$.

To experiment with other examples of antipodal points, see MScript 34 in Appendix A.5. ■

5.2 Borsuk–Ulam Theorem

Borsuk–Ulam Theorem: Basic Idea.
Every continuous map $f : S^n \longrightarrow \mathbb{R}^n$ must identify a pair of antipodal points [9, p. 121]. See, for example, in Fig. 5.3, the antipodal points on the circle perimeter S^2 mapped to a single point in the Euclidean plane $\mathbb{R}^2$. For recent applications of the Borsuk–Ulam Theorem in the study of brain activity, see [10] and in quantum entanglement on a hypersphere [11]. ■

The Borsuk–Ulam Theorem is given in the following form by M.C. Crabb and J. Jaworowski [12].

Theorem 5.3 Borsuk–Ulam Theorem [1].
Let $f : S^n \longrightarrow \mathbb{R}^n$ *be a continuous map. There exists a point* $x \in S^n \subseteq \mathbb{R}^{n+1}$ *such that* $f(x) = f(-x)$.

Proof A proof of this form of the Borsuk–Ulam Theorem follows from a result given in 1930 by L.A. Lusternik and L. Schnirelmann [13]. For the backbone of the proof, see Remark 3.4 in [12]. □

Let S^n denote the unit n-sphere in $\mathbb{R}^{n+1}$, i.e., all points at distance 1 from the origin. Then the Borsuk–Ulam Theorem is stated by F.E. Su [14] in the following way.

Theorem 5.4 Borsuk–Ulam Theorem [1].
Let $f : S^n \longrightarrow \mathbb{R}^n$ *be a continuous map. There exists a pair of antipodal points on* S^n *that are mapped to the same point in* $\mathbb{R}^n$.

Proof A direct proof of the Borsuk–Ulam Theorem is given by F.E. Su [14]. □

A digital form of the Borsuk–Ulam Theorem is given and proved by G. Burak and I. Karaca [15]. For related work on a fixed point theorem for digital images, see O. Ege and I. Karaca [16].

Theorem 5.5 Digital Borsuk–Ulam Theorem.
If $f : (S^n, \kappa) \longrightarrow \mathbb{Z}^n$ *is continuous for* $n = 1, 2$, *for* $\kappa = 4$ *for* S^1, $\kappa = 6$ *for* S^2, *then there exists* $x \in S^n$ *with* $f(x) = f(-x)$.

> In a point-free geometry, the focus shifts from antipodal points to antipodal regions. ■

Let A be a region in a topological space X. The **antipode of a region** A is its complement, namely, $A^c = X \setminus A$. That is, A and A^c are antipodal regions. In the sequel, this leads to new forms of the Borsuk–Ulam Theorem. This also leads to a new perspective on object recognition. Objects belonging to antipodal regions can either be different or similar, depending on the features of objects.

The antipodal region paradigm gives a means of separating objects in terms of the antipodal regions they belong to. This paradigm gives rise to antipodal objects. **Antipodal objects** are objects that belong to antipodal regions.

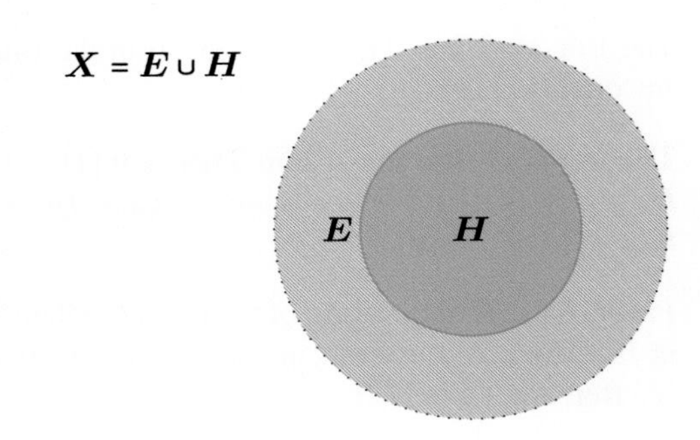

Fig. 5.5 Concentric concrete 2-spaces with shades of *green*

Example 5.6 **Antipodal Regions**.
A pair of antipodal regions is shown in Fig. 5.5, namely,

Antipodal region: $H \subset X$.
Antipodal region: $E = X \setminus H = H^c$.

To experiment with other examples of antipodal regions, see MScript 35 in Appendix A.5. ■

Theorem 5.7 *Antipodal regions are disconnected.*

Let X, Y be topological spaces. Recall that a **function** or **map** $f : X \longrightarrow Y$ on a set X to a set Y is a subset $X \times Y$ so that for each $x \in X$ there is a unique $y \in Y$ such that $(x, y) \in f$ (usually written $y = f(x)$). The mapping f is defined by a rule that tells us how to find $f(x)$. For a good introduction to mappings, see S. Willard [17, Sect. 1.6, pp. 2–3].

A mapping $f : X \longrightarrow Y$ is **continuous**, provided, when $A \subset Y$ is open, then the inverse $f^{-1}(A) \subset X$ is also open. For more about this, see [18, Sect. 3]. In this chapter, a proximal view of the Borsuk–Ulam Theorem is given based on strong proximal continuity.

5.3 Homotopic Maps and Deformation Retracts

An interest in continuous mappings from object spaces to feature spaces leads into homotopy theory and the study of shapes.

Let $f, g : X \longrightarrow Y$ be continuous mappings from X to Y. The continuous map $H : X \times [0, 1] \longrightarrow Y$ is defined by

$$H(x, 0) = f(x),$$
$$H(x, 1) = g(x), \text{ for every } x \in X.$$

The mapping H is a **homotopy**, provided there is a continuous transformation (called a deformation) from f to g. The continuous maps f, g are called **homotopic maps**, provided $f(X)$ continuously deforms into $g(X)$ (denoted by $f(X) \searrow g(X)$). The sets $f(X), g(X)$ are called shapes. For more about this, see M. Manetti [19, p. 121] and the introduction by M.M. Cohen [20]. ■

For the mapping $H : X \times [0, 1] \longrightarrow \mathbb{R}^n \setminus 0$, $H(X, 0)$ and $H(X, 1)$ are homotopic, provided $f(X)$ and $g(X)$ have the same shape. That is, $f(X)$ and $g(X)$ are homotopic, provided

$$\|f(X) - g(X)\| < \|f(X)\|, \text{ for all } x \in X.$$

It was K. Borsuk who first associated the geometric notion of shape and homotopies. This leads into the geometry of shapes and shapes of space [21]. A good introduction to homotopy theory is given by J.R. Munkres [5].

A pair of connected planar subsets in $\mathbb{R}^2$ have equivalent shapes, provided the planer sets have the same number of holes [19, p. 121]. For example, the letters

e, O, P and numerals 6, 9

belong to the same equivalence class of single-hole shapes.

Example 5.8 **Homotopy on Shapes With Holes.**
Let $f, g : X \longrightarrow Y$ be continuous mappings from X to Y. The continuous map $H : X \times [0, 1] \longrightarrow Y$ is defined by

$$H(X, 0) = f(X) = P,$$
$$H(X, 1) = g(X) = 6.$$

In this case,

P and 9

are shape equivalent. Hence, H is a homotopy and f, g are homotopic mappings. See MScript 36 in Appendix A.5 for the implementation of a homotopic mapping from the question mark ? to J. ■

Example 5.9 **Shapes Without Holes.**
This example is derived from [19, Ex. 4.8]. A subset A in Euclidean space $\mathbb{R}^2$ is **star shaped** with respect to the point $a \in A$, provided the straight segment $\overline{ab}$ joining a and b is contained in A for every $b \in A$. The letter X (drawn on the Euclidean plane) and star-shape A are examples of equivalent shapes, since both shapes have no holes. ■

Problem 5.10 🚲 Give a pair of planar shapes that have no points in common and that are not equivalent, i.e., non-equivalent, disconnected shapes drawn on a plane surface. ■

Problem 5.11 🚲 Give a pair of planar shapes that have no points in common and that are equivalent, i.e., equivalent, disconnected shapes drawn on a plane surface. ■

Problem 5.12 ☕ Give an example of homotopy on a connected space X (i.e., all subsets in X are connected). ■

The notion of **homotopy type** intuitively corresponds to equivalent shapes. Objects that map to equivalent shapes belong to the same class of objects. In effect, a homotopy type offers an effective means of classifying objects.

Homotopy equivalence is defined in terms of identity maps. In general, an *identity map* $id : X \longrightarrow X$ is defined by $id(x) = x$ for each $x \in X$. The identify map id_X is also denoted by 1_X [20].

Definition 5.13 Homotopy Equivalence [19, Sect. 10.3].
A continuous map $f : X \longrightarrow Y$ is a **homotopy equivalence**, provided there exists a continuous map $g : X \longrightarrow Y$ such that the composition $f \circ g$ is homotopic to the identify map of Y ($id_Y : Y \longrightarrow Y$) and the composition $g \circ f$ is homotopic to the identify map of X ($id_X : X \longrightarrow X$). A pair of topological spaces X, Y are **homotopy equivalent**, provided X and Y have a homotopy equivalence between them. ■

A pair of shapes (closed sets of simplexes) are **homotopic**, provided one shape can be deformed into the other via a homotopic mapping. An interest in homotopic mappings has led to the study of homotopies in 2D digital images [22].

Problem 5.14 ☕ Prove that the letter j and the question mark ? drawn on the Euclidean plane are equivalent shapes. ■

Example 5.15 The pair of 2-spheres (green disks) H, E in Fig. 5.5 are homotopic, since E can be deformed into H. ■

The 2-sphere H in Fig. 5.5 is an example of a deformation retract of E.

Definition 5.16 Deformation Retract [19, Sect. 10.3].
Let I be the closed interval [0,1] in $\mathbb{R}$ (reals). A subspace $Y \subset X$ is a **deformation retract** of X, provided there is a continuous map $H : X \times I :\longrightarrow Y$ (**deformation** of X into Y), and provided the following conditions are satisfied:

1° $R(x,0) \in Y$ and $R(x,1) = x$ for every $x \in X$.
2° $R(y,t) = y$ for every $y \in Y, t \in I$.

Example 5.17 **Star Shape Center is a Deformation Retract.**
From Example 5.9, let the subset A in Euclidean space $\mathbb{R}^2$ be **star shaped** with respect to the center point $p \in A$ so that every straight line segment $\overline{pb}$ joining p and b is contained in A for every $b \in A$. In this case, $\{p\}$ (or just p) is the deformation retract of A into $\{p\}$ is defined by
1° $R(a,0) \in \{p\}$ and $R(a,1) = a$ for every $a \in A$.
2° $R(a,t) = ta + (1-t)p$ for every $a \in A, t \in [0,1]$.
■

Example 5.18 **Deformation Retractable Disks.**
The concentric disks E, H define an object space X (a finite topological space) with a corresponding feature space $\Phi(X)$ (also a finite topological space), the set

Fig. 5.6 A bit more, Punch, 1845

of feature vectors that describe concrete points in X. For example, assume X is a picture containing the overlapping disks.

In this example, each point $x \in X$ is described by a feature vector containing the red, green and blue intensities of x. Let $\{\emptyset, \{g\}, \{r, g, b\}\}$ (red, green, blue colours) be the topology on the feature space Φ_E, which contains the feature space Φ_H. The mappings $f, g : X \longrightarrow \mathbb{R}^3$ (colour space for Φ_E and Φ_H) are continuous.

The homotopy $h : X \times [0, 1] \longrightarrow \mathbb{R}^n$ is defined by $h(x, 0) = \Phi_E(x)$ and $h(x, 1) = \Phi_H(x)$. The pair of mappings f, g for the 2-spheres (green disks) in Fig. 5.5 are homotopic, since the disk defined by Φ_E can be deformed into the disk defined by Φ_H. In fact, disk H is the nucleus of E. ■

Problem 5.19 Mimic the approach in Example 5.17 and give the formulas for a deformation retraction of the green disk Φ_E into the green disk Φ_H from Example 5.18 (Using the same approach, find a deformation retraction from the Serf's foot to one of the Knight's feet in Fig. 5.6). ■

5.4 Object Description, Structured Sets and Shape Features

A nonempty set X endowed with a nearness relation is a **structured set**. The choice of an appropriate nearness relation in defining a structured set is a bit like choosing how far to lower the knight in Fig. 5.7, so that the knight fits in the saddle. A good choice for the knight leads to a good ride. For example, by choosing the descriptive proximity δ_Φ, it would then be possible to ensure that the geometric features of the knight match the geometric features of the knight's saddle by making sure that the shape of the knight and the shape of the saddle are descriptively near. In terms of set structures, there are many choices and a good choice makes it easier to find interesting nearness patterns.

5.4.1 Euclidean Feature Space

A **feature space** is a finite-dimensional normed linear space. The Euclidean space R^n is an example. Let X be the ordinary Euclidean space R^m and let Y be the Euclidean space R^n that is the feature space for X with $m \leq n$. The set X is called an **object space**. For example, planar shapes (regions) in R^2 have descriptions that are region-based feature vectors in R^n. Again, for example, individual points in R^2 have descriptions that are point-based feature vectors in R^n.

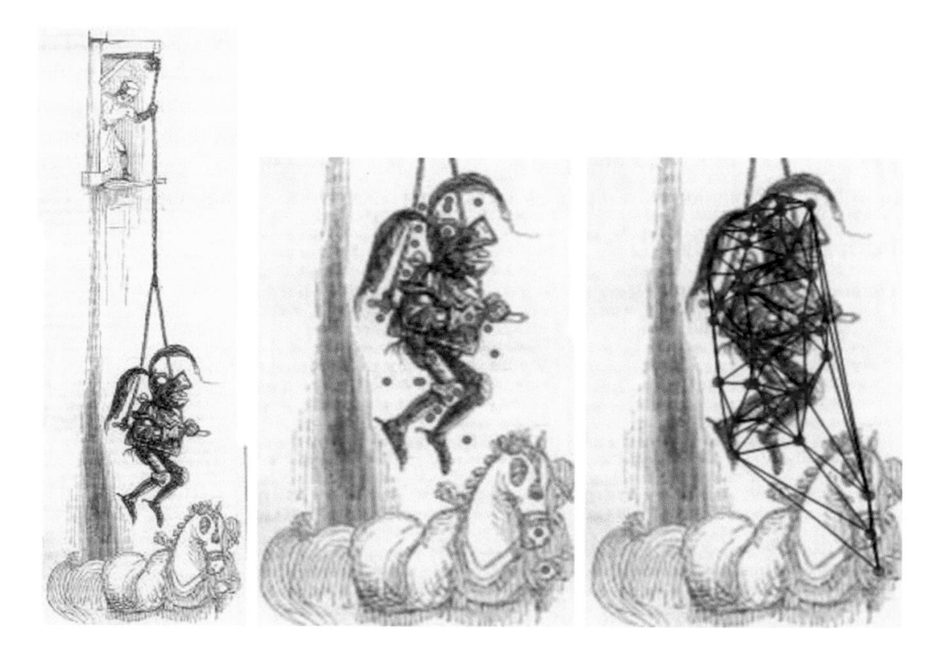

Fig. 5.7 50 knight corners

> **Object Space Axioms.** Every object space has a corresponding feature space. Every region in an object space corresponds to a region-based feature vector in a feature space. Every point in an object space corresponds to a point-based feature vector in a feature space.

Example 5.20 **Pixels and Pixel Features.**
Pixels in a raster image are examples of objects, each with its own location. In most images, there are pixels that are corners, where the edges on either side of a corner pixel change direction, sharply. Cornerness is an example of a pixel feature. The image in Fig. 5.7 highlights (with • bullets) some of the pixels that are corners. ■

Points (regions) in a feature space Y are descriptions of points (regions) in an object space X. Let $x \in X$, $y \in \mathbb{R}^n$. In effect, $x \mapsto y$, i.e., each point $x \in X$ maps to a point $y \in \mathbb{R}^n$ (called a feature vector). Let $\phi : X \to \mathbb{R}$ be a real-valued **point-based probe** such that $\phi(x)$ is a feature value of a point $x \in X$. Let $\varphi : 2^X \to \mathbb{R}$ be a real-valued **region-based probe** such that $\varphi(Re)$ is a feature value of a region

$\text{Re} \subset X$. That is, $\Phi(\text{Re})$ or $y \in \mathbb{R}^n$ is a feature vector of the region $\text{Re} \in 2^X$. The coordinates of a feature vector are probe function values. A feature vector in the feature space $\mathbb{R}^n$ is a **description** of an object in X. Objects are identified by their locations in the object space and known by their descriptions in the feature space. The **location of region** is defined by the coordinates of its centroid (center of mass). For more about region-based probes and region-based descriptions, see Sect. 6.10.

Mscript 5

(*Obtain corners of an image*)

img =  ;

corncoords = ImageCorners[img, MaxFeatures → 50];
HighlightImage[img, corncoords] ■

Example 5.21 **Corner Probe Function.**
Let img be an RGB colour image (a raster image). And let $\phi(p)$ be a probe function that returns an intensity for a corner pixel $p \in img$. **Cornerness** is an image pixel feature. For example, let ϕ be defined by the Mathematica10 function **ImageCorners**, identifies the corner pixels in an image. By way of illustration, consider the Punch drawing of a knight begin lowered onto his horse in Fig. 5.7. The Mscript 5 is used to find up to 50 corners in any raster image. The corners found in the drawing of the knight are shown in Fig. 5.8. In this case, the object space consists of digital image pixels and the vectors in the feature are descriptions of image pixels. ■

5.4.2 Feature Vectors

A **feature vector** is a vector (denoted by $\Phi(y) \in \mathbb{R}^n$ or simply by $y \in \mathbb{R}^n$ in a feature space) with coordinates that are probe function values of a point x or a region Re in a object space X.

Definition 5.22 Point-Based Feature Vector.
A point-based feature vector is defined by

$$\begin{aligned}
X &= \text{ Euclidean object space } \mathbb{R}^m.\\
Y &= \text{ Euclidean feature space } \mathbb{R}^n, m \leq n.\\
&x \in X, y \in Y.\\
&\phi : X \longrightarrow \mathbb{R}, \text{ (objects map to feature values).}
\end{aligned}$$

Fig. 5.8 50 knight corners

$$\phi(x) = \text{ probe function feature value of } x.$$
$$\Phi : X \longrightarrow Y, \text{ (mapping from object space to feature space).}$$
$$\Phi(x) = (\phi_1(x), \ldots, \phi_n(x)) \text{ feature vector for object } x \text{ with } n \text{ features.}$$

■

Object-to-Feature Mapping. To separate data sets effectively, it is common to consider a mapping Φ on a set of objects into a set of feature vectors in a feature space (see, e.g., [23]).

When it is clear that $y = (\phi_1(x), \ldots, \phi_n(x))$ is a feature vector in a feature space $\mathbb{R}^n$, we sometimes write y instead of $\Phi(x)$.

Mscript 6

(*Visualizegradientorientationweightedwiththegradientmagnitudes.*)

***img* = ;**
orientation = GradientOrientationFilter[img, 1]//ImageAdjust;
magnitude = GradientFilter[img, 1]//ImageAdjust;
ColorCombine[{orientation, magnitude, magnitude}, "HSB"]//ColorNegate ■

Example 5.23 **Feature Vectors in $\mathbb{R}^3$ and $\mathbb{R}^4$.**
Let img be the raster image (Punch drawing of a knight being lowered onto his horse) in Fig. 5.7. Let the colour image img have a corresponding greyscale image f. Each point in f has a greyscale intensity $f(x, y)$ in the range [0, 255]. And let $\phi(p)$ be a probe function that returns the gradient orientation (angle) for an edge pixel $p \in img$. *gradient orientation* is an image pixel feature. For example, let ϕ_1, ϕ_2, ϕ_3 be defined by the Mathematica10 functions **GradientFilter, GradientOrientation-Filter**, which return the gradient magnitudes (G_x, G_y in the horizontal and vertical directions) and the gradient orientation of each of the pixels in the greyscale image.

Gx, Gy are defined by the partial derivatives $\frac{\partial f}{\partial x}, \frac{\partial f}{\partial y}$, respectively. The gradient orientation θ of an edge pixel in a greyscale image is defined by

$$\theta = \tan^{-1}\left[\frac{\frac{\partial f}{\partial y}}{\frac{\partial f}{\partial x}}\right] = \tan^{-1}\left[\frac{G_y}{G_x}\right].$$

The Mscript 6 is used to find and visualize the gradients and magnitudes of every pixel in any raster image. As a result, each image pixel has a description given the feature vector

$$\Phi(x) = (\phi_1(x), \phi_2(x), \phi_3(x)) = \left(\theta(x), G_x(x), G_y(x)\right) \in \mathbb{R}^3.$$

The hue, saturation, brightness intensities in the HSB colour space are combined to represent the pixel gradients in the knight image. Let $corner(x)$ denote the cornerness of x (degree that x is a corner pixel).

Combining the gradient orientation and gradient magnitude features with the corner feature in Example 5.21 gives the feature vector $\Phi(x)$ in $\mathbb{R}^4$ defined by

$$\Phi(x) = \left(\theta(x), G_x(x), G_y(x), corner(x)\right) \in \mathbb{R}^4.$$

In Fig. 5.9, edge pixels with the same gradient orientation are represented with the same colour. For example, the top edges of the spurs in have approximately the same gradient orientation and are visualized with magenta •. ■

Fig. 5.9 Knight edge pixel gradients

Without much effort, other pixel features such as RGB pixel intensities (from probe functions R(x), G(x), B(x) that return red, green blue intensities, respectively) can be included in each pixel description, leading to feature vectors in $\mathbb{R}^7$, namely,

$$\Phi(x) = \big(\theta(x), G_x(x), G_y(x), corner(x), R(x), G(x), B(x)\big) \in \mathbb{R}^7.$$

5.5 Strongly Proximal Object and Feature Spaces

A common frame of reference for proximity is some object space that is Euclidean. Often, an *object space* E is a Euclidean topological space $\mathbb{R}^n$ endowed with a proximity δ with the usual Euclidean metric $\|\cdot\|$ (Euclidean norm). This is the space we work in planar objects in $\mathbb{R}^2$ or 3D objects in $\mathbb{R}^3$ such as pixels in 2D and voxels in 3D digital images or 2D sets of nodes in a social network or 3D sets of brain activity signals on the surface of the brain viewed as a hypersphere. Usually, $n = 2$ (Euclidean plane $\mathbb{R}^2$) or $n = 3$ (Euclidean volume space $\mathbb{R}^3$) for common object spaces.

Fig. 5.10 Fisherman object space

Let A^c denote the complement of A (all points of E not in A). The **boundary** of A (denoted bdy$A$) is the set of all points that are near A and near A^c [24, Sect. 2.7, p. 62]. An important structure is the **interior** of A (denoted int$A$), defined by int$A$ = cl$A$ − bdy$A$.

A **proximal object space** is an object space endowed with one or more proximity relations. A **strong proximal object space** is an object space endowed with one or more proximity relations that includes a strong proximity such as $\overset{\wedge\!\!\wedge}{\delta}$ or $\overset{\wedge\!\!\wedge}{\delta}_{\Phi}$.

Example 5.24 **Fisherman Object space**.

Object space: finite Euclidean topological space $X \subset \mathbb{R}^2$ equipped with the proximities $\left\{\delta, \overset{\wedge\!\!\wedge}{\delta}\right\}$ (Lodato-Wallman proximity δ and Lodato strong proximity $\overset{\wedge\!\!\wedge}{\delta}$) is an object space be represented by the fisherman image in Fig. 5.10. The red • dots indicate keypoints in the picture. These dots provide a basis for a Dirichlet tessellation of the picture. Let $A, B \subset X$ be the sets of points in visor (stiff peak) and crown of the fisherman's hat, respectively. Then

$$\mathrm{cl}A = \{x \in X : x\ \delta\ A\}.\ \textbf{(Usual closure of hat peak set)}$$
$$\mathrm{cl}B = \{x \in X : x\ \delta\ B\}.\ \textbf{(Usual closure of hat crown set)}$$

In this case, we can write

$$\mathrm{cl}A\ \overset{\doublewedge}{\delta}\ \mathrm{cl}B,\ \text{since the hat peak and hat crown have a common edge.}$$

Example 5.25 **Interior of a Nonempty Set.**
The interior of a straight line segment $\overline{pq}$ are all of the points in the segment, except the endpoints p and q. Again, for example, let A be the set of all points in one of the knight's spurs in . The interior of A is the set of all points inside the border of the spur. ■

Feature space: finite Euclidean topological space $Y \subset \mathbb{R}^2$ equipped with the proximities $\left\{\overset{\doublewedge}{\delta}, \overset{\doublewedge}{\delta}_{\Phi}\right\}$ is a feature space defined by $\{\emptyset, w, b, wb\}$ for the greyscale colour space (we only consider white and black pixels in the image). Vectors $\Phi(x) \in \mathbb{R}^2$ describe points $x \in X$. In particular,

$$\mathrm{cl}_{\Phi} A = \left\{x \in X : x\ \overset{\doublewedge}{\delta}_{\Phi}\ A\right\}.\ \textbf{(Descriptive closure of hat peak set)},$$
$$i.e., \phi(x) \in \Phi(A).$$
$$\mathrm{cl}_{\Phi} B = \left\{x \in X : x\ \overset{\doublewedge}{\delta}_{\Phi}\ B\right\}.\ \textbf{(Descriptive closure of hat crown set)},$$
$$i.e., \phi(x) \in \Phi(B).$$

In this case, we can write

$$\mathrm{cl}_{\Phi} A\ \overset{\doublewedge}{\delta}\ \mathrm{cl}_{\Phi} B,\ \text{since both sets have white pixels.}$$

$\mathrm{cl}_{\Phi} A\ \overset{\doublewedge}{\delta}\ \mathrm{cl}_{\Phi} B$ means that $\mathrm{cl}_{\Phi} A$ has strong descriptive proximity to $\mathrm{cl}_{\Phi} B$, since there are points $x \in X$ that are common to both sets (e.g., both sets have white pixels). Put another way, $\mathrm{cl}A\ \overset{\doublewedge}{\delta}_{\Phi}\ \mathrm{cl}B$. For example,

$$\Phi(x) = (w, b) = (0.5, 0.5)\ \text{grey tone pixel.}\quad ■$$

Remark 5.26 **Two Different Uses of the Euclidean Plane.**
One of the main things to notice in Example 5.24 is the use finite subsets of the Euclidean plane $\mathbb{R}^2$ in two different ways.

Object space. A finite subset X of the Euclidean plane is the object space for the picture elements in the fisherman image in Fig. 5.10.

Feature space. A second finite subset $\Phi(X)$ of the Euclidean plane is the feature space for the picture elements in the fisherman image in Fig. 5.10. Each 2-dimensional vector in the feature space describes a pixel in the object space. Recall that each $y \in \Phi(X)$ describes a picture element in the object space. ■

Problem 5.27 Do the following:

1° Select digital image img.

2° Select set of sites (generating points) S that can be extracted from img.

3° **Image geometry**: Use Mathematica to tessellate img using the site points in S.

4° **Cell quality**: Compute the average cell quality in the tessellated image. To measure cell quality, use the approach given in Sect. 2.9. **N.B.**: You need to introduce a quality measure for each type of ngon in the tessellated image. For example, introduce a quality measure for 5gons, 6gons and so on. Use Mathematica, to display a list cell quality measurements for the tessellated image and also compute the average cell quality measurement.

5° Briefly, comment on what the quality measurement of each type of cell tells us about the picture region contained in the cell. Do smaller cells (cells with small area) tells us more about the interior of picture regions than bigger cells? Do cells 3 or 4 edges tell us more about the picture regions in the interior of such cells than cells with 5 or more edges?

6° Give a measure of picture region quality. **Hint**: Make the picture region quality a function of cell quality, region features such as Area and Diameter, and region pixel features.

7° Give a mathematical model (formula) that correlates cell quality and picture region quality. Do this by comparing picture region quality measurements with tessellation cell quality measurements.

8° Give a combined plot for (a) cell quality measurements, (b) picture regions quality measurements for img.

9° Define a strong proximal object space X for the tessellated image. **Hint**: Define the object space relative to the objects will be in the Euclidean plane.

10° Indicate subsets $A, B \subset X$ that are strongly near, $A \overset{\wedge\wedge}{\delta} B$.

11° Define a strong proximal feature space X for the tessellated image. **Hint**: Define the feature space relative to pixels features that define feature vectors in the Euclidean plane $\mathbb{R}^n$. This means you must specify features of pixels that you want to describe with feature vectors.

12° Indicate subsets $A, B \subset \mathbb{R}^n$ (feature space) that are strongly descriptively near, $A \overset{\wedge\wedge}{\delta}_{\Phi} B$. ■

Example 5.28 **Digital Image That is a Proximal Object Space**.
Let im be a 2D digital image. Each pixel p in im has a location specified by the coordinates of vector (x, y). The object space for im is a subset X of $\mathbb{R}^2$. The space X endowed with a Wallman proximity δ is an example of a proximal object space (denoted by (A, δ)). Let A, B be subsets in X. Then A is close to B if and only $\text{cl}A \cap \text{cl}B \neq \emptyset$. In other words, A and B have points in common. For example, in

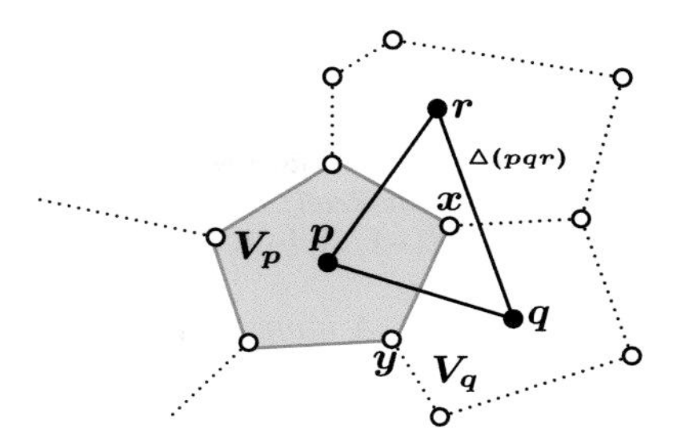

Fig. 5.11 Close voronoï regions

Fig. 5.11, the Voronoï regions V_p, V_q have points in common, namely, the points in the straight line segment $\overline{xy}$. Hence, $V_p \ \delta \ V_q$. ■

Mscript 7

(*Half plane bounded by a line L contains all points on one side L*)

Plot3D[RegionDistance[$\mathcal{R}$2, {x, y}], {x, −1, 1}, {y, −1, 1}, PlotTheme → "Classic", MeshFunctions → {#3&}, Mesh → 0, Exclusions → y == 0] ■

A *half plane* is a planar region consisting of all points on one side of an infinite straight line and no points on the other side of the line [25]. Half planes provide the basic building blocks for Voronoï regions.

Example 5.29 **Proximal Half Planes.**
Let A, B be a pair of half planes in Euclidean 3-space as shown in Fig. 5.12. Since A and B intersect, A and B are proximal half planes, i.e., $A \ \delta \ B$. By experimenting with Mscript 7, it is possible to display many different intersecting half planes. ■

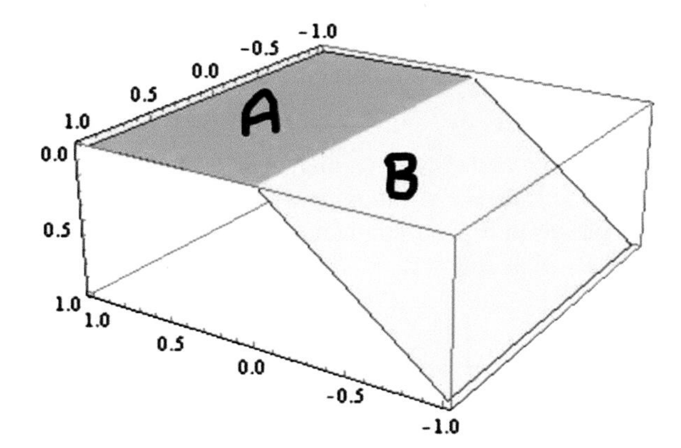

Fig. 5.12 Proximal half planes: $A \ \delta \ B$

Mscript 8

```
Graphics3D[
Table[{Hue[RandomReal[]], Opacity[0.8],
HalfPlane[RandomReal[1, {2, 3}], RandomReal[{−1, 1}, 3]]}, {2}],
PlotRange → {{−1, 1}, {−1, 1}, {−1, 1}}]
```
■

Theorem 5.30 *A nonempty Voronoï region with n vertices contains n intersecting half planes.*

Problem 5.31 (a) Prove Theorem 5.30 and (b) given an example of a pair of half planes that are not near. **Hint**: Use Mscript 8 to produce examples of non-intersecting half planes. ■

5.6 Strong Proximal Borsuk–Ulam Theorem

This section introduces proximal forms of the Borsuk–Ulam Theorem. Instead of the usual form of continuity, we now consider continuous mappings that are strongly proximal. Such mappings preserve proximity structures.

Definition 5.32 Suppose that $(X, \tau_X, \overset{\wedge\!\!\wedge}{\delta}_X)$ and $(Y, \tau_Y, \overset{\wedge\!\!\wedge}{\delta}_Y)$ are topological spaces endowed with strong proximities [26]. We say that the map $f : X \to Y$ is *strongly proximal continuous* and we write **s.p.c.** if and only if, for $A, B \subset X$,

$$A \overset{\wedge\!\!\wedge}{\delta}_X B \Rightarrow f(A) \overset{\wedge\!\!\wedge}{\delta}_Y f(B). \quad ■$$

Theorem 5.33 [26] *Suppose that $(X, \tau_X, \overset{\wedge\!\!\wedge}{\delta}_X)$ and $(Y, \tau_Y, \overset{\wedge\!\!\wedge}{\delta}_Y)$ are topological spaces endowed with compatible strong proximities and $f : X \to Y$ is s.p.c. Then f is an open mapping, i.e., f maps open sets in open sets.*

Example 5.34 This example comes from [26, Sect. 3]. Take $(X, \tau_X, \overset{\wedge\!\!\wedge}{\delta}_X) = (\mathbb{R}^2, \tau_e, \overset{\wedge\!\!\wedge}{\delta}_X)$ and $(Y, \tau_Y, \overset{\wedge\!\!\wedge}{\delta}_Y) = (\mathbb{R}^2, \tau_e, \overset{\wedge\!\!\wedge}{\delta}_Y)$, where τ_e is the Euclidean topology, $A \overset{\wedge\!\!\wedge}{\delta}_X B \Leftrightarrow A \cap B \neq \varnothing$, and $A \overset{\wedge\!\!\wedge}{\delta}_Y B \Leftrightarrow \text{int}A \cap \text{int}B \neq \varnothing$, provided A and B are not singletons; if $A = \{x\}$, then $x \in \text{int}(B)$, and if B too is a singleton, then $x = y$. Consider the triangles as in Fig. 5.14 and let $\mathscr{S}$ be the family of those triangles. Define a function $g : X \to Y$ as follows:

$$g(x) = \begin{cases} x, & \text{if}\, x \leq x_1, \\ R(x_1, 80^\circ), & \text{if}\, x_1 \leq x \leq x_2, \\ R(x_2, 80^\circ) \circ R(x_1, 80^\circ), & \text{if}\, x_2 \leq x \leq x_3, \\ R(x_3, 80^\circ) \circ R(x_2, 80^\circ) \circ R(x_1, 80^\circ), & \text{if}\, x_3 \leq x \leq x_4, \\ \prod_{i \in \{n, n-1.., 1\}} R(x_i, 80^\circ), & \text{if}\, x_n \leq x \leq x_{n+1}. \end{cases}$$

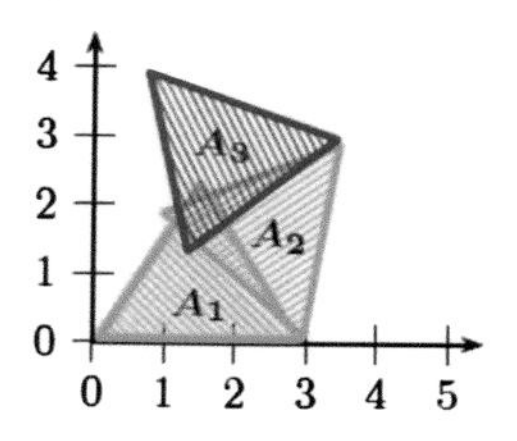

Fig. 5.13 80° rotations

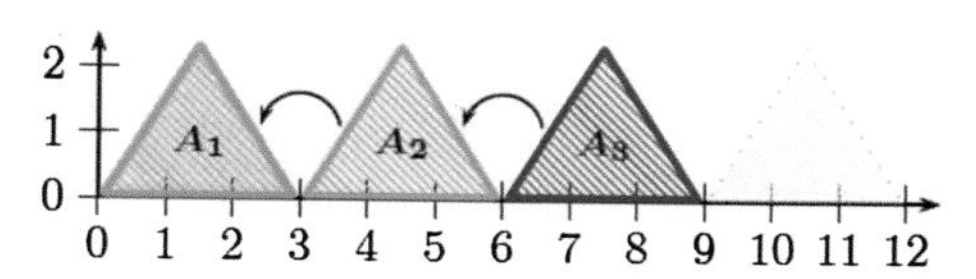

Fig. 5.14 Family of adjacent triangles

where $R(x_i, 80°)$ is the rotation around the point x_i by 80°, and by $\prod_{i\in\{n,n-1..,1\}} R(x_i, 80°)$ we mean the composition of rotations (see Fig. 5.13). We have that g is s.p.c. on $\mathscr{S}$. ■

Recall that antipodal regions are regions that are complements of each other. Antipodal regions can be spatially far and also descriptively near. This situation is possible in a topological space equipped with both Wallman proximity δ and strong descriptive proximity $\overset{\wedge\!\!\wedge}{\delta}_{\Phi}$.

Example 5.35 **Antipodal Region Descriptions.**
Let the object space X (presented in Fig. 5.15) be a finite Euclidean topological space equipped with the relator $\left\{\delta, \overset{\wedge\!\!\wedge}{\delta}_{\Phi}\right\}$ containing the Wallman proximity δ and strong descriptive proximity $\overset{\wedge\!\!\wedge}{\delta}_{\Phi}$. Also, let the points in X be described by feature vectors such that each vector $y =$ (colour, gradient orientation) in the feature space Y. In this case, $C \ \not\delta\ H$, since C and H have no points in common. It is also the case that $C \ \overset{\wedge\!\!\wedge}{\delta}_{\Phi}\ H$, since there points in C that match the description of points in H.

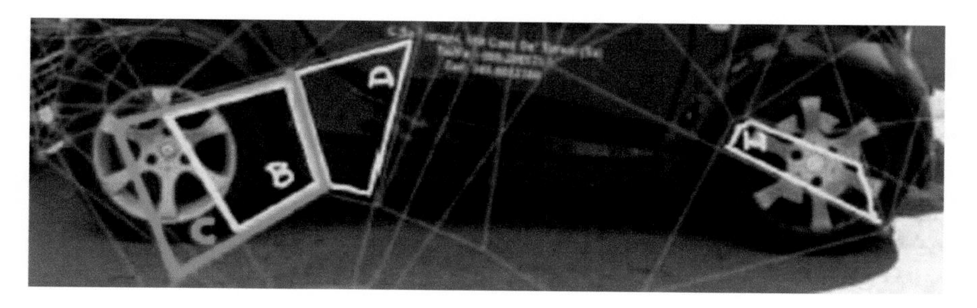

Fig. 5.15 $C \ \not\delta\ H$ and $C \ \overset{\wedge\!\!\wedge}{\delta}_{\Phi}\ H$

Let $f : X \longrightarrow \mathbb{R}^n$ be s.p.c. such that $C \overset{\wedge\wedge}{\delta}_{\Phi} H$ implies $f(C) \overset{\wedge\wedge}{\delta}_{\Phi} f(H)$. In this case, there is $c \in C$ and $h \in H$ with $\{c\} \overset{\wedge\wedge}{\delta} \{h\}$. From $\overset{\wedge\wedge}{\delta}$-Axiom (N6), we know that $c = h$ (both points have the same description). Hence, $f(c) = f(h)$. ■

Let $\neg x$ be any point in $X \setminus \{x\}$ that is not $x \in X$. The Borsuk–Ulam Theorem (BUT) has many different incarnations.

Theorem 5.36 Strong Proximal Borsuk–Ulam Theorem (s.p.c.BUT).
If $f : S^n \longrightarrow \mathbb{R}^n$ is $\overset{\wedge\wedge}{\delta}$-continuous (s.p.c.), then $f(x) = f(\neg x)$ for some $x \in X$.

Proof The mapping f is s.p.c. if and only if $A \overset{\wedge\wedge}{\delta}_{S^n} B$ implies $f(A) \overset{\wedge\wedge}{\delta}_{\mathbb{R}^n} f(B)$. Let $A = \{x\}, B = \{\neg x\}$ for some $x \in X, \neg x \in X \setminus \{x\}$. From $\overset{\wedge\wedge}{\delta}$-Axiom (N6), $\{x\} \overset{\wedge\wedge}{\delta} \{\neg x\} \Leftrightarrow \{x\} = \{\neg x\}$. Hence, $f(x) = f(\neg x)$. □

Since the proof of Theorem 5.36 depends on the domain and range of mapping f being compatible topological spaces equipped with a s.p.c. map and does not depend on the geometry of S^n, we have

Theorem 5.37 S^n-Free Proximal Point-Based Borsuk–Ulam Theorem (S^n-Free s.p.c.BUT).
Let $(X, \tau_X, \overset{\wedge\wedge}{\delta}_X)$ and $(\mathbb{R}^n, \tau_{\mathbb{R}^n}, \overset{\wedge\wedge}{\delta}_{\mathbb{R}^n})$ be topological spaces endowed with compatible strong proximities. If $f : X \longrightarrow \mathbb{R}^n$ is $\overset{\wedge\wedge}{\delta}$-continuous (s.p.c.), then $f(x) = f(\neg x)$ for some $x \in X$.

Problem 5.38 Give an example to illustrate Theorem 5.36. ■

Problem 5.39 Do the following:
1° Prove Theorem 5.37.
2° Give an example to illustrate Theorem 5.37. ■

5.7 Strong Proximal Region-Based Borsuk–Ulam Theorem

We are interested in the case where strongly near regions in an object space are mapped to strongly near regions in the feature space. To arrive at a region-based form of Theorem 5.37, we introduce region-based strong proximal continuity.

Definition 5.40 Region-Based $\overset{\wedge\wedge}{\delta}$-Continuous Mapping.
Let X, Y be nonempty sets. Suppose that $(2^X, \tau_{2^X}, \overset{\wedge\wedge}{\delta}_{2^X})$ and $(Y, \tau_Y, \overset{\wedge\wedge}{\delta}_Y)$ are topological spaces endowed with strong proximities. We say that $f : 2^X \rightarrow Y$ is *region strongly proximal continuous* and we write **Re.s.p.c.** if and only if, for $A, B \in 2^X$,

$$A \overset{\wedge\wedge}{\delta}_X B \Rightarrow f(A) \overset{\wedge\wedge}{\delta}_Y f(B). \quad ■$$

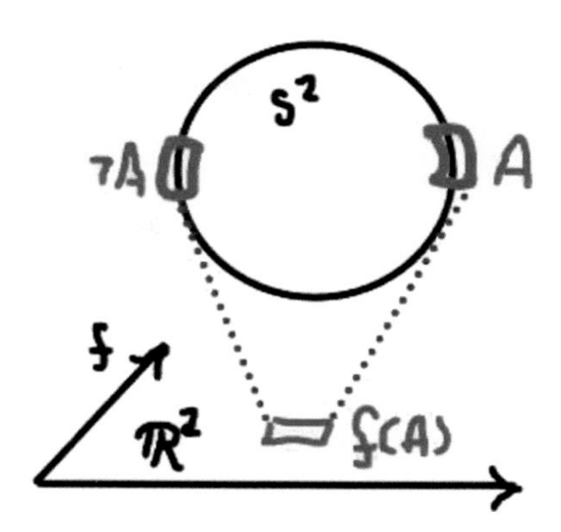

Fig. 5.16 $f : 2^{S^2} \longrightarrow 2^{\mathbb{R}^2}$

Let $\neg A \in 2^X \setminus A$. For a Re.s.p.c. mapping $f : 2^X \to \mathbb{R}^n$ on the collection of subsets 2^X to $\mathbb{R}^n$, the assumption is that 2^X is a region-based object space (each object is represented by a nonempty region) and $\mathbb{R}^n$ is a feature space (each region A in 2^X maps to a feature vector y in $\mathbb{R}^n$ such that y is a description of region A). Then we have

Theorem 5.41 Proximal Region-Based Borsuk–Ulam Theorem.
Suppose that $(X, \tau_X, \overset{\wedge\wedge}{\delta}_X)$ and $(\mathbb{R}^n, \tau_{\mathbb{R}^n}, \overset{\wedge\wedge}{\delta}_{\mathbb{R}^n})$ are topological spaces endowed with compatible strong proximities. If $f : 2^X \longrightarrow \mathbb{R}^n$ is $\overset{\wedge\wedge}{\delta}$-Re.s.p continuous, then $f(A) = f(\neg A)$ for some $A \in 2^X$.

Proof The proof is symmetric to the proof of Theorem 5.36. □

Example 5.42 **Antipodal S^2 regions mapped to a region in $\mathbb{R}^2$.**
Let $\neg A, A \in 2^{S^2}$ be antipodal regions in a topological space on S^2 equipped with the strong proximity $\overset{\wedge\wedge}{\delta}$ and let

$$f : 2^{S^2} \longrightarrow 2^{\mathbb{R}^2}$$

be a $\overset{\wedge\wedge}{\delta}$-continuous mapping on 2^{S^2} into $2^{\mathbb{R}^2}$. From Theorem 5.41, we obtain the result shown in Fig. 5.16. ■

Region-based Borsuk–Ulam Theorem 5.41 provides a means of comparing and classifying object-oriented regions in a topology of digital images. Distinct regions with matching descriptions belong to their own region class. The objects in a region class can be either spatially close or spatially far apart and have matching descriptions in the region feature space. ■

Example 5.43 **Region-Based Feature Vectors and Re-Borsuk–Ulam Theorem at Work.**

Let $(\mathbb{R}^2, \tau_{\mathbb{R}^2}, \overset{\wedge\wedge}{\delta}_{\mathbb{R}^2})$ and $(\mathbb{R}^n, \tau_{\mathbb{R}^n}, \overset{\wedge\wedge}{\delta}_{\mathbb{R}^n})$ be Euclidean topological spaces endowed with the Lodato $\overset{\wedge\wedge}{\delta}$-proximity. The space $\mathbb{R}^2$ is an object space and the space $\mathbb{R}^n$ is a feature space (each region in the object space has a description that is a feature vector in $\mathbb{R}^n$). For simplicity, assume each $A \in 2^{\mathbb{R}^2}$ is described by the feature vector (r, g, b) that contains the intensities of the red, green, blue colours of region A. Let the object space $\mathbb{R}^2$ be represented by the image in Fig. 5.15. Assume $f : 2^{\mathbb{R}^2} \longrightarrow \mathbb{R}^n$ is $\overset{\wedge\wedge}{\delta}$-continuous.

From Theorem 5.41, $f(A) = f(\neg A)$ for some region $A \in 2^{\mathbb{R}^2}$. For example, the feature vector $f(C)$ that describes region C in Fig. 5.15 matches the feature vector $f(H)$ that describes region H (portions of the Salerno Poste wheels in the picture). In this example, H plays the role of $f(\neg C)$ such that $H := \neg C \in 2^{\mathbb{R}^2} \setminus C$. There are many other regions in Fig. 5.15 that satisfy Theorem 5.41. ■

Problem 5.44 Prove Theorem 5.41. Give an example to illustrate Theorem 5.41. ■

5.8 Complexes, Nuclei and Applied Homotopy

In the study of homotopy theory, the focus is on deforming (mapping) one set of sets (called a complex) into another complex. In the language of Di Concilio-Gerla point-free geometry [27, 28], a **complex** is a structure called a region. A pair of complexes are equivalent, provided they have the same nucleus. The nucleus we have in mind here is a **Whiteheadean nucleus** [29], which is a sub-region of a complex that can be transformed into a subregion in another complex.

Algorithm 8: Deform **j** into **?**

Input : Read digital image *img*.
Output: Homotopy $H : pointData \times [0, 1] \longrightarrow Y$.

1 $img \longmapsto edgeBoundaries$;
2 /* $edgeBoundaries$. */ ;
3 $pointData \leftarrow edgeBoundaries$;
4 /* $pointData$ contains edge pixel coordinates used to map **j** to **?**. */ ;
5 $pointData \longmapsto H(x, 0), H(x, 1)$;
6 /* $x \in pointData$. */ ;

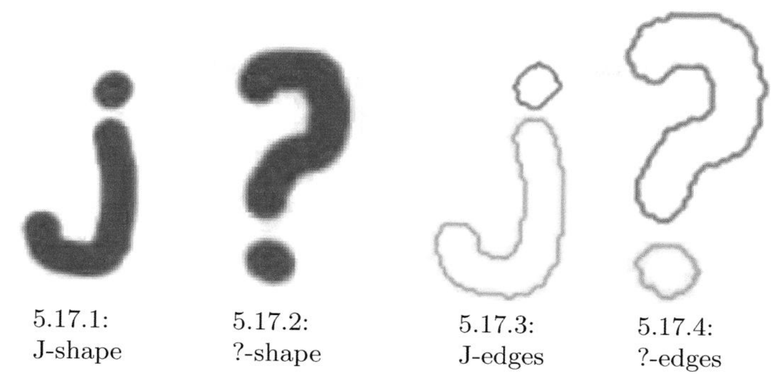

Fig. 5.17 Shape J deformed into shape?

Example 5.45 **Sample Homotopy on a Pair of Fat Characters.**
This example illustrates a deformation mapping of the set of points on the edges of the letter **j** to the set of points on the edges of the question mark **?**.

We start with the set of points in the fat letter **j** in Fig. 5.17.1 and the set of points in the fat question mark **?** in Fig. 5.17.2. Next, the edges on this pair of characters are found. This results in a pair of silhouettes for **j** and **?** shown in Fig. 5.17.3, 4. A set of edge points X ($=pointData$ in Algorithm 8) for the pair of characters is then found.

A homotopy $H : pointData \times [0, 1] \longrightarrow Y$ is defined by $H(x, 0)$ (= **j**) and $H(x, 1)$ (= **?**). The dot blob on the **j** edges in Fig. 5.17.3 is an example of what J.H.C. Whitehead [29, Sect. 6, p. 259] calls a *nucleus*, which is equivalent to the dot blob on the **?** edges in Fig. 5.17.4. By a clever form of deformation mapping, the set of points in the **j** edges are transformed into the set of points in the **?** edges.

These steps leading to a homotopy between the pair of characters are given in Algorithm 8. The implementation of Algorithm 8 is given in the Mathematica script 36 in Appendix A.5. The homotopy widget produced by the Mathematica script 36 is shown in Fig. 5.18. By moving the slider in Fig. 5.18 back and forth slowly, the continuous transformation of the **j** into **?** can be observed. ■

Problem 5.46 Use Mathematica to demonstrate that the shape in Fig. 5.19.1 and the shape in Fig. 5.19.2 are equivalent, i.e., show that can be deformed into (an 8 on its side). ■

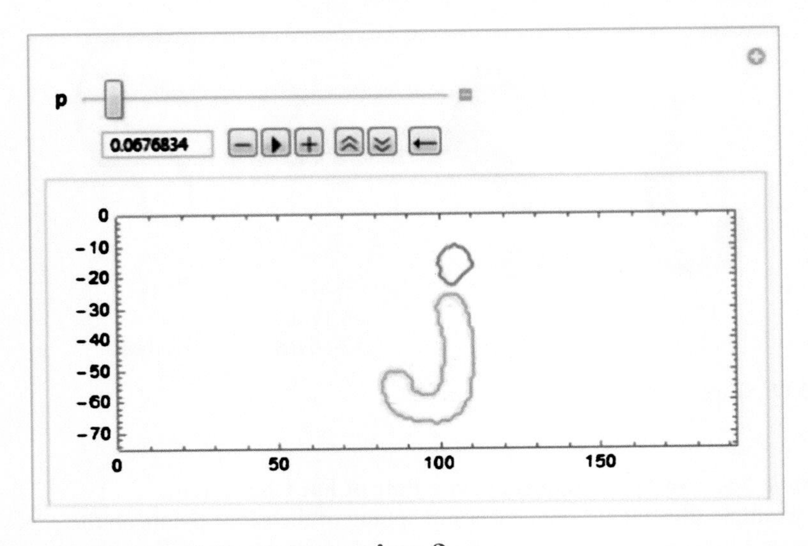

Fig. 5.18 Mathematica widget that deforms **j** into **?**

Fig. 5.19 Equivalent shapes with holes

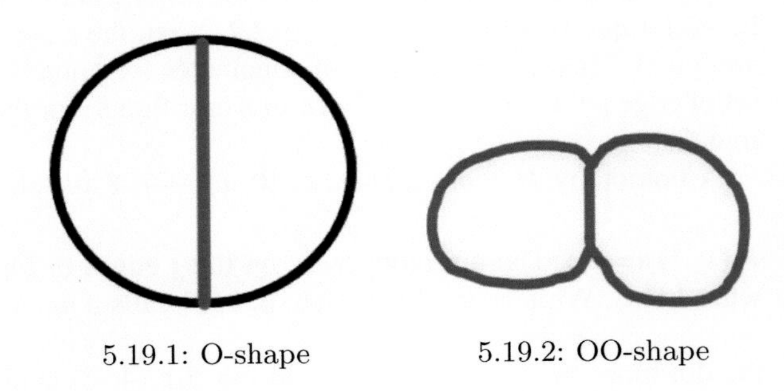

5.19.1: O-shape

5.19.2: OO-shape

5.9 Strong Proximal Homotopy

A region-based, strong proximal form of homotopy can be used to advantage in determining when and to what extent one shape can be deformed into another shape. For this reason, we introduce region-to-region strongly proximal continuous functions.

Definition 5.47 Region-to-Region $\overset{\wedge\wedge}{\delta}$-Continuous Mapping.
Let X, Y be nonempty sets. Suppose that $(X, \tau_{2^X}, \overset{\wedge\wedge}{\delta}_{2^X})$ and $(Y, \tau_{2^Y}, \overset{\wedge\wedge}{\delta}_{2^Y})$ are topological spaces endowed with strong proximities. The strong proximities $\overset{\wedge\wedge}{\delta}_{2^X}, \overset{\wedge\wedge}{\delta}_{2^Y}$ are strong nearness relations on the collections of sets $2^X, 2^Y$, respectively. We say that $f : 2^X \to 2^Y$ is a *region-to-region strongly proximal continuous* map and we write **2Re.s.p.c.** if and only if, for $A, B \in 2^X$ and $f(A), f(B) \in 2^Y$,

$$A \overset{\wedge\!\!\!\wedge}{\delta}_{2^X} B \Rightarrow f(A) \overset{\wedge\!\!\!\wedge}{\delta}_{2^Y} f(B). \quad \blacksquare$$

A 2Re.s.p.c. mapping leads naturally to 2Re.s.p.c. homotopies. That is, we want to consider region-based homotopies instead of the usual point-based homotopies. Informally, this means that the homotopy $H_{Re} : 2^X \times [0, 1] \longrightarrow 2^Y$ is region-based and $H(A, 0) = f(A)$ and $H(A, 1) = g(A)$ are regions (shapes) in Euclidean space such that there is deformation retraction on shape $g(A)$ into shape $f(A)$.

Definition 5.48 Region-based $\overset{\wedge\!\!\!\wedge}{\delta}$-Homotopy.
Let $f, g : 2^X \longrightarrow 2^Y$ be Re.s.p.c. mappings. Suppose that $(X, \tau_{2^X}, \overset{\wedge\!\!\!\wedge}{\delta}_{2^X})$ and $(Y, \tau_{2^Y}, \overset{\wedge\!\!\!\wedge}{\delta}_{2^Y})$ are topological spaces endowed with compatible strong proximities, $A \in 2^X$. The homotopy $H_{Re} : 2^X \times [0, 1] \longrightarrow 2^Y$ is a 2Re.s.p.c. (*region-to-region strongly proximal continuous*) mapping such that

$$H_{Re}(A, 0) = f(A),$$
$$H_{Re}(A, 1) = g(A), \text{ for every } A \in 2^X.$$

The continuous maps f, g are called region-based s.p.c. homotopic maps and $H_{Re}(A, 0)$, $H_{Re}(A, 1)$ are regions (shapes) in 2^Y. $\blacksquare$

Conjecture 5.49 Strongly Deformable Shapes
A shape $H_{Re}(A, 0) = f(A)$ strongly deforms either partially or completely into shape $H_{Re}(A, 1) = g(A)$ if and only if $f(A) \overset{\wedge\!\!\!\wedge}{\delta} g(A)$, i.e., the interior of the two shapes have points in common and there is a $\overset{\wedge\!\!\!\wedge}{\delta}$-continuous mapping on $f(A)$ into $g(A)$ that results in a partial or complete deformation of shape $f(A)$ into shape $g(A)$. $\blacksquare$

Theorem 5.50 *If $H_{Re}(A, 1) \subset H_{Re}(A, 0)$, then Conjecture 5.49 is satisfied.*

Proof The result follows from the fact that $\text{int}H_{Re}(A, 0) \cap \text{int}H_{Re}(A, 1) \neq \emptyset$ and H is 2Re.s.p.c. $\square$

Problem 5.51 Give a counterexample to disprove Conjecture 5.49 for shapes that share points but one shape is not deformable into the other shape. $\blacksquare$

The feature space for region-to-region homotopies leads to a promising approach in determining when one shape is deformable into another shape. In this approach, the assumption is that the collection of sets 2^X is a finite topological space equipped with the Lodato $\overset{\wedge\!\!\!\wedge}{\delta}$ proximity.

Let $f_\Phi, g_\Phi : 2^X \longrightarrow \mathbb{R}^n$ be Re.s.p.c. mappings on 2^X into $\mathbb{R}^n$. This means, for example, for $A \in 2^X$, $f_\Phi(A)$ is a feature vector that describes region A. That is, the assumption made here is that 2^X is a collection of regions Re that represent objects as shapes and that $f_\Phi(Re) \in \mathbb{R}^n$ is an n-dimensional feature vector describing

$Re \in 2^X$. The region-based homotopy H_{Re_Φ} on a set of regions maps to descriptions of the regions, i.e.,

$$\begin{aligned} H_{Re_\Phi} : 2^X \times [0, 1] \longrightarrow \mathbb{R}^n, &\text{ is a 2Re.s.p.c map defined by} \\ H_{Re_\Phi}(A, 0) &= f_\Phi(A), \\ H_{Re_\Phi}(A, 1) &= g_\Phi(A). \end{aligned}$$

Theorem 5.52 *1°* $H_{Re_\Phi}(A, 0) = H_{Re_\Phi}(\neg A)$.
2° $H_{Re_\Phi}(A, 1) = H_{Re_\Phi}(\neg A)$.
3° $f_\Phi(A) \ \overset{\wedge\!\!\wedge}{\delta}\ g(A)$ *if and only if* $f_\Phi(\neg A) \ \overset{\wedge\!\!\wedge}{\delta}\ g_\Phi(\neg A)$
4° $H_{Re_\Phi}(A, 0) \ \overset{\wedge\!\!\wedge}{\delta}\ H_{Re_\Phi}(A, 1)$ *if and only if* $f_\Phi(A) \ \overset{\wedge\!\!\wedge}{\delta}\ g_\Phi(A)$.

Proof $1^o, 2^o, 3^o$: Immediate from Theorem 5.41 (Re-based Borsuk–Ulam Theorem). 3°: $H_{Re_\Phi}(A, 0) \ \overset{\wedge\!\!\wedge}{\delta}\ H_{Re_\Phi}(A, 1)$ means $H_{Re_\Phi}(A, 0)$ overlaps with $H_{Re_\Phi}(A, 1)$ if and only if $f(A) \ \overset{\wedge\!\!\wedge}{\delta}\ g(A)$. □

The assumption made here is that 2^X is a collection of regions that represent objects and that $Re \in 2^Y$ is a collection of descriptions for simplexes (subsets in a complex or region). This means that $h(A, 0) = f(A), h(A, 1) = g(A)$ are subsets of the feature space $2^{\mathbb{R}^n}$ and $f(A), g(A)$ contain features vectors that describe the simplices contained in region A. Similarly, $h(B, 0) = f(B), h(B, 1) = g(B)$ are subsets of the feature space $2^{\mathbb{R}^n}$ and $f(B), g(B)$ contain features vectors that describe the simplices contained in region B. These two sets of descriptions need not be the same.

The question arises When are the two set of descriptions near enough for us to conclude that the shapes described by $f(A), g(A)$ are deformable into the shapes described by $f(B), g(B)$?

Obviously, shape A can be deformed into shape B, provided $A \subset B$. Let $\phi(A) \in \mathbb{R}$ be a feature value of a region A. A description $\Phi(A)$ of a region A is a feature vector:

$$\begin{aligned} \Phi : A &\longrightarrow \mathbb{R} \times \mathbb{R} \times \cdots \times \mathbb{R}, \\ \phi : A &\longrightarrow \mathbb{R}, \\ \Phi(A) &= (\phi_1(A), \phi_2(A), \ldots, \phi_n(A)) \text{ (description of } A). \end{aligned}$$

When does **region description** resemble the description of another region? One region A has a description $\Phi(A)$ that **resembles** the description $\Phi(B)$ of a region B, provided $\|\Phi(A) - \Phi(B)\| \neq 0$. The description $\Phi(A)$ that **strongly resembles** the description $\Phi(B)$ of a region B, provided $\Phi(A), \Phi(B)$ have more than one $\phi_i(A)$ value in common.

Example 5.53 Let A, B, C be a nonempty sets of points in a picture, $\Phi = \{r, g, b\}$, where $r(A)$ equals the intensity of the redness of A, $g(A)$ equals the intensity of the greenness of A and $b(A)$ equals the intensity of the blueness of A. Then

$$\Phi(A) = (1, 1, 1) \text{ (description of } A)$$
$$\Phi(B) = (0, 0, 1) \text{ (description of } B)$$
$$\Phi(C) = (0, 1, 1) \text{ (description of } C)$$
$$\Phi(D) = (0.4, 0.4, 0.4) \text{ (description of } D)$$
$$\Phi(E) = (0.4, 0.2, 0.3) \text{ (description of } E). \quad \blacksquare$$

From this, region A resembles region B, since both regions have equal blue intensity. Region A strongly resembles (is strongly near) region C, since both regions have equal green and blue intensities. By contrast, the description of region A is far from the descriptions of regions D and E.

Conjecture 5.54 Feature-Based Deformable Shapes
Let $H : 2^X \times [0, 1] \longrightarrow 2^Y$ be a region-base Re.s.p.c. homotopy, $A \in 2^X$. A shape $H(A, 0)$ deforms to shape $H(A, 1)$ if and only if $H_{Re_\Phi}(A, 0) = H_{Re_\Phi}(A, 1)$, i.e., the description of region $H(A, 0)$ matches or strongly resembles the description of region $H(A, 1)$. ■

Example 5.55 Let r,g,b colours and gradient orientation be features of the points in a nonempty picture set X. The letter j and question mark ? have $\overset{\pitchfork}{\delta}$ descriptions (both have similar curves and identical colours. ■

Problem 5.56 Let $\Phi = \{r, g, D\}$ be the red, green colours and diameter D of any region in a picture space X. Do the following:

1° Give an example of a pair of shapes A, B with either matching or strongly similar descriptions $\Phi(A)$, $\Phi(B)$ so that one shape can deformed into the other shape.
2° Write a Mathematica script to illustrate Part 1.
3° Give a counterexample to disprove Conjecture 5.49 that, for all pairs of shapes A, B with matching or strongly similar descriptions $\Phi(A)$, $\Phi(B)$, one shape can be deformed (mapped into) another shape, provided the shapes have matching descriptions.
4° Write a Mathematica script to disprove Conjecture 5.49. ■

5.10 Borsuk–Ulam Theorem Extended to Hyperbolic Surfaces by A. Tozzi

The Borsuk–Ulam theorem (BUT) tells us that there exists a pair of antipodal points on a convex n-sphere that are mapped to the same point in n-dimensional Euclidean space $\mathbb{R}^n$. In this section,[3] let S^n be a convex manifold containing a pair of antipodal points that map to a single point in $\mathbb{R}^n$.

[3]Many thanks to A. Tozzi for contributing this section.

So far, the features of geometric regions in 2D and 3D digital images have features such as oblong, concave, convex, diameter, area, perpendicular, circular that are mapped to real numbers that quantify region features. In this section, we consider image regions equipped with hyperbolic geometry characterized by sectional curvature -1, i.e., regions with a concave shape. The question to consider now is whether the antipodal points on flat or on curved surfaces with positive curvature (predicted by BUT) can also be found on image surfaces with negative curvature. In other words, is it possible to transport antipodal points found by BUT to corresponding vectors on a Riemannian manifold M^n with negative curvature?

The answer to this question is positive, provided we map (parallel transport) the antipodal points identified by BUT to points on a hyperbolic manifold. Although parallel transport requires the solution of a second-order differential equation, analysis shows us that we are allowed to have a first-order approximation of the parallel transport. In effect, we pursue parallel transport by solving geodesic equations for sufficient statistics [30] (see, also, an overview of the important aspects of geometric numeric integration by E. Hairer [31]). To solve the parallel transport problem, a number of options are available.

Parallel Transport Options

1° Ehresmann connection [32].

2° Levi-Civita connection [33].

3° We can formulate the Hessian operator on the Riemannian manifold in terms of the Laplace-Beltrami operator [34].

4° We can also retain a first-order approximation and formulate descent directions that are orthogonal to the previous descent ones, through numerical analysis with the conjugate gradient-descent algorithm [35]. Routinely used in optimization, conjugate gradient descent methods have been utilized for gradient descent on manifolds traced out by energy functions such as the variational free-energy [36]. ■

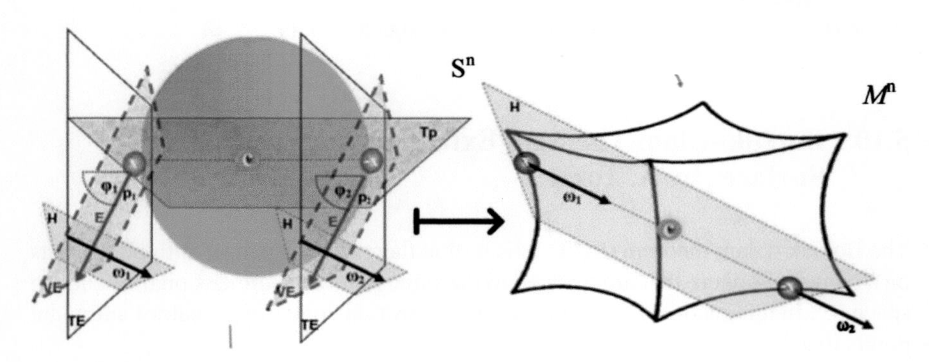

Fig. 5.20 $S^n \mapsto M^n$

Example 5.57 walks through the steps for parallel transport using the Ehresmann connection.

Example 5.57 **Antipodal points on $S^n \mapsto$ Riemannian manifold M^n.**
Parallel transport of antipodal points (predicted by BUT) on convex S^n to antipodal points on a concave Riemannian manifold M^n is illustrated in Fig. 5.20. This is accomplished with an Ehresmann connection (for the details, see C. Ehresmann [32]). After the Ehresmann connection has been performed and the vectors ω_1, ω_2 are found, then use the Riemannian exponential map $exp(\omega_i), i = 1, 2$ from S^n to M^n and the logarithmic map $log(\omega_i), i = 1, 2$ for the opposite mapping $M^n \mapsto S^n$. Let H in Fig. 5.20 be the plane on the Riemannian manifold M^n, containing the vectors ω_1, ω_2 that represent the antipodal points $x, -x$ on S^n. After this has been achieved, use the ham-sandwich theorem to find the center of the hyperbolic manifold (denoted by Ⓒ). ■

Theorem 5.58 Ulam's Ham Sandwich Theorem.
For any three given sets in Euclidean space, each of finite outer Lebesgue measure, there exists a plane that bisects all three sets, i.e., separates each of the given sets into two sets of equal measure.

Proof Assume that the Borsuk–Ulam Theorem holds and then the proof of the Ham-Sandwich Theorem follows (for the details, see [37]). □

In summary, the Borsuk–Ulam Theorem can be extended to concave surfaces, making it possible to look for antipodal points in images equipped with a negative curvature.

References

1. Borsuk, K.: Drei sätze über die n-dimensionale euklidische sphäre. Fundamenta Mathematicae **XX**, 177–190 (1933)
2. Weisstein, E.: Antipode. WolframMathWorld (2015). http://mathworld.wolfram.com/Antipode.html
3. Aichholzer, O., Caraballo, L., Díaz-Bánez, J., Fabila-Monroy, R., Ochoq, C., Nigsch, P.: Characterization of extremal antipodal polygons. Graphs Comb. **31**, 321–333 (2015)
4. Weisstein, E.: Hypersphere. Wolfram MathWorld (2015). http://mathworld.wolfram.com/Hypersphere.html
5. Munkres, J.: Topology, 2nd edn. Prentice-Hall, Englewood Cliffs (2000). Xvi + 537 pp. 1st edn. in 1975, MR0464128
6. Borsuk, K.: Concerning the classification of topological spaces from the stand-point of the theory of retracts. Fundamenta Mathematicae **XLVI**, 177–190 (1958–1959)
7. Borsuk, K.: Fundamental retracts and extensions of fundamental sequences. Fundamenta Mathematicae **64**(1), 55–85 (1969)
8. Borsuk, K., Gmurczyk, A.: On homotopy types of 2-dimensional polyhedra. Fundamenta Mathematicae **109**(2), 123–142 (1980)
9. Dodson, C., Parker, P.: A User's Guide to Algebraic Topology. Kluwer, Dordrecht (1997). (Xii+405 pp. ISBN: 0-7923-4292-5, MR1430097)

10. Tozzi, A., Peters, J.: Our thoughts follow a donut-like trajectory in the brain. ResesarchGate Preprint, pp. 1–13 (2015). doi:10.13140/RG.2.1.3305.8008
11. Peters, J., Tozzi, A.: The Borsuk-Ulam theorem explains quantum entanglement. Resesarch-Gate Preprint, pp. 1–7 (2015). doi:10.13140/RG.2.1.3860.1685
12. Crabb, M., Jaworowski, J.: Aspects of the borsuk-ulam theorem. J. Fixed Point Theory Appl. **13**, 459–488 (2013). doi:10.1007/s11784-013-0130-7
13. Lusternik, L., Schnirelmann, L.: Topological methods in calculus of variations [Russian]. Gosudarstv. Izdat. Tehn.-Teor, Lit (1930)
14. Su, F.: Borsuk-Ulam implies Brouwer: A direct construction. Am. Math. Mon. **104**(9), 855–859 (1997). MR1479992
15. Burak, G., Karaca, I.: Digital borsuk-ulam theorem. Bull. Iran. Math. Soc. (2015) (to appear)
16. Ege, O., Karaca, I.: Banach fixed point theorem for digital images. J. Nonlinear Sci. Appl. **8**(3), 237–245 (2015)
17. Willard, S.: General Topology. Dover Pub. Inc, Mineola (1970). Xii + 369pp, ISBN: 0-486-43479-6 54-02, MR0264581
18. Krantz, S.: A Guide to Topology. The Mathematical Association of America, Washington, D.C. (2009). Ix + 107 pp
19. Manetti, M.: Topology. Springer, Heidelberg (2015). doi:10.1007/978-3-319-16958-3. Xii+309 pp
20. Cohen, M.: A Course in Simple Homotopy Theory. Springer, New York (1973). X+144 pp., MR0362320
21. Collins, G.: The shapes of space. Sci. Am. **291**, 94–103 (2004)
22. Malgouyres, R.: Homotopy in 2-dimensional digital images. Theor. Comput. Sci. **230**(1–2), 221–233 (2000)
23. Murrell, H., Hashimoto, K., Takatori, D.: Fisher discrimination with kernels. Mathematica J. **11**, 1–17 (2011)
24. Naimpally, S., Peters, J.: Topology with Applications. Topological Spaces via Near and Far. World Scientific, Singapore (2013). Xv + 277 pp, Am. Math. Soc. MR3075111
25. Renze, J., Uznanski, D., Weisstein, E.: Half-plane. Wolfram MathWorld (2015). http://mathworld.wolfram.com/Half-Plane.html
26. Peters, J., Guadagni, C.: Strongly proximal continuity & strong connectedness. arXiv **1504**(02740), 1–11 (2015)
27. Di Concilio, A., Gerla, G.: Quasi-metric spaces and point-free geometry. Math. Struct. Comput. Sci. **16**(1), 115–137 (2006). MR2220893
28. Di Concilio, A.: Point-free geometries: Proximities and quasi-metrics. Math. Comp. Sci. **7**(1), 31–42 (2013). MR3043916
29. Whitehead, J.: Simplicial spaces, nuclei and m-groups. Proc. Lond. Math. Soc. **45**, 243–327 (1939)
30. Hairer, E., Lubich, C., Wanner, G.: Geometric numerical integration. Structure-Preserving Algorithms for Ordinary Differential Equations. Springer Series in Computational Mathematics, vol. 31, 2nd edn. Springer, Heidelberg (2006). Xviii+644 pp. ISBN: 978-3-642-05157-9, MR2840298
31. Hairer, E.: Important aspects of geometric numerical integration. J. Sci. Comput. **25**(1–2), 67–81 (2005). MR2231943
32. Ehresmann, C.: Les connexions infinitésimales dans un espace fibré différentiable. (french) [infinitesimal connections in a differentiable fiber space]. Séminaire Bourbaki **1**(24), 153–168 (1995). MR1605161
33. Boothby, W.: An introduction to differentiable manifolds and Riemannian geometry. Pure and Applied Mathematics, vol. 120, 2nd edn. Academic Press Inc, Orlando (1986). Xvi+430 pp. ISBN: 0-12-116052-1, MR0861409
34. Jost, J.: Riemannian geometry and geometric analysis. Pure and Applied Mathematics, vol. 120, 6th edn. Springer, Heidelberg (2002). Xiv+611 pp. ISBN: 978-3-642-21297-0, MR2829653
35. Snyman, J.: Practical mathematical optimization. An introduction to basic optimization theory and classical and new gradient-based algorithms. Applied Optimization, vol. 97. Springer, New York (2005). Xx+257 pp. ISBN: 0-387-24348-8, MR2120543

36. Sengupta, B., Friston, K., Penny, W.: Efficient gradient computation for dynamical models. Neuroimage **98**, 521–527 (2014). doi:10.1016/j.neuroimage.2014.04.040
37. Beyer, W., Zardecki, A.: The early history of the ham sandwich theorem. Am. Math. Mon. **111**(1), 58–61 (2004). MR2212093

Chapter 6
Visibility, Hausdorffness, Algebra and Separation Spaces

Fig. 6.1 Activity separation pattern, 1870 Punch

This chapter introduces visibility and separation spaces, useful in the study of set patterns. The notion of visibility stems from our daily experience in being able to detect spatially as well as descriptively near and strongly near shapes in each visual field with our vision (see, e.g., the activity separation pattern in Fig. 6.1). In a sense, visibility is the opposite of separation of sets. In topology, disjoint sets are separated. In effect, disjoint sets are **separated** from each other by intervening points. In the language of visibility, separated sets are invisible to each other. The notion of visibility comes from art gallery theorems (see, e.g., [1]) and computational geometry [2].

The solution of spatial visibility problems is one of the hallmarks of computational geometry and is especially useful in pinpointing regions-of-interest in various forms of object recognition and analysis as well as in pattern recognition and analysis, e.g.,

J.F. Peters, *Computational Proximity*, Intelligent Systems Reference Library 102, DOI 10.1007/978-3-319-30262-1_6

placement of objects (walls, windows, doors, furniture) in the design of buildings, identifying the proximity of objects in digital image, determining the closeness of nodes in social networks.

6.1 Visibility

Sets in a proximity space are **spatially visible** to each other, provided the sets have at least one common point. In other words, strongly near sets are visible to each other.

Example 6.1 In Fig. 6.2, adjacent triangles are visible to each other. For example, the orange triangle ▶ and the brown triangle ▶ are visible to each other, since they have a common edge. ■

Let A, B be subsets in a Delaunay mesh, $\triangle(pqr) \in B, \triangle(qrt) \in A$. Subsets A, B in a Delaunay mesh are *visible* to each other (denoted $A v B$), provided at least one triangle vertex $x \in \mathrm{cl}A \cap \mathrm{cl}B$. A, B are *strongly visible* to each other (denoted $A \overset{\wedge\wedge}{v} B$), provided at least one triangle edge is common to A and B.

Example 6.2 **Visibility in Delaunay Meshes**.
In the Delaunay mesh in Fig. 4.7, $A\ v\ D$ (i.e., D is visible from A), since A and D have one triangle vertex in common, namely, vertex r. Sets B and D in Fig. 4.7 are strongly visible (i.e., $B \overset{\wedge\wedge}{v} D$), since edge $\overline{wx}$ is common to B and D. In Fig. 4.7, let $C = \{\triangle(pqu)\}$. Then $C \overset{\wedge\wedge}{v} B$, since $C \subset B$. In Fig. 6.3, edge $\overline{qr}$ is common to A and B. $\overline{qr}$ is visible from $p \in B$ and from $t \in A$. Hence, $A \overset{\wedge\wedge}{v} B$ ■

Subsets A, B in a Delaunay mesh are *invisible* to each other (denoted $A\ \underline{v}\ B$), provided $\mathrm{cl}A \cap \mathrm{cl}B = \emptyset$, i.e., A and B have no triangle vertices in common. A, B are *strongly invisible* to each other (denoted $A \overset{v}{\mathrm{w}} B$), provided $C\ \underline{v}\ A$ for all sets of mesh triangles $C \subset B$.

Example 6.3 **Invisible and Strongly Invisible Subsets in a Delaunay Mesh**.
In the Delaunay mesh in Fig. 4.7, A and B are not visible to each other, since $\mathrm{cl}A \cap$

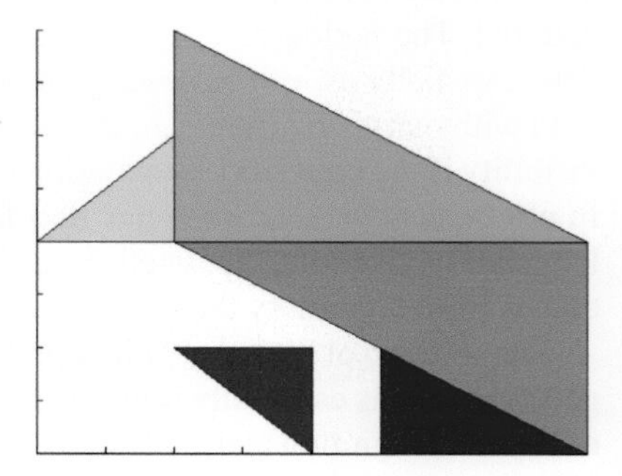

Fig. 6.2 Visible sets in the plane

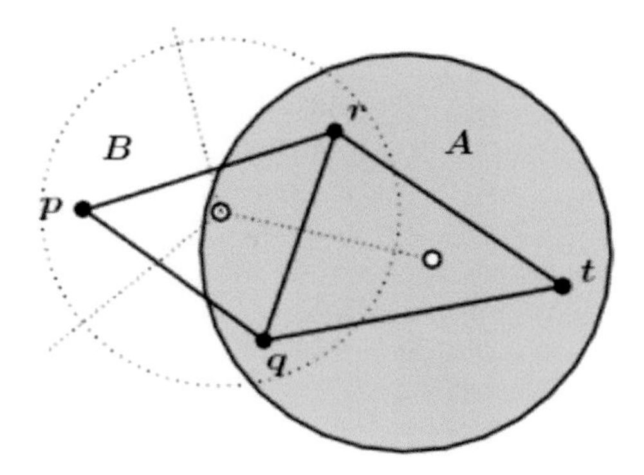

Fig. 6.3 Strongly visible sets $A \overset{\wedge\!\wedge}{v} B$

cl$B$ $= \emptyset$, i.e., A and B have no triangle vertices in common. In Fig. 4.7, let $C = \{\triangle(pqu)\}$. Then $A \overset{v}{\curlyvee} B$ (A and B are strongly invisible to each other), since $C \ \underline{v}\ A$ for all sets of mesh triangles $C \subset B$. ■

Problem 6.4 Identify two sets of strongly near triangles in Fig. 6.4. Use false-colouring to highlight and label the sets of strongly near triangles that you identify. ■

Problem 6.5 Identify three longest edges in Fig. 6.4. Use false-colouring to highlight and label the longest edges in the triangles that you identify. ■

Problem 6.6 Identify two sets of strongly near Voronoï regions in Fig. 6.5. Use false-colouring to highlight and label the sets of strongly near Voronoï regions that you identify. ■

Fig. 6.4 Delaunay mesh

Fig. 6.5 Voronoï mesh

Problem 6.7 Identify three longest edges in Fig. 6.5. Use false-colouring to highlight and label the longest edges in the Voronoï regions that you identify. ■

6.2 Separation Axioms

Various forms of separation in topological spaces are defined by what are known as separation axioms. The main purpose of a separation axiom is to make the points and sets in a space topologically distinguishable [3, Sect. 14.1]. The earliest of such spaces comes from F. Hausdorff, where distinct points belong to disjoint neighbourhoods [4, Sect. 40.II]. The corresponding space with *pairs of distinct points belonging to disjoint neighbourhoods* [4, Sect. 40.II] is now named after him and is called the T_2 or Hausdorff space. There is a descriptive counterpart of a traditional T_2 space (denoted by T_2^{Φ}), introduced in [5] (see, also, [6]). In a T_2^{Φ} space, one can observe that descriptively distinct points belong to disjoint descriptive neighbourhoods. In this chapter, traditional separation spaces are extended to description-based separation spaces. The practical benefit of considering descriptive separation spaces is the generation of multiple patterns that are descriptively distinguishable. In a Hausdorff space, for example, a pair of descriptively distinct points become generators of distinguishable set patterns.

R_0 (Symmetric) topological space.

$$(*) \qquad \{x\} \text{ is near } \{y\} \Rightarrow \{y\} \text{ is near } \{x\}.$$

6.3 R0 Symmetric Space

The R_0 space arose during a consideration of symmetric proximities compatible with a topology. For example, ($*$) is satisfied as a result of partitioning a digital image into equivalence classes. The R_0 axiom was discovered by A.S. Davis [7].

***R*₀ Axiom** . Let x, y be points in a topological space X endowed with the Lodato proximity δ.

$$x\ \delta\ y \Leftrightarrow y\ \delta\ x (x \text{ and } y \text{ are spatially close to each other}).$$

6.4 Descriptive R0 Space

This section introduces an extension of the R_0 space that is descriptive. Before we introduce this extension, we first consider L-proximity (Figs. 6.6 and 6.7).

6.4.1 Near Sets in L-Proximity Spaces

A proximity relation δ called an **L-proximity** is introduced in this section. Let X be a nonempty set, $A, B, C \in \mathcal{P}(X)$, then $A\ \delta\ B$ (A is near B), provided δ satisfies the following axioms [1] from S. Leader [10] and M.W. Lodato [11].

Lodato Axioms Revisited

L(1) $A\ \delta\ B$ implies $B\ \delta\ A$.
L(2) $A\ \delta\ X$, if and only if, A is nonempty.
L(3) $A \cup B\ \delta\ C$, if and only if, A or B is near C.
L(4) If, for every $E \in \mathcal{P}(X)$, $A\ \delta\ E$ or $B\ \delta\ X \setminus E$, then $A\ \delta\ B$.
L(5) $\{x\}\ \delta\ \{y\}$ implies $x = y$.
L(6) $A\ \delta\ B$ and $b\ \delta\ C$ for every $b \in B$ implies $A\ \delta\ C$ (LO axiom [11]).

Example 6.8 **Instance of Axiom L(4)**.
Let X and $\mathbf{X}\backslash\mathbf{E}$ in Fig. 6.8 be endowed with an L proximity relation δ. Sets X and $\mathbf{X}\backslash\mathbf{E}$ in Fig. 6.8 illustrate Axiom L(4). In the set $\mathbf{X}\backslash\mathbf{E}$, the set B has at least one point in common with A. Hence, $A\ \delta\ B$. The basic idea in looking for instances of axiom L(4) is to identify image segments that have points in common, hearkening back to the idea of nearness of segments as in P. Pták and W. Kropatsch [12]. In other words, given a pair of image segments $A, B \in \mathbf{X}\backslash\mathbf{E}$, $A\ \delta\ B$, provided $A \cap B \neq \emptyset$. ■

[1] Leader's axioms L(1), L(2), L(3) and L(5) are a variation of the proximity space axioms P1,P2,P3,P4 in Ju.M. Smirnov [8], introduced by V.A. Efremovič [9].

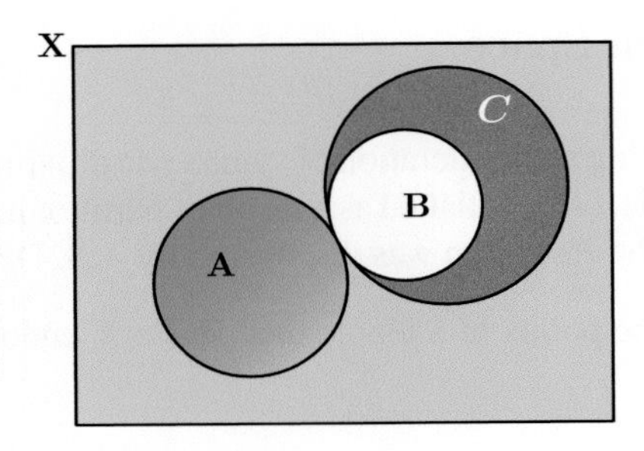

Fig. 6.6 $A\ \delta$ **B** and **B** $\ll$ **C** implies **A** δ **C**

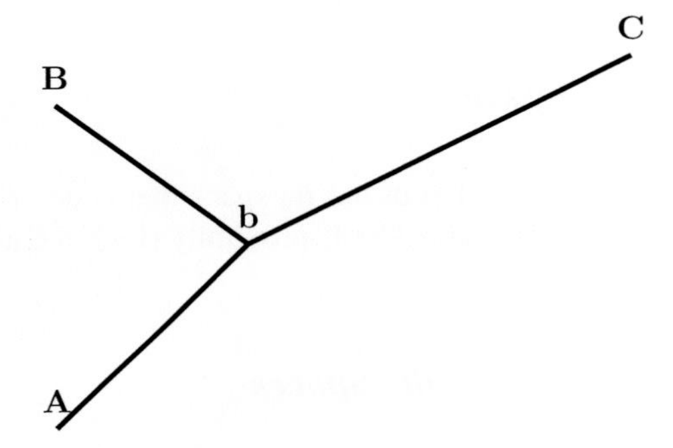

Fig. 6.7 $\text{cl}A\ \delta\ b$ and $b\ \delta\ \text{cl}C$ for some $b \in \text{cl}B$ implies $\text{cl}A\ \delta\ \text{cl}C$

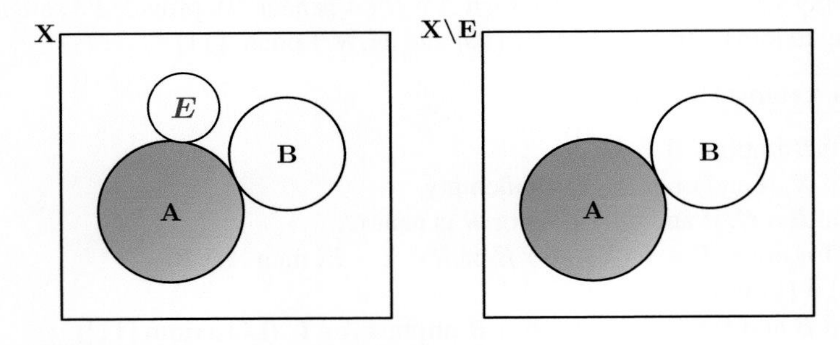

Fig. 6.8 $A\ \delta\ E$ for every $E \in \mathcal{P}(X)$ or $B\ \delta\ X \setminus E \implies A\ \delta\ B$

The following (wLO.7) is a weak form of the Lodato axiom (LO.6).
(**wLO**.7) $\text{cl}A\ \delta\ b$ and $b\ \delta\ \text{cl}C$ for some $b \in \text{cl}B$ implies $\text{cl}A\ \delta\ \text{cl}C$ (weak LO axiom).

Example 6.9 **Instance of (wLO.7)**.
Assume that X is an L-proximity space and that $A, B, C \in \mathcal{P}(X)$. Let A, B, C be represented by the line segments in Fig. 6.7 that have the point $b \in \text{cl}B$ in common. Observe that $A\ \delta\ b$ and $b\ \delta\ C$. Hence, $A\ \delta\ C$. ■

For a set endowed with an L-proximity and containing proximal neighbourhoods, we obtain the result in Theorem 6.10.

Theorem 6.10 *$A\ \delta\ B$ and $B \ll C$ implies $A\ \delta\ C$.*

Proof Assume $A\ \delta\ B$ and $B \ll C$. Since C is a proximal neighbourhood of B, then $B \subset C$. Hence, $b\ \delta\ C$ for every $b \in B$. Then, from the LO axiom, $A\ \delta\ C$. See Fig. 6.6 for a picture proof. □

6.4.2 Descriptive L-Proximity

Next, consider a **descriptive L-proximity space**, obtained by an extension of Leader's L-proximity space axioms. Let Φ be a set of probe functions that represent features of $x \in X$. One set A is **descriptively near** another set B, provided one or more parts of A matches the description of one or more parts of B (e.g., in Fig. 6.9, A partly resembles B, so we write $A\ \delta_{\Phi}\ B$ in terms of the features represented by the probe functions in Φ). The relation δ_{Φ} is a descriptive L-proximity, provided the following axioms are satisfied.

Descriptive Lodato Axioms Revisited

dL(1) $A\ \delta_{\Phi}\ B$ implies $B\ \delta_{\Phi}\ A$.
dL(2) $A\ \delta_{\Phi}\ X$, if and only if, A is nonempty.
dL(3) $A \cup B\ \delta_{\Phi}\ C$, if and only if, A or B is descriptively near C.
dL(4) If, for every $E \in \mathcal{P}(X)$, $A\ \delta_{\Phi}\ E$ or $B\ \delta_{\Phi}\ X \setminus E$, then $A\ \delta_{\Phi}\ B$.
dL(5) $\{x\}\ \delta_{\Phi}\ \{y\}$ implies $\Phi(x) = \Phi(y)$, i.e., the description of x matches the description of y.
dL(6) $A\ \delta_{\Phi}\ B$ and $b\ \delta_{\Phi}\ C$ for *for every* $b \in B$ implies $A\ \delta_{\Phi}\ C$ (descriptive LO axiom (dLO)).

Example 6.11 **Instance of Axiom dL(4)**.
Let X and $\mathbf{X \backslash E}$ in Fig. 6.10 be endowed with a descriptive L-proximity relation δ_{Φ}. Choose Φ to be a set of probe functions that represent colours and greylevel intensities of image points in an appropriate colour space such as rgb. Sets X and $\mathbf{X \backslash E}$ in Fig. 6.8 illustrate Axiom dL(4). In the set $\mathbf{X \backslash E}$, some of the points in set B descriptively match

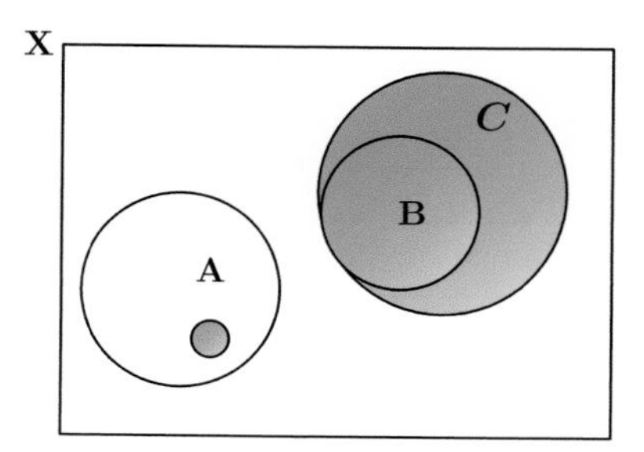

Fig. 6.9 $\mathbf{A}\ \delta_{\Phi}\ \mathbf{B}$ and $\mathbf{B} \ll \mathbf{C}$ implies $\mathbf{A}\ \delta_{\Phi}\ \mathbf{C}$

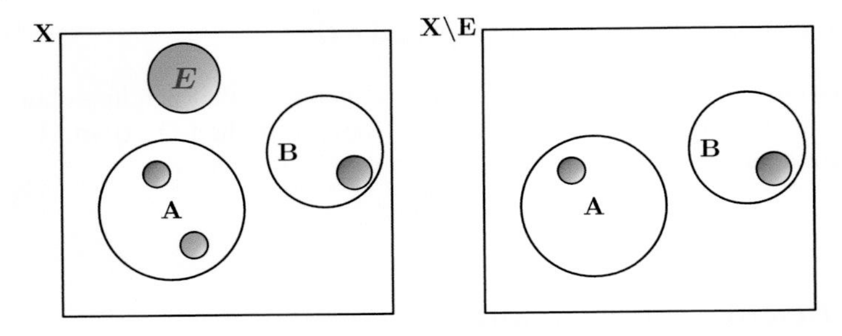

Fig. 6.10 $A\ \delta_{\Phi}\ E$ for every $E \in \mathcal{P}(X)$ or $B\ \delta_{\Phi}\ X \setminus E$ implies $A\ \delta_{\Phi}\ B$

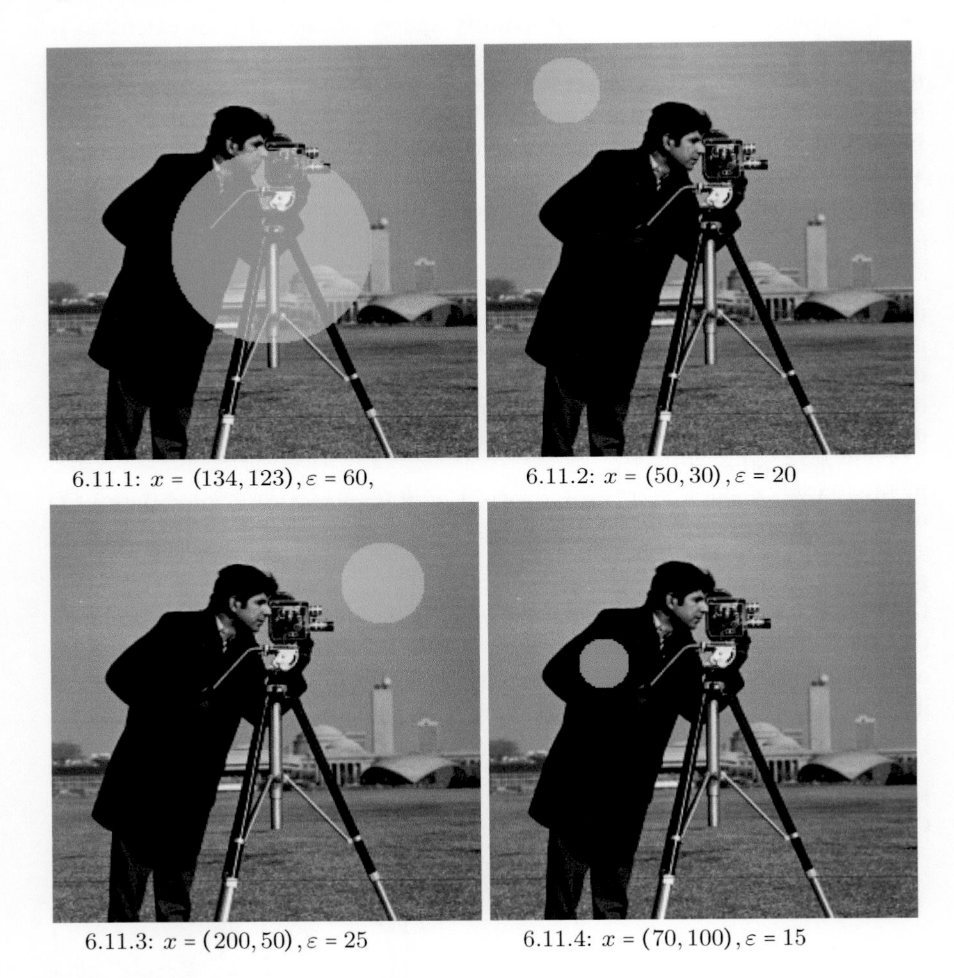

6.11.1: $x = (134, 123), \varepsilon = 60,$

6.11.2: $x = (50, 30), \varepsilon = 20$

6.11.3: $x = (200, 50), \varepsilon = 25$

6.11.4: $x = (70, 100), \varepsilon = 15$

Fig. 6.11 Sample disjoint neighbourhoods

some of the points in A. In other words, $A \underset{\Phi}{\cap} B \neq \emptyset$. Hence, $A\ \delta_\Phi\ B$. For example, one can profitably consider disjoint descriptive neighbourhoods such as those in Fig. 6.11 in looking for instances where axiom dL(4) is satisfied. The basic idea in looking for instances of axiom dL(4) is to identify image segments so that some of the points in each segment descriptively match some of the points in the other segments. ■

6.4.3 Descriptive Proximity Space Revisited

Choose Φ to be a set of probe functions that represent features of points in a topological space and let X be endowed with a descriptive proximity. Notice that X is a topological space, if we consider the descriptive uniform topology on X, where each nonempty subset A determines a descriptive motif pattern (collection of subsets near A). Then there is a descriptive form of the R_0 axiom (R_0^Φ).

R_0^Φ (Descriptively Symmetric) space.

$$\{x\} \text{ is descriptively near } \{y\} \Rightarrow \{y\} \text{ is descriptively near } \{x\},$$

$i.e.,\ \{x\}\ \delta_\Phi\ \{y\} \Rightarrow \{y\}\delta_\Phi\ \{x\}.$

Example 6.12 **Descriptively Symmetric R_0^Φ Space.**
After choosing a set Φ of probe functions representing pixel features for a digital image X endowed with a descriptive L proximity relation δ_Φ, the set of points forms a $\boldsymbol{R_0^\Phi}$ space.

Proof Immediate from axiom dL(1). □

6.5 T0 Anti-Symmetric Space

Earlier separation axioms had been discovered and were called *Trennungsaxiome* (*Trennung* is German for *separation*) by P. Alexandroff and H. Hopf [13, p. 58ff, Sect. 4]. Hence, these axioms are named with a subscripted $\boldsymbol{T}$ as $T_n, n = 0, 1, 2, 3, 4$. Often these axioms have alternate names such as Hausdorff, normal, regular, Tychonoff, and so on and there is no unanimity in the nomenclature. It is helpful to construct counterexamples to delineate various separation axioms (see, e.g., [14, Pt. 2, Sect. 87]).

The focus here is on the **separation of neighbourhoods**, i.e., disjoint sets have neighbourhoods with distinct properties. This directly relates to the study of image segments and the classification of segmented images.

Remark 6.13 **Distinct Points**.
Let X be a nonempty set endowed with a proximity relation δ. Points x, y in a digital image are **spatially distinct**, provided the closures of x and y are not near, i.e., $\text{cl}\{x\} \underline{\delta} \text{ cl}\{y\}$. ■

In contrast to the symmetry axiom R_0, the anti-symmetric axiom T_0 (discovered by A. Kolmogorov) is defined as follows.

T_0: (a) For every pair of distinct points, at least one of them is far from the other, or
(b) For every pair of distinct points in a topological space X, there exists an open set containing one of the points but not the other point (cf. [13, p. 58]).
Note: The discovery of T_0 topologies in digital images hinges on what is meant by the observation that points are distinct.

The discovery of T_0 topologies in digital images hinges on what is meant by the observation that points are descriptively distinct.

Remark 6.14 **Descriptively Distinct Points**.
Let Φ be a set of probe functions that represent features of points x in a nonempty set X. Then let X be endowed with a descriptive proximity relation δ_Φ. Points x, y are **descriptively distinct**, provided x and y are spatially distinct and the feature vectors $\Phi(x)$ and $\Phi(y)$ are not equal. For example, points x, y in a digital image X are descriptively distinct (*descriptively far*), provided x and y are spatially distinct and have different descriptions, i.e., $x \not\delta_\Phi y$. ■

Let Φ be a set of probe functions representing features of members of a set and let $\varepsilon > 0$. There is a descriptive form of T_0 space (denoted by T_0^Φ), defined in terms of descriptive open neighbourhoods. Recall that a *descriptive open neighbourhood* $N_{\Phi(x)}$ is defined by

$$N_{\Phi(x)} = \{y \in X : \Phi(x) = \Phi(y) \text{ and } |x - y| < \varepsilon\}.$$

That is, the description of each point in $N_{\Phi(x)}$ matches the description of x. Due to the spatial restriction $|x - y| < \varepsilon$, $N_{\Phi(x)}$ is also called a *bounded descriptive neighbourhood*.

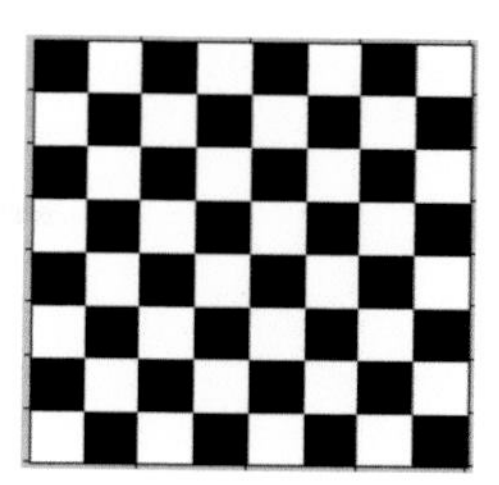

Fig. 6.12 Sample visual space

> T_0^{Φ}: For every pair of descriptively distinct points in a topological space X, there exists a descriptive open neighbourhood containing one of the points but not the other point.

Example 6.15 T_0^{Φ} **Visual Space.**
Let X be represented by the checkerboard in Fig. 6.12 and let x, y be black and white points in X. It is easily verified that X is a topological space. Then let $N_{\Phi(x)}$ be a descriptive open neighbourhood of x. The point y is excluded from $N_{\Phi(x)}$, since $\Phi(x) \neq \Phi(y)$. This is true for every pair of descriptively distinct points in X. Hence, X is a T_0^{Φ} space. ■

6.6 T1 Space

Next, consider a topological space (called a T_1 space) where distinct points are not near.

> T_1: A topological space is T_1 if, and only if, distinct points are not near.

In a T_1 space, singleton sets, and consequently, all finite sets are closed. Recall that an **indiscrete topology** on a nonempty set X contains only X and the empty set. An indiscrete topological space τ is not T_0, since X is the only set containing points and the empty set contains no points. A T_0 space that is not R_0 is $X = \{a, b\}$ with open sets $\emptyset$, X, $\{a\}$. Thus, R_0 and T_0 are independent. Computable versions of the separation axioms T_0, T_1, and T_2 are considered in [15]. The next separation axiom is $T_1 = T_0 + R_0$ (discovered by M. Fréchet and F. Riesz).

Let Φ be a set of probe functions representing features of members of a set. There is a descriptive form of T_1 space (denoted by T_1^{Φ}). For a unified theory for topologies on the closed sets of a metrizable space, see G. Beer and R. Lucchetti [16].

Fig. 6.13 Sample T_1^{Φ} visual space

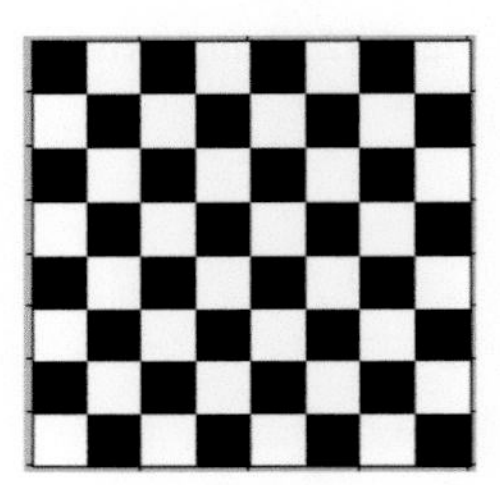

T_1^{Φ}: A topological space is T_1^{Φ} if, and only if, descriptively distinct points are not near.

Example 6.16 **A visual T_1^{Φ} space.**
Choose Φ to be a set of probe functions that represent greyscale and colour intensities of points in an image. Let a topological space X be represented by the checkerboard in Fig. 6.13. X is an example of a visual T_1^{Φ} space. To see this, let $x, y \in X$ be points in black and white squares, respectively. The points x and y are descriptively distinct. In general, black and white pixels in X are both spatially distinct and not near, descriptively. Hence, $x \underline{\delta}_{\Phi}\ y$ and the checkerboard is an example of a T_1^{Φ} space.
■

Lemma 6.17 *A digital image X that is a finite topological space endowed with the Lodato descriptive proximity δ_{Φ} such that X contains two or more descriptively distinct points is a T_1^{Φ} space.*

Proof Let X be a digital image (a set of points called pixels) endowed with a descriptive proximity δ_{Φ}. Choose Φ, a set of probe functions that represent features of points in X. Let points $x, y \in X$ be descriptively distinct. Then $x \underline{\delta}_{\Phi}\ y$, i.e., x is descriptively not near y. Hence, X is a T_1^{Φ} space. □

From Lemma 6.17 and the definition of a segmented image, we obtain the following result.

Theorem 6.18 *A digital image that is a T_1^{Φ} space is segmented.*

A topology on a set X has been compared to a load of gravel. The finer the gravel, the greater the number of open sets that can be used to construct a topology.

6.7 Hausdorff T2 Space

Hausdorff observed that it is possible for a pair of distinct points to have distinct neighbourhoods and used this axiom in his work. The corresponding space such that *pairs of distinct points belong to disjoint neighbourhoods* [4, Sect. 40.II] is now named after him and is called the T_2 or Hausdorff space.

> T_2: A topological space is T_2 if, and only if, distinct points have disjoint neighbourhoods (distinct points live in disjoint *houses*). From an application perspective, especially in solving object recognition problems, the Hausdorffness of finite spaces is important. For example, the Hausdorffness of a digital image makes it possible to separate image regions occupied by distinct objects.

In effect, a T_2 space defines a partition of the space[2]. It has also been observed that every T_1 space is a T_0 space [17, Sect. 1.5]. Putting these observations together, $T_2 \Rightarrow T_1 \Rightarrow T_0$ [15, Sect. 3].

There is a descriptive counterpart of a traditional T_2 space (denoted by T_2^{Φ}), introduced in [5] (see, also, [6]). In a T_2^{Φ} space, one can observe that descriptively distinct points belong to disjoint neighbourhoods.

> T_2^{Φ}: A topological space is T_2^{Φ} if, and only if, descriptively distinct points have disjoint neighbourhoods.

Example 6.19 **A visual T_2^{Φ} space.**
Choose Φ to be a set of probe functions that represent greyscale and colour intensities of points in an image. Let a topological space X again be represented by the checkerboard in Fig. 6.13. X is an example of a visual T_2^{Φ} space. To see this, let $x, y \in X$ be points in black and white squares, respectively. Then consider a pair of descriptive neighbourhoods $N_{\Phi(x)}, N_{\Phi(y)}$ of x and y, respectively. Neighbourhood $N_{\Phi(x)}$ contains only points with descriptions that match the description of x, i.e., $N_{\Phi(x)}$ contains only black points. Similarly, neighbourhood $N_{\Phi(y)}$ contains only points with descriptions that match the description of y, i.e., $N_{\Phi(y)}$ contains only white points. Hence, $N_{\Phi(x)}, N_{\Phi(y)}$ are disjoint. ■

Observe that a T_2^{Φ} space is also a T_1^{Φ} space, since, by definition, descriptively distinct points are not near. Also observe that a T_1^{Φ} space is also a T_0^{Φ} space, since, for every pair of descriptively distinct points, one can find a descriptive open set containing of the points and not containing the other point. The penultimate example of a T_1^{Φ} space that is also a T_0^{Φ} space is a space where descriptively distinct points

---

[2]A partition is a T_2 space, provided every class has no more than one point, i.e., every class is single tenant 'house'.

Fig. 6.14 Manitoba dragonfly

belong to open descriptive neighbourhoods. From these observations, observe that $T_2^{\Phi} \Rightarrow T_1^{\Phi} \Rightarrow T_0^{\Phi}$.

Recall that a **bounded descriptive neighbourhood** $N_{\Phi(x)}$ of a point x in a set X is defined by

$$N_{\Phi(x)} = \{y \in X : d(\Phi(x), \Phi(y)) = 0 \text{ and } |x - y| < \varepsilon\},$$

where d is the taxicab distance between the descriptions of x and y, i.e.,

$$d(\Phi(x), \Phi(y)) = \sum_{i=1}^{n} |\phi_i(x) - \phi_i(y)| : \phi_i \in \Phi.$$

Observe that a T_2^{Φ} space is also a T_1^{Φ} space, since, by definition, descriptively distinct points are not near. The dragonfly in Fig. 6.14 provides an illustration of a biology-based T_2^{Φ} space (see Example 6.21 for details). Also observe that a T_1^{Φ} space is also a T_0^{Φ} space, since, for every pair of descriptively distinct points, one can find a descriptive open set containing of the points and not containing the other point. The penultimate example of a T_1^{Φ} space that is also a T_0^{Φ} space is a space where descriptively distinct points belong to open descriptive neighbourhoods. From these observations, observe that $T_2^{\Phi} \Rightarrow T_1^{\Phi} \Rightarrow T_0^{\Phi}$.

Theorem 6.20 *A digital image X endowed with a descriptive proximity δ_{Φ} such that X contains two or more descriptively distinct points is a T_2^{Φ} space.*

Proof Let X be a digital image (a set of points called pixels) endowed with a descriptive proximity δ_{Φ}. Choose Φ, a set of probe functions that represent features of points in X. Let points $x, y \in X$ be descriptively distinct. Let $N_{\Phi(x)}, N_{\Phi(y)}$ be descriptive neighbourhoods of x, y, respectively. If $a \in N_{\Phi(x)}$, then $d(\Phi(a), \Phi(x)) = 0$, i.e., each member of $N_{\Phi(x)}$ must descriptively match x. Similarly, each $b \in N_{\Phi(y)}$ descriptively matches y. Then, $N_{\Phi(x)} \cap N_{\Phi(y)} = \emptyset$. Hence, X is a T_2^{Φ} space. $\square$

Fig. 6.15 Manitoba dragonfly

Example 6.21 **Dragonfly T_2^{Φ} Shape Space.**
Choose Φ to be a set of probe functions that represent the gradient orientation of the points in an image. Let a topological space X be represented by the dragonfly in Fig. 6.14, endowed with a descriptive proximity relation δ_{Φ}. X is an example of a complex visual T_2^{Φ} shape space. To see this, let $x, y \in X$ be points along the edges of the filtered dragonfly image in Fig. 6.15. The points x and y are descriptively distinct, since these points have different gradient orientations. In addition, points x, y are centers of disjoint descriptive neighbourhoods $N_{\Phi(x)}, N_{\Phi(y)}$, respectively, in a T_2^{Φ} Shape Space.

Proposition 6.22 *The set X represented by the image in Fig. 6.14 is a T_2^{Φ} Shape Space.*

Proof We assume that $\Phi(x) \neq \Phi(y)$, i.e., x and y have different gradient orientations in Fig. 6.15. The descriptive neighbourhood $N_{\Phi(x)}$ of point x (with no spatial restriction) is defined by

$$N_{\Phi(x)} = \{a \in X : \Phi(x) = \Phi(a) \text{ and } |x - a| < \varepsilon\},$$

i.e., the gradient orientation of x matches the gradient orientation of each point a in $N_{\Phi(x)}$. Hence, $y \notin N_{\Phi(x)}$, since the gradient orientation of y does not match the gradient orientation of x. Similarly, observe that $x \notin N_{\Phi(y)}$. Then $N_{\Phi(x)}, N_{\Phi(y)}$ are disjoint. This is true of every pair of points in X that have unequal gradient orientations. Hence, X is an example of a descriptive T_2^{Φ} shape space. □

6.8 Hausdorffness and Visibility

The **Hausdorffness of a space** means that distinct points in the space belong to disjoint neighbourhoods that are not visible to each other. In other words, Hausdorffness implies the absence of visibility between separated sets, i.e., sets that have no points

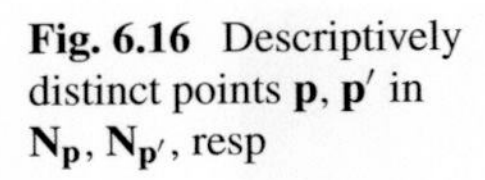

Fig. 6.16 Descriptively distinct points $\mathbf{p}, \mathbf{p}'$ in $\mathbf{N_p}, \mathbf{N_{p'}}$, resp

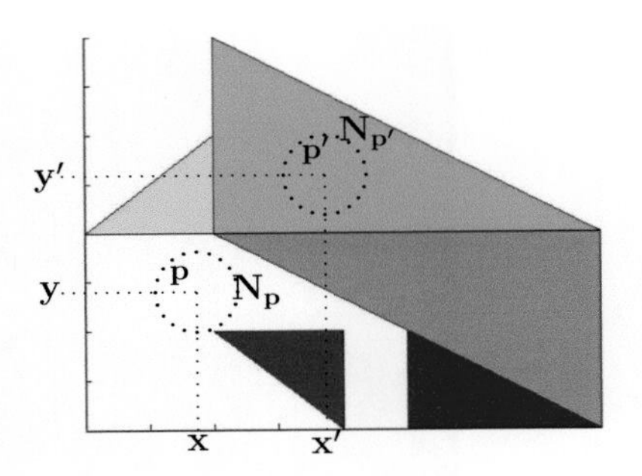

in common. Between any two separated sets, one can always finds points in between the sets that block the view between the sets. Separated sets in a Hausdorff space are analogous to shorelines on opposite sides of a very large ocean. Standing on one shoreline of a very large lake or sea, the opposite shoreline is not visible.

The descriptive Hausdorffness of a space endowed with a proximity relation means that descriptively distinct points belong to disjoint descriptive neighbourhoods that are descriptively not visible to each other. This form of invisibility may seem a bit strange at first but later it becomes apparent that descriptive invisibility has utility in separating objects in a scene. Descriptively separate sets are analogous to canine vision vs. human vision in which dogs cannot see as many colours as human because dogs have only two kinds colour-detecting cells (cones), whereas human have three kinds of cones in the retinas of their eyes. Descriptively separated sets are analogous to canine vs. human vision. For example, what is blue-green for human vision is gray for canine vision.

The descriptive Hausdoffness of such things as digital images leads to a natural separation of image regions in terms of their descriptions. There are many different forms of separation of sets. Let X be a proximity space, $A, B \subset X$ and let $A \not\delta_{\Phi} B$ denote the descriptive separation of A and B (A, B have no points with matching descriptions), $A \underset{\Phi}{\overset{\wedge\!\wedge}{\not\delta}} B$ denotes the strong descriptive separation of A and B. Strangely enough, A can be spatially near B and yet A can be descriptively remote from B (see, e.g., Fig. 6.17).

Example 6.23 **Spatially near, descriptively remote sets.**
In Fig. 6.16, the orange triangle ▶ and the brown triangle ▶ are spatially near each other, since they have a common edge but these filled in triangles are descriptively remote. ■

An overview of the different forms of spatial and descriptive separation is given next.

Spatial separation $A \not\delta B$, i.e., A and B have no points in common. In a Hausdorff space, disjoint neighbourhoods are spatially separated.

Example 6.24 In Fig. 6.17, consider Voronoï regions $V_{a_1} \not\delta V_{a_8}$, $V_{a_{11}} \not\delta_{\Phi} V_{a_8}$ and $V_{a_{12}} \not\delta_{\Phi} V_{a_8}$. ■

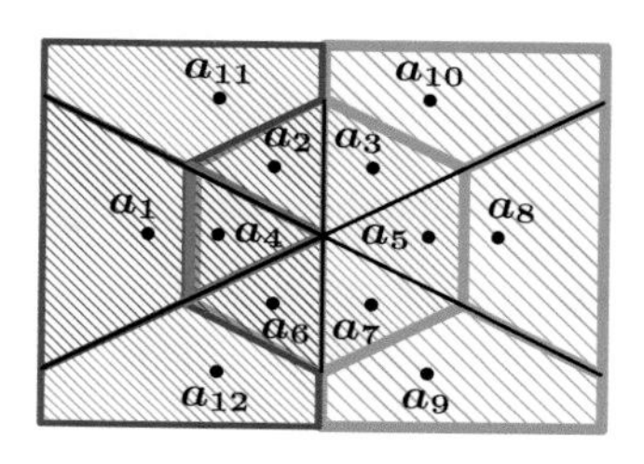

Fig. 6.17 Voronoï regions: $V_{a_5} \overset{\wedge\!\wedge}{\delta} V_{a_8}$ and $V_{a_5} \not\delta_{\Phi} V_{a_8}$

Strong spatial separation $A \underset{\Join}{\not\delta} B$, i.e., A and B are strongly far from each other, which means not only that A, B have no points in common but also that A, B are strongly included in other sets, which are disjoint. In a Hausdorff space, disjoint neighbourhoods A, B are strongly spatially separated, provided $\mathrm{cl}A \ll E \Longleftrightarrow E^c \ll \mathrm{cl}A^c$ for some $E \subset X$. For more about this, see [6, p. 15].

Descriptive separation $A \not\delta_{\Phi} B$, i.e., A and B have different descriptions.

Example 6.25 In Fig. 6.17, consider Voronoï regions $V_{a_{10}} \not\delta_{\Phi} V_{a_3}$ and $V_{a_1} \not\delta_{\Phi} V_{a_9}$. ■

Strong descriptive separation $A \underset{\Join_{\Phi}}{\not\delta} B$, i.e., A and B have different descriptions and A, B are descriptive strongly included in other sets, which are disjoint. In a Hausdorff space, disjoint neighbourhoods A, B are strongly spatially separated, provided $\mathrm{cl}A \ll_{\Phi} E \Longleftrightarrow E^c \ll_{\Phi} \mathrm{cl}A^c$ for some $C \subset X$. For more about this, see [6, p. 115].

Example 6.26 In Fig. 6.17, consider Voronoï regions $V_{a_1} \underset{\Join_{\Phi}}{\not\delta} V_{a_8}$ such that $V_{a_1} \subset (V_{a_1} \cup V_{a_4})$. ■

Spatial separation and Descriptively Near A, B are spatially separated and descriptively near, provided A, B have no points in common and A, B contain points with matching descriptions. That is, $A \not\delta B$ and $A \delta_{\Phi} B$. In the physical world, spatially separate, descriptively near sets are common. For example, a pair of paintings of waterscapes in an art gallery are spatially separated but descriptively near.

Example 6.27 In Fig. 6.17, consider Voronoï regions $V_{a_1} \underset{\Join}{\not\delta} V_{a_8}$ and $V_{a_5} \delta_{\Phi} V_{a_9}$. ■

Spatial separation and Descriptively Remote A, B are spatially separated and descriptively remote, provided A, B have no points in common and A, B contain no points with matching descriptions. ■

Example 6.28 In Fig. 6.17, consider Voronoï regions $V_{a_1} \underset{\Join}{\not\delta} V_{a_8}$ and $V_{a_5} \not\delta_{\Phi} V_{a_8}$. ■

Lemma 6.29 *Nonempty sets are spatially separated, if and only if the sets are not visible to each other.*

Proof Immediate from the definitions of spatial separation and spatial visibility. □

Theorem 6.30 *In a Hausdorff space, distinct points belong to disjoint neighbourhoods that are invisible to each other.*

Proof Let x, y be distinct points in a Hausdorff space X. From the Hausdorfness of X, x, y belong to disjoint neighbourhoods A, B, respectively. From Lemma 6.29, A, B are not visible to each other. □

Definition 6.31 Spatial Visibility Between Sets.
Let A, B be nonempty sets in a proximity space endowed with a spatial visibility relation v. A and B are descriptively visible to each other (denoted by $A \text{ v } B$), if and only if the interiors int$A$ and int$B$ have common points, i.e.,

$$A \text{ v } B \Leftrightarrow \text{int}A \cap \text{int}B \neq \emptyset. \quad \blacksquare$$

Definition 6.32 Descriptive Visibility Between Sets.
Let A, B be nonempty sets in a descriptive proximity space endowed with a descriptive visibility relation v_Φ. A and B are descriptively visible to each other (denoted by $A \text{ v}_\Phi B$), if and only if the sets of descriptions of the interiors $\Phi(\text{int}A)$ and $\Phi(\text{int}B)$ have common points, i.e.,

$$A \text{ v}_\Phi B \Leftrightarrow \Phi(\text{int}A) \cap \Phi(\text{int}B) \neq \emptyset. \quad \blacksquare$$

Definition 6.33 Descriptive invisibility.
Nonempty sets A, B are descriptive invisible, provided there is no point in A that has the same description as a point in B, i.e., $\Phi(A) \cap \Phi(B) = \emptyset$. Let A, B be nonempty sets in a descriptive proximity space (X, δ_Φ). A and B are descriptively invisible, provided $A \not\delta_\Phi B$. ■

Lemma 6.34 *Nonempty sets are descriptively separated, if and only if the sets are not descriptively visible to each other.*

Proof Immediate from the definitions of descriptive separation and descriptive invisibility. □

Theorem 6.35 *In a Hausdorff space endowed with a descriptive proximity δ_Φ, descriptively distinct points belong to descriptively disjoint neighbourhoods that are invisible to each other.*

Proof Let x, y be distinct points in a descriptive Hausdorff space (X, δ_Φ). From the Hausdorfness of X, x, y belong to descriptively disjoint neighbourhoods A, B, respectively. From Lemma 6.34, $\Phi(A), \Phi(B)$ are not visible to each other. That is, $\Phi(A) \not\delta_\Phi \Phi(B)$, if and only if $\Phi(A) \cap \Phi(B) = \emptyset$. By definition, $\Phi(A), \Phi(B)$ are not visible to each other. Hence, A and B are descriptively invisible to each other. □

Fig. 6.18 Unbounded descriptive neighbourhood of pixel edge of righthand side of dish

Fig. 6.19 Unbounded descriptive neighbourhood of pixel closest to the Nhbd in Fig. 6.18

To see this, consider the following examples of separated descriptive neighbourhoods.

Example 6.36 **Descriptive Visibility** The Matlab script A.3 in Appendix A.6 was used to produce the unbounded descriptive neighbourhoods shown in Fig. 6.18. The neighbourhoood (denoted by *nhdb*) in Fig. 6.18 is an unbounded descriptive nhdb A of a pixel on the upper righthand edge of the dish holding the planter. The nhdb in Fig. 6.19 is also an unbounded descriptive nhdb B of a pixel that is closest to the nhdb in Fig. 6.18. In this case we can observe that

1° $A \ \delta \ B$, i.e., A and B are close to each other (both sets share points).

2° $A \ \mathrm{v} \ B$, i.e., A and B are spatially visible to each other, since A, B have common points.

3° $A \ \not\delta_{\Phi} \ B$, i.e., A and B are descriptively remote from each other, since A, B do not have matching descriptions. ■

Theorem 6.37 *Let $A(p)$, $B(q)$ be neigbourhoods of points p, q, respectively, in a digital image such that $A(p)$, $B(q)$ overlap (have common points). Then*

1° $A(p)$ is descriptively near $B(q)$, i.e., $A(P)\ \delta_\Phi\ B(q)$.

2° $A(p)$ and $B(q)$ are spatially visible to each other, i.e., $A(p)\ v\ B(q)$.

3° $A(p)$ and $B(q)$ are descriptively visible to each other, i.e., $A(p)\ v_\Phi\ B(q)$.

Problem 6.38 Do the following:

1° Prove all parts of Theorem 6.37.

2° Write a Mathematica script to illustrate Part 1 of Theorem 6.37. That is, the script should make it possible to click on an image pixel p and produce a nhdb $A(p)$ for a given radius ε so that

$$A(p) = \{x \in X : \|x - p\| \leq \varepsilon\}. \text{ (Neighbourhood of } p\text{).}$$

Then your script should display a neighbourhood B of a point q with the same ε so the nhdb $B(q)$ overlaps with $A(p)$. Use false colouring to shade with green the pixels in the overlapping region of the two neighbourhoods. ■

Problem 6.39 The Matlab script A.3 in Appendix A.6 displays unbounded descriptive neighbourhoods (UBdNhbds) of points so that, for each given UBdNhbd A, the script find the spatially closest UBdNhbd B. Do the following

1° The modified Matlab script A.3 makes it possible to display

(a) Bounded descriptive neighbourhoods.

2° Write a Mathematica script to illustrate Part 1 of Theorem 6.37. ■

6.9 Partial Descriptive Groupoids in Hausdorff Spaces

This section introduces algebraic structures in Hausdorff spaces. The emphasis is on partial descriptive groupoids derived from descriptive neighbourhoods of points in descriptive T_2^Φ Spaces.

Remark 6.40 **Partial Descriptive Groupoids in T_2^Φ Spaces**
Let X be a descriptive Hausdorff space T_2^Φ, and let $A(x) \subset X$ be a descriptive neighbourhood of a point in $x \in X$. Also, let nhbd denote the neighbourhood of a point and let $\Phi(x)$ denote the description of x, i.e., a feature vector that describes x. Observe that every point in A will have the same description as the neighbourhood center x. Let $\circ_\Phi : A \times A \to A$ be defined by

$$\circ_\Phi(a, b) = \Phi(x), \text{ for } a, b \in \text{nhbd}A(x), \text{nbhd center } x \in X.$$

Since every point in neigbourhood A has the same description, $\circ_\Phi$ is a partial binary operation and $A(\circ_\Phi)$ is a **partial descriptive groupoid**. ■

Theorem 6.41 *Let $\circ_\Phi$ be the partial binary operation defined in Remark 6.40 and let X be a descriptive Hausdorff space T_2^Φ. Every descriptive neighbourhood in X is a partial descriptive groupoid.*

Proof Immediate from the observations in Remark 6.40. □

Remark 6.42 **Near Partial Descriptive Groupoids in T_2^Φ Spaces**
Let X be a descriptive Hausdorff space T_2^Φ, and let $A \subset X$ is a descriptive neighbourhood of a point in $x \in X$. Further, assume that a metric d is defined on X. Let A, B be partial descriptive groupoids defined by the partial binary operation $\circ_\Phi$ from Remark 6.40. Then the descriptive distance $D_\Phi(A, B)$ is defined

$$D_\Phi(A, B) = inf \{d(\Phi(a), \Phi(b)) : a \in A, b \in B\}.$$

Let $\Phi = \{\phi_i, ..., \phi_n\}$, a set of probe functions that represent features of points in X. Then we can define metric d to be the Manhattan distance between $\Phi(a), \Phi(b)$, i.e.,

$$d(\Phi(a), \Phi(b)) = \sum_{i=1}^{n} |\phi_i(a) - \phi_i(b)|.$$

Further, let $\varepsilon > 0$. Then the partial descriptive groupoid A is descriptively near the partial descriptive groupoid B, provided $D_\Phi(A, B) < \varepsilon$. ■

Theorem 6.43 *Let $\mathfrak{P}_\Phi(A)$ be a collection of partial descriptive groups that are near the partial descriptive groupoid A. If B is a partial descriptive groupoid in $\mathfrak{P}_\Phi(A)$,*
(1) $D_\Phi(A, B) < \varepsilon$ for $\varepsilon > 0$,
(2) The partial descriptive groupoid $A(\circ_\Phi)$ is a descriptive pattern generator,
(3) $\mathfrak{P}_\Phi(A)$ is a set pattern.

Proof (1): Immediate from the observations in Remark 6.42. (2)–(3): Immediate from the definitions of pattern generator and set pattern. □

Remark 6.44 **Partial Descriptive Semigroups in T_2^Φ Spaces**
If $\circ_\Phi$ in Remark 6.40, then by extension of Theorem 6.41, every descriptive neighbourhood in a T_2^Φ space is a semigroup. ■

6.10 Set-Based Probes and Proximal Delaunay Groupoids

Set-based probes serve as a means of describing regions in computational point-free geometry as well as in computational geometry in general. A **region** is a set with cardinality greater than one in an n-dimensional space. Set-based probe functions are used to define set descriptions, instead of the usual point descriptions. In this work, a *probe* is a real-valued function that extracts a feature value from a nonempty set (the notion of a probe was introduced by M. Pavel [18, Sect. 2.3, p. 9]). Let X be a topological space endowed with the proximity δ, 2^X the powerset of X, $A, B \in 2^X$.

Definition 6.45 Set-Based Probe
Let A be a region in an n-dimensional space X. A **set-based probe** $\phi : 2^X \longrightarrow \mathbb{R}$ is defined by $\phi(A) \in \mathbb{R}$. ■

Example 6.46 Let A be a solid polygon in the Eucludean plane, i.e., int$A \neq \emptyset$. Here are sample probes on A.

1° $\phi_1(A) =$ area of A.
2° $\phi_2(A) =$ perimeter of A.
3° $\phi_3(A) =$ diameter of A (see, e.g., [19, 20]).
4° $\phi_4(A)$ uses cl$A$ to probe the density of A such that

$$\phi_4(A) = \begin{cases} 1, & \text{if } A \text{ is dense, } i.e., \text{cl}A = X; \\ 0, & \text{if } A \text{ is not dense.} \end{cases}$$

5° $\phi_5(A)$ uses int(cl$A$) to probe the regular openness of A such that

$$\phi_5(A) = \begin{cases} 1, & \text{if } A \text{ is regular open, } i.e., A = \text{int}(\text{cl}A); \\ 0, & \text{if } A \text{ is not regular open.} \end{cases}$$

6° $\phi_6(A)$ uses $\alpha A + (1 - \alpha)A$ to probe the convexity of A such that

$$\phi_6(A) = \begin{cases} 1, & \text{if } A \text{ is convex, } i.e., \alpha A + (1-\alpha)A \subset A \forall \alpha \in [0,1] [13, \text{Sect. 1.1}, p.4]; \\ 0, & \text{if } A \text{ is not convex.} \quad ■ \end{cases}$$

7° $\phi_7(A)$ uses a descriptive strong connectedness relation

$$\overset{\pitchfork,\Phi}{\text{conn}}(A) = \left\{ B \in Re : A \overset{\pitchfork,\Phi}{\text{conn}} B \right\},$$

(also written $\overset{\pitchfork,\Phi}{\text{conn}}(A, Re)$ or $A \overset{\pitchfork,\Phi}{\text{conn}} Re$) for a collection of regions $Re \in 2^{2^X}$ in a proximity space X (analogous to the connection relation in [19]). $(Re, \overset{\pitchfork,\Phi}{\text{conn}})$ is called a descriptive strong connectedness space. See, also, [20]), to probe the connectedness of A in relation to regions $B \in Re$ such that

$$\phi_7(A) = \begin{cases} 1, & \text{if } \overset{\pitchfork,\Phi}{\text{conn}}(A) \neq \emptyset, i.e., A \text{ is descriptively strongly connected to Re}; \\ 0, & \text{if } \overset{\pitchfork,\Phi}{\text{conn}}(A) = \emptyset. \quad ■ \end{cases}$$

8° $\phi_8(A)$ uses $\underset{\Phi}{\cap}$ to determine if A is descriptively near nerve nucleus $B \subset X$ such that

$$\phi_8(A) = \begin{cases} 1, & \text{if } A \underset{\Phi}{\cap} B \neq \emptyset; \\ 0, & \text{if } A \underset{\Phi}{\cap} B = \emptyset. \quad ■ \end{cases}$$

Strong descriptive connectedness is defined in terms of descriptive strong proximity.

Definition 6.47 ($\overset{\pitchfork}{\text{conn}}$-connectedness[21, Sect. 4, Definition 4.2], [22]).
Let X be a topological space with $\overset{\pitchfork}{\delta}$ a strong proximity on X. We say that X is $\overset{\pitchfork}{\delta}$ −connected (denoted by $\overset{\pitchfork}{\text{conn}}$) if and only if $X = \bigcup_{i \in I} X_i$, where I is a countable subset of $\mathbb{N}$, X_i and int(X_i) are connected for each $i \in I$, and $X_{i-1} \overset{\pitchfork}{\delta} X_i$ for each $i \geq 2$. ■

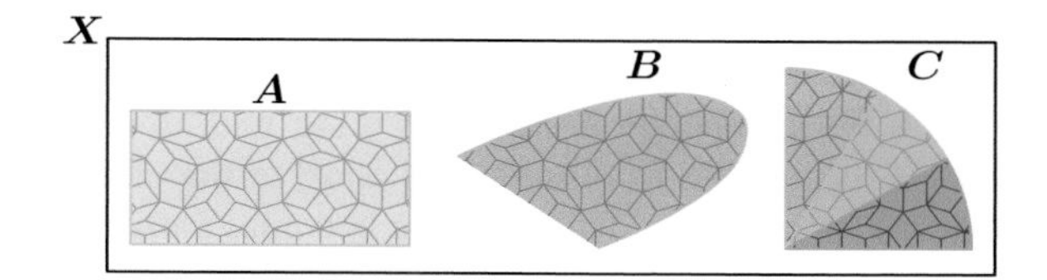

Fig. 6.20 $\Phi(A)$ descriptively strongly near $\Phi(C)$

Definition 6.48 ($\overset{\wedge\wedge,\Phi}{\text{conn}}$-connectedness).
Let X be a topological space with $\overset{\wedge\wedge}{\delta}$ a strong proximity on X. We say that X is $\overset{\wedge\wedge}{\delta}$ −connected (denoted by $\overset{\wedge\wedge,\Phi}{\text{conn}}$) if and only if $X = \bigcup_{i \in I} X_i$, where I is a countable subset of $\mathbb{N}$, X_i and $\text{int}(X_i)$ are connected for each $i \in I$, and $X_{i-1} \overset{\wedge\wedge}{\delta}_{\Phi} X_i$ for each $i \geq 2$. ■

Example 6.49 **Descriptive Strong Connectedness** [21].
Let X be a space of picture points represented in Fig. 6.20. with red, green or blue colours and let $\Phi : X \to \mathbb{R}$ a description on X representing the colour of a picture point, where 0 stands for red (r), 1 for green (g) and 2 for blue (b). Suppose the range is endowed with the topology given by $\tau = \{\emptyset, \{r, g\}, \{r, g, b\}\}$.

The space $X = A \cup B \cup C$ is descriptively $\overset{\wedge\wedge,\Phi}{\text{conn}}$-connected. In fact $\Phi(A) = \{g\}$, $\text{int}(\Phi(A)) = \emptyset$, $\Phi(C) = \{r, g, b\} = \text{int}(\Phi(C))$, $\Phi(B) = \{r\}$, $\text{int}(\Phi(B)) = \emptyset$ and they are all connected in τ. Furthermore $\Phi(A) \overset{\wedge\wedge}{\delta} \Phi(C)$ and $\Phi(C) \overset{\wedge\wedge}{\delta} \Phi(B)$. ■

Remark 6.50 **Probing Descriptive Nearness of a Region to a Nucleus-Based Nerve**.
Let B be the nucleus of a nerve in a descriptive proximity space X.

$$\phi_8(A) = 1 \Leftrightarrow A \ \overset{\wedge\wedge}{\delta}_{\Phi} \ B, \text{ for some } A \in 2^X. \quad ■$$

Definition 6.51 Set-Based Description
Let A be a subset of a nonempty set X, set-based probes $\phi_1, \ldots, \phi_n$ such that $\phi_i(A) \in \mathbb{R}, 1 \leq i \leq n$. A **set-based description** $\Phi : 2^X \longrightarrow \mathbb{R}^n$ is defined by a feature vector of a subset A, i.e.,

$$\Phi(A) = (\phi_1(A), \ldots, \phi_n(A)) \in \mathbb{R}^n, \text{ feature vector describing } A. \quad ■$$

Each probe function value is the value of a feature of a set. A *feature of a set* is a characteristic (something typical) of the set. Probes on a nonempty set describe the set.

The focus here is on the description of sets in, for example, Euclidean space $\mathbb{R}^2$ (the plane). For example, let A be triangle in a Delaunay triangulation of a set of points $S \subset X$ that covers a set X in the plane. The set of descriptions of members of X equipped with a binary operation $\circ$ leads naturally to a groupoid.

Remark 6.52 **Groupoids in Proximity Spaces**.
Groupoids offer a means of observing the behaviour of members of a set equipped with a binary operation. Experimenting with different binary operations on a set makes to possible to observe and analyze the effect of each binary operation on pairs of set elements. This behavior becomes particular interesting when we operate on pairs of sets in the context of a descriptive proximity space in the case where the operation on the sets has an *indeterminate* character. For example, let $U, V \in 2^X$ be neighbourhoods of points in 2^X, $\Phi(U), \Phi(V)$ descriptions of U and V and $\circ(\Phi(U), \Phi(V)) = \frac{0}{0}$ for descriptions that str both zero vectors. The indeterminancy of a binary operation depends on the behaviour the two set-based probe functions that varies with the variation of the neighbourhoods U, V. ■

Definition 6.53 A **Delaunay groupoid** is a nonempty set of Delaunay triangles equipped with a binary relation $\circ$. ■

Definition 6.54 A **proximal Delaunay groupoid** is a Delaunay groupoid in a proximity space. ■

Example 6.55 **Delaunay Groupoids**.
Let X be a finite set of triangles in a Delaunay triangulation of a set S in the plane, A a triangle in X, $\mathfrak{G}$ is a set of descriptions of triangles in X. A single Delaunay triangle is nonempty set of points contained in the edges and vertices of the triangle. Here are some examples of Delaunay groupoids.

$$
\begin{aligned}
A, B \in X &= \text{ Delaunay triangles},\\
\Phi(A) &= (\phi_1(A), \phi_2(A)) \in \mathbb{R}^2,\\
\phi_i(A) \in \mathbb{R} &= \text{ feature value of } A, \text{ where}\\
\phi_1(A) &= \text{ area of } A,\\
\phi_2(A) &= \text{ perimeter of } A,\\
G &\subseteq X, \text{ set of Delaunay triangles},\\
\mathfrak{G} &= \left\{(\phi_1(A), \phi_2(A)) : (\phi_1(A), \phi_2(A)) \in \mathbb{R}^2 \text{ for all } A \in G\right\}\\
\Phi(A), \Phi(B) &\in \mathfrak{G},\\
(\mathfrak{G}, \circ_1)&, \text{ feature space groupoid, where}\\
\Phi(A) \circ_1 \Phi(B) &= \begin{cases} \Phi(A), & \text{if } \|\Phi(A) - \Phi(B)\| = 0;\\ \Phi(B), & \text{otherwise.} \end{cases}\\
(\mathfrak{G}, \circ_2) &= \text{ feature space groupoid, where}\\
\Phi(A) \circ_2 \Phi(B) &= \begin{cases} \Phi(B), & \text{if } \|\Phi(A) - \Phi(B)\| > 0;\\ \Phi(A), & \text{otherwise.} \quad \blacksquare \end{cases}\\
(\mathfrak{G}, \circ_3) &= \text{ feature space groupoid, where}\\
\Phi(A) \circ_3 \Phi(B) &= \begin{cases} \Phi(B), & \text{if } \Phi(A) = \Phi(B);\\ \Phi(A), & \text{otherwise.} \quad \blacksquare \end{cases}
\end{aligned}
$$

Definition 6.56 Delaunay Feature Vector.
Let (X, δ) be a finite Lodato proximity space of triangles in a Delaunay triangulation of a set X in the Euclidean plane, A, B a pair of triangles in X, $\mathfrak{G}$ is a set of descriptions of triangles in X. A **Delaunay feature vector** is a description of a Delaunay triangle. ■

Theorem 6.57 *Let $\mathfrak{G}$ is a set of descriptions of Delaunay triangles in the Euclidean plane. The feature space groupoid $(\mathfrak{G}, \circ_1)$ is a Delaunay feature space groupoid.*

Proof Immediate from the definition of the feature space groupoid $(\mathfrak{G}, \circ_1)$. □

Theorem 6.58 *Let $\mathfrak{G}$ be a set of descriptions of Delaunay triangles in the Euclidean plane, $\Phi(A), \Phi(B) \in \mathfrak{G}$, $\circ : \mathfrak{G} \times \mathfrak{G} \longrightarrow \mathfrak{G}$ a binary operation. The feature space groupoid $(\mathfrak{G}, \circ)$ is a Delaunay feature space groupoid.*

Proof Immediate from the definition $\circ$. □

Remark 6.59 **Near Delaunay Feature Vectors.**
Let (X, δ) be a finite proximity space of triangles in a Delaunay triangulation of a set X in the Euclidean plane, A, B a pair of triangles in X, $\mathfrak{G}$ is a set of descriptions of triangles in X. Descriptions $\Phi(A), \Phi(B) \in \mathfrak{G}$ are near each other, provided $\phi(A) = \phi(B)$ for at least one probe function ϕ used to define the descriptions of $\Phi(A), \Phi(B)$. That is, $\phi(A)\ \delta\ \phi(B)$, provided $\phi(A) = \phi(B)$ for some $\phi(A) \in \Phi(A), \phi(B) \in \Phi(B)$. In other words, Delaunay feature vectors are **near** (**proximal**), provided the feature vectors have one or more matching coordinates. ■

Next, let $\Phi(A), \Phi(B)$ be descriptions of Delaunay triangles A, B that cover a subset of the Euclidean plane. The binary operation $\circ_{\overset{\wedge\!\!\!\wedge}{\delta}}$ be defined by

$$\Phi(A) \circ_{\overset{\wedge\!\!\!\wedge}{\delta}} \Phi(B) = \begin{cases} \Phi(A), & \text{if } \Phi(A) \overset{\wedge\!\!\!\wedge}{\delta} \Phi(B); \\ \Phi(B), & \text{otherwise.} \end{cases}$$

The binary operation $\circ_{\overset{\wedge\!\!\!\wedge}{\delta}}$ has a bias towards the set of descriptions $\Phi(A)$ that are strongly near $\Phi(B)$.

Theorem 6.60 *Let $\mathfrak{G}$ be a set of descriptions of Delaunay triangles in the Euclidean plane, equipped with the strong proximity $\overset{\wedge\!\!\!\wedge}{\delta}$. The feature space groupoid $(\mathfrak{G}, \circ_{\overset{\wedge\!\!\!\wedge}{\delta}})$ is a strong proximal Delaunay feature space groupoid.*

Proof Immediate from the definition of $\circ_{\overset{\wedge\!\!\!\wedge}{\delta}}$. □

Nonempty sets of Delaunay triangles G, H are **spatially near sets**, provided every member of G has a vertex in common every member of H. Groupoids are **spatially near groupoids**, provided the groupoids have at least one point in common.

Example 6.61 **Spatially near Delaunay triangles.**
Let X be a set of Delaunay triangles endowed with proximity δ, $G, H \subset X$ such that

1° $G = \{A_1, A_2\}$.
2° $H = \{B_1, B_2, B_3, B_4\}$.
$G\ \delta\ H$, since the triangles in G and H have a common vertex. Let G, H be equipped with any binary operation $\circ$. Then the groupoids $(G, \circ), (H, \circ)$ are spatially near. ■

Theorem 6.62 *Let X be a nonempty set of spatially near Delaunay triangles that is a topological space endowed with a Lodato proximity δ, $G, H \subset X$ equipped with a binary relation $\circ$. Then G is spatially near H.*

Proof By definition, all triangles in X have a common vertex. $G\ \delta\ H$, since $G, H \subset X$. Hence, the Delaunay groupoids $(G, \circ), (H, \circ)$ are spatially near. □

Corollary 6.63 *Let X be a nonempty set of spatially near Delaunay triangles, 2^X collection of all subsets of triangles in X. For all $G, H \in 2^X$ equipped with any binary operations $\circ$ in the proximity space (X, δ), G and H are spatially near groupoids.*

Proof Immediate from Theorem 6.62. □

Nonempty sets of Delaunay triangles G, H are **descriptively near sets** (denoted $G\ \delta_\Phi\ H$), provided there is a member of G that has a description in common with a member of H. Groupoids are **descriptively near groupoids**, provided the descriptions of the groupoids have at least one feature vector in common.

Example 6.64 **Descriptively near Delaunay groupoids.**
Let G, H be sets of Delaunay triangles as defined in Example 6.61 (see Fig. 6.21). Let $(G, \circ), (H, \circ)$ be Delaunay groupoids. Since triangles $B_4 \in H, A_1 \in G$ have the same area and perimeters of equal length, then $\Phi(B_4) = \Phi(A_1)$. Consequently, $\Phi(B_4)\ \delta_\Phi\ \Phi(A_1)$, triangles B_4, A_1 are desciptively near. Hence, by definition, $(G, \circ), (H, \circ)$ are descriptively near. ■

Theorem 6.65 *Let X be a nonempty set of Delaunay triangles that is a topological space endowed with a descriptive Lodato proximity δ_Φ, $G, H \subset X$ equipped with a binary relation $\circ$. If $A \in G, B \in H$ are descriptively near triangles, then G is descriptively near H.*

Proof Let $A \in G, B \in H$ be descriptively near triangles, i.e, $A\ \delta_\Phi\ B$. By definition, $G\ \delta_\Phi\ H$, since A, B have matching descriptions. Hence, the Delaunay groupoids $(G, \circ), (H, \circ)$ are descriptively near. □

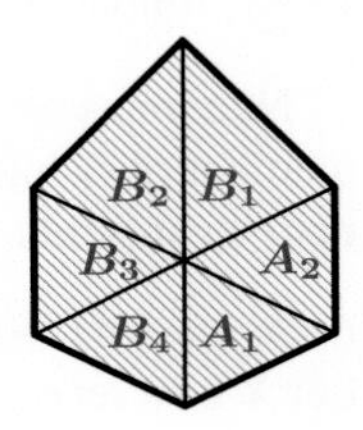

Fig. 6.21 $X = \{$Spatially near Delaunay triangles$\}$

Conjecture 6.66 Let X be a finite set of triangles in a Delaunay triangulation of a set of points S in the Euclidean plane, X endowed with the descriptive proximity δ_Φ, $G \subset X$. Every Delaunay groupoid G is a semigroup.

Counterexample 6.67 *Let H be a set of Delaunay triangles as defined in Example 6.61 (see Fig. 6.21). For $A \in H$, let $\Phi(A) =$ (area of A), i.e., the description of triangle A is defined by its area. Let $H = \{B_1, B_2, B_3, B_4\}$ be equipped with the binary relation $\circ$ defined by*

$$A, B \in H,$$

$$\Phi(A) \circ \Phi(B) = \begin{cases} \Phi(B), & \text{if } \Phi(A) = \Phi(B); \\ \Phi(A), & \text{otherwise.} \end{cases} \quad \blacksquare$$

H is not a semigroup, since

$$(\Phi(B_4) \circ \Phi(B_3)) \circ \Phi(B_2) = \Phi(B_3) \circ \Phi(B_2) = \Phi(B_3),$$
$$\Phi(B_4) \circ (\Phi(B_3) \circ \Phi(B_2)) = \Phi(B_4) \circ \Phi(B_2) = \Phi(B_4). \quad \blacksquare$$

6.11 Voronoï Groupoids

Voronoï Groupoids are the counterpart of Delaunay groupoids. Such groupoids are important enough to merit an intense view.

Definition 6.68 Voronoï-Hausdorff Convex Groupoids.
Let X be a Hausdoff space, $A \subset X$ that is a set of Voronoï regions in a Dirichlet tessellation of A, $\circ_{T_2} : A \times A \longrightarrow A$, a binary operation on subsets of X. The pair $(A, \circ_{T_2})$ is a **Voronoï Groupoid**. Since each of the Voronoï regions is a convex set in a Hausdorff space, the pair $(A, \circ_{T_2})$ is called a **Voronoï-Hausdorff Convex Groupoids**. ■

Definition 6.69 Voronoï Feature Vector.
Let (X, δ) be a finite Lodato proximity space of Voronoï regions in a Dirichlet tessellation of a subset X in the Euclidean plane, A, B a pair of triangles in X, $\mathfrak{G}$ is a set of descriptions of Voronoï regions in X. A **Voronoï feature vector** is a description of a Voronoï region. ■

Definition 6.70 Feature Space Groupoid.
Let X be a descriptive Hausdoff space, $A \subset X$ that is a set of Voronoï regions in a Dirichlet tessellation of A, $\circ^{\Phi}_{T_2} : \Phi(A) \times \Phi(A) \longrightarrow \Phi(A)$, a binary operation on subsets of $\Phi(X)$. The pair $(A, \circ^{\Phi}_{T_2})$ is a **feature space Groupoid**. ■

Theorem 6.71 *Let $\mathfrak{G}$ be a set of descriptions of Voronoï regions on the tessellation of a subset in the Euclidean plane. The feature space groupoid*

$$(\mathfrak{G}, \circ_{T_2}^{\Phi})$$

is a Voronoï feature space groupoid.

Proof Immediate from the definition of the feature space groupoid $(A, \circ_{T_2}^{\Phi})$. □

Theorem 6.72 *Let $\mathfrak{G}$ be a set of descriptions of Voronoï regions in a Dirichlet tessellation of a subset in the Euclidean plane, $\Phi(A), \Phi(B) \in \mathfrak{G}$, $\circ : \mathfrak{G} \times \mathfrak{G} \longrightarrow \mathfrak{G}$ a binary operation. The feature space groupoid $(\mathfrak{G}, \circ)$ is a Dirichlet feature space groupoid.*

Proof Immediate from the definition $\circ$. □

Remark 6.73 **Near Voronoï region Feature Vectors**.
Let (M, δ) be a finite proximity space of Voronoï region in a Dirichlet tessellation of a subset in the Euclidean plane, A, B a pair of triangles in X, and let $\mathfrak{G}$ be a set of descriptions of Voronoï region in M. Descriptions $\Phi(A), \Phi(B) \in \mathfrak{G}$ are near each other, provided $\phi(A) = \phi(B)$ for at least one probe function ϕ used to define the descriptions of $\Phi(A), \Phi(B)$. That is, $\phi(A)\ \delta\ \phi(B)$, provided $\phi(A) = \phi(B)$ for some $\phi(A) \in \Phi(A), \phi(B) \in \Phi(B)$. In other words, Voronoï region feature vectors are **near (proximal)**, provided the feature vectors have one or more matching coordinates. ■

Next, let $\Phi(A), \Phi(B)$ be descriptions of Voronoï regions A, B that cover a subset of the Euclidean plane. The binary operation $\circ_{\overset{\wedge\!\!\wedge}{\delta}}$ is defined by

$$\Phi(A) \circ_{\overset{\wedge\!\!\wedge}{\delta}}^{V} \Phi(B) = \begin{cases} \Phi(A), & \text{if } \Phi(A) \overset{\wedge\!\!\wedge}{\delta} \Phi(B); \\ \Phi(B), & \text{otherwise.} \end{cases}$$

The binary operation $\circ_{\overset{\wedge\!\!\wedge}{\delta}}^{V}$ has a bias towards the set of descriptions $\Phi(A)$ that are strongly near $\Phi(B)$.

Theorem 6.74 *Let $\mathfrak{G}$ be a set of descriptions of Voronoï regions in the Euclidean plane, equipped with the strong proximity $\overset{\wedge\!\!\wedge}{\delta}$. The feature space groupoid $(\mathfrak{G}, \circ_{\overset{\wedge\!\!\wedge}{\delta}}^{V})$ is a strong proximal Dirichlet feature space groupoid.*

Proof Immediate from the definition of $\circ_{\overset{\wedge\!\!\wedge}{\delta}}^{V}$. □

Remark 6.75 **Set-Based Probes or Point-Based Probes**.
Notice that a Voronoï feature vector can either be defined by set-based probes or point-based probes. A **set-based probe** returns a real number that is a feature value of a region. For example, let $A(p)$ be a Voronoï region of a point p and let $\phi(A) :=$ area of A. By contrast, a **point-based probe** returns a real number that is a feature value of a point in a region. For example, let X be a picture covered by a Voronoï diagram and let $A(p)$ be a Voronoï region of a picture point p and let $\phi(p) :=$

intensity of p, a feature value of a picture point. In other words, point-based probes are restricted to concrete points such as picture points, whereas region-based probes are not restricted to concrete regions such as subsets of picture points. For this reason, region-based descriptive groupoids are more general and have a broad range of applications. ■

Theorem 6.76 *Let $\mathfrak{G}$ be a set of descriptions of Voronoï regions in the Euclidean plane, $\circ_\Phi : \mathfrak{G} \times \mathfrak{G} \longrightarrow \mathfrak{G}$. The feature space groupoid $(\mathfrak{G}, \circ_\Phi)$ is a* Voronoï feature space groupoid.

Proof Immediate from the definition of the feature space groupoid $(\mathfrak{G}, \circ_\Phi)$. □

Remark 6.77 **Near Sets of Voronoï Feature Vectors.**
Let (X, δ) be a finite proximity space of triangles in a Delaunay triangulation of a set S in the plane, A, B a pair of triangles in X, and let $\mathfrak{G}$ be a set of descriptions of triangles in X. Descriptions $\Phi(A), \Phi(B) \in \mathfrak{G}$ are near each other, provided $\phi(A) = \phi(B)$ for at least one probe function ϕ used to define the descriptions of $\Phi(A), \Phi(B)$. That is, $\phi(A)\ \delta\ \phi(B)$, provided $\phi(A) = \phi(B)$ for some $\phi(A) \in \Phi(A), \phi(B) \in \Phi(B)$. In other words, Delaunay feature vectors are **near** (**proximal**), provided the feature vectors have one or more matching coordinates. ■

Next, let $\Phi(A), \Phi(B)$ be descriptions of Delaunay triangles A, B that cover a subset of the Euclidean plane. The binary operation $\circ_{\overset{\wedge\!\!\!\wedge}{\delta}}$ is defined by

$$\Phi(A) \circ_{\overset{\wedge\!\!\!\wedge}{\delta}} \Phi(B) = \begin{cases} \Phi(A), & \text{if } \Phi(A) \overset{\wedge\!\!\!\wedge}{\delta} \Phi(B); \\ \Phi(B), & \text{otherwise.} \end{cases}$$

The binary operation $\circ_{\overset{\wedge\!\!\!\wedge}{\delta}}$ has a bias towards the set of descriptions $\Phi(A)$ that are strongly near $\Phi(B)$.

Theorem 6.78 *Let $\mathfrak{G}$ be a set of descriptions of Delaunay triangles in the Euclidean plane, equipped with the strong proximity $\overset{\wedge\!\!\!\wedge}{\delta}$. The feature space groupoid $(\mathfrak{G}, \circ_{\overset{\wedge\!\!\!\wedge}{\delta}})$ is a strong proximal Delaunay feature space groupoid.*

Proof Immediate from the definition of $\circ_{\overset{\wedge\!\!\!\wedge}{\delta}}$. □

6.12 T3 Space

The discovery of separation axioms continues. L. Vietoris [23] defined a stronger axiom than T_2.

T_3 (**regular**): A T_1 space is T_3 or **regular** if, and only if, a point a does not belong to a closed set B implies a and B have disjoint neighbourhoods. Since the complement of B is an open neighbourhood of a, an alternate description of T_3 can be made, i.e., *each neighbourhood of a point contains a closed neighbourhood.*

Obviously, every T_3 space is Hausdorff or T_2. There is a descriptive equivalent of a T_3 space (denoted by T_3^{Φ}). Let X be endowed with a descriptive proximity relation δ_{Φ}. Then a **descriptively closed set** B in X results from the fact that, for each pair of points $x, y \in \text{cl}(B)$, the description of x is near the description of y, i.e., $x\ \delta_{\Phi}\ y$.

T_3^{Φ} (**descriptively regular**): A T_1^{Φ} space is T_3^{Φ} or **descriptively regular** if, and only if, a point a does not belong to a descriptively closed set B implies a and B have disjoint neighbourhoods so that *each descriptive neighbourhood of a point contains a descriptive closed neighbourhood.*

Example 6.79 $\boldsymbol{T_3^{\Phi}}$ **Descriptively Regular Space**.
Choose Φ to be a set of probe functions that represent greyscale and colour intensities of points in an image. Again, let a topological space X again be represented by the checkerboard in Fig. 6.13. X is also an example of a visual T_3^{Φ} descriptively regular space. To see this, let B be a set of black points in a black checkerboard tile and let $N_{\Phi(x)} \subset B$ be a closed descriptive neighbourhood for a black point $x \in B$ and let y be a white point that is the center of the descriptive neighbourhood $N_{\Phi(y)}$. The only points in $N_{\Phi(y)}$ are white points. Hence, $N_{\Phi(x)}, N_{\Phi(y)}$ are disjoint. ■

6.13 T4 Space

The next separation axiom (T_4 or **normal**) was discovered by H. Tietze in 1923.

T_4 (**normal**): A topological space that is T_1 space in which disjoint closed sets have disjoint neighbourhoods. Alternatively, if a closed set A is contained in an open set U, then there is an open set V such that

$$A \subset V \subset clV \subset U (T_4).$$

Obviously, T_4 is stronger than all of the previous separation axioms and we have the following scheme exhibiting the relation between T_4 and the other axioms.

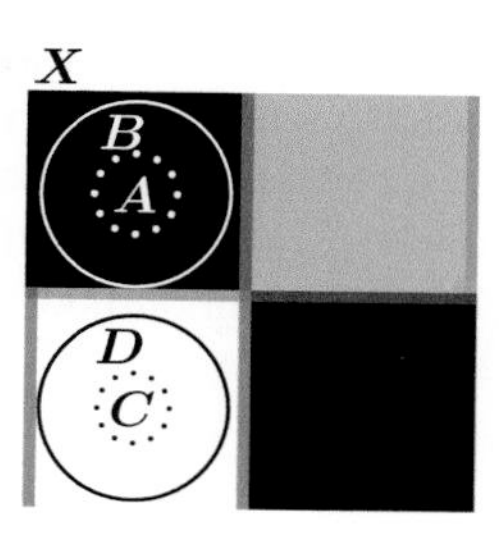

Fig. 6.22 Sample T_4^{Φ} visual space

$$T_4 \Rightarrow T_3 \Rightarrow T_2 \Rightarrow T_1 \Rightarrow \begin{Bmatrix} R_0 \\ T_0 \end{Bmatrix}.$$

These considerations lead to the following interesting result.

Theorem 6.80 *T_1 topological space X is normal, if and only if, its compatible fine L-proximity δ_0 is Efremovič.*

Proof Recall that $A\ \delta_0\ B$, if and only if $clA \cap clB \neq \emptyset$. Suppose A, B are disjoint closed sets of a normal space. Then they are far and so there is a subset $E \subset X$ such that A is far from E and E^c is far from B. Then E^c and E are disjoint neighbourhoods of A, B, respectively. The converse follows easily. □

It is a straightforward task to formulate a descriptive form of T_4 space (denoted by T_4^{Φ}).

> T_4^{Φ} (**descriptively normal**): A descriptive topological space that is a T_1^{Φ} space is T_4^{Φ}, provided disjoint closed sets have descriptively disjoint neighbourhoods. Alternatively, if a closed set A is contained in an open set U, then there is an open set V such that
>
> $$\Phi(A) \subset \Phi(V) \subset \Phi(clV) \subset \Phi(U),$$
>
> i.e., the set of descriptions of points in A is a subset of the set of descriptions of points in V, which is a subset of the set of descriptions of points in cl$V$, and so on.

Example 6.81 **T_4^{Φ} Descriptively Normal Space**.
Let the set of points X in Fig. 6.22 be endowed with a descriptive proximity relation δ_{Φ}. Choose Φ to be a set of probe functions that represent greyscale and colour features of points in X. Further, let B, D represent disjoint closed sets and let A, C represent disjoint open sets.

Proposition 6.82 *The set X is an example of a descriptive T_4^{Φ} space.*

Proof First, observe that B is a proximal neighbourhood of A, which is an example of a bounded descriptive neighbourhood. Similarly, D is a proximal neighbourhood of C, which is a bounded descriptive neighbourhood. For points a, c in A, B, respectively, $a\ \underline{\delta}_{\Phi}\ c$ (the description of a (black) does not match the description of c (white)). Hence, neighbourhoods A and C are descriptively disjoint. In effect, X is a T_4^{Φ} space. □

6.14 Visual Set Patterns in Descriptive Separation Spaces

Visual patterns arise naturally from the different forms of descriptive separation spaces. We illustrate this in terms of the T_1^{Φ} and T_2^{Φ} spaces (see, e.g., Fig. 6.23).

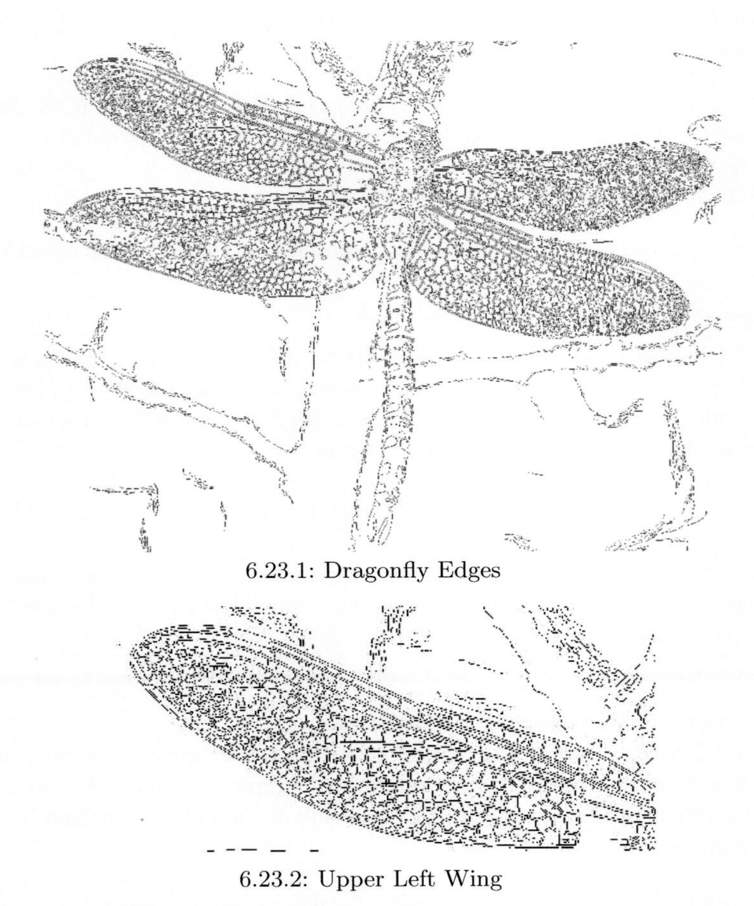

6.23.1: Dragonfly Edges

6.23.2: Upper Left Wing

Fig. 6.23 Descriptive separation space: dragonfly edges

6.14.1 Visual Patterns in Descriptive T1 Spaces

To find visual patterns in descriptive T_1 spaces, do the following:

(1) Choose Φ, a set of probe functions representing features of points in a T_1^{Φ} space X.
(2) Select a pair of descriptively distinct points $x, y \in X$. By definition, $x \not\delta_{\Phi} y$. Hence, the T_1^{Φ} space property is satisfied.
(3) Let M_1, M_2 denote point sets $\{x\}, \{y\}$, respectively.
(4) Determine all subsets of X containing points that descriptively match M_1 and then determine all subsets of points that descriptively match M_2.

As a result of the above steps, we can identify a pair of descriptive motif patterns $\mathfrak{P}_{\Phi}(M_1), \mathfrak{P}_{\Phi}(M_2)$ in a T_1^{Φ} space X. A welcome side-effect of this approach is a resulting descriptive uniform topology on X. In addition, each such a motif pattern is a member of a descriptive Herrlich topology ξ_{Φ} defined on X.

Let X be endowed with a proximity δ_{Φ} such that X is a T_1^{Φ} space and let $M = \{x\}$ be a motif containing a single point $x \in X$, which defines a descriptive motif pattern $\mathfrak{P}_{\Phi}(M)$. If $A, B \in \mathfrak{P}_{\Phi}(M)$, then $A \underset{\Phi}{\cap} B \neq \emptyset$. From this, we obtain the following result.

Theorem 6.83 *Let (X, δ_{Φ}) be a T_1^{Φ} space with nearness structure ξ_{Φ} on X and let $\mathfrak{P}_{\Phi}(M)$ be a descriptive motif pattern determined by a motif M containing a single point x in X. Then $\mathfrak{P}_{\Phi}(M) \in \xi_{\Phi}$.*

Example 6.84 **Point Motif Pattern in T_1^{Φ} space.**
Let X be a set of edge points in Fig. 6.24 extracted from the 1870 Punch drawing of a delivery boy in Fig. 6.1 (not all of the points, just those of interest). Choose Φ to be a set of probe functions representing the orientation (gradient direction) of pixels in X. Observe that *if* each pair of pixels x, y in X have different orientations (i.e., x and y are descriptively distinct), then $x \underline{\delta}_{\Phi}\ y$. Hence, X is an example of T_1^{Φ} space.

The points x_1, y_1 in Fig. 6.24 have different orientations. Then let M_1, M_2 denote motif point sets $\{x_1\}, \{y_1\}$, respectively. An indication of the descriptive motif patterns $\mathfrak{P}_{\Phi}(M_1), \mathfrak{P}_{\Phi}(M_2)$ determined by M_1, M_2 is suggested by the bounded regions containing points x_1, y_1. In the bounded region representing $\mathfrak{P}_{\Phi}(M_1)$, for example,

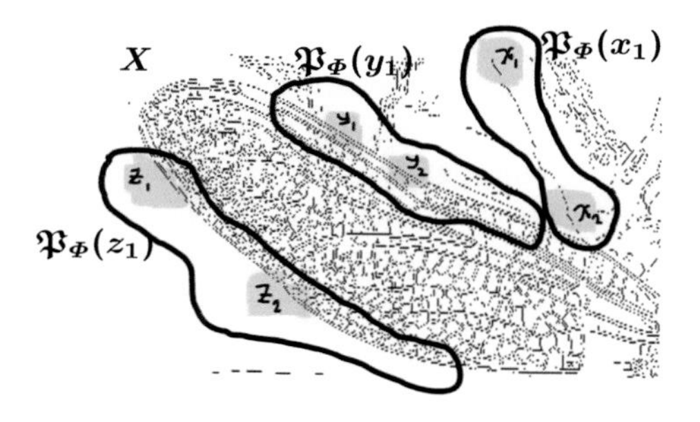

Fig. 6.24 Sample T_1^{Φ} visual patterns

notice that x_2 is the centre of bounded descriptive neighbourhood $N_{\Phi(x_2)}$ containing points with matching orientations. Similarly, there is a descriptive neighbourhood $N_{\Phi(y_2)}$ in the bounded region representing $\mathfrak{P}_{\Phi}(M_2)$. Continuing this process, one can observe many other descriptive motif patterns (see, e.g., the bounded region containing z_1, representing $\mathfrak{P}_{\Phi}(M_3)$, where $M_3 = \{z_1\}$). ■

Thought Problem 1 Do the following:

(1) Give an example of a descriptive Herrlich nearness structure containing two or more collections of descriptively near sets in a digital image that is a T_1^{Φ} space.
(2) Give an example of a visual motif pattern that is a member of a descriptive nearness structure in a digital image that is a T_1^{Φ} space. ■

Thought Problem 2 Naimpally Family T_1^{Φ} space.
Let X be a set containing at least two separate traditional families[3] (father, mother, sisters, brothers, aunts, uncles and grandparents) and let x in X be a member of a particular family, i.e., x is a point in X. Then

(a) Prove that X is T_1^{Φ} space.
(b) Give an example of a descriptive motif pattern in X, where the motif M is a point (family member). ■

6.14.2 Visual Patterns in Descriptive T2 Spaces

To find visual patterns in descriptive T_2 spaces, do the following:

(1) Choose Φ, a set of probe functions representing features of points in a T_2^{Φ} space X.
(2) Select a pair of descriptively distinct points $x, y \in X$. By definition, $N_{\Phi(x)} \not\delta_{\Phi} N_{\Phi(y)}$, since the description of each point in a descriptive neighbourhood matches the description of the point at the centre of the neighbourhood and $x \underline{\delta}_{\Phi}\ y$. That is, neighbourhoods $N_{\Phi(x)} \not\delta_{\Phi} N_{\Phi(y)}$ are descriptively disjoint. Hence, the T_2^{Φ} space property is satisfied.
(3) Let M_1, M_2 denote neighbourhoods $N_{\Phi(x)} \not\delta_{\Phi} N_{\Phi(y)}$, respectively.
(4) Determine all subsets of X that are descriptively near M_1 and then determine all subsets of X such that descriptively near M_2.

As a result of the above steps, we can identify a pair of descriptive motif patterns $\mathfrak{P}_{\Phi}(M_1), \mathfrak{P}_{\Phi}(M_2)$ in a T_2^{Φ} space X. A welcome side-effect of this approach is a resulting descriptive uniform topology on X. However, in general, each such a motif pattern is not a member of a descriptive Herrlich topology ξ_{Φ} defined on X. To see this, consider a pair of neighbourhoods $N_{\Phi(x')}, N_{\Phi(x'')}$ that are descriptively near $N_{\Phi(x)}$. We know that $N_{\Phi(x)}\ \delta_{\Phi}\ N_{\Phi(x')}$ but it is possible that $N_{\Phi(x')} \not\delta_{\Phi} N_{\Phi(x'')}$, if, for example, we compare pixel colours. $N_{\Phi(x)}$ may have a mixture of red and green

[3]This example is named after S.A. Naimpally, who first suggested the possibility of non-spatially near sets of family members [24].

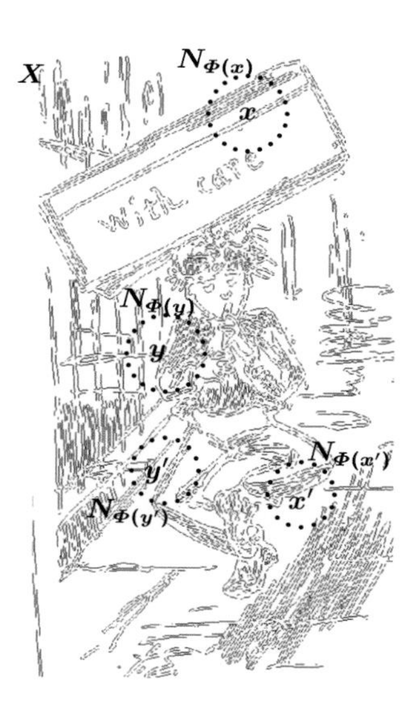

Fig. 6.25 Sample T_2^{Φ} descriptive neighbourhoods

colours, where $N_{\Phi(x')}$ has pixels with red colours but no green colours and $N_{\Phi(x'')}$ has pixels with green colours but no red colours.

Theorem 6.85 *A digital image containing two or more descriptively distinct points is a T_2^{Φ} space.*

Example 6.86 **Edge Motif Pattern in T_2^{Φ} space.**
Let X be the set of edge points in Fig. 6.25. Choose Φ to be a set of probe functions representing the orientation (gradient direction) of edge pixels in X. Observe that if pixels x, y in X have different orientations (i.e., x and y are descriptively distinct), then $x \ \underline{\delta}_{\Phi} \ y$. Then $N_{\Phi(x)}, N_{\Phi(y)}$ are descriptively disjoint neighbourhoods. Hence, X is an example of T_2^{Φ} space.

Then let M_1, M_2 denote motif neighbourhood edge point sets $N_{\Phi(x)}, N_{\Phi(y)}$ of points x, y, respectively. An indication of the descriptive motif patterns $\mathfrak{P}_{\Phi}(M_1)$, $\mathfrak{P}_{\Phi}$ (M_2) determined by M_1, M_2 is suggested by the edge regions containing points x', y'. In the pattern representing $\mathfrak{P}_{\Phi}(M_1)$, for example, notice that x' is the centre of bounded descriptive neighbourhood $N_{\Phi(x')}$ containing points with matching orientations. And the M_1 edge point neighbourhood is descriptively near $N_{\Phi}(x')$, since the orientation of one or more edges in $N_{\Phi(x)}$ match the orientation of one or more edges in $N_{\Phi(x)}$, i.e.,

$$N_{\Phi(x)} \ \delta_{\Phi} \ N_{\Phi(x')} : M_1 \doteqdot N_{\Phi(x)}.$$

Similarly, there is a descriptive neighbourhood $N_{\Phi(y')}$ in the edge pattern $\mathfrak{P}_{\Phi}(M_2)$ so that

$$N_{\Phi(y)} \ \delta_{\Phi} \ N_{\Phi(y')} : M_2 \doteqdot N_{\Phi(y)}.$$

Continuing this process, one can observe many other edge motif patterns in this T_2^{Φ} space. ■

Theorem 6.87 *A descriptive T_2^{Φ} space contains distinct descriptive motive patterns.*

Proof Immediate from Lemma 6.17 and the definition of descriptive motif patterns. □

6.14.3 Set Pattern Generators

The study of set patterns became prominent, thanks to T. Pavlidis [25], U.Grenander [26, Sect. 17.5], leading recently to [5, 27] and [6, Sect. 14.5]. One can generate multiple set patterns in an asymmetric space X such as a $T_1(T_1^{\Phi})$ or $T_2(T_2^{\Phi})$ space by endowing X with a proximity relation δ (δ_{Φ}). In addition, generated patterns can be derived so that they are remote from each other. We consider the spatial case, first. For a given subset M in X, one can find a set pattern by finding all sets B that are near M, all sets $A\ \delta\ M$. The collection of subsets near M is an example of a nearness set pattern (denoted by $\mathfrak{P}(M)$). Let *motif* M be a set used to generate a set pattern in an asymmetric space.

Example 6.88 **Multiple Patterns in a T_1 Space.**
For a T_1 space X, let $x, y \in X$ be distinct such that $\text{cl}\{x\} \cap \text{cl}\{y\} = \emptyset$. Then let $M_1 = \text{cl}\{x\}$ and find

$$\mathfrak{P}(M_1) = \{A \in \mathcal{P}(X), x, y \in X : \text{cl}A \cap\ \text{cl}\{x\} \neq \emptyset \& \text{cl}A \cap \text{cl}\{y\} = \emptyset\}.$$

Similarly, Then let $M_2 = \text{cl}\{y\}$ and find

$$\mathfrak{P}(M_2) = \{B \in \mathcal{P}(X), x, y \in X : \text{cl}B \cap\ \text{cl}\{y\} \neq \emptyset \& \text{cl}B \cap \text{cl}\{x\} = \emptyset\}.$$

In effect, multiple, disjoint spatial patterns can be derived in a T_1 Space. The choice of $M_1 = \text{cl}\{x\}$, $M_2 = \text{cl}\{y\}$ (instead of $M_1 = \{x\}$, $M_2 = \{y\}$) is important. Otherwise, the boundaries of distinct points may have nonempty intersection. ■

One can generate multiple descriptive set patterns in an asymmetric space X by endowing X with a descriptive proximity relation δ_{Φ}. Descriptive motifs are sets that are descriptively near each member of a set pattern. Let x be a point in X and let Φ be a set of probe functions ϕ used to represent features of each x. Then $\Phi(x)$ denotes a feature vector that provides a description of x. Sets A, B are near each other, provided there is at least one pair of points $a \in A, b \in B$ with matching descriptions. Motif set patterns containing descriptively near sets are denoted by $\mathfrak{P}_{\Phi}(M)$. Motif set patterns containing sets that are descriptively remote are also considered (denoted by $\underline{\mathfrak{P}}_{\Phi}(M)$), i.e., each $A \in \underline{\mathfrak{P}}_{\Phi}(M)$ has no points that descriptively match any point in motif M.

Example 6.89 **Multiple Patterns in a T_2^{Φ} Space.**
For a T_2^{Φ} space X, let $x, y \in X$ be distinct points with disjoint neighbourhoods N_x, N_y, respectively, such that $\mathrm{cl}N_x \underset{\Phi}{\cap} \mathrm{cl}N_y = \emptyset$. Then let $M_1 = \mathrm{cl}N_x$ and generate the descriptive pattern

$$\mathfrak{P}_{\Phi}(M_1) = \left\{A \in \mathcal{P}(X) : \mathrm{cl}A \underset{\Phi}{\cap} \mathrm{cl}N_x \neq \emptyset \& \mathrm{cl}A \underset{\Phi}{\cap} \mathrm{cl}N_y = \emptyset\right\}.$$

Similarly, Then let $M_2 = \mathrm{cl}N_y$ and find

$$\mathfrak{P}(M_2) = \left\{B \in \mathcal{P}(X) : \mathrm{cl}B \underset{\Phi}{\cap} \mathrm{cl}N_y \neq \emptyset \& \underset{\Phi}{\cap} \mathrm{cl}A\ \mathrm{cl}N_x = \emptyset\right\}.$$

In effect, multiple, disjoint descriptive patterns can be derived in a T_2^{Φ} Space. ■

In a given asymmetric space, one can usually find many motifs that serve as set pattern generators. Let M, M_1, M_2 be motifs. From this, we obtain the following result.

Theorem 6.90

(1) M generates a spatial set pattern $\mathfrak{P}(M)$.
(2) M generates a spatial set pattern $\underline{\mathfrak{P}}(M)$ (collection of sets remote from M).
(3) The pair M, Φ generates a descriptive set pattern $\mathfrak{P}_{\Phi}(M)$.
(4) The pair M, Φ generates a descriptive set pattern $\underline{\mathfrak{P}}_{\Phi}(M)$ (collection of sets descriptively remote from M).
(5) $M_1 \cup M_2$ (union of motifs) generates a spatial set pattern $\mathfrak{P}(M_1 \cup M_2)$ (collection of sets spatially near $M_1 \cup M_2$).
(6) $M_1 \cup M_2$ generates a spatial set pattern $\underline{\mathfrak{P}}_{\Phi}(M_1 \cup M_2)$ (collection of sets remote from $M_1 \cup M_2$).
(7) $M_1 \cup M_2, \Phi$ (union of motifs, Φ) generates a set pattern $\mathfrak{P}(M_1 \cup M_2)$ (collection of sets descriptively near $M_1 \cup M_2$).
(8) $M_1 \cup M_2, \Phi$ (union of motifs, Φ) generates a set pattern $\underline{\mathfrak{P}}(M_1 \cup M_2)$ (collection of sets descriptively remote from $M_1 \cup M_2$).
(9) $M_1 \cap M_2$ (intersection of motifs) generates a set pattern $\mathfrak{P}(M_1 \cap M_2)$ (collection of sets near $M_1 \cap M_2$).
(10) $M_1 \cap M_2$ (intersection of motifs) generates a set pattern $\underline{\mathfrak{P}}(M_1 \cap M_2)$ (collection of sets descriptively remote from $M_1 \cap M_2$).
(11) $M_1 \cap M_2, \Phi$ (intersection of motifs plus Φ) generates a descriptive set pattern $\mathfrak{P}_{\Phi}(M_1 \cap M_2)$(collection of sets descriptively near $M_1 \cap M_2$).
(12) $M_1 \cap M_2, \Phi$ (intersection of motifs plus Φ) generates a descriptive set pattern $\underline{\mathfrak{P}}_{\Phi}(M_1 \cap M_2)$(collection of sets descriptively remote from $M_1 \cap M_2$).

Proof Let (X, δ), (X, δ_{Φ}) be spatial and descriptive EF-proximity spaces, respectively. Assume that the base space is a Hausdorff space. We prove only (1) and (3).
(1)**Spatial case**: Let M be a neighbourhood of a point x in X. Since M is near itself,

$\mathfrak{P}(M)$ is nonempty. Find $A \subset X$ such that $A\ \delta\ M$, i.e., there is at least one $a \in A$ so that $a\ \delta\ M$ (the point a is spatially near M, i.e., $D(\{a\}, M) = 0$). Then $A \in \mathfrak{P}(M)$. Continuing this process, motif M generates the spatial motif pattern $\mathfrak{P}(M)$. Let B, C be disjoint neighbourhoods of distinct points $b, c \in X$, then B and C serve as generators of spatial motif patterns $\mathfrak{P}(B), \mathfrak{P}(C)$.

(3)**Descriptive case**: Choose Φ to be a set of probe functions that represent features of each x in X. Since M is descriptively near itself, $\mathfrak{P}_\Phi(M)$ is nonempty. Find $A \subset X$ such that $A\ \delta_\Phi\ M$, i.e., there is at least one $a \in A$ and at least one $m \in M$ so that $a\ \delta_\Phi\ m$. In effect, $A \underset{\Phi}{\cap} M$ is nonempty. Then $A \in \mathfrak{P}_\Phi(M)$. Continuing this process, motif M generates the descriptive motif pattern $\mathfrak{P}_\Phi(M)$. Let B, C be disjoint bounded neighbourhoods of descriptively distinct points $b, c \in X$ (i.e., $\Phi(b) \neq \Phi(c)$), then B and C serve as generators of descriptive motif patterns $\mathfrak{P}_\Phi(B), \mathfrak{P}_\Phi(C)$. □

Thought Problem 3 Naimpally Family T_2^Φ space.
Let X be a set containing at least two separate traditional families (father, mother, sisters, brothers, aunts, uncles and grandparents) and let x in X be a member of a particular family, i.e., x is a point in X. Then

(a) Prove that X is a T_2^Φ space.
(b) Give an example descriptive motif pattern in X, where the motif M is a neighbourhood of a point (family member). ■

References

1. O'Roorke, J.: An alternative proof of the rectilinear art gallery theorem. J. Geom. **11**, 118–130 (1983)
2. Ghosh, S.: Visibility Algorithms in the Plane. Cambridge University Press, Cambridge (2007). Xiii + 318 pp
3. Thron, W.: Topological Spaces. Holt, Rinehart and Winston, New York (1966). Xi + 240 pp
4. Hausdorff, F.: Set Theory, trans. by J.R. Aumann. AMS Chelsea Publishing, Providence (1957). 352 pp
5. Peters, J.: Local near sets: pattern discovery in proximity spaces. Math. Comp. Sci. **7**(1), 87–106 (2013). doi:10.1007/s11786-013-0143-z, MR3043920
6. Naimpally, S., Peters, J.: Topology with applications. Topological spaces via near and far. World Scientific, Singapore (2013). Xv + 277 pp. Am. Math. Soc. MR3075111
7. Davis, A.: Indexed systems of neighborhoods for general topological spaces. Am. Math. Mon. **68**, 886–893 (1961)
8. Smirnov, J.M.: On proximity spaces. Math. Sb. (N.S.) 31(73), 543–574 (1952). English translation: Am. Math. Soc. Trans. Ser. 2(38), 5–35 (1964)
9. Efremovič, V.: The geometry of proximity I (in Russian). Mat. Sb. (N.S.) 31(73)(1), 189–200 (1952)
10. Leader, S.: On clusters in proximity spaces. Fundam. Math. **47**, 205–213 (1959)
11. Lodato, M.: On topologically induced generalized proximity relations, Ph.D. thesis. Rutgers University (1962). Supervisor: S. Leader
12. Pták, P., Kropatsch, W.: Nearness in digital images and proximity spaces.In: Proceedings of the 9th International Conference on Discrete Geometry, LNCS, vol. 1953, pp. 69–77 (2000)

13. Alexandroff, P., Hopf, H.: Topologie. Springer, Berlin (1935). Xiii + 636pp
14. Steen, L., Seebach, J.: Counterexamples in Topology. Springer, New York (1970). Dover Edition (1995)
15. Weihrauch, K.: Computable separation in topology, from t_0 to t_2. J. Univ. Comp. Sci. **16**(18), 2733–2753 (2010)
16. Beer, G., Lucchetti, R.: Weak topologies for the closed subsets of a metrizable space. Trans. Am. Math. Soc. **335**(2), 805–822 (1993)
17. Engelking, R.: General Topology. Revised and completed edition. Heldermann Verlag, Berlin (1989)
18. Pavel, M.: Fundamentals of Pattern Recognition, 2nd edn. Marcel Dekker Inc, New York (1993). Xii + 254 pp. ISBN: 0-8247-8883-4, MR1206233
19. Di Concilio, A., Gerla, G.: Quasi-metric spaces and point-free geometry. Math. Structures Comput. Sci. **16**(1), 115–137 (2006). MR2220893
20. Di Concilio, A.: Point-free geometries: proximities and quasi-metrics. Math. Comp. Sci. **7**(1), 31–42 (2013). MR3043916
21. Peters, J., Guadagni, C.: Strong proximities on smooth manifolds and Voronoï diagrams. Adv. Math. Sci. J. **4**(2), 91–107 (2015)
22. Peters, J., Guadagni, C.: Strongly proximal continuity and strong connectedness, pp. 1–11 (2015). arXiv:1504.02740v1
23. Vietoris, L.: Stetige mengen. Monatshefte für Mathematik und Physik 31, pp. 173–204 (1921). MR1549101
24. Naimpally, S.: Near and far. A centennial tribute to Frigyes Riesz. Siberian Electronic Mathematical Reports 2, pp. 144–153 (2009)
25. Pavlidis, T.: Analysis of set patterns. Pattern Recog. **1**, 165–178 (1968)
26. Grenander, U.: General pattern theory. A Mathematical Study of Regular Structures. Oxford University Press, Oxford (1993). Xxi + 883 pp
27. Peters, J.: Nearness of local admissible covers. Theory and application in micropalaeontology. Fund. Inform. **126**, 433–444 (2013). doi:10.3233/FI-2013-890

Chapter 7
Strongly Near Sets and Overlapping Dirichlet Tessellation Regions

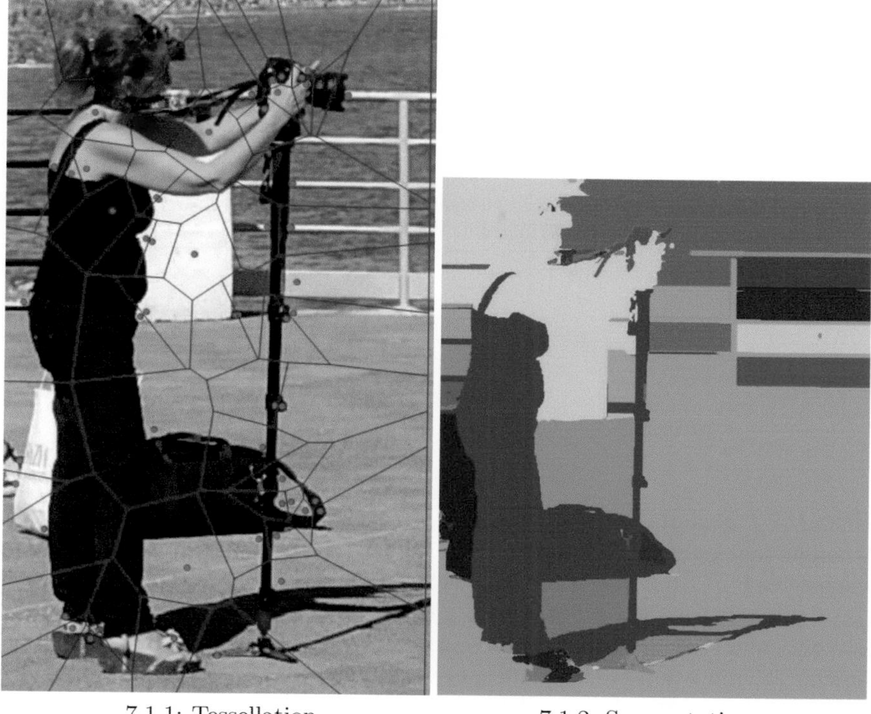

7.1.1: Tessellation

7.1.2: Segmentation

Fig. 7.1 Keypoint-based Dirichlet tessellation versus segmentation

This chapter introduces various forms of sites (seed or generating points, including hybrid generating points) used in tessellating digital images. A natural outcome of an image tessellation is an approximate image segmentation. The segments in this case

J.F. Peters, *Computational Proximity*, Intelligent Systems Reference Library 102, DOI 10.1007/978-3-319-30262-1_7

are the byproduct of various forms of Voronoï and Delaunay mesh cells containing all of those image pixels nearest to each mesh generating point. A digital image is ***segmented*** by partitioning the image into a set of regions whose union is the image. Recall that the ***partition of a set*** X is a collection of disjoint subsets of X which cover X [1, Sect. 29.14]. Strictly speaking, an image segmentation is not a partition of the image, since each pair of adjacent image segments has a common border. Similarly, a tessellated image is a tiled image. A ***plane tiling*** is a collection of disjoint open sets and the union of the closures of the open sets cover the plane [2]. The interior of each tessellation cell approximates the usual notion of an image segment, since the interiors of each pair of cells are disjoint, i.e., the interior of each cell does not have any points in common with the interior of any other tessellation cell. Each cell interior is an example of an open set. And the union of the closures of the image cell interiors covers the image.

Algorithm 9: Keypoint-based Dirichlet Tessellation versus Cell-based Segmentation

Input : Read digital image *img*.
Output: Keypoint-based Dirichlet Tessellation + Cell-based Segmentation.

1 $img \longmapsto sites$;
2 /* Use *sites* = keypoints to construct tessellation regions (= Dirichlet cells). */ ;
3 $VoronoiMesh \leftarrow sites$;
4 /* *sites* are used to construct Voronoï *Regions* (= Dirichlet cells). */ ;
5 $sites \longmapsto ImageForestingComponents\ (segmentation)$;
6 /* Image segments include *sites* as markers. */ ;

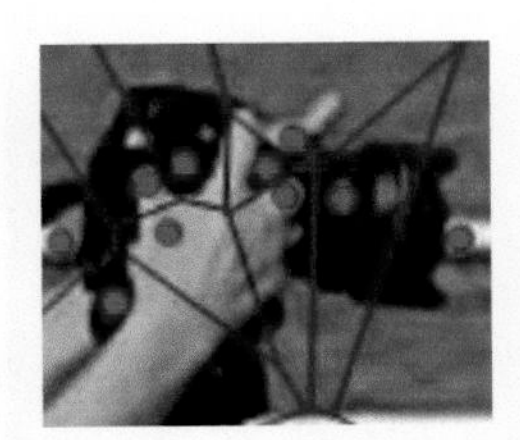

Example 7.1 **Sample Tessellation Cells**:
A sample segmentation of a digital image induced by a keypoint-based tessellation is shown in Fig. 7.1.1. The interior of each tessellation cell is an image region that has no points in common with the interior of any other tessellation cell in Fig. 7.1.1. Examples of tessellations cells are those cells covering the photographer's hand and camera. The red • dots are keypoint sites used to generate the tessellated image.

The steps to obtain the tessellated image as well as the segmented image are given in Algorithm 9. The Mathematica ImageForestingComponents method is used in MScript 37 in Appendix A.7 to obtain the image segmentation in Fig. 7.1.2. In this application of the ImageForestingComponents method, the keypoint sites serve as markers that are included in image segments. A very different segmentation is obtained in MScript 38 in Appendix A.7 so that the segments are obtained without the use of sites as markers. An important thing to notice is that with image segmentation there is a loss of image information, since only some of the shapes (including interior content) in the original image are carried over into the image segments. ■

7.1 Known Mesh Seed or Generating Points

Here is a list of *possible* mesh generating points.

1° **centroid**. The earliest known mesh generating point.
2° **corner**.
3° **intensity**.
4° **edge pixel**.
5° **key point**. Key feature in an image (cf. **ImageKeypoints** in Mathematica).
6° **critical point**. Also called an interesting point.
7° **salient point**. cf. **ImageSaliencyFilter** in Mathematica.
8° hybrid centroid-corner. A centroid that is also corner.
9° hybrid centroid-edge. A centroid that is also an edge.
10° hybrid intensity-centroid. A centroid that is a distinguished point in a set of intensities.
11° hybrid intensity-corner. A corner that is a distinguished point in a set of intensities.
12° hybrid intensity-edge. An edge pixel that is a distinguished point in a set of intensities.
13° hybrid intensity-keypoint. An keypoint that is a distinguished point in a set of intensities.
14° hybrid intensity-critical point. An critical point that is a distinguished point in a set of intensities.
15° hybrid intensity-salient point. An salient point that is a distinguished point in a set of intensities.
16° hybrid keypoint-centroid. A keypoint that is also a centroid.
17° hybrid keypoint-corner. A keypoint that is also a corner.
18° hybrid keypoint-edge. A keypoint that is also an edge pixel.
19° hybrid corner-centroid. A corner that is also a centroid.
20° hybrid corner-edge. A corner that is also an edge pixel.
21° hybrid critical point-centroid. A critical point that is also a centroid.
22° hybrid critical point-corner. A critical point that is also a corner.
23° hybrid critical point-corner. A critical point that is also an edge point.
24° hybrid critical point-corner. A critical point that is also a key point.
25° hybrid salient point-corner. A salient point that is also a centroid.
26° hybrid salient point-corner. A salient point that is also a corner.
27° hybrid salient point-corner. A salient point that is also an edge point.
28° hybrid salient point-corner. A salient point that is also a key point.
29° hybrid salient point-corner. A salient point that is also a critical point.

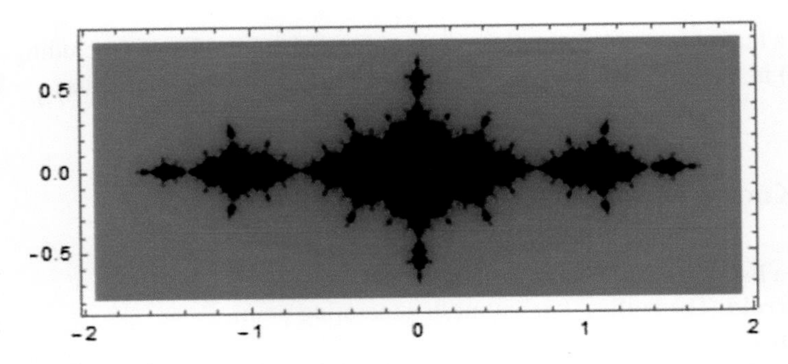

Fig. 7.2 Quadratic julia set image (see mathematica script 9)

7.2 Julia Set Image Example

For simplicity, the segmentation of an image is illustrated here in Fig. 7.2 with a quadratic Julia set image.

Let c be a real number, z a complex number. A quadratic Julia set is generated with the mapping

$$z_{n_1} = z_n^2 + c.$$

The Mathematica function **JuliaSetPlot[]** can be used to generate various quadratic Julia set plots (some of them are known fractals). The beauty of using such plots to experiment with image segmentation is that we know a lot about a Julia set plot image before we start the segmentation process.

Mathematica script 9 generates a Julia set image.

Mscript 9 Julia Set Plot.

***JuliaSetPlot*[−1.2];**

***Options*[*JuliaSetPlot*[−1.2]]** ■

Example 7.2 Experiment with Mathematica script 9. Try varying the JuliaSetPlot argument to produce other Julia sets. Use **Options[]** to learn the options available for this or another Mathematica function. ■

7.3 Strongly Near Regions and Mesh Cells

In this section, we revisit the notion of strongly near sets [3, 4] in terms of sets of polygons in a Voronoï diagram. The concept of ***strongly near*** proximity arises from the need to introduce a relation that yields information about the interiors of pairs of subsets that have at least have non-empty intersection.

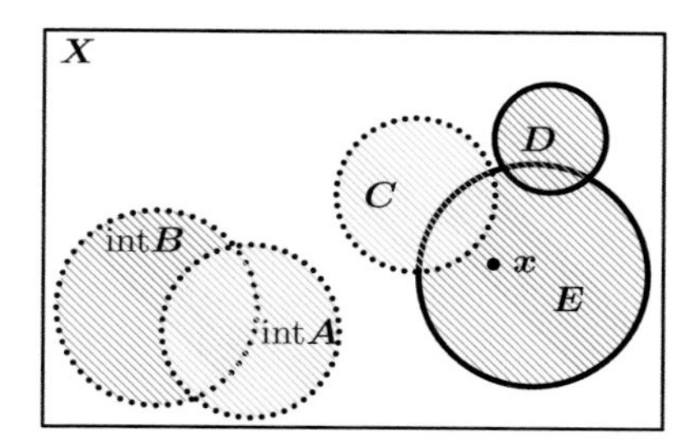

Fig. 7.3 $A \overset{\doublewedge}{\delta} B$, $E \overset{\doublewedge}{\delta} (C \cup D)$

For the sake of clarity, the axioms for the $\overset{\doublewedge}{\delta}$ proximity are revisited. Let X be a topological space equipped with the $\overset{\doublewedge}{\delta}$ proximity [4].

We say that the relation $\overset{\doublewedge}{\delta}$ on the family of subsets of X (denoted 2^X) is a *strongly near proximity*, provided it satisfies the following axioms. Let $A, B, C \subset X$ and $x \in X$.

(N0) $\varnothing \not\overset{\doublewedge}{\delta} A, \forall A \subset X$, and $X \overset{\doublewedge}{\delta} A, \forall A \subset X$

(N1) $A \overset{\doublewedge}{\delta} B \Leftrightarrow B \overset{\doublewedge}{\delta} A$

(N2) $A \overset{\doublewedge}{\delta} B \Rightarrow A \cap B \neq \varnothing$

(N3) If int(B) and int(C) are not equal to the empty set, $A \overset{\doublewedge}{\delta} B$ or $A \overset{\doublewedge}{\delta} C \Rightarrow A \overset{\doublewedge}{\delta} (B \cup C)$

(N4) $\text{int}A \cap \text{int}B \neq \varnothing \Rightarrow A \overset{\doublewedge}{\delta} B$

For each *strong proximity*, we also assume the following point-based axioms.

(N5) $x \in \text{int}(A) \Rightarrow x \overset{\doublewedge}{\delta} A$

(N6) $\{x\} \overset{\doublewedge}{\delta} \{y\} \Leftrightarrow x = y$ ■

Example 7.3 **Strongly Near Sets and Intersecting Interiors.**

Let X in Fig. 7.3 be a finite planar Euclidean topological space equipped with the $\overset{\doublewedge}{\delta}$ proximity. Then observe the following strongly near sets.

1° $\text{int}A \cap \text{int}B \neq \varnothing \Rightarrow A \overset{\doublewedge}{\delta} B$ (Axiom (N4)).

2° $A \overset{\doublewedge}{\delta} B \Rightarrow A \cap B \neq \varnothing$ (axiom N4) That is, sets A and B share points.

3° $E \overset{\doublewedge}{\delta} C$ or $E \overset{\doublewedge}{\delta} D \Rightarrow E \overset{\doublewedge}{\delta} (C \cup D)$ (axiom N3). This case is interesting, since int$E$ and int$C$ are nonempty disjoint sets but the union of the E and C shares points with D. Hence, the set D is strongly near the set $C \cup E$.

4° $x \in \text{int}(E) \Rightarrow x \overset{\doublewedge}{\delta} E$. In fact, every point in the interior of a set is strongly near the set. ■

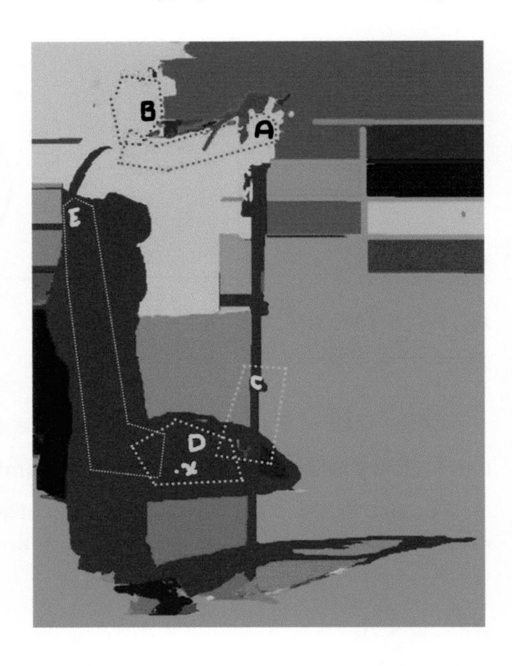

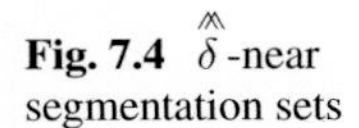
Fig. 7.4 $\stackrel{\wedge\!\!\!\wedge}{\delta}$-near segmentation sets

Example 7.4 **Strongly Near Image Segment Interiors**.

Let a finite planar Euclidean topological space X equipped with the $\stackrel{\wedge\!\!\!\wedge}{\delta}$ proximity be represented by the image segmentation in Fig. 7.4. And let $A, B, C, D, E \in 2^X$ be segmentation subsets as shown in Fig. 7.4. Then observe the following strongly near sets.

1° $A \stackrel{\wedge\!\!\!\wedge}{\delta} B \Rightarrow A \cap B \neq \emptyset$ (axiom N4) That is, sets A and B share points.

2° $D \stackrel{\wedge\!\!\!\wedge}{\delta} C$ or $D \stackrel{\wedge\!\!\!\wedge}{\delta} E \Rightarrow D \stackrel{\wedge\!\!\!\wedge}{\delta} (C \cup E)$ (axiom N3). This case is interesting, since int$E$ and int$C$ are nonempty disjoint sets but the union of the E and C shares points with D. Hence, the set D is strongly near the set $C \cup E$.

3° $x \in \text{int}(D) \Rightarrow x \stackrel{\wedge\!\!\!\wedge}{\delta} D$. In fact, every point in the interior of a set is strongly near the set. ■

When we write $A \stackrel{\wedge\!\!\!\wedge}{\delta} B$, we read A is *strongly near*. For a source of strongly near sets in a digital image, see Fig. 7.5. This is explained in Example 7.5.

Example 7.5 **Strongly Near Dirichlet Tessellation Regions**.

Let a finite planar Euclidean topological space X that is mapped onto the keypoint-based Dirichlet tessellation of X in Fig. 7.6.2 equipped with the $\stackrel{\wedge\!\!\!\wedge}{\delta}$ proximity. And let $A, B, C, D, E \in 2^X$ be the tessellation cells as shown in Fig. 7.7.2. To make it easier to view the tessellation cells without the underlying digital image, the unlabelled Voronoï diagram is shown in Fig. 7.6.1 and labelled Voronoï diagram cells are shown in Fig. 7.7.1.

Fig. 7.5 Photographer taking a picture of the sea

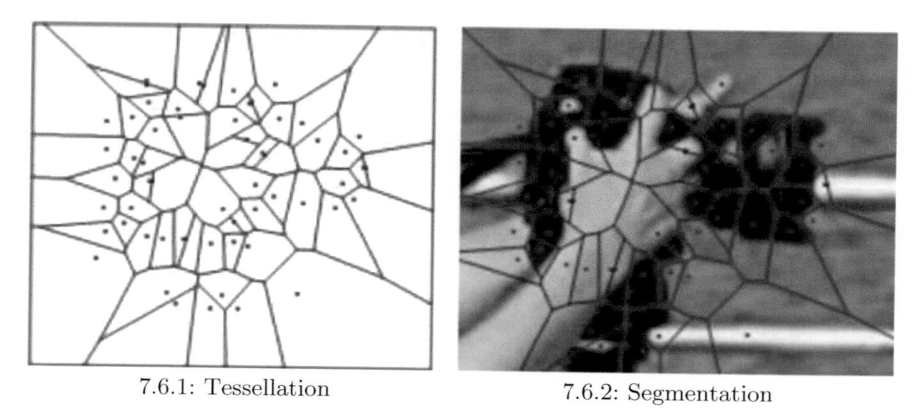

7.6.1: Tessellation 7.6.2: Segmentation

Fig. 7.6 Photographer hand keypoint-based Voronoï mesh cells and diagram

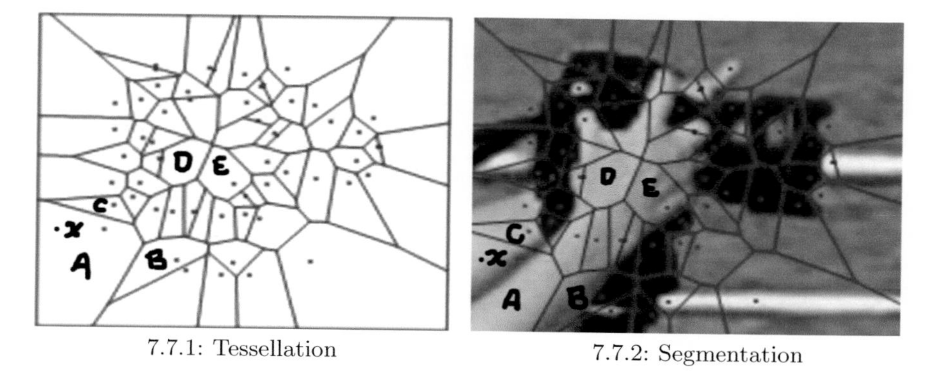

7.7.1: Tessellation 7.7.2: Segmentation

Fig. 7.7 Labelled keypoint-based Voronoï mesh cells and diagram

Then observe the following strongly near sets.

1° $A \stackrel{\wedge\wedge}{\delta} B \Rightarrow A \cap B \neq \varnothing$ (axiom N4) That is, mesh cells A and B are strongly near, since these cells have a common edge. Similarly,

2° $D \stackrel{\wedge\wedge}{\delta} E \Rightarrow D \cap E \neq \varnothing$ (axiom N4) That is, mesh cells D and E are strongly near, since these cells share points along their common edge.

3° $A \stackrel{\wedge\wedge}{\delta} B$ or $A \stackrel{\wedge\wedge}{\delta} C \Rightarrow A \stackrel{\wedge\wedge}{\delta} (B \cup C)$ (axiom N3). This is the case, since cl$B$ and cl$C$ are nonempty disjoint sets but the union of the B and C shares points with A. Hence, the set A is strongly near the set $B \cup C$.

4° $x \in \text{int}(A) \Rightarrow x \stackrel{\wedge\wedge}{\delta} A$. In fact, every point in the interior of region A is strongly near the region. ■

$\tau^{\wedge\wedge}$ is the *strongly hit and far-miss* topology [5] (see, also, [6]) having as subbase the sets of the form:

- $V^{\wedge\wedge} = \{E \in \text{CL}(X) : E \stackrel{\wedge\wedge}{\delta} V\}$, where V is an open subset of X,
- $A^{++} = \{ E \in \text{CL}(X) : E \not\delta X \setminus A \}$, where A is an open subset of X,

where in both cases δ is a Lodato proximity compatible with the topology on X.

Theorem 7.6 *In a corner-based Voronoï mesh,*

1° Corners are found along varying image edges.
2° Close together, small polygons will cover image edges.
3° The set of polygons covering off-edge image regions will be coarser than the set of polygons covering on-edge image regions.
4° The set of edge-polygons is finer than the set of off-edge polygons.
5° If edge corners are close together, then the corner-generated Voronoï regions are smaller the off-edge polygons.

Theorem 7.7 *In a corner-based Voronoï mesh,*

1° Self-similar edge corners generate self-similar Voronoï regions.

Theorem 7.8 *In a corner-based Voronoï mesh,*

1° A Voronoï mesh endowed with a strongly near proximity induces a mesh topology.
2° A strongly hit and far miss topology $V_{\mathbb{W}}^{-}$ is induced by a Voronoï mesh.

7.4 Back to Earth

Mathematica script 10 is used to detect key points in a Julia set image.

Mscript 10 Image Keypoint-Generated Mesh.

im0 = 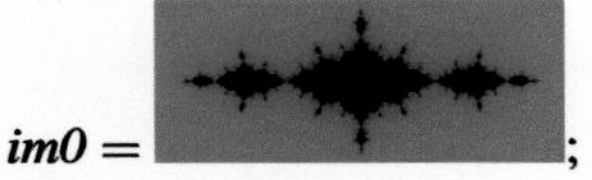;
im = ImageKeypoints[im0, MaxFeatures → **100]**

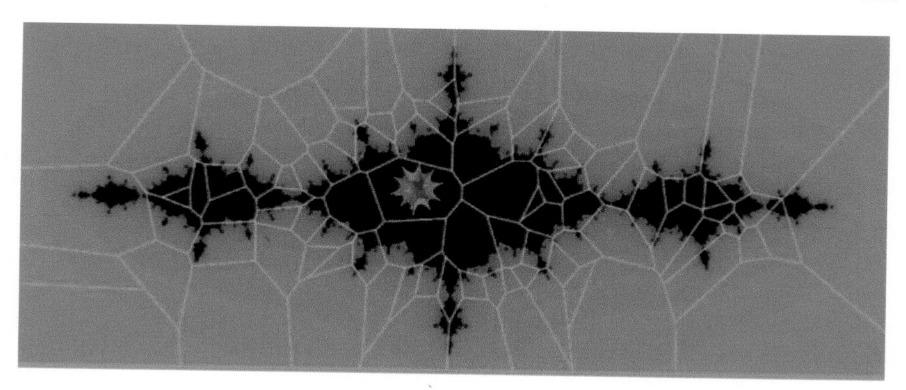

Fig. 7.8 Labelled image polygon

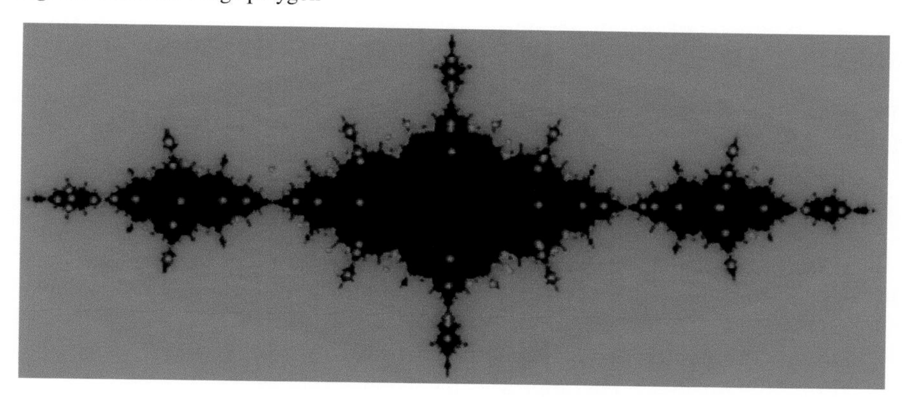

Fig. 7.9 Julia set image key points

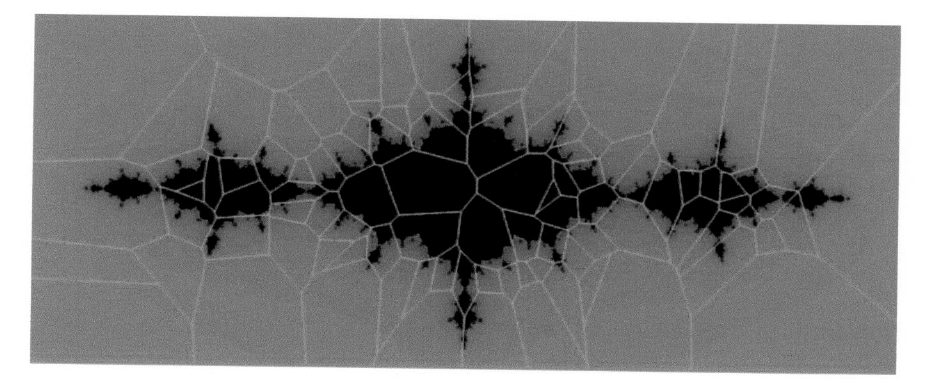

Fig. 7.10 Julia set key points-generated mesh

***voronoi* = *HighlightMesh*[*VoronoiMesh*[*im*], {*Style*[1, *Green*], *Style*[2, *Opacity* [0.0]]}];**

***im3* = *Show*[*im0*, *voronoi*]**

***ImageCompose*[*im3*, ✸, {361.3, 173.0}]** ■

Example 7.9 **Corner Generated Voronoï Image Mesh**.
The Mathematica Script 10 finds the keypoints in an image, e.g., the key points for a sample Julia set image in Fig. 7.9. The key points are then used to segment an image with a Voronoï mesh, e.g., the mesh in the sample Julia set image in Fig. 7.10. In keeping with interest in identifying mesh polygons that are strongly near other mesh polygons, a sample labelled polygon that has edge points in common with many other polygons is shown in Fig. 7.8. The coordinates of the ✸ identify a point in the neighbourhood of one of the image key points (assigned to the variable **im** in Script 10). ■

Problem 7.10 Edge-Corner-Based Voronoï Mesh.
Do the following:

- i^o Assign a given image to a variable, **im0**. Display im0.
- ii^o Find image edges. Display edge image. Assign edge point coordinates to a variable, e.g., im2.
  Hint: Use **im2 = ImageValuePositions[EdgeDetect[im0,...]];**
- iii^o Use im2 as generating points for a Voronoï mesh. Assign mesh to a variable, e.g., voronoi.
- iv^o Display voronoi (by itself, not on image).
  Hint: Use **voronoï = HighlightMesh[VoronoiMesh[im2], ...]**
- v^o Display Voronoi mesh on im0.
  Hint: Use **Show[im0,voronoi];**
- vi^o For three images of your own choosing, superimpose an edge-corner based Voronoï mesh on three colour images of your own choosing.

For a solution, see Appendix A.39.

Problem 7.11 Edge-Corner-Based Delaunay Mesh.
Do the following:

- i^o Repeat the steps in Problem 7.10 to generate a Delaunay mesh on im0.
- ii^o For three images of your own choosing, superimpose an edge-corner based Delaunay mesh on three colour images of your own choosing.

Problem 7.12 Key Points-Based Voronoï Mesh.
Do the following:

- i^o Assign a given image to a variable, **im0**. Display im0.
- ii^o Find key points in an image. Display key point image. Assign key point coordinates to a variable, e.g., im2.
  Hint: Use **im5 = ImageKeypoints[im0, MaxFeatures → 100];**.
- iii^o Vary MaxFeatures for 3 cases of your own choosing and repeat the following steps for each case, e.g., MaxFeatures → 50, 100, 150.

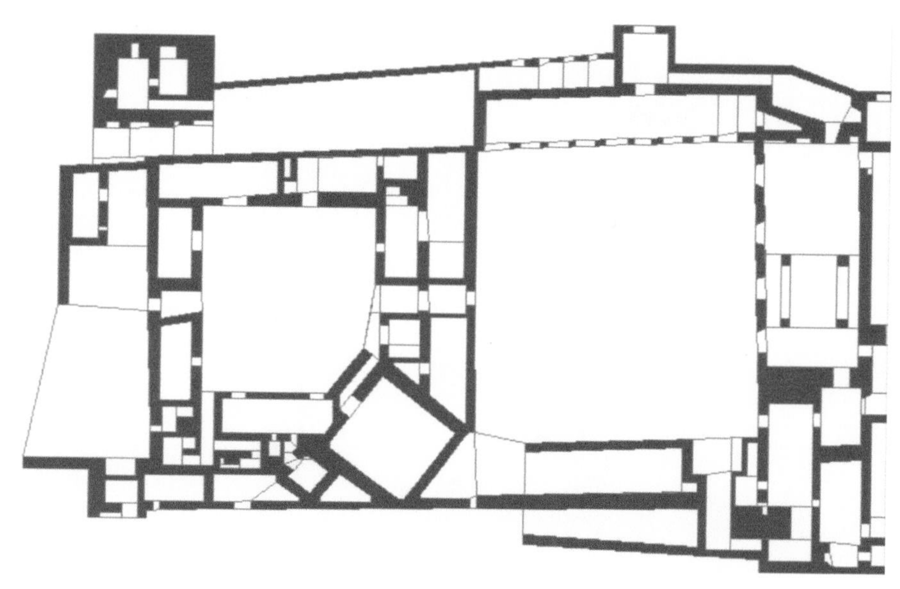

Fig. 7.11 Alhambra floorplan

iv^o Use im2 as generating points for a Voronoï mesh. Assign mesh to a variable, e.g., voronoi.

v^o Display voronoi (by itself, not on image).

vi^o Display key point Voronoï mesh on im0.

vii^o For three images of your own choosing, superimpose an key point-based Voronoï mesh on three colour images of your own choosing. ■

Problem 7.13 Key Points-Based Delaunay Mesh.
Do the following:

i^o Repeat the steps in Problem 7.12 to generate a Delaunay mesh on im0.

ii^o For three images of your own choosing, superimpose an edge-corner based Delaunay mesh on three colour images of your own choosing. ■

Problem 7.14 Edge-Key-Points-Based Voronoï Mesh.
Do the following:

i^o Assign a given image to a variable, **im0**. Experiment with an image like the Alhambra floorplan[1] in Fig. 7.11.

ii^o Find the edges in im0 (see Fig. 7.12). Assign edge image to a variable, e.g., im2.

iii^o Find edge key points in im2 (edges). Use MaxFeatures → 100. Assign edge key points to variable, e.g., **im5**.

iv^o Display edge key points on im0. See, for example, Fig. 7.13.

---

[1]Many thanks to Bin Jang, University of Gävie, for the Alhambra floorplan image.

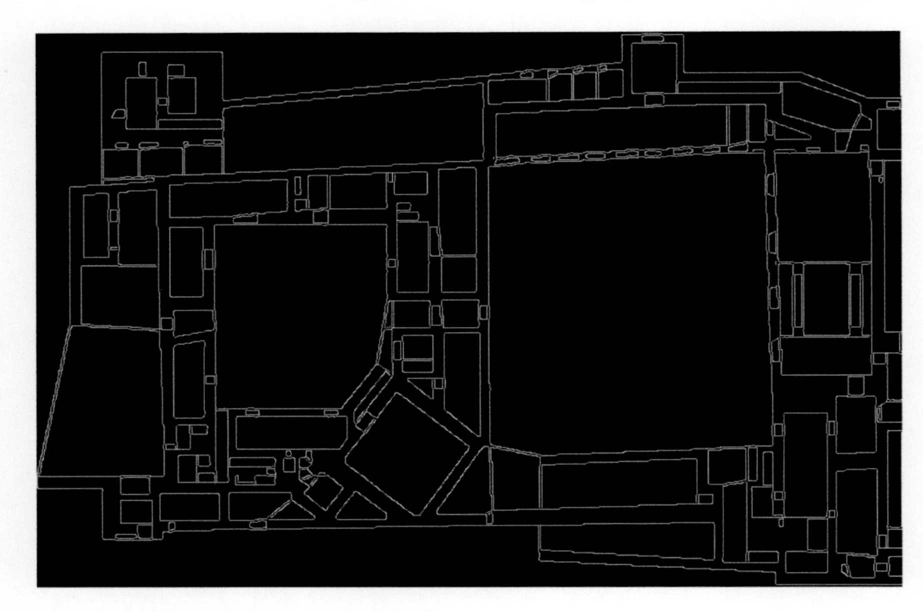

Fig. 7.12 Edges in the Alhambra floorplan

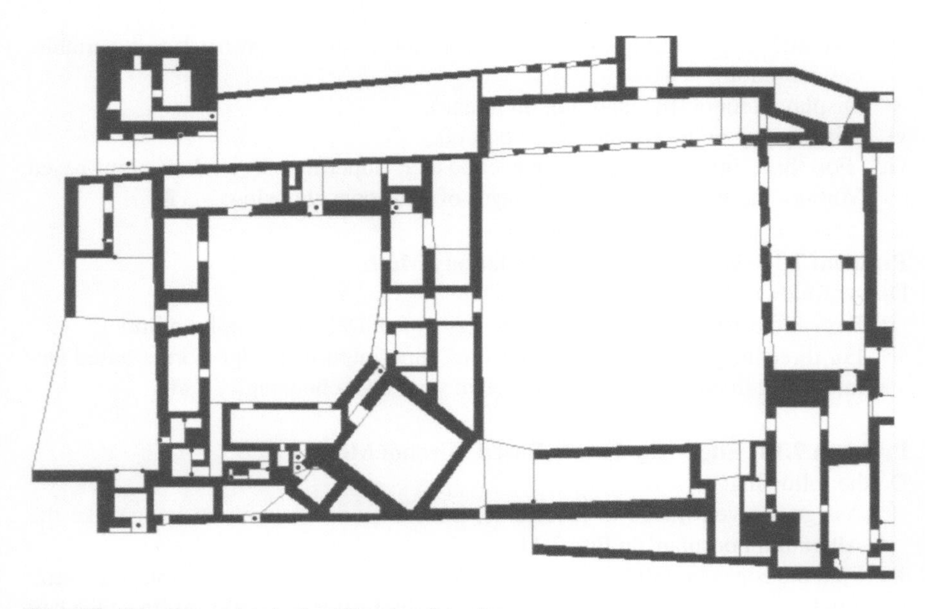

Fig. 7.13 Edge-key-points in the Alhambra floorplan

v^o Vary MaxFeatures for 3 cases of your own choosing and repeat the following steps for each case, e.g., MaxFeatures → 50, 100, 150.

vi^o Use **edge key points** im5 to generate a Voronoï mesh image. Assign mesh to a variable, e.g., **voronoi**.

vii^o Display Voronoi mesh, using variable voronoï.

viii^o Show the mesh on im0. See, for example, Fig. 7.14.

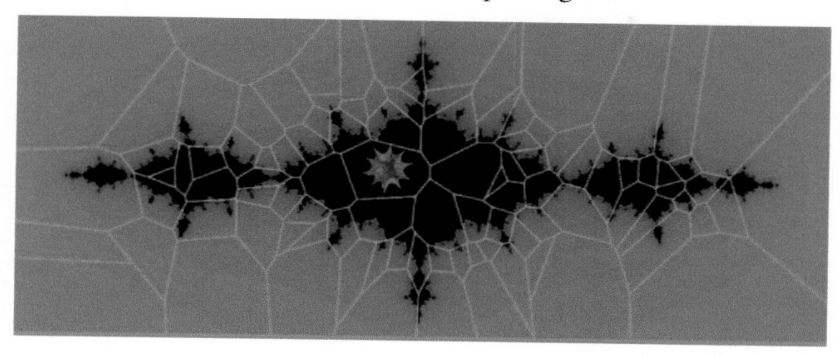

ix^o Identify polygons with at least 7 adjacent, strongly near mesh polygons. See, for example, Fig. 7.15. Label (highlight) 3 of the mesh polygons with high edge adjacency to other polygons in the mesh. **Hint:** Use the coordinates of the edge key point of the selected polygon as the location of a label such as × . Then use **ImageCompose[im3,symbol,position]**. See sample labelled key points generated mesh.

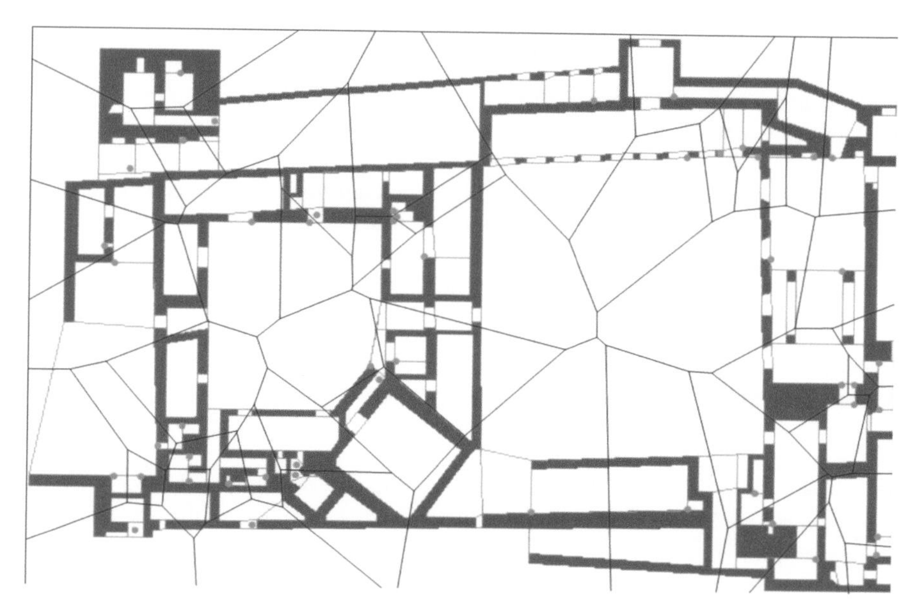

Fig. 7.14 Edge-key-points Voronoï mesh on the Alhambra floorplan

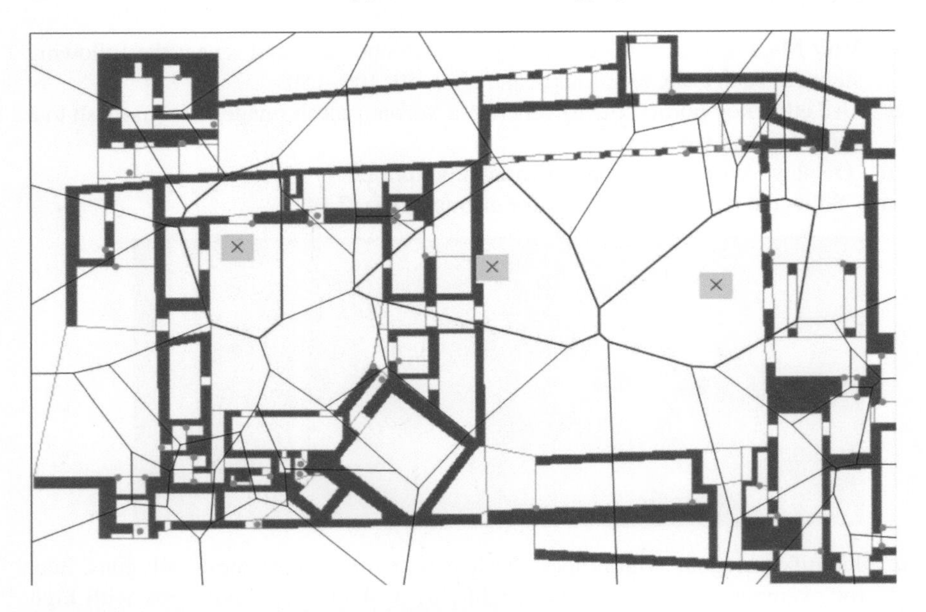

Fig. 7.15 Strongly near polygons in the Alhambra floorplan

x^o For three images of your own choosing, superimpose an edge key points-based Voronoï mesh on the images. ■

Problem 7.15 Edge-Key-Points-Based Voronoï Mesh.
Do the following:

i^o Repeat the steps in Problem 7.14 to generate a Delaunay mesh on im0.

ii^o Identify triangles that have a vertex common with at least 7 other, near mesh triangles. Label (highlight) 3 of the mesh triangles with high vertex adjacency to other triangles in the mesh. **Hint:** Use the coordinates of the edge key point of the selected polygon as the location of a label such as ×.

iii^o For three images of your own choosing, superimpose an edge key points-based Delaunay mesh on the images. See Figs. 7.16, 7.17, 7.18, 7.19, 7.20 and 7.21 for examples. ■

Problem 7.16 Voronoï Meshes Over Changing Scenes.
Do the following:

i^o Suggest how Voronoï meshes can be used to identify objects in a image. Give 3 examples.

ii^o Given a pair of images in a sequence of images showing a changing scene. Suggest how Voronoï meshes can be used to detect changes in the images. Give 3 examples.

iii^o Suggest how Voronoï meshes can be used to identify patterns (e.g., repeated edges, colours, polygons with the same shape, polygons with the same area) in a image. Give 3 examples. ■

Fig. 7.16 Rainbow on a plant

Fig. 7.17 Edges in the rainbow plant floorplan

Fig. 7.18 Edge-key-points in the rainbow plant

Fig. 7.19 Edge-key-points Delaunay mesh on the rainbow plant

Fig. 7.20 Strongly Delaunay mesh near triangles in the rainbow plant

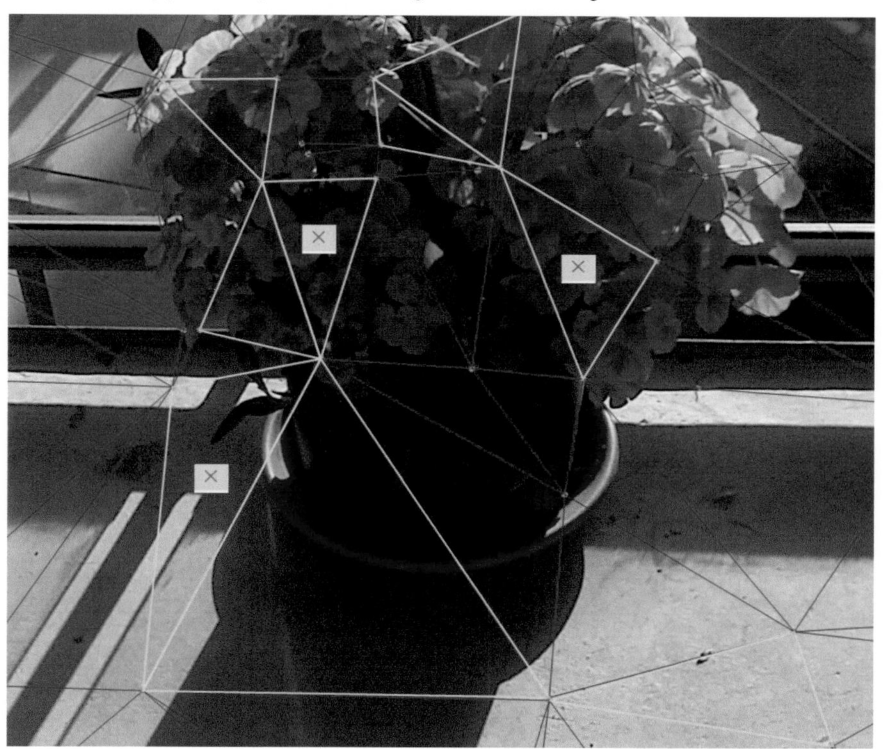

Fig. 7.21 High Delaunay mesh common vertices in the rainbow plant

Problem 7.17 Delaunay Meshes Over Changing Scenes.
Do the following:
i^o Repeat the steps in Problem 7.16 for Delaunay meshes on an image. ■

Example 7.18 **Object Detection Using Image Key Points**.
There are quite a number of ways to do object detection on digital images using computational geometry. This is especially true in cases where hybrid generating points are used to construct either Voronoï or Delaunay meshes on digital images. For example, consider the use of edge-keypoint based meshes in the Mathematica script suggested by S. Younas (see Appendix A.41). Mesh-based detection of image objects results from high polygon clusters in parts of an image where keypoints are close together. This happens in cases where keypoints are extracted from detected image edges. The result is a collection of segments that reveal image objects such as the sparrow in Fig. 7.22. From the detection edge key points, it is then possible to examine the clusters of points that reveal the segments found in a particular object (see, e.g., the clusters in Fig. 7.27). A Delaunay mesh works well in object identification, since a Delaunay mesh is only constructed over generating points that belong to an object region. This can been seen, for example, in Fig. 7.29. The conjecture is that the groupings of mesh polygons with the smallest area provide a basis for object recognition. See Figs. 7.23, 7.24, 7.25, 7.26, 7.27, 7.28 and 7.29 for the details. ■

Fig. 7.22 Sparrow

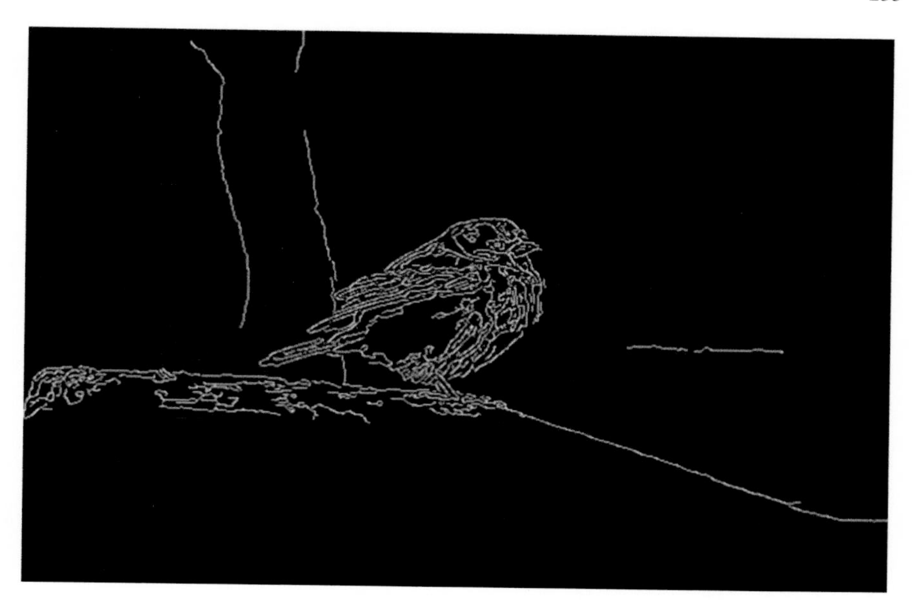

Fig. 7.23 Edges in the sparrow

Fig. 7.24 Edge-key-points in the sparrow

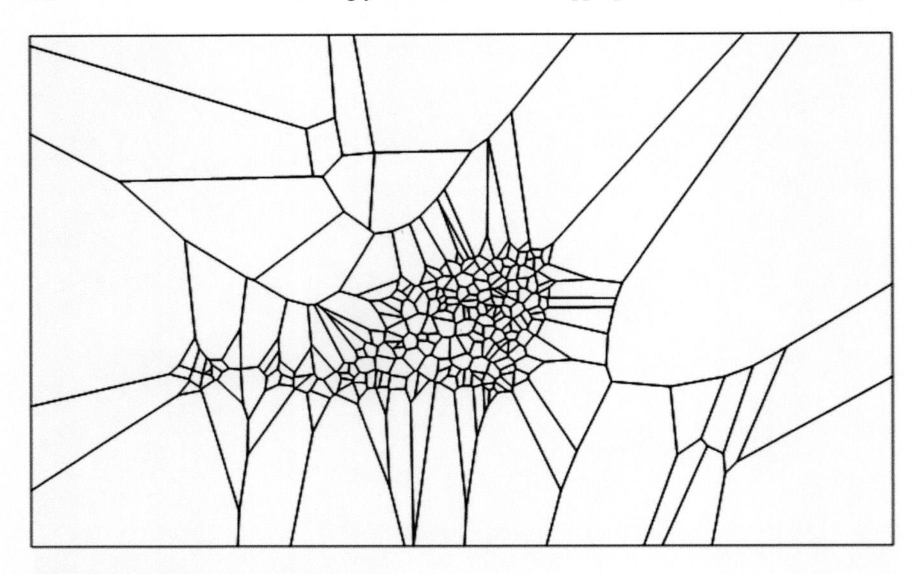

Fig. 7.25 Edge-Keypoint Voronoï mesh

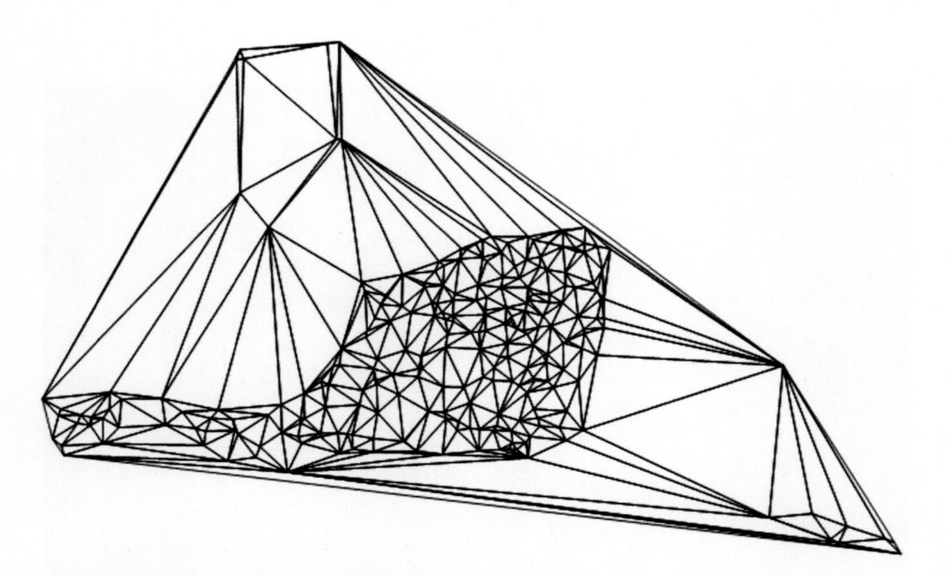

Fig. 7.26 Edge-Keypoint Delaunay mesh

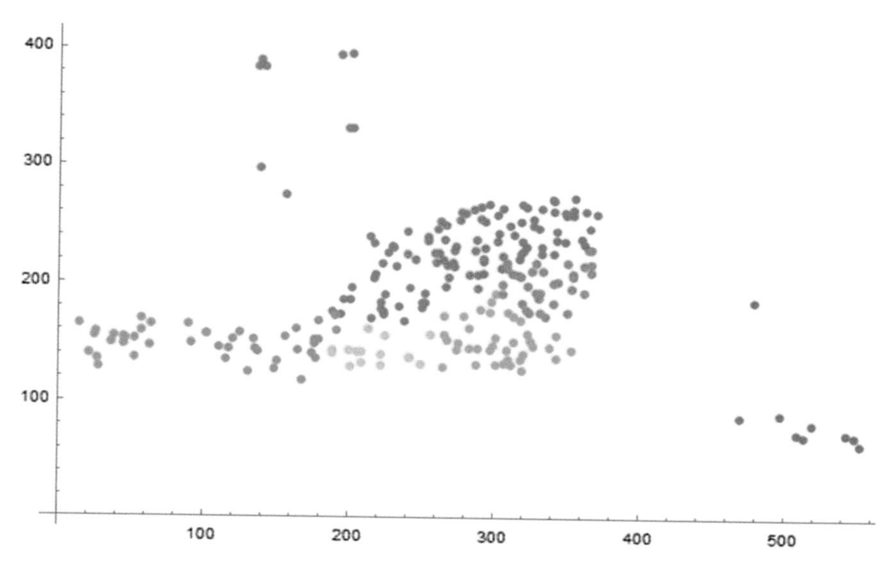

Fig. 7.27 Edge-Key-points clusters in the sparrow

Fig. 7.28 Edge-Keypoint Voronoï mesh on sparrow

Fig. 7.29 Edge-keypoint Delaunay mesh on sparrows

References

1. Willard, S.: General Topology, Xii + 369pp. Dover Publisher Inc., Mineola (1970). ISBN: 0-486-43479-6 54-02, MR0264581
2. Weisstein, E.: Tiling. Wolfram MathWorld (2015). http://mathworld.wolfram.com/Tiling.html
3. Peters, J.: Visibility in proximal delaunay meshes, pp. 1–5 (2015). arXiv:1501.02357
4. Peters, J., Guadagni, C.: Strongly near proximity and hyperspace topology, 1–6 (2015). arXiv:1502.05913
5. Peters, J., Guadagni, C.: Strongly hit and far miss hypertopology and hit and strongly far miss hypertopology, pp. 1–8 (2015). arXiv:1503.02587
6. Guadagni, C.: Bornological convergences on local proximity spaces and ω_μ-metric spaces. Ph.D. thesis, Università degli Studi di Salerno, Salerno, Italy (2015). Supervisor: A. Di Concilio, 79 pp

Chapter 8
Proximal Manifolds

8.1.1: Colour Image Manifold

8.1.2: Greyscale Image

Fig. 8.1 Colour image manifold $\mathbb{R}^5$ mapped to greyscale image $\mathbb{R}^3$

Topological Manifold.
A **topological manifold** is a Hausdorff space with a countable basis where each point has a neighbourhood homeomorphic to some Euclidean space [1]. J. Milnor looks at whether a topological manifold can be embedded in a high dimensional Euclidean space so as to have a continuously turning tangent plane [1, 2]. ■

The notion of a manifold results from a specialization of topological spaces. The simplest manifold is an n-dimensional topological space (called an **topological n-manifold**) that locally looks like some Euclidean space $\mathbb{R}^n$. For a topological manifold M with special properties, each point $x \in M$ has an open neighbourhood

J.F. Peters, *Computational Proximity*, Intelligent Systems Reference Library 102, DOI 10.1007/978-3-319-30262-1_8

homeomorphic to an open set in $\mathbb{R}^n$. That is, there is a continuous, bijective mapping $f : M \longrightarrow 2^{\mathbb{R}^n}$ on M onto N, a subset in the collection of subsets $2^{\mathbb{R}^n}$. Let $Y \in 2^{\mathbb{R}^n}$. A mapping $f : M \longrightarrow Y$ is **continuous**, provided, when $A \subset Y$ is open, then the inverse $f^{-1}(A) \subset M$ is also open. For more about this, see [3, Sect. 3].

Recall that a mapping $f : M \longrightarrow Y$ is **bijective**, provided f is 1-1, onto and has an inverse f^{-1} on Y into M. This form of manifold has two properties, namely, Hausdorff (distinct points belong to disjoint neighbourhoods) and M has a countable basis for the topology on M, i.e., M is second countable. Recall that **basis** for a topology is a collection of subsets $\mathscr{B} \in 2^M$ of a topological space M such that

$$M = \left\{ \bigcup B : B \in \mathscr{B} \right\}.$$

In other words, every member of a topological manifold M equals the union of subcollections in the basis $\mathscr{B}$ [4, Sect. 5.1]. A topological basis is also called a **topological base**.

Example 8.1 This example comes from S. Krantz [3, Sect. 2.1]. Let X be the real line that is also a topological space with a basis that is the collection of open intervals $(a, b) : a, b \in \mathbb{R}$. ■

Common examples of topological manifolds are Euclidean spaces

$$\mathbb{R}^1, \mathbb{R}^2, \mathbb{R}^3, \dots, \mathbb{R}^n, n \in \mathbb{N}.$$

as well as smooth plane curves such as circles and parabolas, and smooth surfaces such as spheres, tori, paraboloids, ellipsoids and hyperboloids. Other examples include n-spheres, $n \geq 2$ and graphs of smooth maps between Euclidean spaces. An n-sphere is a set of n-tuples of vectors $(x_1, x_2, \dots, x_n)$ such that

$$n - \text{sphere with radius } R : x_1^2 + x_2^2 + \cdots + x_n^2 = R^2.$$

See Sect. 5.1 for examples of n-spheres and their use in the Borsuk-Ulam Theorem. Manifolds are useful in solving dimensionality reduction problems.

Example 8.2 **Digital Image Manifold**.
Let X be subset of $\mathbb{R}^5$ be represented by the colour image in Fig. 8.1.1, $p \in X$. Let $x, y \in \mathbb{Z}^{0+}$ (set of integers greater than or equal to 0) that are the coordinates of a pixel in X and let $r, g, b \in [0, 1]$ be the intensities of the red, green and blue colour channels for each image pixel. Every member of X is a vector (x, y, r, g, b). X is an example of a *finite Euclidean topological space* that is Hausdorff and which has a countable **basis** equal to the set $\{\varnothing, (x, y, r, g, b)\}$.

Let Y be a subset of $\mathbb{R}^3$ be represented by the greyscale image in Fig. 8.1.2. Let $x, y \in \mathbb{Z}^{0+}$ (set of integers greater than or equal to 0) that are the coordinates of a pixel in $q \in Y$ and let $gr \in [0, 1]$ is the greyscale intensity of the red, green and blue colour channels of pixel q. A homeomophic map $f : M \longrightarrow Y, Y \subset \mathbb{R}^3$ is defined by

$$f(p) = \left(x, y, gr_{r,g,b}(p) = \frac{r+g+b}{3}\right) \in \mathbb{R}^3, \text{ for } p \in M.$$

The notation $gr_{r,g,b}(p)$ indicates that each q knows (can recall) the r, g, b values used to compute its greyscale value. That is, for each pixel $p \in X$, f maps $p \in M$ to a point $q \in Y$ that has the same coordinates as p and maps the colour channels intensities r, g, b to an average intensity, which is a greyscale value. Since the coordinates for p and q are the same and q can recall the r, g, b intensities used to compute its greyscale value, $f^{-1}(p)$ exists. Hence, X is a digital image manifold. To experiment with other image manifolds, try MScript 43 in Appendix A.8. ■

The introduction of manifolds in a computational proximity framework is motivated by its utility in a broad spectrum of applications. One such application is manifold learning that leads to dimensionality reduction by mapping high dimensional spaces such as feature spaces to tractable lower dimensional spaces (see, e.g., [5]). For more examples smooth manifolds, see J.M. Lee [6, Sect. 1, starting on p. 17].

8.1 Manifolds: Basic Notions

This section introduces the basic notions underlying manifolds.

Definition 8.3 Manifold.
A **manifold** is a topological space, provided it is Hausdorff, second countable and locally Euclidean of dimension n so that each point p in $\mathscr{M}$ has a neighbourhood U that is homeomorphic to an open set in $\mathbb{R}^n$. (i.e., around every point in $\mathbb{R}^k$, $k \geq 1$, there is a neighborhood that is topologically equivalent to an open unit ball in $\mathbb{R}^n$, $n \geq 1$). ■

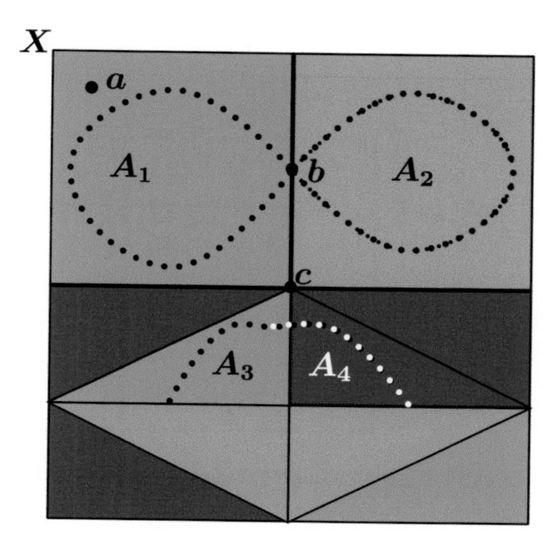

Fig. 8.2 Euclidean topology on a tessellation of $\mathbb{R}^2$

Example 8.4 **Sample Topological Basis for an RGB Colour Space.**
Let X be the tessellation of a subspace of $\mathbb{R}^2$ endowed with the Euclidean topology as shown in Fig. 8.2. Let $Y = \{g, \ r, \ b\}$ be the set of colour green, red and blue and take $\mathscr{P}(Y)$ endowed with the topology τ having as basis $\mathscr{B}$ defined by

$$\begin{aligned}\mathscr{B} &= \{\{g, r\}, \ \{b, r\}, \ \{g, b\}, \ \{r, g, b\}, \\ &= \{\{r, g, b\}, \{r, g\}, \{r\}\}, \ \{\{r, g, b\}, \{r, g\}, \{g\}\}, \\ &= \ \{\{r, g, b\}, \{r, g\}, \{b\}\}\}. \ (\tau\text{-basis})\end{aligned}$$

Every subset in τ equals the union or intersection of members of $\mathscr{B}$. A penultimate example of such a space is a RGB digital image. Let every member of $\mathscr{B}$ be a blend of colours that are common feature of pixels in an RGB image. Think of τ as a painter's palette with dabs of paint on it, with some dabs having only one colour and other dabs containing a mixture of colours. For example, let $\{c\} \in \tau$ be a singleton containing a mixture of colours in Fig. 8.2 and observe that

$$\{c\} = \{r\} \cup \{g\} \cup \{b\}.$$

In fact, every subset in τ equals the union or intersection of members of $\mathscr{B}$. For more about this, see [7]. ■

Families of curves provide a good hunting ground for examples of manifolds.

Example 8.5 **Curves Manifold.**
Let $\mathscr{M}$ be a manifold represented in Fig. 8.3. For each point P_i in $\mathscr{M}$ consider a family of curves $\{\alpha_k^i : k \in K_i\}$ and fix a specific curve α_1^i as reference curve. Take $\theta_{\{1,k\}}^i$ as the angle between the curves α_1^i and α_k^i; $A_{P_i} = \{\theta_{\{1,k\}}^i : k \in K_i \setminus \{1\}\}$. We can talk about descriptive strong connectedness for family of curves. In this case, our space X is represented by all the curves for all the points of $\mathscr{M}$. Our description maps each curve α_k^i, for $k \neq 1$, in $\theta_{\{1,k\}}^i$, and α_1^i in some of the already found values among $\theta_{\{1,k\}}^i$ with $k \neq 1$.

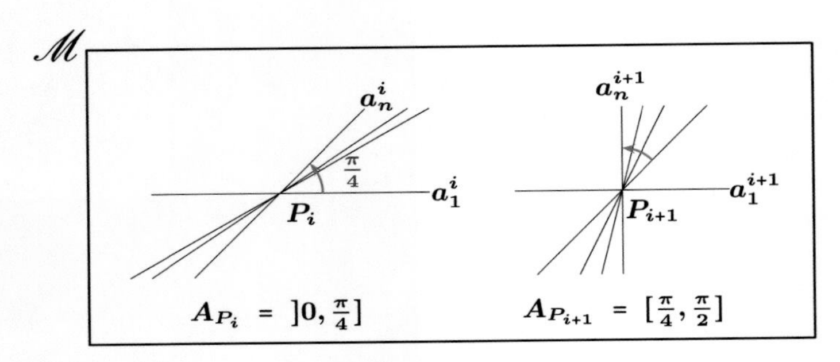

Fig. 8.3 $A_{P_{n-1}}$ strongly near A_{P_n} for each $n \geq 2$

In particular, we have that the set of all the curves is descriptively $\overset{\wedge\wedge}{\delta}$ −connected if we can find a countable subfamily of points P_n such that A_{P_n} is connected and $A_{P_{n-1}} \overset{\wedge\wedge}{\delta} A_{P_n}$ for each $n \geq 2$. To better understand this look at figure. We have $A_{P_i} =]0, \frac{\pi}{4}]$ and $A_{P_{i+1}} = [\frac{\pi}{4}, \frac{\pi}{2}]$. They are connected and, if we take $A \overset{\wedge\wedge}{\delta} B \Leftrightarrow A \cap B \neq \varnothing$, they are also strongly near. This means that, fixed α_1^i and α_1^{i+1} there is at least one curve through P_i and one through P_{i+1} such that they form the same angle with α_1^i and α_1^{i+1} respectively. We could require more by choosing a stronger strong proximity. In the previous way we obtained a sort of angle connectedness for families of curves. ■

8.2 Charts and Atlases

Start with a manifold $\mathscr{M}$. Consider a homeomorphic mapping on a neighbourhood $U \subset \mathscr{M}$ of a point p onto a open set in $\mathbb{R}^n$. In particular, let $\varphi : U \longrightarrow \mathbb{R}^n$ be that homeomorphism on U.

Definition 8.6 Chart.
A *chart* on $\mathscr{M}$ is a pair (U, φ). When the meaning is clear from the context, we write chart U instead of (U, φ). ■

Definition 8.7 Atlas.
An *atlas* $\mathcal{A}$ for manifold $\mathscr{M}$ is a collection of charts whose domain covers $\mathscr{M}$. ■

Given a pair of charts (U, φ), (V, ψ), the composite map $\psi \circ \varphi^{-1} : \varphi(U \cap V) \longrightarrow \psi(U \cap V)$ is called a transition map from φ to ψ.

A pair of charts is *smoothly compatible*, provided $U \cap V \neq \varnothing$ and the transition map $\psi \circ \varphi^{-1}$ is a C^∞-diffeomorphism on $\varphi(U \cap V)$. An atlas $\mathcal{A}$ is *smoothly compatible*, provided any pair of charts in $\mathcal{A}$ is smoothly compatible [6, Sect. 1, p. 12]. By replacing the requirement that charts be smoothly compatible with the weaker requirement that each transition map $\psi \circ \varphi^{-1}$ and its inverse are C^r-differentiable, $\mathscr{M}$ is called a C^r-manifold.

Suppose that $\mathscr{M}$ is an n-dimensional C^r-manifold. We can endow it with a proximity that is strictly connected with its structure. For example, if $\mathcal{A} = \{(U_i, \phi_i) : i \in I\}$ is an atlas on $\mathscr{M}$, we can define a proximity on $\mathcal{A} \times \mathcal{A}$ in the following way:

$$U_i \; \delta \; U_j$$
$$\Leftrightarrow$$
$$\exists C \subseteq U_i, D \subseteq U_j \; f : \phi_i(C) \to \phi_j(D),$$

such that f is C^r-diffeomorphism.

Theorem 8.8 ([8, Theorem 5.1]) *The relation δ is a proximity on $\mathcal{A} \times \mathcal{A}$.*

On a manifold, it is possible to define also a stronger kind of proximity, called a *manifold strong proximity*. As before, take an atlas $\mathcal{A} = \{(U_i, \phi_i) : i \in I\}$.

Definition 8.9 Manifold Strong Proximity.
Let $\overset{\wedge\!\!\wedge}{\delta}_{\mathcal{A}}$ be a relation on $\mathcal{A} \times \mathcal{A}$. It is called *manifold strong proximity*, if the following axioms hold:

$(M0)$ $\varnothing \not\overset{\wedge\!\!\wedge}{\delta}_{\mathcal{A}} U_i \forall i \in I$,

$(M1)$ $U_i \overset{\wedge\!\!\wedge}{\delta}_{\mathcal{A}} U_j \Leftrightarrow U_j \overset{\wedge\!\!\wedge}{\delta}_{\mathcal{A}} U_i \ \forall i, j \in I$

$(M2)$ $U_i \overset{\wedge\!\!\wedge}{\delta}_{\mathcal{A}} U_i \Rightarrow \phi_i(U_i) \cap \phi_j(U_j) \neq \varnothing$,

$(M3)$ If $\{U_h\}_{h \in H \subseteq I}$ is an arbitrary family of domains in $\mathcal{A}$ and $U_i \overset{\wedge\!\!\wedge}{\delta}_{\mathcal{A}} U_j$ for some $j \in H \setminus \{i\} \subseteq I$, then $U_i \overset{\wedge\!\!\wedge}{\delta}_{\mathcal{A}} \bigcup_{h \in H \setminus \{i\}} U_h$.

$(M4)$ $\text{int}(\phi_i(U_i)) \cap \text{int}(\phi_j(U_j)) \neq \varnothing \Rightarrow U_i \overset{\wedge\!\!\wedge}{\delta}_{\mathcal{A}} U_j$

Define the following relation on $\mathcal{A} \times \mathcal{A}$:

$$U_i \ \overset{\wedge\!\!\wedge}{\delta}_{\mathcal{A}} \ U_j \Leftrightarrow \phi_i(U_i) \cap \phi_j(U_j) \neq \varnothing.$$

That is, chart U_i is strongly near chart U_j, if and only if the chart descriptions $\phi_i(U_i) \cap \phi_j(U_j)$ have nonempty intersection. Moreover, if $U_j \cup U_k$ is not a domain in $\mathcal{A}$, define

$$\text{Property } (*) \ U_i \overset{\wedge\!\!\wedge}{\delta}_{\mathcal{A}} (U_j \cup U_k) \Leftrightarrow \phi_i(U_i) \cap (\phi_j(U_j) \cup \phi_k(U_k)) \neq \varnothing.$$

Theorem 8.10 ([8, Theorem 5.3]) *The relation $\overset{\wedge\!\!\wedge}{\delta}_{\mathcal{A}}$ is a manifold strong proximity on $\mathcal{A} \times \mathcal{A}$, if $U_i \cup U_j$ is not a domain in $\mathcal{A}$ for all i and j with $i \neq j$.*

Proof $(M0)$ That $\varnothing \not\overset{\wedge\!\!\wedge}{\delta}_{\mathcal{A}} U_i$ for each $i \in I$ is straightforward. $(M1)$ Symmetry is obvious. $(M2)$ The descriptive form of $A \delta B \Rightarrow A \cap B \neq \varnothing$ holds by definition. $(M3)$ This holds because we know that $U_i \cup U_j$ is not a domain in $\mathcal{A}$ for all i and j with $i \neq j$. So we refer to $(*)$. $(M4)$ This is easily observed. □

In terms of the proximity relation δ on $\mathcal{A} \times \mathcal{A}$ from Theorem 8.8, we obtain the following result.

Theorem 8.11 ([8, Theorem 5.4]) *Let $U_i, U_j \in \mathcal{A}$ be charts in manifold atlas $\mathcal{A}$. $U_i \ \overset{\wedge\!\!\wedge}{\delta}_{\mathcal{A}} \ U_j \Rightarrow U_i \ \delta \ U_j$.*

We want to define a strong proximity acting on Voronoï regions. We say that two Voronoï regions are strongly near, and we write $V_p \ \overset{\wedge\!\!\wedge}{\delta} \ V_q$, if and only if they share more than one point.

Fig. 8.4 Voronoï Regions $V_{a_i}, i \in \{1, 2, 3, 4, 5, 6, 7, 8\}$

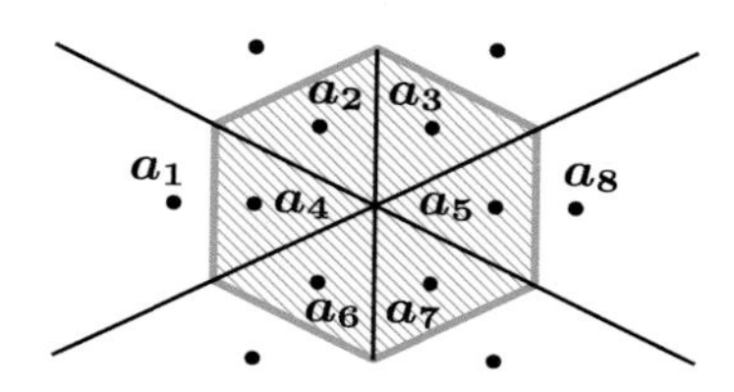

$$\boldsymbol{V(a_4)\ \overset{\wedge\!\!\wedge}{\delta}\ V(a_2)} \text{ and } \boldsymbol{V(a_4)\ \overset{\wedge\!\!\wedge}{\not\delta}\ V(a_5)}$$

Theorem 8.12 ([8, Theorem 6.3]) *Let $(X, \|\cdot\|)$ be a finite-dimensional normed vector space and S a collection of points in X. The relation defined by saying $V_p \overset{\wedge\!\!\wedge}{\delta} V_q$, if and only if V_p, V_q share more than one point, is a strong proximity on $\mathscr{V}(S)$, the class of Voronoï regions generated by S.*

Theorem 8.12 is illustrated in Example 8.13.

Example 8.13 **Strongly Proximal Voronoï Regions.**
Let X be a space covered with a Voronoï diagram $\mathscr{V}(S)$, S, a set of sites. A partial view of $\mathscr{V}(S)$ is shown in Fig. 8.4, where

$$V_{a_i} \in \mathscr{V}(S), i \in \{1, 2, 3, 4, 5, 6, 7, 8\}.$$

From Theorem 8.12, observe that

$$V_{a_4} \overset{\wedge\!\!\wedge}{\delta} V_{a_2},\ V_{a_4} \overset{\wedge\!\!\wedge}{\delta} V_{a_6} \text{ and } V_{a_4} \overset{\wedge\!\!\wedge}{\not\delta} V_{a_5},\ V_{a_2} \overset{\wedge\!\!\wedge}{\not\delta} V_{a_5},\ V_{a_6} \overset{\wedge\!\!\wedge}{\not\delta} V_{a_5},$$

since $\{V_{a_2}, V_{a_4}\}, \{V_{a_4}, V_{a_6}\}$ have a common edge. Further, $V_{a_2}, V_{a_4}, V_{a_6}$ are not strongly near V_{a_5}. $V_{a_2}, V_{a_4}, V_{a_6}$ share only one point with V_{a_5}. Similarly,

$$V_{a_5} \overset{\wedge\!\!\wedge}{\delta} V_{a_3}, V_{a_5} \overset{\wedge\!\!\wedge}{\delta} V_{a_7}, V_{a_5} \overset{\wedge\!\!\wedge}{\delta} V_{a_8},$$

since, taken pairwise, these Voronoï regions have a common edge. There are also Voronoï regions in Fig. 8.4 that are near but not strongly near, e.g., $V_{a_3} \overset{\wedge\!\!\wedge}{\not\delta} V_{a_6}, V_{a_7} \overset{\wedge\!\!\wedge}{\not\delta} V_{a_2}$. ■

Remark 8.14 **Hit and Strongly Far-Miss and Strongly Hit and Far-Miss Hypertopologies.**
Let CL(X) be the hyperspace of all non-empty closed subsets of a space X. *Hit and miss* and *hit and far-miss* topologies on CL(X) are obtained by the join of two halves. Well-known examples are Vietoris topology [9–12] (see, also, [13–25]) and Fell topology [26]. In [27, 28], the following new hypertopologies are introduced.

♣ $\tau_{\mathbb{W}}$ is the *hit and strongly far-miss* topology having as subbase the sets of the form:

- $V^{-} = \{E \in \mathrm{CL}(X) : E \cap V \neq \varnothing\}$, where V is an open subset of X,
- $A_{\mathbb{W}} = \{\ E \in \mathrm{CL}(X) : E \underset{\mathbb{W}}{\not\delta} X \setminus A\ \}$, where A is an open subset of X

♣ $\tau^{\wedge\wedge}$ is the *strongly hit and far-miss* topology having as subbase the sets of the form:

- $V^{\wedge\wedge} = \{E \in \mathrm{CL}(X) : E \overset{\wedge\wedge}{\delta} V\}$, where V is an open subset of X,
- $A^{++} = \{\ E \in \mathrm{CL}(X) : E \not\delta X \setminus A\ \}$, where A is an open subset of X,

where in both cases δ is a Lodato proximity compatible with the topology on X. For more about these and other hypertopologies, see [28, 29] (see, also, [25]). ■

From Theorem 8.12, we can define a strongly hit and miss hypertopology, $\tau^{\wedge\wedge}$, on the space of Voronoï regions generated by S, $\mathscr{V}(S)$, to which we add the empty set, [28]. The hypertopology $\tau^{\wedge\wedge}$ has as subbase the elements of the following form:

- $\mathrm{int}(V_p)^{\wedge\wedge} = \{V_q \in \mathscr{V}(S) : V_q \overset{\wedge\wedge}{\delta} \mathrm{int}(V_p)\} = \{V_q \in \mathscr{V}(S) : V_q \overset{\wedge\wedge}{\delta} V_p\}$,
- $V_s^{+} = \{V_q \in \mathscr{V}(S) : V_q \cap V_s = \varnothing\}$,

where Voronoï regions $V_p, V_s \in \mathscr{V}(S)$.

Theorem 8.15 ([8, Theorem 5.4]) *Let* $(X, \|\cdot\|)$ *be a finite-dimensional normed vector space and* S *a collection of points in* X*. For any* $p \in S$ *let* $\{a_i\}_{i \in I}$ *the family of points in* S *such that* $V_{a_i} \overset{\wedge\wedge}{\delta} V_p$*, and* $\{b_j\}_{j \in J}$ *the family of points in* S *such that* $V_{b_j} \cap V_p = \varnothing$*. Then* $\mathscr{A} = (\bigcap_{i=1}^{n} \mathrm{int}(V_{a_i})^{\wedge\wedge}) \cap (\bigcap_{j=1}^{m} V_{b_j}^{+})$ *is the smallest open set in* $\tau^{\wedge\wedge}$ *containing* V_p*.*

Example 8.16 **Smallest Open Set in a Hypertopology Containing a Voronoï Region.**
Consider the situation in Fig. 8.5. Take, for example, the Voronoï region V_{a_4}. The smallest open set in $\tau^{\wedge\wedge}$ containing V_{a_4} is given by

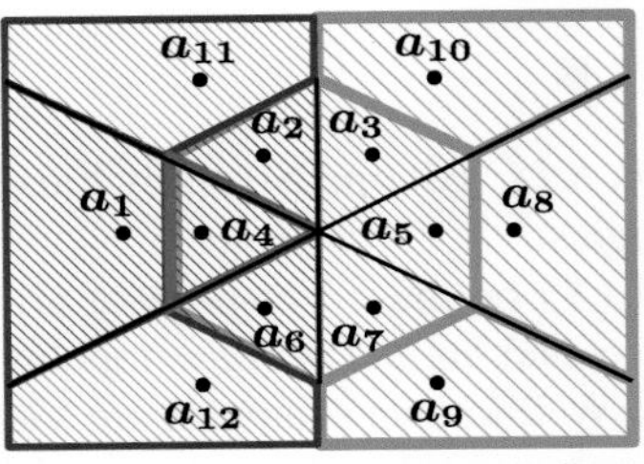

Fig. 8.5 Smallest open set containing Voronï region of a4

$V(a_4) \overset{\wedge\wedge}{\delta} V(a_2)$ and $V(a_4) \overset{\wedge\wedge}{\not\delta} V(a_5)$

$$\mathscr{A} = (\bigcap_{i=1,2,4,6} \text{int}(V_{a_i})^{\Cap}) \cap \left(\bigcap_{q=8,9,10} V_q^+ \right). \quad \blacksquare$$

Theorem 8.17 ([8, Theorem 5.4]) *Let $(X, \|\cdot\|)$ be a finite-dimensional normed vector space and S a collection of points in X. If $p \in S$ and $\mathscr{A}$ is the smallest open set containing V_p, then $\mathscr{A}$ cannot contain any other region in $\mathscr{V}(S)$.*

8.3 Proximal Voronoï Manifolds, Atlases and Charts

Let $\mathcal{M}$ be a manifold, that is a topological space which is Hausdorff, second countable, locally Euclidean of dimension n. This means that for each point there is a neighbourhood U of $\mathcal{M}$ with a homeomorphism $\varphi : U \longrightarrow \hat{U} = \varphi(U) \subseteq \mathbb{R}^n$. $\mathcal{M}$ is a Voronoï manifold, provided $\varphi(U)$ is a Voronoï diagram. The pair (U, φ) is called a *Voronoï chart* on $\mathcal{M}$. The collection $\mathcal{A}$ of all Voronoï charts on $\mathcal{M}$ is called a *Voronoï atlas*.

Let $\mathcal{M}_1, \mathcal{M}_2$ be Voronoï manifolds and let $S_1 \subset \mathcal{M}_1$, $S_2 \subset \mathcal{M}_2$ be nhbds (neighbourhoods) of points in $\mathcal{M}_1$ and $\mathcal{M}_2$ respectively, φ, ψ homeomorphisms from S_1, S_2 to subsets $\varphi(S_1), \psi(S_2) \subseteq \mathbb{R}^n$ such that $\text{cl}(\varphi(S_1)), \text{cl}(\psi(S_2))$ are Voronoï diagrams. From what has been observed about manifolds, we make the following observations. Define

$$\mathcal{M}_1 \overset{\Cap}{\delta} \mathcal{M}_2 \Leftrightarrow \exists (S_1, \varphi), (S_2, \psi) : \text{cl}(\varphi(S_1)) \overset{\Cap}{\delta} \text{cl}(\psi(S_2)) (\text{see Theorem 8.12}).$$

Example 8.18 **Strongly Near Charts.**
Let $\mathcal{M}_1, \mathcal{M}_2$ be Voronoï manifolds in the plane. From Fig. 8.6, let

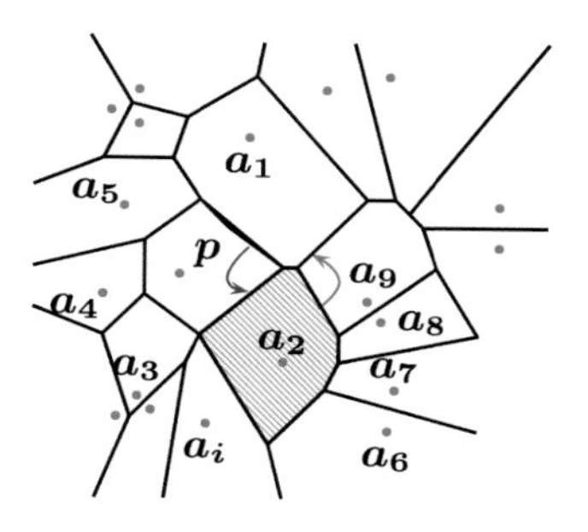

Fig. 8.6 Voronoï diagram with respect to Theorem 8.17

$$\begin{aligned}
\mathcal{M}_1 &= \text{the portion of the plane containing the regions}\\
&\quad\ \text{associated with } a_1, a_3, a_4, a_5, p.\\
\mathcal{M}_2 &= \text{the portion of the plane containing the regions}\\
&\quad\ \text{associated with } a_2, a_6, a_7, a_8, a_9.\\
\varphi(S_1) &= \text{the interior of the portion of the plane containing the regions}\\
&\quad\ \text{associated with } p, a_1.\\
\psi(S_2) &= \text{the interior of the portion of the plane containing the regions}\\
&\quad\ \text{associated with } a_2, a_8.\\
\text{cl}(\varphi(S_1)) &= \mathscr{V}(S_1)\ \text{(Voronoïdiagram)},\ V_p \in \mathscr{V}(S_1).\\
\text{cl}(\psi(S_2)) &= \mathscr{V}(S_2)\ \text{(Voronoïdiagram)},\ V_{a_2} \in \mathscr{V}(S_1).
\end{aligned}$$

In this simple case, the homeomorphisms correspond to the identity map. In Fig. 8.6, $\mathscr{V}(S_1), \mathscr{V}(S_2)$ share the edge between Voronoï regions V_p and V_{a_2}. Hence, $\text{cl}(\varphi(S_1)) \stackrel{\doublewedge}{\delta} \text{cl}(\psi(S_2))$. So $\mathcal{M}_1 \stackrel{\doublewedge}{\delta} \mathcal{M}_2$. ■

In [7], new kind of connectedness related to strong proximities is introduced.

Definition 8.19 Let X be a topological space and $\stackrel{\doublewedge}{\delta}$ a strong proximity on X. We say that X is $\stackrel{\doublewedge}{\delta}$ −connected if and only if $X = \bigcup_{i \in I} X_i$, where I is a countable subset of $\mathbb{N}$, X_i and $\text{int}(X_i)$ are connected for each $i \in I$, and $X_{i-1} \stackrel{\doublewedge}{\delta} X_i$ for each $i \geq 2$. ■

Remark 8.20 $X_{i-1} \stackrel{\doublewedge}{\delta} X_i$ in Definition 8.19 can be formulated in descriptive terms. Let X be a set, Φ a description that maps X to $\mathbb{R}^n$, $\stackrel{\doublewedge}{\delta}$ a strong proximity on $\mathbb{R}^n$ endowed with the Euclidean topology. We say that two subsets A, B are *descriptively strongly near*, and we write

$$A \stackrel{\doublewedge}{\delta}_{\Phi} B, \text{ if and only if } \Phi(A) \stackrel{\doublewedge}{\delta} \Phi(B). \text{ (descriptive strongly near)} \quad ■$$

In terms of descriptively near manifolds $\mathcal{M}_1, \mathcal{M}_2, S_1 \subset \mathcal{M}_1, S_2 \subset \mathcal{M}_2$ with corresponding descriptively near charts $(S_1, \varphi), (S_2, \psi)$, we have

$$\mathcal{M}_1 \stackrel{\doublewedge}{\delta}_{\Phi} \mathcal{M}_2 \Leftrightarrow \exists (S_1, \varphi), (S_2, \psi) : \varphi(S_1) \stackrel{\doublewedge}{\delta}_{\Phi} \psi(S_2), \text{ (in the sense of Remark 8.20)},$$

where $\Phi : \mathcal{M}_1 \cup \mathcal{M}_2 \to \mathbb{R}^n$.

Example 8.21 Continuing Example 8.18, assume

$$\begin{aligned}
&x \in \varphi(S_1),\ y \in \psi(S_2).\\
\Phi(x) &= (\text{colour of } x, \theta_x \text{ gradient angle}),\ \text{feature vector for } x.\\
\Phi(y) &= (\text{colour of } y, \theta_y \text{ gradient angle}),\ \text{feature vector for } y.
\end{aligned}$$

Assume x, y have matching feature vectors, then

$$\mathcal{M}_1 \overset{\wedge\!\!\wedge}{\delta}_{\Phi} \mathcal{M}_2 \Leftrightarrow \exists (S_1, \varphi), (S_2, \psi) : \Phi(\varphi(S_1)) \overset{\wedge\!\!\wedge}{\delta} \Phi(\psi(S_2)). \quad \blacksquare$$

Let $\mathcal{A}_1 = \left\{(U_i, \varphi_i) : i \in \mathbb{N}^+\right\}, \mathcal{A}_2 = \left\{(V_j, \psi_j) : j \in \mathbb{N}^+\right\}$ be atlases on smooth manifolds $\mathcal{M}_1, \mathcal{M}_2$, respectively, $\hat{U}_i = \varphi_i(U_i), \hat{V}_j = \psi_j(V_j)$, and define the *descriptive intersection of the disjoint charts* by

$$\hat{U}_i \underset{\Phi,\hat{A}}{\cap} \hat{V}_j = \left\{x \in \hat{U}_i \cup \hat{V}_j : \Phi(x) \in \Phi(\hat{U}_i),\ \Phi(x) \in \Phi\left(\hat{V}_j\right)\right\}.$$

Then define the relation $\underset{\hat{A},\Phi}{\overset{\wedge\!\!\wedge}{\delta}}$ on $\mathcal{A}_1 \times \mathcal{A}_2$ by

$$U_i \underset{\hat{A},\Phi}{\overset{\wedge\!\!\wedge}{\delta}} V_j \Leftrightarrow \hat{U}_i \underset{\Phi,\hat{A}}{\cap} \hat{U}_j \neq \varnothing.$$

Theorem 8.22 ([8, Theorem7.3]).
Let $\mathcal{A}_1 = \left\{(U_i, \varphi_i) : i \in \mathbb{N}^+\right\}, \mathcal{A}_2 = \left\{(V_j, \psi_j) : j \in \mathbb{N}^+\right\}$ be atlases on smooth manifolds $\mathcal{M}_1, \mathcal{M}_2$, respectively. Then

1. $\hat{U}_i \overset{\wedge\!\!\wedge}{\delta}_{\Phi} \hat{V}_j \Rightarrow U_i \underset{\hat{A},\Phi}{\overset{\wedge\!\!\wedge}{\delta}} V_j$
2. $\mathcal{M}_1 \overset{\wedge\!\!\wedge}{\delta}_{\Phi} \mathcal{M}_2 \Rightarrow \exists (U_i, \varphi_i) \in \mathcal{A}_1, (V_j, \psi_j) \in \mathcal{A}_2 : U_i \underset{\hat{A},\Phi}{\overset{\wedge\!\!\wedge}{\delta}} V_j$

8.4 Mean Shift Manifold

Examples of image manifolds are given in this section. For simplicity, we consider colour image manifolds in Euclidean space $\mathbb{R}^3$ and greyscale image manifolds in Euclidean space $\mathbb{R}^2$. The basic approach here is to map the neighbourhoods of an image onto a subset of an Euclidean space.

The basic example of a manifold is Euclidean space, and many of its properties carry over to manifolds. In addition, any smooth boundary of a subset of Euclidean space, like the circle or the sphere, is a manifold. Manifolds are useful in solving dimensionality reduction problems (mapping on a high dimensional space to a lower dimensional space) and solving object recognition, pattern recognition and image classification problems. Manifolds are therefore of interest in the study of digital images as well as in geometry, topology, and mathematical analysis. Manifolds are defined in terms of homeomorphic mappings between topological spaces.

Definition 8.23 Homeomorphically Topological Spaces.
Topological spaces are *homeomorphically* or *topologically equivalent*, provided there

is a continuous 1-1, onto mapping (bijection) from one space onto the other whose inverse is also continuous (cf. [30, Sect. I.2, p. 9]). ■

Example 8.24 **Open Disk in $\mathbb{R}^2 \longrightarrow \mathbb{R}^2$.**
Let $N(0, 1)$ be an open disk centered around the origin in the Euclidean plane with a radius equal to 1, i.e.,

$$N(0,1) = \left\{x \in \mathbb{R}^2 : \|x\| < 1\right\}.$$

The open disk $N(0, 1)$ is homeomorphic to a subset of $\mathbb{R}^2$. That is, the homeomorphism $f : N(0, 1) \longrightarrow \mathbb{R}^2$ is defined by

$$f(x) = \frac{x}{(1 - \|x\|)}, x \in N(0,1).$$

This example is based on a homeomorphism introduced in [30, Sect. II.1, p. 27]). ■

Example 8.25 **Open Disk in $\mathbb{R}^2 \longrightarrow$ mean-shifted disk in $\mathbb{R}^2$.**
Let $N(p, r)$ be an open disk centered around a point p in the Euclidean plane with a radius equal to $r > 0$, i.e.,

$$N(p,r) = \left\{x \in \mathbb{R}^2 : \|x - p\| < r\right\}.$$

The open disk $N(p, r)$ is homeomorphic to a subset of $\mathbb{R}^2$. Let $i(x)$ denote the intensity of a pixel x in a digital image. Then, the homeomorphism $f : N(p, r) \longrightarrow \mathbb{R}^2$ is defined by

$$f(x) = \frac{i(x)}{\sum\limits_{i=1}^{|N(p,r)|} i(y)}, x, y \in N(p,r).$$ ■

Definition 8.26 Manifold.
A topological space X is a *n-manifold* if and only if every $x \in X$ has a neighbourhood homeomorphic to a subset in $\mathbb{R}^n$ (cf. [30, Sect. III.2, p. 27])

Mscript 11 Mean Shift Manifold.

```
(*Manifold resulting from amean shift filter of an image :
MeanShifFiter[image, r, d]replaceseachpixelwith the mean of the pixelsin arange−

rneighborhood and whose value is within adistance d *)
```

Fig. 8.7 Result of mean-shifting image r-neighbourhoods

im = ;
***MeanShiftFilter*[*im*, 3, .1, *MaxIterations* → 50]** ■

8.5 Digital Image Manifolds

A 2D digital image is an example of a finite topological space on the Euclidean space $\mathbb{R}^2$. This is the case, since the intersection and union of every pair of image subsets belongs to the image. Each pair of image pixels is separated by points in the Euclidean plane. For this reason, it is always the case that distinct image points belong to disjoint nhbds. For this reason, a 2D image is a Hausdorff space. Recall that a topological space is Hausdorff, provided every pair of distinct points belong to disjoint neighbourhoods.

Also recall that a *topological basis* (base) is a subset $\mathcal{B}$ of a space X in which all open sets of the space can be written as unions or finite intersections of basis elements. For example, the family of subsets of a digital image form a basis for the image. A topological space is a *second countable* space, provided the space has a countable base. This the case with a digital image, since every subset of an image belongs to the union or intersection of image subsets.

In effect, every image is a topological space that is Hausdorff and second countable.

Let f be a mapping on a set X into a set Y. The mapping is continuous, provided every pair of points $x, y \in X$ that are near (i.e., $\|x - y\| < \varepsilon, \varepsilon > 0$) implies that $f(x)$ is near $f(y)$. Assuming that a proximity δ has been given, then a mapping f is proximally continuous, provided

$$x \ \delta \ y \Rightarrow f(x) \ \delta \ f(y).$$

A 1-1 mapping f on an image onto itself is a homeomorphism. If we define a continuous, homeomorphic mapping on an image into itself, then the image is a manifold. A homeomorphic mapping on a Euclidean space onto either itself or another Euclidean space guarantees that the mapping is continuous. So, in the context of digital images, we drop the the reference to continuity.

A nonempty set X with a continuous homeomophic mapping on every neighbourhood around every point of X onto a subset of the Euclidean space $\mathbb{R}^n$ is a **manifold**.

In general, a subset X of a Euclidean plane is a manifold, provided there a homeomorphic mapping from X onto itself.

Theorem 8.27 *A 2D digital image X is a manifold, provided there a homeomorphic mapping from X onto itself.* ■

Proof
Every digital image is a finite topological space that is Hausdorff and second countable. Hence, a digital image X with a homeomorphic mapping on X onto itself is, by definition, a manifold. □

An image manifold that satisfies Theorem 8.27 is called a selfie image manifold.

Example 8.28 **Selfie Image Manifold.**
In general, a subset X of a Euclidean plane is a manifold, provided there a homeomorphic mapping from X onto itself. ■

Given a pair of digital images X and Y, define a homeomorphic mapping on X onto Y. Then X is a manifold that is locally Euclidean. This holds true for both 2D and 3D images.

Given a digital image X in the plane, define a homeomorphic mapping on X onto Y in the Euclidean space $\mathbb{R}^n$. Then X is a manifold. The same holds for 3D images.

8.8.1: $\mathbb{R}^3$ Shoe

8.8.2: $\mathbb{R}^3$ MeanShiftShoe

Fig. 8.8 Rainbow-on-a-shoe colour image manifold in $\mathbb{R}^3$

8.6 Mean-Shift Image Manifolds

In this section, each of the neighbourhood disks in an image is mean-shifted onto neighbourhood disks, where all intensities each manifold neighbourhood disk are replaced by the average neighbourhood disk intensity.

Example 8.29 **Image Disks in $\mathbb{R}^3 \longrightarrow$ mean-shifted disks in $\mathbb{R}^3$.**
This example is a continuation of Example 8.25. In this example, the manifold is a colour image in $\mathbb{R}^3$ that is a topological space homeomorphic to a subset in $\mathbb{R}^3$. This manifold results from replacing each intensity in each r-neighbourhood in a colour image im with the average value of the neighbourhood pixel intensities within a distance d. This means that every neighbourhood pixel intensity is mapped to an average neighbourhood intensity. In effect, there is a homeomorphic mapping on each neighbourhood $N(p, r)$ in $\mathbb{R}^3$ onto a subset of $\mathbb{R}^3$ that contains the average neighbourhood intensities.

The Mathematica script 11 illustrates a mean-shift manifold that results from an application of the **MeanShifFiter** function on the colour image shown in Fig. 8.8.1. The subset of $\mathbb{R}^3$ containing the mean-shifted intensities in the neighbourhoods of Fig. 8.8.1 is shown in Fig. 8.7. ■

Example 8.30 **Image Disks in $\mathbb{R}^2 \longrightarrow$ mean-shifted disks in $\mathbb{R}^2$.**
In this example, the manifold is a greyscale satellite image of the Alhambra[1] in $\mathbb{R}^2$ that is a topological space homeomorphic to a subset of $\mathbb{R}^2$. This manifold results from replacing each intensity in each r-neighbourhood in the greyscale image with the average value of the neighbourhood pixel intensities. This means that every neighbourhood pixel intensity is mapped to an average neighbourhood intensity. In effect, there is a homeomorphic mapping on each neighbourhood $N(p, r)$ in $\mathbb{R}^2$ onto a subset of $\mathbb{R}^2$ that contains the average neighbourhood intensities.

[1] Many thanks to Bin Jang, University of Gävie, for the Alhambra satellite image.

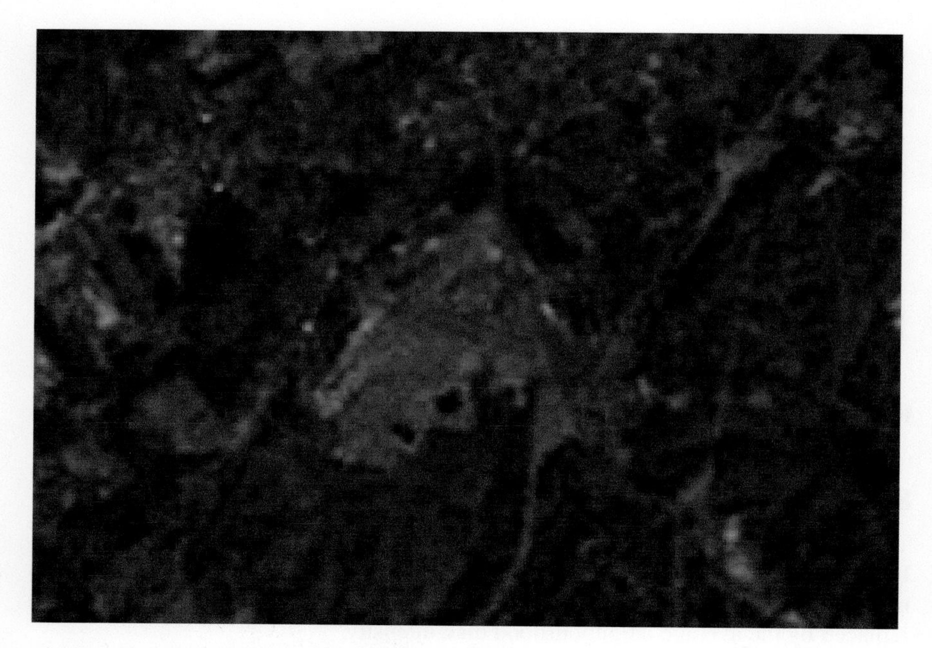

Fig. 8.9 Alhambra satellite image manifold

The Mathematica script 11 is now used on the greyscale image shown in Fig. 8.9. The subset of $\mathbb{R}^2$ containing the mean-shifted intensities in the neighbourhoods of Fig. 8.9 is shown in Fig. 8.10. ■

Remark 8.31 **What the Mean-Shift Manifold Reveals.**
The Alhambra satellite image manifold introduced in Example 8.30 results in a clearcut segmentation of the satellite image. The uneven, blurred parts of the original satellite image have been replaced by sharply defined segments. These segments facilitate image analysis and object recognition. Objects in the satellite image manifold are now revealed more clearly in the homeomorphic image containing the mean-shifted neighbourhoods. ■

Mscript 12 Image Manifold Saliency Map.

(* Saliency Manifold *)

im2 = 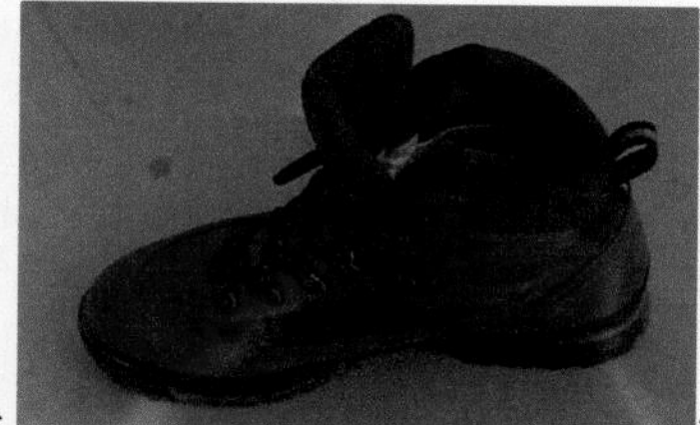 ;
ImageSaliencyFilter[im2]//ImageAdjust ■

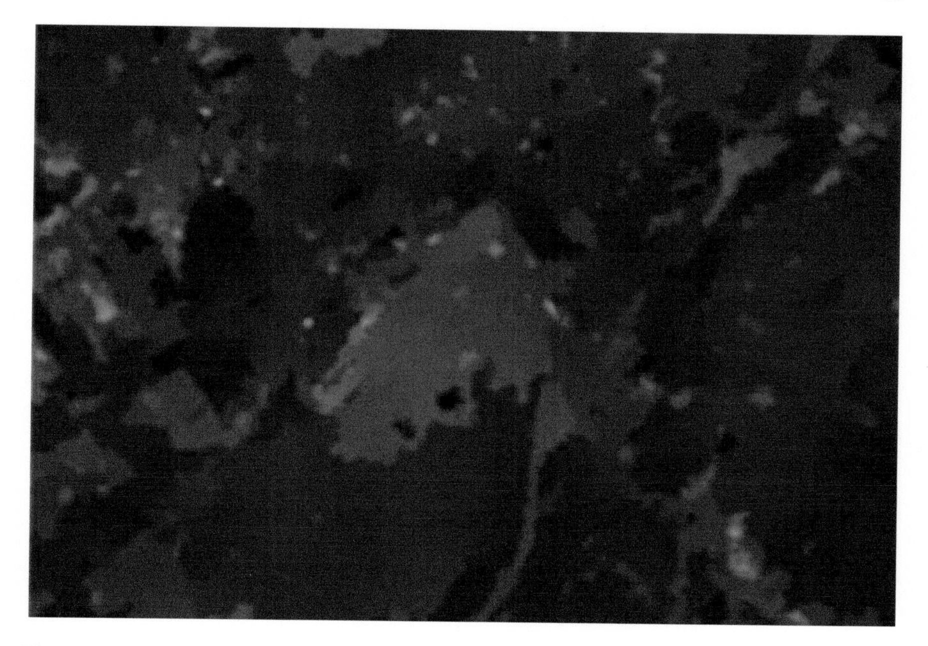

Fig. 8.10 Mean-shift filtered Alhambra Satellite image

8.7 Image Saliency Manifolds

A *saliency map* predicts the focus of attention in an input image. A saliency value is nonlinearly emphasized or suppressed to simulate the lateral inhibition mechanism [31, Sect. 4.3.1, p. 661]. The saliency value is found using the Gabor pyramid-based orientation detection method. The end result of a saliency mapping is a segmentation of an image into salient image regions. The salient regions of an image can be extracted by considering the image intensities that match a given intensity value within a subimage containing a maximum intensity in a saliency filtered image. Each salient region can serve as a manifold, provided some form of a homeomorphic mapping is defined on the salient region onto a subset of $\mathbb{R}^n$.

Example 8.32 **Image Disks in $\mathbb{R}^3$ $\longrightarrow$ mean-shifted disks in $\mathbb{R}^3$.**
In this example, the manifold is the colour image in $\mathbb{R}^3$ in Fig. 8.8.1 that is a topological space homeomorphic in $\mathbb{R}^3$ to a subset in $\mathbb{R}^3$. This manifold results from replacing each intensity in each r-neighbourhood in the colour image im with a saliency value. This means that every neighbourhood pixel intensity is mapped to a salient intensity. In effect, there is a homeomorphic mapping on each neighbourhood $N(p, r)$ in $\mathbb{R}^3$ onto a subset of $\mathbb{R}^3$ that contains the neighbourhood salient intensities.

The Mathematica script 12 illustrates an application of the **ImageSaliencyFilter** function on the colour image shown in Fig. 8.8.1. The subset of $\mathbb{R}^3$containing the salient intensities in the neighbourhoods of Fig. 8.8.1 is shown in Fig. 8.11. ■

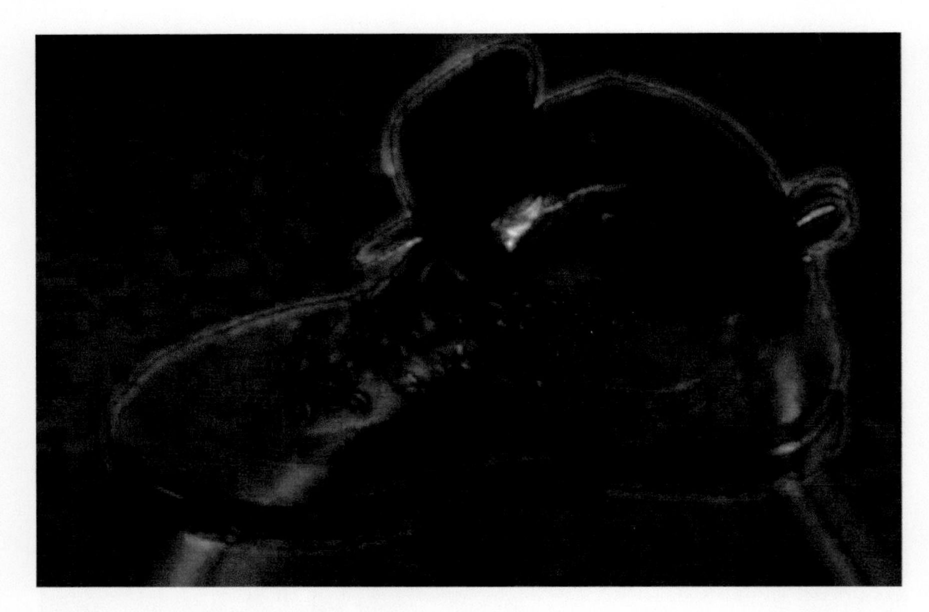

Fig. 8.11 Shoe saliency manifold map

8.8 Selfie Voronoï Manifold

Generating a Voronoï mesh on any digital image leads to a member of a class of image mesh manifolds. This class of image manifolds (called selfie Voronoï Manifolds) is particularly interesting, since each polygon in the mesh is a convex image segment that yields information about image object(s) covered by the convex segment.

Theorem 8.33 *A homeomorphic mapping on a digital image onto a Voronoï mesh on the same image, defines an image manifold.*

Proof Let X be a digital image. Each polygon P in a Voronoï mesh on X is a closed set with a nonempty interior. Each of the points in P has a 1-1 correspondence with a point in X. A Voronoï mesh covers an entire image with convex polygons. Hence, there is a homeomorphic mapping f on X onto itself. By definition, X is a topological space that is Hausdorff and second countable. Consequently, X is an image manifold. □

The following example of a selfie Voronoï manifold is based on a colour image of the hand from Binglin Li.

Example 8.34 **Binglin Li Mesh Manifold**.
Each pixel in the hand image Im in Fig. 8.12 corresponds to a point in a Voronoï mesh V superimposed on the image. The Voronoï mesh on the hand image is shown in Fig. 8.12. The fact that each pixel in Im corresponds to a mesh pixel implies that Im

8.12.1: $\mathbb{R}^3$ Binglin Li Image

8.12.2: $\mathbb{R}^3$ Mesh

Fig. 8.12 Selfie Voronoï manifold

is homeomorphic to V. Hence, from Theorem 8.33, Im is a manifold (called a mesh manifold). The mesh itself originally created by Binglin Li has been modified so that the polygon edges are more easily detected on the darker background. Because the edges are shown as — straight line segments, the generating points are shown as ●, displayed as tiny disks. ■

8.9 Manifolds with a Proximity

Proximal manifolds were introduced in [32] (see, also, [33]). A *proximal manifold* is a manifold endowed with one or more proximity relations.

Let M be a proximal manifold endowed with the Wallman proximity δ (denoted (M, δ)).

Theorem 8.35 *Let $A, B \subset M$. $A\ \delta B$ if and only if A and B have at least one point in common.*

Proof Since δ is a Wallman proximity, $A\ \delta\ B \ \Leftrightarrow\ \mathrm{cl}A \cap \mathrm{B} \neq \emptyset$. Hence, A and B have at least one point in common. □

Example 8.36 **Voronoï Regions with a Common Boundary are Near.**
Polygons A, B with an edge in common in a Voronoï mesh manifold satisfy Theorem 8.35. Consequently, $A\ \delta\ B$, i.e., polygons A and B are near polygons. ■

Recall that nonempty sets are *strongly near*, provided the sets have more than one point in common.

Theorem 8.37 *Let $A, B \subset M$, a manifold. $A \stackrel{\wedge\!\!\wedge}{\delta} B$ if and only if A and B have more than one point in common.*

Proof Immediate from the definition of a strongly near proximity. □

Theorem 8.38 *Let $A, B \subset M$, an image manifold. $A \stackrel{\wedge\!\!\wedge}{\delta} B$ if and only if A and B have more than one point in common.*

Proof Immediate from Theorem 8.37. □

Example 8.39 **Strongly Near Sets of Voronoï Regions in a Mesh Manifold.**
Let A, B be sets of polygons in a Voronoï mesh. Assume that $\text{int}A \cap \text{int}B \neq \emptyset$. From Theorem 8.37, $A \stackrel{\wedge\!\!\wedge}{\delta} B$, i.e., polygons A and B are strongly near sets. ■

Example 8.40 **Strongly Near Sets of Voronoï Regions in an Image Mesh Manifold.**
Let A, B be sets of polygons in an image Voronoï mesh manifold. Assume that $\text{int}A \cap \text{int}B \neq \emptyset$. From Theorem 8.38, $A \stackrel{\wedge\!\!\wedge}{\delta} B$, i.e., A and B are strongly near sets of sets of image mesh polygons. ■

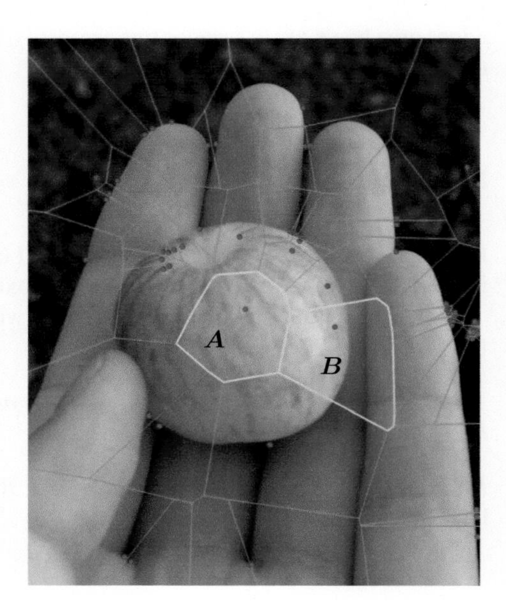

Fig. 8.13 $A \stackrel{\wedge\!\!\wedge}{\delta} B$

For example, there are many Voronoï regions in Fig. 8.12 that are strongly near.

Example 8.41 **Strongly Near Voronoï Regions in the Binglin Li Mesh Manifold.** Polygons A, B in the mesh manifold in Fig. 8.13 have an edge in common. In effect, A and B have more than one point (pixel) in common. Hence, these polygons satisfy Theorem 8.35. Consequently, $A \overset{\wedge\wedge}{\delta} B$, i.e., polygons A and B are strongly near polygons. ■

References

1. Milnor, J.: Topological manifolds and smooth manifolds. Proceedings of the International Congress Mathematicians (Stockholm, 1962), pp. 132–138. Institute Mittag-Leffler, Djursholm (1962). MR0161345
2. Milnor, J.: Microbundles. I. Topology 3(suppl. 1), 53–80 (1964). MR0161346
3. Krantz, S.: A Guide to Topology. The Mathematical Association of America, Washington (2009). Ix + 107 pp
4. Willard, S.: General Topology. Dover Pubisher Inc., Mineola (1970). Xii + 369 pp. ISBN: 0-486-43479-6 54-02, MR0264581
5. Hettiarachchi, R., Peters, J.: Multi-manifold LLE learning in pattern recognition. Pattern Recognit. Elsevier **48**(9), 2947–2960 (2015)
6. Lee, J.: Introduction to Smooth Manifolds. Graduate Texts in Mathematics vol. 218. Springer, Berlin (2016). Xvi + 708 pp., MR2954043
7. Peters, J., Guadagni, C.: Strongly proximal continuity and strong connectedness, 1–11 (2015). arXiv:1504.02740
8. Peters, J., Guadagni, C.: Strong proximities on smooth manifolds and Voronoï diagrams. Adv. Math.: Sci. J. **4**(2), 91–107 (2015)
9. Vietoris, L.: Stetige mengen. Monatshefte für Mathematik und Physik **31**, 173–204 (1921). MR1549101
10. Vietoris, L.: Bereiche zweiter ordnung. Monatshefte für Mathematik und Physik **32**, 258–280 (1922). MR1549179
11. Vietoris, L.: Kontinua zweiter ordnung. Monatshefte für Mathematik und Physik **33**, 49–62 (1923). MR1549268
12. Vietoris, L.: Über den höheren zusammenhang kompakter räume und eine klasse von zusammenhangstreuen abbildungen. Math. Ann. **97**(1), 454–472 (1927). MR1512371
13. Beer, G.: Topologies on Closed and Closed Convex Sets. Kluwer Academic Publishers, The Netherlands (1993)
14. Beer, G., Lucchetti, R.: Weak topologies for the closed subsets of a metrizable space. Trans. Am. Math. Soc. **335**(2), 805–822 (1993)
15. Beer, G., Di Concilio, A., Di Maio, G., Naimpally, S., Pareek, C., Peters, J.: Somashekhar Naimpally, 1931–2014. Topol. Appl. **188**, 97–109 (2015). http://dx.doi.org/10.1016/j.topol.2015.03.010, MR3339114
16. Di Concilio, A.: Uniformities, hyperspaces, and normality. Monatshefte für Math. **107**, 303–308 (1989). MR1012462
17. Di Concilio, A.: Point-free geometries: proximities and quasi-metrics. Math. Comput. Sci. **7**(1), 31–42 (2013). MR3043916
18. Di Concilio, A.: Action on hyperspaces. Topol. Proc. **41**, 85–98 (2013)
19. Di Concilio, A., Naimpally, S.: Proximal set-open topologies. Boll. Unione Mat. Ital. Sez. B Artic Ric. Mat. **8**(1), 173–191 (2000). MR1755708
20. Di Concilio, A.: Proximal set-open topologies on partial maps. Acta Math. Hungar. **88**(3), 227–237 (2000). MR1767801

21. Di Maio, G., Kočinac, L.: Some covering properties of hyperspaces. Topol. Appl. **155**(17–18), 1959–1969 (2008). MR2457978
22. Di Maio, G., Holá, L.: On hit-and-miss topologies. Rendiconto dell'Accademia delle Scienze Fisiche e Matematiche **4**(LXII), 103–124 (1995). MR1419286
23. Di Maio, G., Naimpally, S.: Comparison of hypertopologies. Rend. Istit. Mat. Univ. Trieste **22**(1–2), 140–161 (1992). MR1210485
24. Di Maio, G., Meccariello, E., Naimpally, S.: Uniformly discrete hit-and-miss hypertopology. a missing link in hypertopologies. Appl. Gen. Topol. **7**(2), 245–252 (2006). MR2295173
25. Guadagni, C.: Bornological convergences on local proximity spaces and ω_μ-metric spaces. Ph.D. thesis, Università degli Studi di Salerno, Salerno, Italy (2015) (Supervisor: A. Di Concilio, 79 pp.)
26. Fell, J.: A Hausdorff topology for the closed subsets of a locally compact non-Hausdorff space. Proc. Am. Math. Soc. **13**, 472–476 (1962). MR0139135
27. Peters, J., Guadagni, C.: Strongly far proximity and hyperspace topology, 1–6 (2015). arXiv:1502.02771
28. Peters, J., Guadagni, C.: Strongly near proximity and hyperspace topology, 1–6 (2015). arXiv:1502.05913
29. Peters, J., Guadagni, C.: Strongly hit and far miss hypertopology and hit and strongly far miss hypertopology, 1–8 (2015). arXiv:1503.02587
30. Edelsbrunner, H.: Geometry and Topology of Mesh Generation. Cambridge University Press, Cambridge (2001)
31. Hata, M., Toyoura, M., Mao, X.: Automatic generation of accentuated pencil drawing with saliency map and lic. Vis. Comput. **28**, 657–668 (2012)
32. Peters, J., Hettiarachchi, R.: Proximal manifold learning via descriptive neighbourhood selection. Appl. Math. Sci. **8**(71), 3513–3517 (2014)
33. Peters, J., Hettiarachchi, R.: Dimensionality reduction via proximal manifolds. Gen. Math. Notes **22**(2), 1–10 (2014)

Chapter 9
Watershed, Smirnov Measure, Fuzzy Proximity and Sorted Near Sets

9.1.1: Hand Image

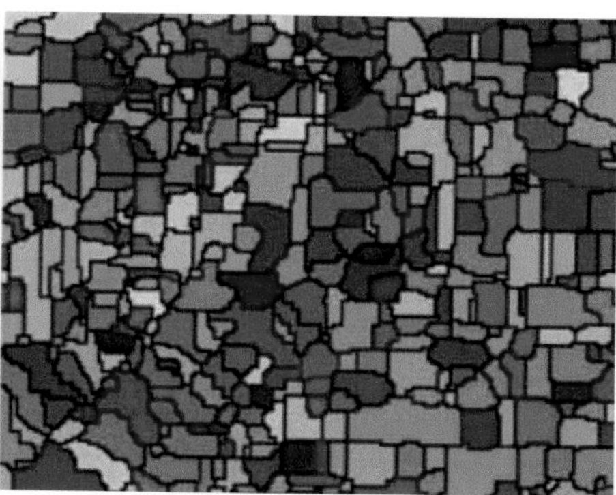

9.1.2: Watershed Segments

Fig. 9.1 Watershed segments

This chapter introduces proximal watershed image segments and Voronoï diagrams. The watershed image segmentation method is a centerpiece in mathematical morphology (NM). Basically, watershed segmentation is a separation of digital image pixels into regions analogous to catchment basins. An image catchment basin is a region in which pixel intensities descend to lower values analogous to the way water flows from a catchment into a lower area such as a river or lake or reservoir. This very interesting approach to image segmentation was introduced by G. Matheron (founder of geostatistics [1]) and J. Serra [2] during the 1960s (Matheron was Serra's Ph.D. supervisor).

Example 9.1 **Sample Image Watershed Catchment Basins.**
Let *img* be the hand image in Fig. 9.1.1 and let X be the watershed segmentation of *img* shown in Fig. 9.1.2. This segmentation is obtained using MScript 44 in Appendix A.9. ■

J.F. Peters, *Computational Proximity*, Intelligent Systems Reference Library 102, DOI 10.1007/978-3-319-30262-1_9

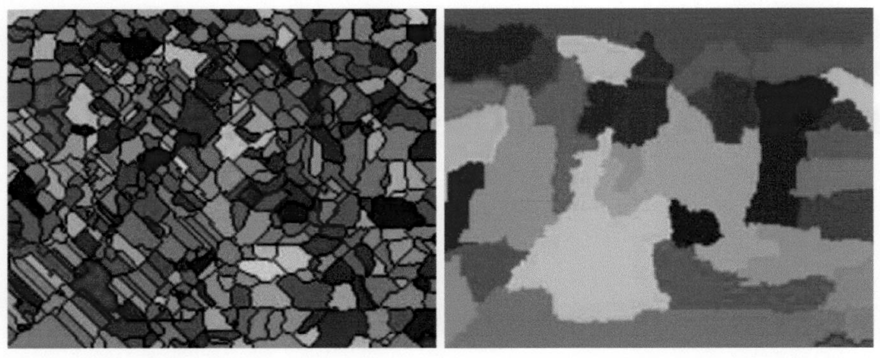

9.2.1: Hand Image 9.2.2: Watershed Segments

Fig. 9.2 Watershed salient boxed segments

Proximity enters into the picture thanks to an interest in comparing catchment basins, highlighting those that are near and those that are far apart. This is made possible by introducing a finite topology on a segmented image and equipping the topology with one or more proximities. From this, we obtain a watershed proximity space susceptible to a variety of methods, e.g., topological sorting of close watershed basins, separation of image regions containing watershed basins that are far apart. A practical benefit of this approach is the classification of watershed segments (grouping basins in terms of their proximity or remoteness from each other) and a heightened understanding of each original image that has been segmented.

Example 9.2 **Sample Watershed Proximity Space**.
Let *img* be the boxed segmented hand image in Fig. 9.2.1 and let X be the watershed salient boxed segmentation of *img* shown in Fig. 9.2.2. The boxed segmentation in Fig. 9.2.1 is obtained by considering only horizontally and vertically adjacent pixels as neighbours and excluding corner neighbours. The salient boxed segmentation in Fig. 9.2.2 is obtained ignoring corner neighbours and choosing a minimum saliency level as a means of merging watershed segments. The **CornerNeighbours** condition and the **MinimuSaliency** method are both implemented in Mathematica 10.

Let X be the watershed segmented hand image *img*. This is also a finite Euclidean topological space as defined in Example 8.2. By equipping X with the descriptive strong Lodato proximity $\overset{\wedge\!\!\wedge}{\delta}_{\Phi}$, X becomes a proximity space. This watershed salient boxed segmentation is also obtained using MScript 44 in Appendix A.9. ■

This chapter also introduces some basic mathematical morphology operations. Digital images are viewed as sets of objects (dark areas) that are either thinned (eroded) or expanded (dilated). Combinations of erosion and dilation operations lead to **closing** and **opening** operations. In addition, **watershed** segmentation is introduced as a means of delineating the boundaries of image objects as closed contours.

9.1 See-Through MM Segments Via Dirichlet Tessellation on an Image

This section introduces an approach to tessellating the morphological components (segments) derived from a digital image. The centroids of the segments are used to construct a Voronoï diagram on the image. The end result is the production of see-through image segments. The steps to do this are given in Algorithm 10.

Algorithm 10: See-Through MM Segments Via Centroid-Based Dirichlet Tessellation

---

Input : Read digital image *img*.
Output: See-Through MM Segments Via Centroid-Based Dirichlet Tessellation.
1 $img \longmapsto watershedComponents$;
2 /* Use $watershedComponents$ to find centroids. */ ;
3 $watershedComponents \leftarrow centroidCoordinates$;
4 /* $centroidCoordinates$ are used to construct Voronoï $Regions$ (= Dirichlet cells). */ ;
5 $centroidCoordinates \longmapsto VoronoiDiagram\ (on\ segments)$;
6 /* Superimpose Voronoi Diagram on image. */ ;
7 $img \leftarrow VoronoiDiagram$;

Example 9.3 **Centroids on the MM components of an Image**.
The production of mathematical morphology components (segments) for the photographer image in Fig. 9.3.1 is accomplished using MScript 45 in Appendix A.9. This MScript implements the steps in Algorithm 10, which finds the centroids in collections of close MM segments. The black • dots indicate the locations of centroids found by this algorithm are shown in Fig. 9.3.2. ■

Example 9.4 **Centroid-Based See-Through Digital Image Segments**.
The construction of a centroid-based Voronoï diagram on mathematical morphology components (segments) for the photographer image in Fig. 9.3.1 is also accomplished using MScript 45 in Appendix A.9. This MScript also implements the steps in Algorithm 10, which uses the MM segment centroids as sites to generate a Voronoï diagram on the collection of image morphological components (segments). For the result, see Fig. 9.4.1. The red • dots in Fig. 9.4.2 indicate the locations of centroids of the MM segments.

A see-thru segmentation of the original image is obtained by superimposing the Voronoï diagram on the image. This is shown in Fig. 9.4.2. ■

Notice that most of the Voronoï regions in Fig. 9.4.1 contain more than one MM segment and the MM segments in each regions overlap with adjacent regions in varying amounts. Hence, in this instance, the Voronoï diagram is a very rough approximation of the MM segmentation of the image. Depending on the choice of generating points (or sites), the accuracy of the Voronoï diagram approximation of an MM segmentation will vary. To see this, consider the following example in which a keypoint-based Voronoï diagram is a closer approximation of the MM watershed segmentation of an image.

9.3.1: Photographer 9.3.2: MM Centroids

Fig. 9.3 Segment centroids

9.4.1: MM centroidsMesh 9.4.2: See-thru Segments

Fig. 9.4 See-thru mesh via centroids on MM segments

Algorithm 11: See-Through MM Segments Via Keypoint-Based Dirichlet Tessellation

Input : Read digital image *img*.
Output: See-Through MM Segments Via Keypoint-Based Dirichlet Tessellation.
1 *Repeat lines 1 and 2 in Algorithm 10*;
2 /* Use *watershedComponents* to find keypoints. */ ;
3 *watershedComponents* ← *keypointCoordinates*;
4 /* *keypointCoordinates* are used to construct Voronoï *Regions* (= Dirichlet cells). */ ;
5 *keypointCoordinates* ⟼ *VoronoiDiagram (on segments)*;
6 /* Superimpose Voronoi Diagram on image. */ ;
7 *img* ← *VoronoiDiagram*;

Example 9.5 **Centroidal- Versus Keypoints-Based Mesh on Photographer MM Segments**.

The construction of a keypoints-based Voronoï diagram on mathematical morphology components (segments) for the photographer image in Fig. 9.3.1 is also accomplished using MScript 45 in Appendix A.9. This MScript also implements the steps in Algorithm 11, which uses the MM segment keypoints as sites to generate a Voronoï diagram on the collection of image morphological components (segments). For a comparison of the results of centroidal- versus keypoints-based Voronoï diagram on the mathematical morphology components (segments) for the photographer image in Fig. 9.3.1, see Fig. 9.5. For the keypoints-based mesh on the MM segments, see Fig. 9.5.2. The black • dots in Fig. 9.5.1 indicate the locations the centroids found and in Fig. 9.5.2 indicate the locations the keypoints found on the segmentation of the photographer image.

Notice that now that there is less overlap between many of the Voronoï regions, i.e., in many cases, the MM segments are predominately in one region. For this reason, a keypoints-based Voronoï diagram provides a fairly good approximation of the MM segmentation of an image. ■

A Voronoï diagram that is a good approximation of a MM segmentation provides us with means of seeing where the MM watershed basin are in an image. The convexity of the Voronoï regions gives a powerful means of analyzing that comparing image segments.

Example 9.6 **Centroidal- Versus Keypoints Mesh MM Segments on Photographer Image**.

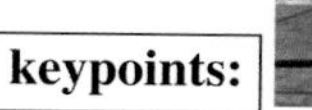

centroids:

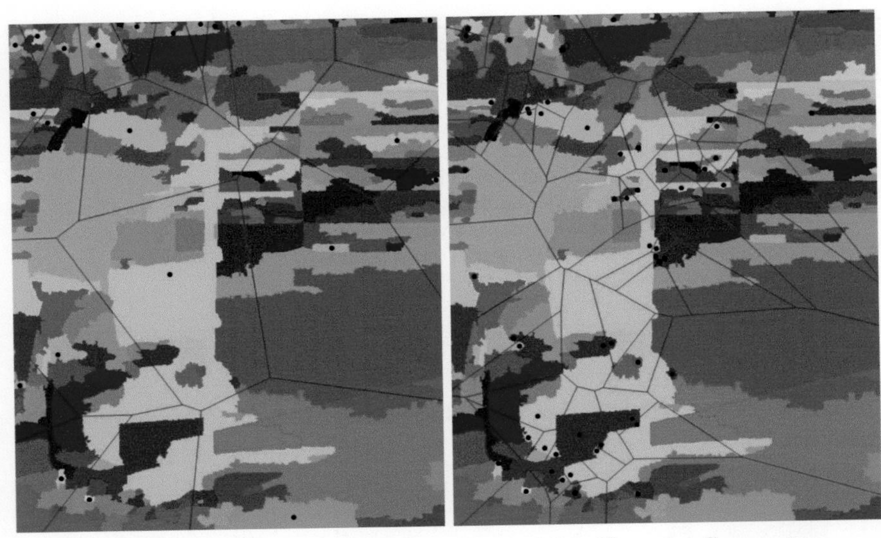

9.5.1: Centroidal Segments 9.5.2: Keypoint Segments

Fig. 9.5 Centroidal- versus keypoints mesh on MM segments

For a comparison of the results of centroidal- versus keypoints-based Voronoï diagram on the mathematical morphology components (segments) on the photographer image, see Fig. 9.6. For this image, keypoints are a better choice of sites in generating a Voronoï diagram derived from the MM segmentation of the image. This can be seen, for example, in Voronoï regions surrounding the head and camera in the keypoints-based segmentation of the image. By contrast, the centroidal-based Voronoï regions are bigger (each region contains more MM segments). This means that keypoints-based segmentation of the image is a better approximation of the basins in an MM image segmentation. ■

Conjecture 9.7 Keypoint-Based Approximation of an MM Segmentation.

1º Keypoint-based Voronoï regions on an MM segmentation of an image provide a more accurate MM watershed approximation than centroidal-based Voronoï regions on an MM segmentation of the same image.

2º Keypoint-based Voronoï regions on an MM segmentation of an image provide a more accurate MM watershed approximation than any other form of Voronoï regions on an MM segmentation of the same image. ■

Problem 9.8 The goal in this problem is to find the centroids for as many of the Mathematical Morphology segments as possible. This will ensure that the regions in the corresponding Voronoï diagram closely correspond to the MM segments.

9.6.1: MM centroidsMesh 9.6.2: MM keypointsMesh

Fig. 9.6 Centroidal- versus keypoints mesh segments on an image

Motivation for doing this is that Voronoï regions are convex polygons, whereas MM segments in general are not convex. Do the following:

1° 🚲 The Binarize operation in MScript 45 in Appendix A.9 uses 2/3 to force the Mathematica Binarize operation to get the high intensity pixels. To change the number of centroids found, experiment with different options for Binarize, e.g., Cluster, Entropy, Mean, Median. Include in your experiments a count of the number of centroids found.

2° 🚲 The Dilation operation in MScript 45 in Appendix A.9 uses 1, which calls for a box structuring element with MaxFilter[img2,1] == Dilation[img2,1]. To change the number of centroids found, experiment with different structure elements used by Dilation, e.g., BoxMatrix and DiskMatrix. ■

Problem 9.9 Do the following:

1° ☕ Look for counterexamples that disprove Conjecture 9.7.1. To do this, test this conjecture on different images for both centroid-based and keypoint-based tessellations of MM segmentations of the chosen images.

2° ☕ Look for counterexamples that disprove Conjecture 9.7.2. ■

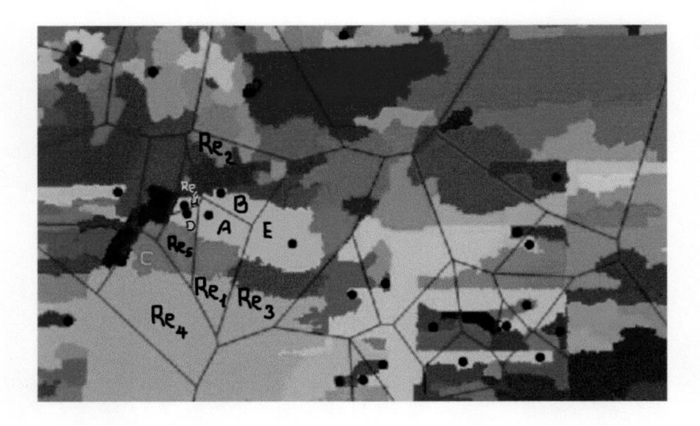

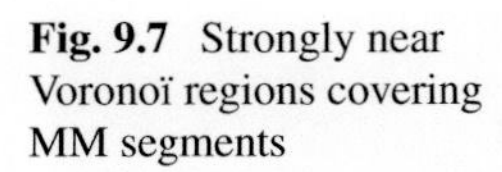
Fig. 9.7 Strongly near Voronoï regions covering MM segments

9.2 Strongly Near Voronoï Region Interiors on MM Segmentations

This section briefly looks at the overlap between Voronoï Regions in a tessellation of an MM segmentation of a digital image. By identifying the extent that neighbouring Voronoï Regions have parts in common, it is then possible to sort the regions in terms of the extent that the regions share points, some more and some less.

Example 9.10 **Strongly near Voronoï regions on an MM segmentation**. Figure 9.7 shows a fragment of a Voronoï diagram covering MM segments of a digital image. Let X be a finite Euclidean space of picture points represented in Fig. 9.7 with red (r), green (g), blue (b) colours and let $\Phi : X \longrightarrow \mathbb{R}^3$ be a description a point x in X that is a feature vector of the form $(\phi_r(x), \phi_g(x), \phi_b(x))$ defined by

$$\phi_r(x), \phi_g(x), \phi_b(x) \in [0, 1], \text{ (r, g, b colour intensities for} x \in X).$$

For instance, $\phi_r : X \longrightarrow \mathbb{R}$ is a picture point colour probe function defined by $\phi_r(x) \in [0, 1]$, which a red channel intensity value for the picture point x. Assume that the range of Φ is endowed with the topology τ with the

$$\tau - basis = 2^{\{(\phi_r(x), \phi_g(x), \phi_b(x))\}}.$$

That is, the topology τ is a collection of all unions of elements of the τ-basis, which consists of sets of feature vectors that describe picture points in X.

Also assume that X is equipped with the proximity relations in the proximal relator $\mathscr{R} = \left\{\delta, \delta_\Phi, \overset{\wedge\!\!\wedge}{\delta}, \overset{\wedge\!\!\wedge}{\delta}_{\Phi}\right\}$, i.e., Wallman proximity, Lodata descriptive proximity, strongly near proximity and Lodato descriptive strongly near proximity. Putting these structures together, we obtain the proximal relator space $(X, \tau, \mathscr{R})$.

In addition, in Fig. 9.7, let $Re_1, Re_2, Re_3 \subset X$ be Voronoï Regions with $A \subset Re_1$, $B \subset Re_2$, $E \subset Re_3$. In addition, let

$$\begin{aligned}
\Phi(A) &= \{\Phi(a) = (0, 0.8, 0) : a \in A\} \text{ (green segment).}\\
&= \Phi(B) = \Phi(E) \text{ (green segments), and observe that}\\
\Phi(A) &\subset \Phi(\text{int}(\text{cl}Re_1)),\\
\Phi(B) &\subset \Phi(\text{int}(\text{cl}Re_2)),\\
\Phi(E) &\subset \Phi(\text{int}(\text{cl}Re_3)).
\end{aligned}$$

For instance, the description of the picture set A is the set of feature vectors $(0, 0.8, 0)$ that describe picture point with a 0.8 green intensity and zero red and blue intensities. And the descriptions of the picture sets B, E match the description of the picture set A. In these sets of descriptions of picture points, for example, $\Phi(A)$ is a set of colour feature vectors that describe the interior of region Re_1. Then observe

1° $\text{cl}Re_1\ \delta\ \text{cl}Re_2$ and $\text{cl}Re_1\ \delta\ \text{cl}Re_3$, since the Voronoï Regions have a common vertex.

2° $Re_1\ \overset{\wedge\wedge}{\delta}\ Re_2$ and $Re_1\ \overset{\wedge\wedge}{\delta}\ Re_3$, since the Voronoï Regions have common edges.

3° $\text{int}(\text{cl}Re_1)\ \delta_\Phi\ \text{int}(\text{cl}Re_2)$ and $\text{int}(\text{cl}Re_1)\ \delta_\Phi\ \text{int}(\text{cl}Re_3)$, since parts of the interiors of these Voronoï Regions have matching descriptions.

4° $A\ \overset{\wedge\wedge}{\delta}_\Phi\ B$ and $A\ \overset{\wedge\wedge}{\delta}_\Phi\ E$, since the interiors of these sets have matching descriptions.

5° $\text{int}(\text{cl}Re_1)\ \overset{\wedge\wedge}{\delta}_\Phi\ \text{int}(\text{cl}Re_2)$ and $\text{int}(\text{cl}Re_1)\ \overset{\wedge\wedge}{\delta}_\Phi\ \text{int}(\text{cl}Re_3)$, since these interiors are both green. ■

9.3 Extension of Smirnov Proximity Measure

This section introduces two extensions of the measure of proximity δ introduced by Ju. M. Smirnov [3]. A **proximity measure** is a measure of the closeness of a pair of nonempty sets. Notice that a proximity measure is not a distance metric but instead a proximity measure is an **set inclusion** measure, i.e., a measure of the extent (degree) that one set is included in another set. Let A, B be nonempty subsets in a proximity space X. The **Smirnov proximity measure** $\delta(A, B) \in \{0, 1\}$ be defined by

$$\delta(A, B) = \begin{cases} 1, & \text{if } A \text{ is close to B},\\ 0, & \text{if } A \text{ is far from B}. \end{cases}$$

Example 9.11 **Smirnov Proximity Measure**
Let the subsets of picture points Re_1, Re_2 be Voronoï regions in the proximal relator space $(X, \tau, \mathscr{R})$ in Example 9.10. In that case, $\delta(Re_1, Re_2) = 1$, since Re_1, Re_2 have an edge in common. ■

Let $\varepsilon > 0$, $\upsilon(A, B) = \frac{|A \cap B|}{|X|}$. For a relator space like the one in Example 9.10, the **spatial Smirnov proximity measure** $\delta_{\varepsilon,\upsilon}(A, B) \in [0, 1]$ is defined by

$$\delta_{\varepsilon,v}(A, B) = \begin{cases} \frac{|A \cap B|}{|X|}, & \text{if } \varepsilon < v(A, B) \leq 1, \\ 0, & \text{if } v(A, B) \leq \varepsilon. \end{cases}$$

Example 9.12 **Smirnov Spatial Inclusion Measure**,
Let the subsets of picture points Re_1, Re_2 be Voronoï regions in the proximal relator space $(X, \tau, \mathscr{R})$ in Example 9.10. Put $\varepsilon > 0$. Assume $v(Re_1, Re_2) > \varepsilon$. In that case, $\delta_{\varepsilon,v}(Re_1, Re_2) = 1$, since Re_1, Re_2 share more than one point, namely, the points in the common edge. ■

9.4 Fuzzy Lodato Proximity and Fuzzy Lodato–Smirnov Membership Function

There are many different forms of fuzzy proximity,e.g., A.P. Šostak [4, Sect. 3.1.1], Y.-M. Liu and M.-K. Luo [5, Sect. 13.2], J. Brennan and E. Martin [6, Sect. 3, p. 945], V. Çetkin, A.P. Šostak and H. Aygün [7], K.C. Chattopadhyay, H. Hazra and S.K. Samanta [8]. In this section, we introduce the axioms that need to be satisfied by a membership function on a fuzzy Lodato proximity space. We also introduce the fuzzy Lodato–Smirnov membership function.

Let A, B be nonempty subsets in a space X. Recall that a L.A. Zadeh membership (characteristic) function $f_A : X \longrightarrow [0, 1]$ defined by $f_A(x)$ gives the degree of membership of x in A [9]. For recent work on membership functions, see H. Fashandi on inclusion measures in image processing [10] and A.H. Meghdadi on fuzzy tolerance neighbourhoods in content-based image retrieval [11].

Definition 9.13 Fuzzy Lodato Proximity.
Let X be a topological space equipped with the Lodato proximity δ, $A, B, C \subset X$. A membership function $\mu_\delta : 2^X \times 2^X \longrightarrow [0, 1]$ on the space X is a **fuzzy Lodato proximity**, provided it satisfies the following axioms.

$N_\mu.1$ $\mu_\delta(\emptyset, A) = 0$ for all $A \subset X$.
$N_\mu.2$ $\mu_\delta(A, B) = \mu_\delta(B, A)$.
$N_\mu.3$ $\mu_\delta(A, B) \neq 0$ implies $A\ \delta\ B$.
$N_\mu.4$ $\mu_\delta(A, (B \cup C)) \neq 0$ implies $\mu_\delta(A, B) \neq 0$ and $A\ \delta\ B$ or $\mu_\delta(A, C) \neq 0$ and $A\ \delta\ C$.
$N_\mu.5$ $\mu_\delta(A, B) \neq 0$ and $\mu_\delta(\{b\}, C) \neq 0$ for all $b \in B$ implies $\mu_\delta(A, C) \neq 0$ and $A\ \delta\ C$. ■

For a fuzzy Lodato–Smirnov proximity membership function μ_δ (briefly, *fuzzy Lodato–Smirnov proximity*), a **fuzzy Lodato–Smirnov proximity space** is denoted by (X, δ, μ_δ). Unlike the Zadeh membership function $f_A(x)$ for the degree of membership of a point in a nonempty set A, the **fuzzy Lodato–Smirnov proximity membership function** $\mu_\delta : 2^X \times 2^X \longrightarrow [0, 1]$ maps a pair of nonempty sets on a Lodato proximity space to a number in $[0, 1]$, which is the degree of membership of one set in the other set. Let $\delta_{\varepsilon,v}$ be the Smirnov proximity measure introduced in Sect. 9.3.

Theorem 9.14 *The Smirnov proximity measure $\delta_{\varepsilon,v}$ on the Lodato proximity space is a fuzzy Lodato–Smirnov proximity membership function.*

Proof By definition, $\delta_{\varepsilon,v} \longrightarrow [0, 1]$ on the Lodato proximity space. Here, we prove Axiom N_μ.3. Let X be a topological space equipped with the Lodato proximity δ, nonempty $A, B \subset X$. Hence,

$$\begin{aligned}\delta_{\varepsilon,v}(A, B) > \varepsilon &\Rightarrow \frac{|A \cap B|}{|X|} > \varepsilon \text{ (By def. of } \delta_{\varepsilon,v}) \\ &\Rightarrow |A \cap B| \neq 0 \\ &\Rightarrow A\ \delta\ B.\end{aligned}$$

□

From Theorem 9.14, $\delta_{\varepsilon,v}$ is called a **fuzzy Lodato–Smirnov proximity**.

Example 9.15 **Sample Fuzzy Lodato Space.**
Let the Voronoï mesh in Fig. 9.8 represent a fuzzy Lodato proximity space $(X, \delta, \delta_{\varepsilon,v})$ with labelled regions A, B, C, D, E and point x. Then we have

1° $\delta_{\varepsilon,v}(B, E) = 0$, since regions B, E have no points in common.

2° 

:

$\delta_{\varepsilon,v}(D, E) > 0$, since regions D, E share the points along the common green edge.

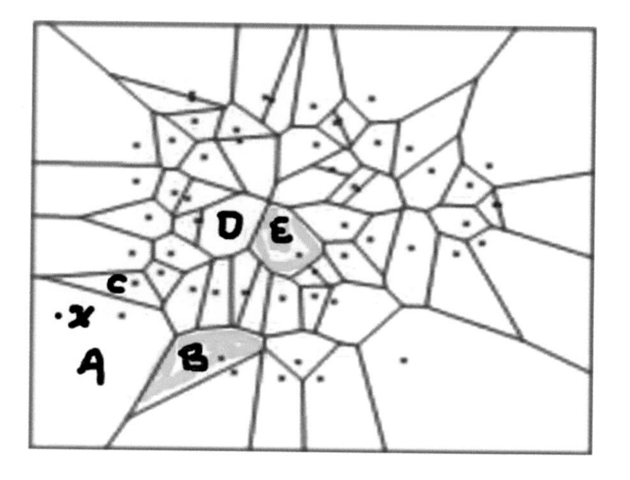

Fig. 9.8 Fuzzy Lodato proximity space on a Voronoï mesh

3° 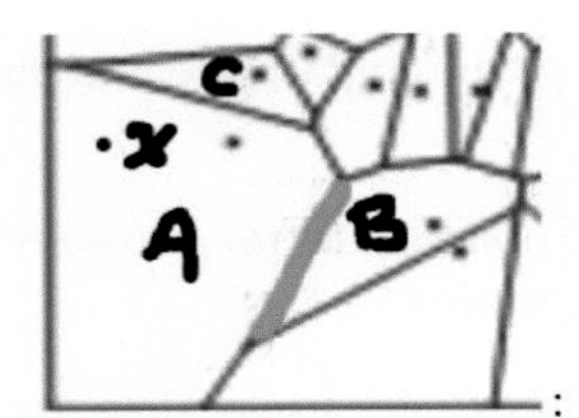

:

$\delta_{\varepsilon,v}(A, B) > 0$, since regions A, B share the points along the common green edge.

4° $\delta_{\varepsilon,v}(\{x\}, A) > 0$, since $\{x\}\ \delta\ A$, i.e., x is a member of region A.

5° $\delta_{\varepsilon,v}(\{x\}, E) = 0$, since $\{x\}\ \not\delta\ E$, i.e., x is not a member of region E. ■

Let X be a finite set of picture points. In that case, $(X, \delta, \delta_{\varepsilon,v})$ is a **digital fuzzy Lodato proximity space**.

Example 9.16 **Sample Digital Fuzzy Lodato Space**.
Let the Voronoï mesh on hand-camera image in Fig. 9.9 represent a digital fuzzy Lodato proximity space $(X, \delta, \delta_{\varepsilon,v})$ with labelled regions A, B, C, D, E and point x. Then we have

1° $\delta_{\varepsilon,v}(B, E) = 0$, since regions B, E have no points in common.

2° 

:

$\delta_{\varepsilon,v}(D, E) > 0$, since regions D, E share the points along the common blue **edge**.

Fig. 9.9 Fuzzy Lodato proximity space on a Voronoï mesh

3° 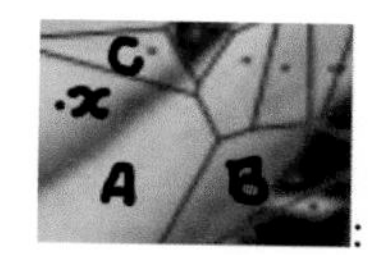

:

$\delta_{\varepsilon,v}(A, B) > 0$, since regions A, B share the points along the common blue **edge**.

4° $\delta_{\varepsilon,v}(\{x\}, A) > 0$, since $\{x\}\ \delta\ A$, i.e., x is a member of region A.

5° $\delta_{\varepsilon,v}(\{x\}, E) = 0$, since $\{x\}\ \not\delta\ E$, i.e., x is not a member of region E. ■

Problem 9.17

☕ Complete the proof of Theorem 9.14, i.e., show that $\delta_{\varepsilon,v}$ satisfies the remaining axioms for a fuzzy Lodato proximity membership function. ■

We obtain a fuzzy Wallman–Lodato proximity space by adding the following axiom to those in Definition 9.13.

$N_{\mu}.6$ $\mu_{\delta}(\mathrm{cl}A, \mathrm{cl}B) \Leftrightarrow \mathrm{cl}A \cap \mathrm{cl}B \neq \emptyset \Leftrightarrow A\ \delta\ B$. ■

Theorem 9.18 *The spatial Smirnov proximity measure $\delta_{\varepsilon,v}$ on Wallman–Lodato proximity space is spatial Wallman–Lodato fuzzy proximity.*

Problem 9.19

☕ Prove Theorem 9.18, i.e., show that $\delta_{\varepsilon,v}$ satisfies Axiom $N_{\mu}.6$. ■

The following little theorems flow from the definitions of fuzzy Lodato proximity and strong proximity.

Theorem 9.20 *Let A, B be subsets in fuzzy Lodato topological space equipped with the strong proximities $\overset{\wedge\!\!\wedge}{\delta}$ and Wallman–Lodato proximity δ. Then*

1° $\mu_{\delta}(A, B) \neq 0$, if and only if $A\ \overset{\wedge\!\!\wedge}{\delta}\ B$.

2° $\mu_{\delta}(A, B) \neq 0$, if and only if $A\ \delta\ B$.

Proof

1°:

$$
\begin{aligned}
\mu_{\delta}(A, B) \neq 0 &\Leftrightarrow \mathrm{int}A \cap \mathrm{int}B \neq \emptyset \\
&\Leftrightarrow A\ \overset{\wedge\!\!\wedge}{\delta}\ B
\end{aligned}
$$
□

Problem 9.21

☕ Prove Part 2° for Theorem 9.20. ■

Conjecture 9.22 *There are at least 100 little theorems about the fuzzy proximity membership functions $\mu_{\delta}, \delta_{\varepsilon,v}$ and ordinary proximities.* ■

Remark 9.23 **Distance Between Euclidean Plane Points Versus Distance Between Picture Points**.
In Problem 9.27, the set X contains the points in a subset of the Euclidean plane. For the implementation, the assumption is that the distance between points in X is a real number. For a 2D digital image, the picture points (pixels) are also a subset of the Euclidean Plane. The **Euclidean plane** is the 2-dimensional Euclidean space denoted by $\mathbb{R}^2$, i.e., the Euclidean plane is the space of all 2-tuples of real numbers (x_1, x_2) [12, Sect. II.1] (see, also, M. Berger [13]). And there are gaps between picture points. From a plane geometry perspective, there are no gaps between the points in the Euclidean plane. For a 3D digital image, the picture points are a subset of the 3-dimensional Euclidean space $\mathbb{R}^3$, also with gaps between the picture points. In the next problem, the set X is limited to the pixels in a 2D digital image. In that case, the distance between pixels in X is an integer. ■

The Euclidean plane has its own fascination, even nowadays. Here is K. Borsuk's conjecture for the plane .

Conjecture 9.24 Borsuk's conjecture [14].
In the Euclidean space $\mathbb{R}^2$*, any bounded set of points A can be decomposed into 3 parts of diameter strictly less than the diameter of A. This also holds true for Euclidean space* $\mathbb{R}^n$, $N \geq 2$. ■

Example 9.25 **Based on Borsuk's conjecture**.
For a sample decomposition of a plane bounded region A split three parts, namely, A_1, A_2, A_3, see Fig. 9.10. ■

There is a digital counterpart of Borsuk's conjecture for digital images. Call it the **Digital Borsuk Conjecture**.

Problem 9.26 Digital Borsuk Conjecture.
☕ State Borsuk's conjecture for bounded subsets in digital images in the Euclidean plane. ■

Problem 9.27 Fuzzy Lodato Proximity.
Do the following: Let M be a Voronoï mesh that covers a nonempty subset X of the Euclidean plane equipped with the Lodato proximity δ. For implementations, use Mathematica.

Fig. 9.10 Decomposition of 2D bounded set into 3 parts

1° Select a set of points X in the Euclidean plane.
2° Select a set of sites $S \subset X$ and $\varepsilon > 0$.
3° Construct a Voronoï mesh M using S.
4° Select $A, B \in M$.
5° Implement

$$v(A, B) = \frac{|A \cap B|}{|X|}$$

for nonempty Voronoï regions $A, B \in M$.
6° Implement the Smirnov proximity measure $\delta_{\varepsilon,v}(A, B)$ defined by

$$\delta_{\varepsilon,v}(A, B) = \begin{cases} \frac{|A \cap B|}{|X|}, & \text{if } \varepsilon \leq v(A, B) \leq 1, \\ 0, & \text{if } v(A, B) < \varepsilon. \end{cases}$$

7° Display the values of $v(A, B)$, $\delta_{\varepsilon,v}(A, B)$.
8° Repeat Problem 9.27, beginning with Step 4 for 10 choices of A, B. ■

Problem 9.28 Digital Fuzzy Lodato Proximity.
🚲 Let M be a Voronoï mesh that covers a digital image X in the Euclidean plane equipped with the Lodato proximity δ. Repeat the steps in Problem 9.27. ■

Problem 9.29 Fuzzy Wallman–Lodato Proximity.
Do the following: Let M be a Voronoï mesh that covers a nonempty subset X of the Euclidean plane equipped with the Wallman–Lodato proximity δ. For implementations, use Mathematica.
1° Select a subset X of the Euclidean plane.
2° Select a set of sites $S \subset X$ and $\varepsilon > 0$.
3° Construct a Voronoï mesh M using S.
4° Select $A, B \in M$.
5° Implement

$$v(\text{cl}A, \text{cl}B) = \frac{|\text{cl}A \cap \text{cl}B|}{|X|}$$

for nonempty Voronoï regions $A, B \in M$.
6° Implement the fuzzy Wallman–Smirnov proximity measure $\delta_{\varepsilon,v}(A, B)$ defined by

$$\delta_{\varepsilon,v}(\text{cl}A, \text{cl}B) = \begin{cases} \frac{|\text{cl}A \cap \text{cl}B|}{|X|}, & \text{if } \varepsilon \leq v(\text{cl}A, \text{cl}B) \leq 1, \\ 0, & \text{if } v(\text{cl}A, \text{cl}B) < \varepsilon. \end{cases}$$

7° Display the values of $v(\text{cl}A, \text{cl}B)$, $\delta_{\varepsilon,v}(\text{cl}A, \text{cl}B)$.
8° Repeat Problem 9.29, beginning with Step 4 for 10 choices of A, B. ■

Problem 9.30 Digital Fuzzy Wallman–Lodato Proximity.
Let M be a Voronoï mesh that covers a digital image X in the Euclidean splane equipped with the Wallman–Lodato proximity δ. Repeat the steps in Problem 9.29. ■

Problem 9.31
Give the axioms for a fuzzy strong proximity space $(X, \overset{\wedge\wedge}{\delta}, \mu_{\overset{\wedge\wedge}{\delta}})$. ■

Let $v_{\overset{\wedge\wedge}{\delta}}$ be defined by

$$v_{\overset{\wedge\wedge}{\delta}}(A, B) = \frac{|\text{int}A \cap \text{int}B|}{|X|}$$

For a relator space like the one in Example 9.10, the **strong Smirnov proximity measure** $\delta_{\varepsilon, v_{\overset{\wedge\wedge}{\delta}}} : 2^X \times 2^X \longrightarrow [0, 1]$ is defined by

$$\delta_{\varepsilon, v_{\overset{\wedge\wedge}{\delta}}}(A, B) = \begin{cases} \frac{|\text{int}A \cap \text{int}B|}{|X|}, & \text{if } \varepsilon < v_{\overset{\wedge\wedge}{\delta}} \leq 1, \\ 0, & \text{if } v_{\overset{\wedge\wedge}{\delta}} \leq \varepsilon. \end{cases}$$

Problem 9.32
Prove that $\underset{\varepsilon, v}{\overset{\wedge\wedge}{\delta}}$ is a fuzzy strong proximity measure. ■

9.5 Descriptive Smirnov Proximity

The extended Smirnov proximity measure $\delta_{\varepsilon, v}(A, B) \in [0, 1]$ has a descriptive counterpart. Let $\varepsilon_\Phi > 0$, $v(A, B) = \frac{|\Phi(A) \cap \Phi(B)|}{|\Phi(X)|} \in [0, 1]$. The **descriptive Smirnov proximity measure** $\delta_{\varepsilon_\Phi, v}(A, B) \in [0, 1]$ is defined by

$$\delta_{\varepsilon_\Phi, v}(A, B) = \begin{cases} \frac{|\Phi(A) \cap \Phi(B)|}{|\Phi(X)|}, & \text{if } \varepsilon_\Phi \leq v(A, B) \leq 1, \\ 0, & \text{if } v(A, B) \leq \varepsilon_\Phi. \end{cases}$$

Example 9.33 **Smirnov Descriptive Inclusion Measure.**
Let the subsets of picture points $A \subset \text{int}(\text{cl}R_1), B \subset \text{int}(\text{cl}R_2)$ be subsets in the interiors of the Voronoï regions Re_1, Re_2 in the proximal relator space $(X, \tau, \mathscr{R})$ in Example 9.10. Put $\varepsilon_\Phi > 0$. Assume $v > \varepsilon_\Phi$. In that case, $\delta_{\varepsilon_\Phi, v}(A, B) = 1$, since the descriptions of $\Phi(A), \Phi(B)$ have share more one picture point with the same description, namely, the green points in the two sets. ■

Algorithm 12: Spatially Sorting Near Regions in a Proximity Space

Input : Proximal relator space $(X, \tau, \mathscr{R})$, sets of picture point regions $Re_A, Re_B, Re_E, \ldots, Re_n \subset X, n > 0$, nonempty subsets $A \subset \text{int}(\text{cl}R_A), B \subset \text{int}(\text{cl}R_B), E \subset \text{int}(\text{cl}R_E), \ldots, \varepsilon > 0$.

Output: $\mathscr{S} = \{Re_A, Re_B, Re_E, \ldots\}$ sorted in ascending order.

```
1  /* Select a nonempty region Re_A ⊂ X. */ ;
2  S := Re_A;
3  X := X \ Re_A;
4  Continue ← True;
5  while (Continue and X ≠ ∅) do
6     if ∄Re_B ∈ X \ Re_A then
7        Continue ← False;
8     else
9        Continue ← True;
10       Select Re_B ⊂ X \ Re_A;
11       /* Measure the degree of inclusion of Re_B in Re_A. */ ;
12       v := |Re_A ∩ Re_B| / |X|  /* Compute degree of inclusion */;
13       if δ_{ε,v}(Re_A, Re_B) = 1 then
14          S := S ∪ Re_B;
15          sort S;
16       X := X \ Re_B;
```

9.6 Application: Sorting Near Sets Using Smirnov Proximity Measures

A natural and practical outcome of the introduction of the extensions of the Smirnov proximity measure is a sorting of near sets in a proximity space using a method first suggested in [15].

The extended Smirnov inclusion measures can be used to sort near sets relative to each given near set. This approach to sorting near sets is directly related to S. Leader's approach to constructing a uniform topology on a nonempty set equipped with a proximity relation [16].

Definition 9.34 Spatially Sorted Near Sets.
Let X be a proximity space, $A_1, A_2, A_3, \ldots, A_i, \ldots, A_j, \ldots, A_n \subset X$, $\varepsilon > 0$, $\delta_{\varepsilon,v}(A_i, A_j)$ the extended Smirnov proximity measure. For a selected subset A_1, the subsets A_1, A_i, A_j, A_k are sorted, provided

$$\delta_{\varepsilon,v}(A_1, A_i) \leq \delta_{\varepsilon,v}(A_1, A_j) \leq \delta_{\varepsilon,v}(A_1, A_k).$$

$\delta_{\varepsilon,v}(A_1, A_k)$ has the highest value, indicating that A_k is the subset nearest to A_1. ∎

Example 9.35 **Spatially Sorted Near Sets.**
Let X be the proximal relator space described in Example 9.10. Applying Algorithm 12, we obtain the following spatially sorted near sets in the collection $\mathscr{S}$,

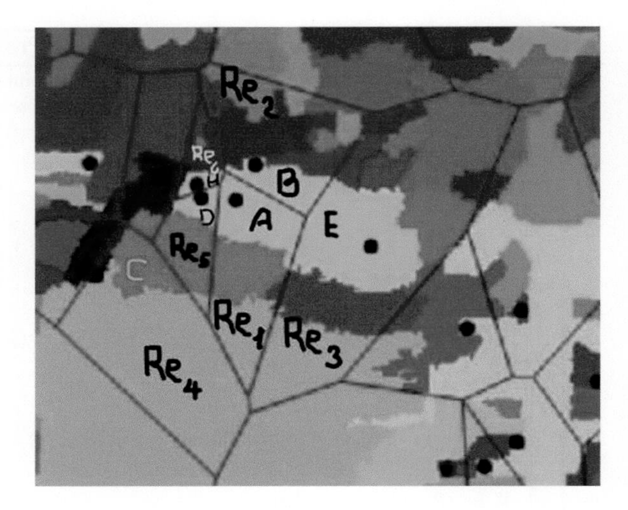

Fig. 9.11 Overlapping MM segments in adjacent Voronoï regions

which are in descending order, starting with Voronoï region Re_1. In this example, let $Re_1, Re_2, Re_3, Re_4, Re_5, Re_6$ in Fig. 9.11, we have

$$\begin{aligned}\delta_{\varepsilon,v}(Re_1, Re_3) &\geq \delta_{\varepsilon,v}(Re_1, Re_5) \geq \delta_{\varepsilon,v}(Re_1, Re_2)\\ &\geq \delta_{\varepsilon,v}(Re_1, Re_4) \geq \delta_{\varepsilon,v}(Re_1, Re_6)\end{aligned}$$

and

$\mathscr{S} = \{$ 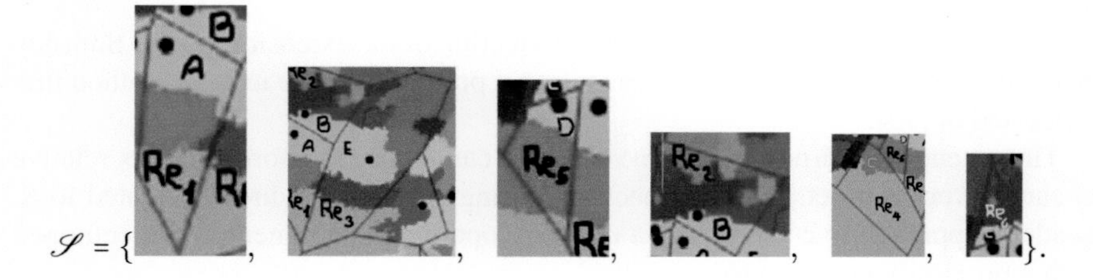 $\}$.

That is, for instance, Re_3 is spatially nearer Re_1 than Re_5, since the common edge between Re_1, Re_3 is longer than the common edge between Re_1, Re_3. ■

Problem 9.36 Do the following:

1° Implement Algorithm 12.

2° ☕ Select three images to illustrate the computations performed by Algorithm 12. ■

Problem 9.37 🚲
Do the following:

1° Select three different digital images in the same class of images such as landscapes, insects and people.
2° Select a value for the nearness lower bound $\varepsilon > 0$.
3° Use Mathematica to obtain a watershed segmentation of each image.
4° Use Mathematica to construct a Voronoï diagram on each image.
5° Select and label Voronoï regions *Re* in each diagram.
6° Select and label subsets in each of the selected *Re* regions.
7° Implement Algorithm 12 to spatially sort each of the selected regions.

(a) Give the degree of nearness of each of the regions to the selected region.
(b) Display the spatially sorted regions as an image sequence like the one Example 9.35. ■

Definition 9.38 Descriptively Sorted Near Sets.
Let $(X, \tau_\Phi, \delta_\Phi)$ be a proximity space with τ_Φ-basis defined by feature vectors determined by a set of probes $\{\phi_1, \dots, \phi_i, \dots, \phi_n\}$ such that $\phi_i : X \longrightarrow \mathbb{R}$. Also, let $A_1, A_2, A_3, \dots, A_i, \dots, A_j, \dots, A_n \in \tau_\Phi$, $\varepsilon_\Phi > 0$, and let $\delta_{\varepsilon_\Phi, v}(A_i, A_j)$ be the extended Smirnov descriptive proximity measure. For a selected subset A_1, the subsets A_1, A_i, A_j, A_k are sorted, provided

---

Algorithm 13: Descriptively Sorting Near Sets in a Proximity Space

---

Input : Proximal relator space $(X, \tau, \mathscr{R})$, sets of picture points $Re_1, Re_2, R_3, \dots, Re_n \subset X$, $n > 0$, nonempty subsets $A \subset \text{int}(\text{cl}Re_1)$, $B \subset \text{int}(\text{cl}Re_2)$, $E \subset \text{int}(\text{cl}Re_3), \dots$, $\varepsilon_\Phi > 0$, probes $\{\phi_1, \dots, \phi_i, \dots, \phi_n\}$ such that $\Phi(x) = (\phi_1(x), \dots, \phi_i(x), \dots, \phi_n(x))$ is a feature vector that describes point $x \in X$.
Output: $\mathscr{S} = \{A, B, E, \dots\}$ sorted in ascending order.

```
1  /* Select a nonempty subset A ⊂ X. */ ;
2  S := A;
3  X := X \ A;
4  Continue ← True;
5  while (Continue and X ≠ ∅) do
6      if ∄B ∈ X \ A then
7          Continue ← False;
8      else
9          Continue ← True;
10         Select B ⊂ X \ A;
11         /* Measure the degree of inclusion of B in A. */ ;
12         v := |Φ(A) ∩ Φ(B)| / |Φ(X)|  /* Compute descriptive degree of inclusion */;
13         if δ_{ε_Φ,v}(A, B) = 1 then
14             S := S ∪ B;
15             sort S;
16         X := X \ B;
```

$$\delta_{\varepsilon_\Phi,v}(A_1, A_i) \le \delta_{\varepsilon_\Phi,v}(A_1, A_j) \le \delta_{\varepsilon_\Phi,v}(A_1, A_k).$$

$\delta_{\varepsilon_\Phi,v}(A_1, A_k)$ has the highest value, indicating that A_k is the subset nearest to A_1. ■

Example 9.39 **Descriptively Sorted Near Sets.**
Let X the proximal relator space described in Example 9.10. Applying Algorithm 13, we obtain the following descriptively sorted near sets in the collection $\mathscr{S}$, which are in descending order, starting with set A in an MM segment partially covered by Voronoï region Re_1. In this example, let A, E, B, D, H be the partial MM segments shown in Fig. 9.11. Then we have

$$\delta_{\varepsilon_\Phi,v}(A, E) \ge \delta_{\varepsilon_\Phi,v}(A, B) \ge \delta_{\varepsilon_\Phi,v}(A, D) \ge \delta_{\varepsilon_\Phi,v}(A, H).$$

and

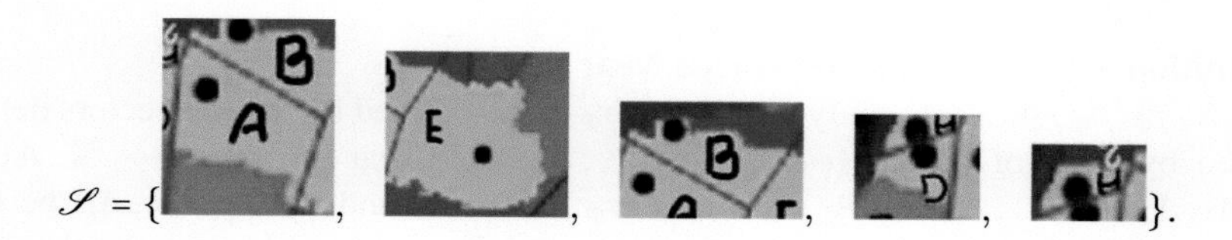

That is, for instance, A is descriptively nearer E than B, since there are more points in E that match the description of points in A that there are points in B that match the description of points in A. ■

Problem 9.40 Do the following:

1° Implement Algorithm 13.
2° Select three images to illustrate the computations performed by Algorithm 13. ■

Problem 9.41
Do the following:

1° Select three different digital images in the same class of images such as landscapes, insects and people.
2° Select probe functions $\{\phi_1, \ldots, \phi_i, \ldots, \phi_n\}$ such that

$$\Phi(x) = (\phi_1(x), \ldots, \phi_i(x), \ldots, \phi_n(x))$$

is a feature vector that describes point $x \in X$.
3° Select a value for the nearness lower bound $\varepsilon_\Phi > 0$.
4° Use Mathematica to obtain a watershed segmentation of each image.
5° Use Mathematica to construct a Voronoï diagram on each image.
6° Select and label Voronoï regions Re in each diagram.

7° Select and label subsets in each of the selected *Re* regions.
8° Implement Algorithm 13 to descriptively sort each of the selected subsets.

(a) Give the degree of nearness of each of the subsets to the selected subset.
(b) Display the descriptively sorted subsets as an image sequence like the one in Example 9.39. ■

9.7 Fuzzy Descriptive Lodato Proximity Space

The counterpart of a fuzzy Lodato proximity space is the space defined with a combination of the descriptive Lodato proximity and an extension of the descriptive Smirnov proximity measure.

Definition 9.42 Fuzzy Descriptive Lodato Proximity.
Let X be a topological space equipped with the descriptive Lodato proximity δ_Φ, $A, B, C \subset X$. A membership function $\mu_{\delta_\Phi} : 2^X \times 2^X \longrightarrow [0, 1]$ on the space X is a **fuzzy descriptive Lodato proximity**, provided it satisfies the following axioms.
$N_{\mu_{\delta_\Phi}}.1$ $\mu_{\delta_\Phi}(\varnothing, A) = 0$ for all $A \subset X$.
$N_{\mu_{\delta_\Phi}}.2$ $\mu_{\delta_\Phi}(A, B) = \mu_{\delta_\Phi}(B, A)$.
$N_{\mu_{\delta_\Phi}}.3$ $\mu_{\delta_\Phi}(A, B) \neq 0$ implies $A\ \delta_\Phi\ B$.
$N_{\mu_{\delta_\Phi}}.4$ $\mu_{\delta_\Phi}(A, (B \cup C)) \neq 0$ implies $\mu_{\delta_\Phi}(A, B) \neq 0$ and $A\ \delta_\Phi\ B$ or $\mu_{\delta_\Phi}(A, C) \neq 0$ and $A\ \delta_\Phi\ C$.
$N_{\mu_{\delta_\Phi}}.5$ $\mu_{\delta_\Phi}(A, B) \neq 0$ and $\mu_{\delta_\Phi}(\{b\}, C) \neq 0$ for all $b \in B$ implies $\mu_{\delta_\Phi}(A, C) \neq 0$ and $A\ \delta_\Phi\ C$. ■

A fuzzy descriptive Lodato proximity space is denoted by $(X, \delta_\Phi, \mu_{\delta_\Phi})$.

Theorem 9.43 *The descriptive Smirnov proximity measure $\delta_{\varepsilon_\Phi, \upsilon}$ on the descriptive Lodato proximity space is a fuzzy descriptive Lodato–Smirnov proximity membership function.*

Proof The proof is symmetric with the proof of Theorem 9.43. □

Problem 9.44
☕ Prove Theorem 9.43. ■

Problem 9.45 Fuzzy Descriptive Lodato Proximity.
Do the following: Let M be a Voronoï mesh that covers a nonempty subset X of the Euclidean plane equipped with the descriptive Lodato proximity δ_Φ. For implementations, use Mathematica.
1° Select a subset X in the Euclidean plane.
2° Select a set of sites $S \subset X$ and $\varepsilon_\Phi > 0$.
3° Construct a Voronoï mesh M using S.
4° Select $A, B \in M$.

5° Implement

$$v_\Phi(A, B) = \frac{\left|A \underset{\Phi}{\cap} B\right|}{|\Phi(X)|}$$

for nonempty Voronoï regions $A, B \in M$.

6° Implement the descriptive Smirnov proximity measure $\delta_{\varepsilon_\Phi, v}(A, B)$ defined by

$$\delta_{\varepsilon_\Phi, v_\Phi}(A, B) = \begin{cases} \frac{\left|A \underset{\Phi}{\cap} B\right|}{|\Phi(X)|}, & \text{if } \varepsilon \leq v_\Phi(A, B) \leq 1, \\ 0, & \text{if } v_\Phi(A, B) < \varepsilon. \end{cases}$$

7° Display the values of $v(A, B)$, $\delta_{\varepsilon, v_\Phi}(A, B)$.

8° Repeat Problem 9.45, beginning with Step 4 for 10 choices of A, B. ■

Problem 9.46 Digital Fuzzy Descriptive Lodato Proximity.
🚲 Let M be a Voronoï mesh that covers a nonempty subset X of digital image in the Euclidean plane equipped with the descriptive Lodato proximity δ_Φ. Repeat the steps in Problem 9.45. ■

Problem 9.47 Borsuk's Conjecture for Digital Fuzzy Descriptive Lodato Proximity Regions.
☕ What does Borsuk's observation in Conjecture 9.24 for the Euclidean plane tell us about a bounded subset $\Phi(A)$ (see of all descriptions of points in the subset A) in a digital image? Do the following:

1° Give a sketch to illustrate your idea.

2° Give an example in a digital image.

3° Use Mathematica to do the following:

- a^o Choose a colour image X.
- b^o For pixel features, choose the red, green, blue colours.
- c^o Select a set of sites in X.
- d^o Superimpose a Voronoï diagram M on X.
- e^o Select Voronoï region A in M.
- f^o Partition A into three subregions containing all pixels of your choice of colour. ■

Problem 9.48
☕ Prove that $\delta_{\varepsilon_\Phi, v}$ is a fuzzy descriptive proximity membership function. ■

Problem 9.49
☕ Give the axioms for a fuzzy descriptive strong proximity space $(X, \overset{\wedge\!\!\!\wedge}{\delta}_\Phi, \mu_{\overset{\wedge\!\!\!\wedge}{\delta}_\Phi})$. ■

Problem 9.50 Let fuzzy descriptive strong proximity space be represented by the collection of regions in a Voronoï mesh M equipped with the descriptive strong proximity $\overset{\wedge\!\!\!\wedge}{\delta}_\Phi$. Let $v_{\overset{\wedge\!\!\!\wedge}{\delta}_\Phi}(A, B)$ for nonempty Voronoï regions $A, B \in M$ be defined by

$$\upsilon_{\overset{\wedge\!\!\wedge}{\delta_\Phi}}(A, B) = \frac{\left|\text{int}A \underset{\Phi}{\cap} \text{int}B\right|}{|\Phi(X)|}.$$

The descriptive Smirnov strong proximity measure $\delta_{\varepsilon_\Phi, \upsilon_{\overset{\wedge\!\!\wedge}{\delta_\Phi}}}(A, B)$ defined by

$$\delta_{\varepsilon_\Phi, \upsilon_{\overset{\wedge\!\!\wedge}{\delta_\Phi}}}(A, B) = \begin{cases} \frac{\left|\text{int}A \underset{\Phi}{\cap} \text{int}B\right|}{|\Phi(X)|}, & \text{if } \varepsilon < \upsilon_{\overset{\wedge\!\!\wedge}{\delta_\Phi}}(A, B) \leq 1, \\ 0, & \text{if } \upsilon_{\overset{\wedge\!\!\wedge}{\delta_\Phi}}(A, B) < \varepsilon. \end{cases}$$

☕ Prove that $\delta_{\varepsilon_\Phi, \upsilon_{\overset{\wedge\!\!\wedge}{\delta_\Phi}}}$ is a fuzzy descriptive strong proximity membership function. ■

Problem 9.51
🚲 Repeat the steps in Problem 9.45 for $\upsilon_{\overset{\wedge\!\!\wedge}{\delta_\Phi}}(A, B)$ and $\delta_{\varepsilon_\Phi, \upsilon_{\overset{\wedge\!\!\wedge}{\delta_\Phi}}}(A, B)$. ■

Mscript 13 Image Edges Dilated and Eroded.

(*morphology*)

im = EdgeDetect[]

im2 = Dilation[***im, DiskMatrix***[3]]

im3 = Erosion[***im2, DiskMatrix***[3]] ■

9.8 Image Dilation

In mathematical morphology (MM), **dilation** is a shape-growing method for binary images. **Dilation** fills in 3×3 neighbourhood pixels near a certain threshold. This leads to the growth of pixel clusters in an image. Dilation is defined in terms of the reflection and translation of a set. The **reflection** of a set of pixels B (denoted $\hat{B}$) is defined by

$$\hat{B} = \{w : w = -b, b \in B\}.$$

The **translation** of a set A by a point z (denoted $(A)_z$) is defined by

$$(A)_z = \{c : c = a + z, a \in A\}.$$

For a sample dilation, see Fig. 9.12. The binary dilation of region A by a structuring element B is a set operation $A \oplus B$ defined by

$$A \oplus B = \left\{z : \hat{B}_z \cap A \neq \varnothing\right\},$$

where the reflected and translated strel (structuring element) B overlaps at least one element of A. Each dilation of an image A by a structuring element B is a pixel $z := 1$ such that A and $\hat{B}_z$ overlap by at least one element. That is, $A \oplus B$ is a set of pixels z such that the translated structuring element overlaps with foreground pixels in A.

Mscript 14 Binarized Rainbow-on-a-plant Image.

(*MorphologicalBinarize[image]withupperandlowerthresholding.*)

im **=** **;**
***MorphologicalBinarize*[*im*]**

***MorphologicalBinarize*[*im*, {0.25, 0.5}]**

■

9.9 MorphologicalBinarize Operation with Lower and Upper Bounds

MM operations are performed on binary images. For this reason, a close look at approaches to binarizing colour or greyscale images is warranted. It is helpful to try the default binarization of an image, first. For sample MM operations on binary images, see Fig. 9.12 and for object and hole pixels in a binary image, see Fig. 9.13.

Example 9.52 **Binarizing a Rainbow-on-a-Plant Colour Image.**
This is done using Mathematica script 14 on the Rainbow-on-a-plant Image in Fig. 9.14.1. The result is shown in Fig. 9.14.2. Notice that most of the details of the plant are lost and the foreground plant shadow dominates. To achieve a better result, Otsu thresholds are set on the lower and upper bounds of the intensities that are mapped to foreground and background intensities in the binarized image. For this example,

$$\{lowthreshold, highthreshold\} = \{0.25, 0.5\}.$$

The choice of lower and upper thresholds will vary, depending on the image. The thresholded binarized plant image is shown in Fig. 9.15. ■

For best results in setting the upper bound on a binarized threshold, use

FindThreshold[*im*];

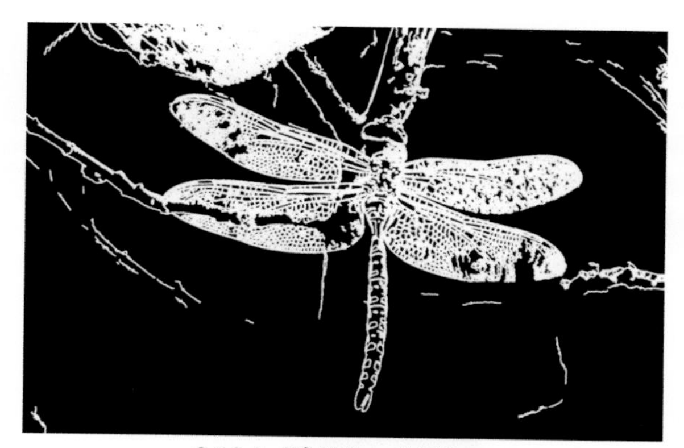

9.12.1: $\mathbb{R}^2$ Dilated Edges

9.12.2: $\mathbb{R}^2$ Eroded Edges

Fig. 9.12 Morphed edges in a DragonFly image

1	1	0	0	1	1	1	0	0	1

Fig. 9.13 White foreground (object) pixels (1), black background pixels (0)

9.14.1: Rainbow on Plant

9.14.2: Default Binarization

Fig. 9.14 Default binarization of the rainbow-on-a-plant image

Fig. 9.15 Bounded binarization of the rainbow-on-a-plant image

$$\mathbf{FindThreshold}[im];$$

Fig. 9.16 Optimal bounded binarization of the rainbow-on-a-plant image

Example 9.53 **Optimal Upper Bounded Binarization.**
For the rainbow-on-the-plant image in Fig. 9.14.1, the optimal upper threshold is 0.439216. Try

MorphologicalBinarize[[, **0.25, 0.439216**]];

The optimal result is shown in Fig. 9.16. ■

9.10 Dilation Method

Let g be a digital image and let se be an $n \times n$ array for a structuring element (usually, n is odd). In MM, the dilation method grows or thickens objects in an image. In a **dilation operation, denoted by** $g \oplus se$, the centre of a structuring element (se) is placed over each background pixel (value $= 0$) in g. If any neighbourhood pixels covered by the se are foreground pixels, then the background pixel (covered by se centre pixel) is switched to a foreground pixel (value $= 1$).

Different structuring elements (kernels) are possible, depending on the appropriate shape to use in either dilating or eroding a binary image. DiskMatrix, BoxMatrix, DiamondMatrix, and GaussianMatrix are examples in Mathematica.

Example 9.54 **Binary Image Dilation.**
For the binarized rainbow-on-the-plant image in Fig. 9.16, try

Dilation[[, ***DiskMatrix*[3]**]];

The result is shown in Fig. 9.17. Notice the foreground pixels dominate in the dilated image. ■

Another tact to try is subtracting a binarized image from the dilated image. The result is a capture of the dominant edges in the binary image.

Example 9.55 **Binary Image Edges** .
For the binarized rainbow-on-the-plant image in Fig. 9.16, try

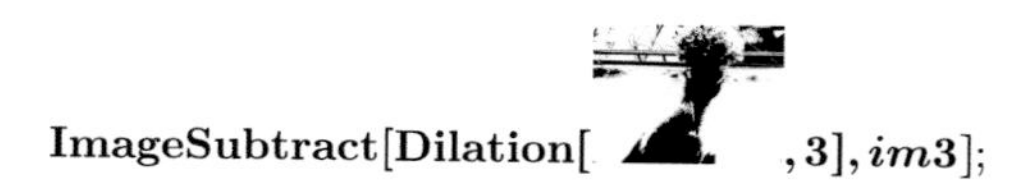

ImageSubtract[**Dilation**[, **3**], ***im3***];

The result is shown in Fig. 9.18. This leads to sharply delineated, dominant edges in the sample plant image. This is remarkable, since the original image is filled with noisy regions. ■

Fig. 9.17 Rainbow-on-a-plant image dilation with disk structuring element

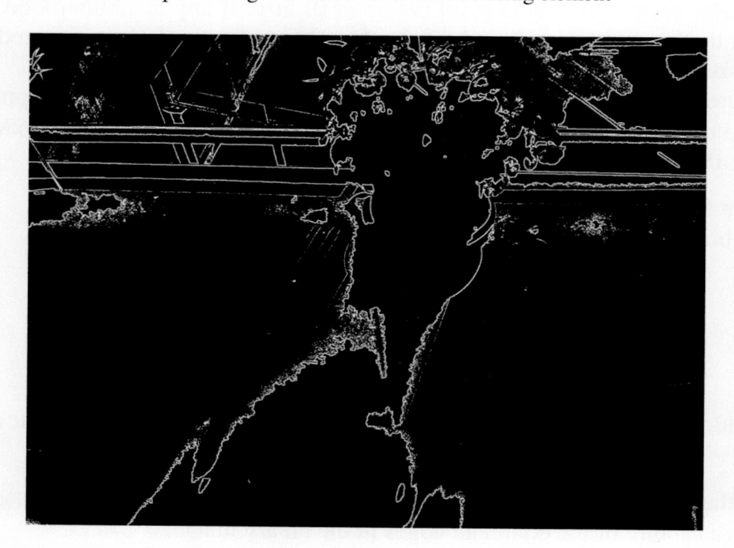

Fig. 9.18 Rainbow-on-a-plant image edges after binarized image subtraction

Mscript 15 Watershed Segmentation of Rainbow-on-a-plant Image.

(* Watershed segmentation *)

i =  ;

***w* = *ColorNegate*[*Image*[*WatershedComponents*[*i*], "*Bit*"]];**
***HighlightImage*[*i*, *Dilation*[*w*, 1]]**

***Image*[*WatershedComponents*[*Threshold*[, 0.18]], "*Bit*"]**

WatershedComponents*[*FillingTransform*[*GradientFilter*[*i*, 1], 0.03, *Padding* → 1]]//*Colorize

***Image*[*WatershedComponents*[*Threshold*[, 0.18]], "*Bit*"]**

***WatershedComponents*[*FillingTransform*[*GradientFilter*[, 1], 0.03, *Padding* → 1]]//**
Colorize ■

9.11 Dilation and Watershed Segmentation

This section briefly introduces watershed segmentation. The basic idea is to view image regions with lower intensities as the analogue of natural **cachement areas** (drainage basins) that rainwater flows into from higher level surface areas. The analogue of cachement areas with lower intensity levels in a digital image (i.e., cachement intensity levels below higher level intensity levels) leads to watershed image segmentation. That is, a **watershed segment** is an image region containing pixels with lower intensities than the pixels in the surrounding region. A good separation of image regions can be achieved by performing a dilation operation on watershed components.

Example 9.56 **Dilation on Watershed Components.**
Here are the steps on the binarized rainbow-on-the-plant image in Fig. 9.16 to obtain a dilated segmented image:
1° Binarize an image i (often a colour image).
2° Find the watershed components w on i.
3° Highlight dilation of w on i.
4° Produce the result as an image.
These steps are carried out at the bit level using Mathematica script 14 +−. The result is shown in Fig. 9.19. ■

Example 9.57 **Watershed Components Without Dilation.**
Here are the steps on the binarized rainbow-on-the-plant image in Fig. 9.16 to obtain a segmented image without dilation:
1° Binarize an image i (often a colour image).
2° Choose an Otsu threshold on i.
3° Find the watershed components w on i.
4° Produce the result as an image.
These steps are carried out again at the bit level using MScript 14. The result is shown in Fig. 9.20. ■

Fig. 9.19 Rainbow-on-a-plant image after dilation of watershed components

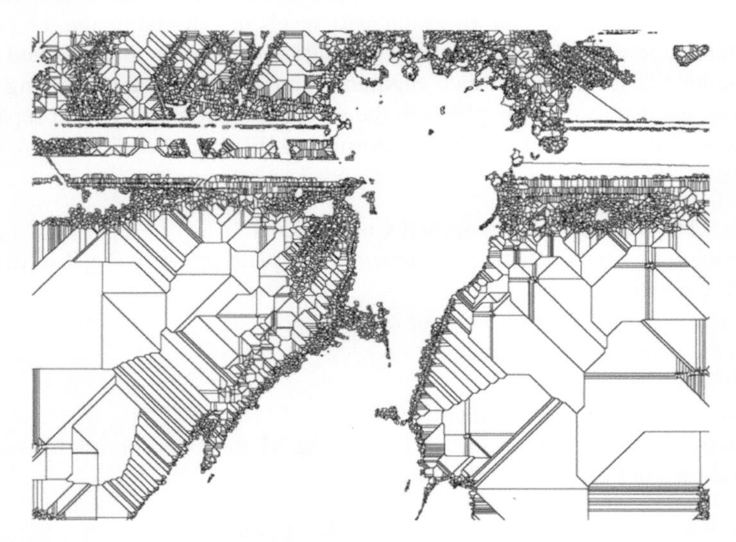

Fig. 9.20 Rainbow-on-a-plant image without dilation of watershed components

9.12 Gradient Filtering Watershed Components

This section briefly introduces gradient filtering watershed components. The basic approach is to compute the gradient magnitude of each pixel using discrete derivatives of a Gaussian of pixel radius r.

Example 9.58 **Watershed Components Without Dilation.**
Here are the steps on the binarized rainbow-on-the-plant image in Fig. 9.16 to obtain a segmented image without dilation:
1° Binarize an image i (often a colour image).
2° Choose pixel radius r.
3° Find the gradient magnitudes of the pixels on i.
4° Choose depth h.
5° Transform the result of gradient filtering with all extended minima of depth h filled.
6° Find the watershed components of the result of the previous step.
7° Produce the result as an image, showing colourized segments.
These steps are carried out again at the bit level using Mathematica script 14. The result is shown in Fig. 9.21. ■

Problem 9.59 ☕ Do the following:

1° Select three digital images.
2° For each of the selected images, follow the steps in Example 9.57 and show the results obtained using MScript 14. ■

Problem 9.60 ☕ Do the following:

1° Select three digital images.
2° For each of the selected images, follow the steps in Example 9.58 and show the results obtained using MScripts 14 and 15. ■

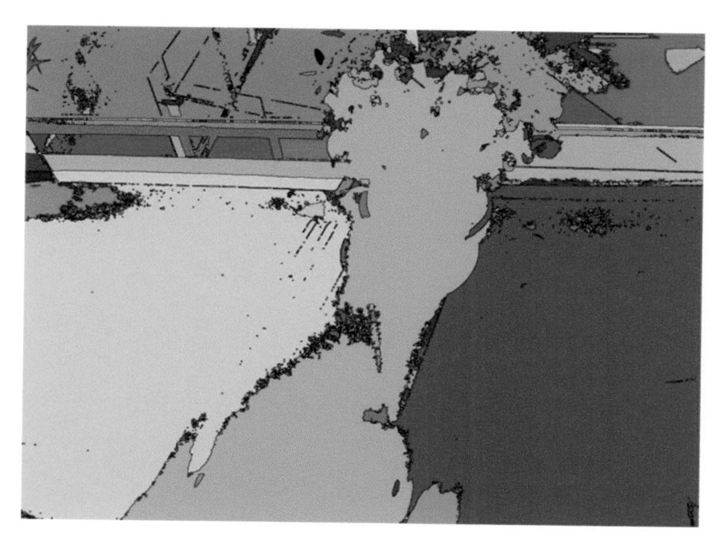

Fig. 9.21 Rainbow-on-a-plant with Gradient Filtering $\mapsto$ Watershed Components

References

1. Matheron, G.: Estimating and Choosing. Springer, Berlin (1989)
2. Serra, J.: Image Analysis and Mathematical Morphology. Academic Press, London (1982)
3. Smirnov, J.M.: On proximity spaces. Math. Sb. (N.S.) **31**(73), 543–574 (1952). English translation: Amer. Math. Soc. Trans. Ser. **2**(38), 5–35 (1964)
4. Šostak, A.: Basic structures of fuzzy topology. J. Math. Sci. **78**(6), 662–701 (1996). MR1384343
5. Liu, Y.M., Luo, M.K.: Fuzzy topology. Advances in Fuzzy Systems-Applications and Theory, 9. World Scientific Publishing Co., Inc, River Edge, NJ (1997). X+353 pp. ISBN: 981-02-2862-7, MR1643076
6. Brennan, J., Martin, E.: Membership functions for spatial proximity. In: A. Sattar, B. Kang (eds.) Lecture Notes in Artificial Intelligence 4304, pp. 942–949. Springer (2006)
7. Çetkin, V., Šostak, A., Aygün, H.: An approach to the concept of soft fuzzy proximity. Abstr. Appl. Anal. Art. ID 782583, 1–9 (2014). MR3191064
8. Chattopadhyay, K., Hazra, H., Samanta, S.: A correspondence between lodato fuzzy proximities and a class of principal type-ii fuzzy extensions. J. Fuzzy Math. **20**(1), 29–46 (2012). MR2934135
9. Zadeh, L.: Fuzzy sets. Information and Control **8**, 338–353 (1965). MR0219427
10. Fashandi, H.: Topology framework for digital image analysis with extended interior and closure operators. Ph.D. thesis, University of Manitoba, Winnipeg, MB, CA (2012). Supervisor: J.F. Peters
11. Meghdadi, A.: Fuzzy tolerance neighborhood approach to image similarity in content-base image retrieval. Ph.D. thesis, University of Manitoba, Winnipeg, MB, CA (2012). Supervisors: J.F. Peters
12. Berger, M.: Geometry revealed. A Jacob's ladder to modern higher geometry. Translated from the French by Lester Senechal. Springer, Heidelberg (2010). Xvi+831 pp. ISBN: 978-3-540-70996-1, MR2724440
13. Berger, M.: Geometry I. Translated from the 1977 French original by M. Cole and S. Levy. Springer-Verlag, Berlin (1977). Xiv+428 pp. ISBN: 978-3-540-11658-5 51-01, MR2724360
14. Borsuk, K.: Drei sätze über die n-dimensionale euklidische sphäre. Fundamenta Mathematicae XX, 177–190 (1933)
15. Dochviri, I., Peters, J.: Topological sorting of finitely many near sets. Math. Comput. Sci. pp. 1–6 (2015). communicated
16. Leader, S.: On clusters in proximity spaces. Fundamenta Mathematicae **47**, 205–213 (1959)

Chapter 10
Strong Connectedness Revisited

Fig. 10.1 Portrait exhibiting strong connectedness

This chapter takes another look at connectedness. The traditional view of connected spaces is augmented with the introduction of strongly near proximity, which ushers in new forms of connectedness, namely,

1° **Strong Connectedness** [1, 2]. Let X be a topological space equipped with the $\overset{\wedge\wedge}{\delta}$ strong proximity. The space X is strongly connected, provided

(i) $X = \bigcup_{i \in \mathbb{N}^+} X_i$ for all $X_i \in 2^X$.

(ii) $\text{int}X_{i-1} \cap \text{int}X_i \neq \emptyset$ for all $i \geq 2$. In that case, we write $A \overset{\wedge\wedge}{\text{conn}} B$ for strongly connected subsets of X.

2° **Descriptive Strong Connectedness** [2]. Let X be a topological space equipped with the $\overset{\wedge\wedge}{\delta}_{\Phi}$ descriptive strong proximity. The space X is descriptively strongly connected, provided

J.F. Peters, *Computational Proximity*, Intelligent Systems Reference Library 102, DOI 10.1007/978-3-319-30262-1_10

Fig. 10.2 Descriptive strong connectedness subsets in a portrait

(i) $\Phi(X) = \bigcup_{i \in \mathbb{N}^+} \Phi(X_i)$ for all $X_i \in 2^X$.

(ii) $X_{i-1} \underset{\Phi}{\cap} X_i \neq \emptyset$ for all $i \geq 2$. In that case, we write $A \overset{\curlywedge,\Phi}{\text{conn}} B$ for descriptively strongly connected subsets of X. ■

We illustrate descriptive strong connectedness with the portrait[1] in Fig. 10.1.

Example 10.1 **Descriptive Strong Connectedness Portrait Space.**
Let X be a topological space with the RGB colour space basis as defined in Example 9.10 and equipped with the $\overset{\curlywedge}{\delta}_{\Phi}$ descriptive strong proximity represented in Fig. 10.2. Assume that $X = A \cup B$. In addition, $A \overset{\curlywedge,\Phi}{\text{conn}} B$, since $\text{int}A \underset{\Phi}{\cap} \text{int}B$, i.e., the description of many points in the interior of A matches the description of many points in the interior of B. Hence, X is a descriptive strongly connected space. ■

10.1 Connectedness: Basic Concepts

This section briefly revisits the notion of connectedness of subsets in various proximity spaces, based on recent work on strong connectedness [1] (see, also, [2]). It appears quite natural to try to generalize this concept using strong proximities. Actually we obtain a strengthening of the standard concept by considering connected sets such as the one in the plot of the Dirichlet L-function shown in (Fig. 10.3). See Example 10.3 for the details.

Recall that a space X is **connected** (denoted by connX), if and only if X is the union of nonempty subsets A and B such that

$$\text{cl}A \cap B \neq \emptyset \text{ and } A \cap clB \neq \emptyset.$$

[1] Many thanks to Maciej Borkowski for contributing this amazing portrait as an illustration of strong connectedness.

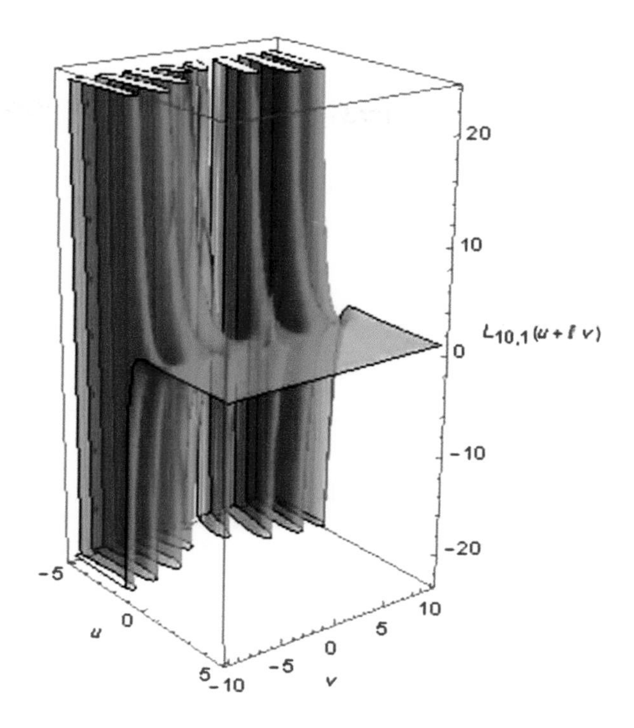

Fig. 10.3 Connected set defined by the Dirichlet L-function in $\mathbb{R}^3$

A subset E in a space X is a **connected subset**, if and only if there are nonempty subsets $A, B \subset X$ such that

$$\text{conn}E = A \cup B \ \text{ and } \ \text{cl}A \cap B \neq \varnothing \text{ and } A \cap clB \neq \varnothing.$$

Let X be a nonempty set endowed with the Wallman proximity δ. Then X is a **connected space**, if and only if, for nonempty subsets $A, B \subset X$,

$$\text{conn}X = A \cup B \text{ and } A \ \delta \ B.$$

Similarly, a subset $E \subset X$ is connected, if and only if, for nonempty subsets $A, B \subset X$,

$$\text{conn}E = A \cup B \text{ and } A \ \delta \ B.$$

Example 10.2 **Topologist's Sine Curve Subset.**
An interesting example of a connected subset T known as the topologist's sine curve in the Euclidean plane $\mathbb{R}^2$ is given in [3, Sect. 1.12, p. 29–30]. The subspace $T \subset \mathbb{R}^2$ defined by

$$\text{conn}T = \{(0, y) : -1 \leq y \leq 1\} \cup \{(x, \sin(\frac{1}{x})) : x > 0\}.$$

Fig. 10.4 Sine curve

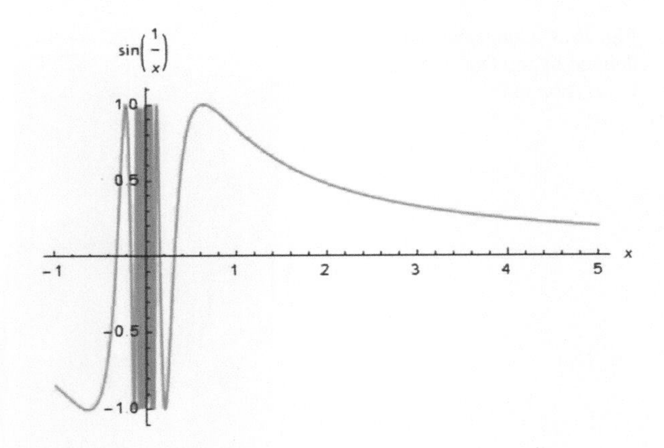

T consists of two parts, namely, a portion of the y-axis and a portion of $\mathbb{R}^2$ to the right of the y-axis (see Fig. 10.4), implemented using MScript 48 in Appendix A.10. Intuitively, T seems to be disconnected but, in fact, from the definition of a connected space, T is connected. ■

Example 10.3 **DirichletL Connected Subset in Euclidean space** $\mathbb{R}^3$.
Let L be defined by

$$\chi_k(n) = \text{ integer function with period } k \text{ (modulus), (index) } n,$$
$$L_k(x, \chi) \equiv \sum_{n=1}^{\infty} \chi_k(n) n^{-x}.$$

Then the connected subset D in Euclidean $\mathbb{R}^3$ is defined, for example, by

$$\begin{aligned} \text{conn} D &= A \cup B \\ &= \{(u, v, L_{10,1}(u + iv)) : -3 \leq u \leq 0, 0 \leq v \leq 1.1\} \cup \\ &\quad \{(u, v, L_{10,1}(u + iv)) : -5 \leq u \leq 5, -10 \leq v \leq 10\}. \end{aligned}$$

Here, the connected subset is defined by two subsets, namely, a portion of the set A defined by

$$A := \{(u, v, L_{10,1}(u + iv)) : -3 \leq u \leq 0, 0 \leq v \leq 1.1\}.$$

(see Fig. 10.5 produced by MScript 46 in Appendix A.10) that has points in common with the set defined by

$$B := \{(u, v, L_{10,1}(u + iv)) : -5 \leq u \leq 5, -10 \leq v \leq 10\}.$$

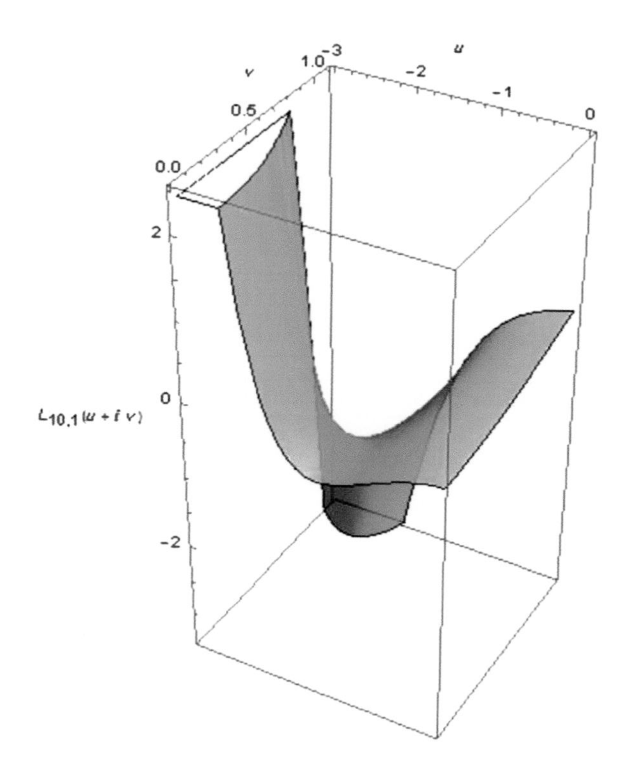

Fig. 10.5 Set defined by the Dirichlet L-function in $\mathbb{R}^3$

See Fig. 10.3 for a plot of the connected subset$D$, implemented using MScript 47 in Appendix A.10 and $A\ \delta\ B$. Since A and B have more than one point in common, it is also the case that $A \stackrel{\wedge\wedge}{\delta} B$ (A is strongly near B). ■

Recall the following result from S. Willard [4].

Theorem 10.4 *If* $X = \bigcup_{n=1}^{\infty} X_n$ *where each* X_n *is connected and* $X_{n-1} \cap X_n \neq \emptyset$ *for each* $n \geq 2$, *then* X *is connected.*

Based on earlier work on strongly near sets, we define strongly proximal connectedness.

Definition 10.5 Let X be a topological space and $\stackrel{\wedge\wedge}{\delta}$ a strong proximity on X. We say that X is $\stackrel{\wedge\wedge}{\delta}$-connected if and only if $X = \bigcup_{i \in I} X_i$, where I is a countable subset of $\mathbb{N}$, X_i and $\text{int}(X_i)$ are connected for each $i \in I$, and $X_{i-1} \stackrel{\wedge\wedge}{\delta} X_i$ for each $i \geq 2$. ■

Theorem 10.6 *Let* X *be a topological space and* $\stackrel{\wedge\wedge}{\delta}$ *a strong proximity on* X. *Then* $\stackrel{\wedge\wedge}{\delta}$*-connectedness implies connectedness.*

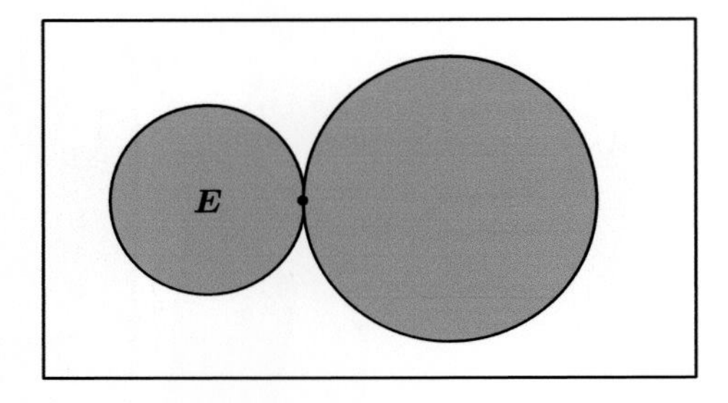

Fig. 10.6 Connected subset E but not $\overset{\wedge\!\!\wedge}{\delta}$-connected

---

Algorithm 14: Strongly Connected Set in a Delaunay Mesh on a Digital Image

---

Input : Read digital image img.
Output: Strongly connected subset connDe of Delaunay triangles D in mesh $\mathscr{M}$ on img.

1 $img \longmapsto cornerCoordinates$;
2 $S \leftarrow cornerCoordinates$;
3 /* S contains corner coordinates used as mesh generating points (sites). */ ;
4 $S \longmapsto$ *DelaunayMesh* $\mathscr{M}$;
5 *DelaunayMesh* $\mathscr{M} \longmapsto img$;
6 *Identify triangles* $A, B \in \mathscr{M}$ *with a common edge* ;
7 $connDe := A \cup B$;

Proof This simply follows from Theorem 10.4 and axiom $(N2)$ in the definition of strong proximities. □

Example 10.7 **Strongly Connected Set.**
Consider the $\overset{\wedge\!\!\wedge}{\delta}$-connected subset $D_1 := A \cup B$ in the Delaunay mesh on the digital image showing a plant in the sunshine in Fig. 10.6 and the strong proximity given by $A \overset{\wedge\!\!\wedge}{\delta} B$. The subset conn$D_1$ is $\overset{\wedge\!\!\wedge}{\delta}$-connected, since the triangular shaped sets A and B have a common edge (Fig. 10.7). ■

The set D_1 in Example 10.7 illustrates the outcome of Algorithm 14. In a Delaunay mesh, it is possible to find many examples of sets of triangles that are strongly connected.

Example 10.8 **Connected but Not Strongly Connected Set.**
Consider the connected subset conn$D_2 := C \cup E$ in the Delaunay mesh on the digital image showing a plant in the sunshine in Fig. 10.6 and Wallman proximity given by $C \; \delta \; E$. The subset D_2 is connected but $\overset{\wedge\!\!\wedge}{\delta}$-connected, since the triangular shaped sets C and E only share a vertex. This shows that the converse of Theorem 10.6 is not always true. ■

The set D_2 in Example 10.8 illustrates the outcome of Algorithm 15. In a Delaunay mesh, it is also possible to find many sets of triangles that are connected but not strongly connected.

Fig. 10.7 Strongly connected subset $D := A \cup B$ in a Delaunay mesh

---

Algorithm 15: Connected subset in a Delaunay Mesh on a Digital Image

---

Input : Read digital image img.
Output: Connected set of Delaunay triangles connDe in mesh $\mathscr{M}$ on img.

1 $img \longmapsto cornerCoordinates$;
2 $S \leftarrow cornerCoordinates$;
3 /* S contains corner coordinates used as mesh generating points (sites). */ ;
4 $S \longmapsto$ *DelaunayMesh* $\mathscr{M}$;
5 *DelaunayMesh* $\mathscr{M} \longmapsto img$;
6 *identify triangles* $A, B \in \mathscr{M}$ *with a common vertex* ;
7 $connDe := A \cup B$;

Problem 10.9 Delaunay connectedness.
Write a Mathematica script that does the following:

1° Select a digital image.
2° Select a set of mesh generating points (sites) S.
3° Constructs a Delaunay mesh $\mathscr{D}(S)$ on the digital image.
4° Implement Algorithm 14. In your implementation, highlight in yellow sets of Delaunay triangles D_1, D_2 that are strongly connected and D_1, D_2 are separated (D_1, D_2 have neither a vertex nor an edge in common), i.e., $D_1 \ \not\delta\ D_2$. Do this for up to 10 strongly connected sets of Delaunay triangles.
5° Implement Algorithm 14. In your implementation, highlight in yellow sets of Delaunay triangles D_1, D_2 that are strongly connected and D_1, D_2 are not separated (D_1, D_2 have either a vertex or an edge in common), i.e., $D_1 \ \delta\ D_2$. Do this for up to 10 strongly connected sets of Delaunay triangles.

6° Implement Algorithm 15. In your implementation, highlight in orange sets of Delaunay triangles D_3, D_4 that are connected but not strongly connected and D_3, D_4 are separated (D_3, D_4 have neither a vertex nor an edge in common), i.e., $D_3 \not\delta D_4$. Do this for up to 10 connected sets of Delaunay triangles.

7° Implement Algorithm 15. In your implementation, highlight in orange sets of Delaunay triangles D_3, D_4 that are connected but not strongly connected and D_3, D_4 are not separated (D_3, D_4 have either a vertex or an edge in common), i.e., $D_3 \ \delta \ D_4$. Do this for up to 10 connected sets of Delaunay triangles.

8° Repeat the preceding steps for three different digital images of your own choosing. **N.B.**: Only use digital images taken by you, not found on the web. ■

The following example also shows that the converse of Theorem 10.6 is not always true.

Example 10.10 **Connected Set That is Not Strongly Connected.**
Consider the subset E of $\mathbb{R}^2$ as shown in Fig. 10.6 and the strong proximity given by $A \overset{\wedge\!\!\!\wedge}{\delta} B \Leftrightarrow \text{int}A \cap \text{int}B \neq \varnothing$ or either A or B is equal to X, provided A and B are not singletons; if $A = \{x\}$, then $x \in \text{int}(B)$, and if B too is a singleton, then $x = y$. The subset E is connected but it is not $\overset{\wedge\!\!\!\wedge}{\delta}$-connected. ■

Algorithm 16: Strongly Connected Set in a Voronoï Mesh on a Digital Image

Input : Read digital image img.
Output: Strongly connected subset of polygons conn$V$ in Voronoï mesh $\mathcal{M}$ on img.

1 $img \longmapsto cornerCoordinates$;
2 $S \leftarrow cornerCoordinates$;
3 /* S contains corner coordinates used as mesh generating points (sites). */ ;
4 $S \longmapsto$ *VoronoiMesh* $\mathcal{M}$;
5 *VoronoiMesh* $\mathcal{M} \longmapsto img$;
6 *Identify polygons* $A, B \in \mathcal{M}$ *with a common edge* ;
7 $connV := A \cup B$;

10.2 Strongly Connected Subsets in a Voronoï Mesh

This section carries forward the observations about connected subsets of triangles in a Delaunay mesh. Here, the focus shifts to sets of polygons in a Voronoï mesh. The basic approach is to find sets of Voronoï mesh polygons that have a common edge (see Algorithm 16).

Example 10.11 **Strongly Connected Set in a Voronoï Mesh on an Image.**
Consider the subset $V_1 := A \cup B$ in the Voronoï mesh $\mathcal{M}$ on the dancing boy shown in Fig. 10.8 and the strong proximity given by $A \overset{\wedge\!\!\!\wedge}{\delta} B$, since polygons A and B have a common edge. Hence, the subset $V_1 \in \mathcal{M}$ is strongly connected. ■

Fig. 10.8 Strongly connected subset $D := A \cup B$ in a Voronoï mesh

The set V in Example 10.11 illustrates the type of output found by Algorithm 16.

Example 10.12 **Multiple Strongly Connected Sets in a Voronoï Mesh on an Image.**
Consider the subsets

$$V_1 := A \cup B \text{ in Fig. 10.8}$$
$$V_2 := C \cup E \text{ in Fig. 10.8}$$

From Example 10.11, $V_1 \in \mathscr{M}$ is strongly connected. $V_2 := C \cup E$ in the Voronoï mesh $\mathscr{M}$ on the dancing boy shown in Fig. 10.8 and the strong proximity given by $C \overset{\wedge\!\!\wedge}{\delta} E$, since polygons C and E have a common edge. Hence, the subset $V_2 \in \mathscr{M}$ is strongly connected.

In addition,

$$A \overset{\wedge\!\!\wedge}{\delta} E, \text{ since } A \in V_1, B \in V_2 \text{ have a common edge.}$$

Hence, V_1 and V_2 are strongly connected. ■

Problem 10.13 Strongly Connected Sets of Polygons in a Voronoï Mesh.
Write a Mathematica script that does the following:

1° Select a digital image.

2° Select a set of mesh generating points (sites) S.

3° Construct a corner-based Voronoï mesh $\mathscr{V}(S)$ on the digital image.

4° Implement Algorithm 16. In your implementation, highlight in yellow sets of Voronoï polygons V_1, V_2 that are strongly connected and V_1, V_2 are separated (V_1, V_2 do not have an edge in common), i.e., $V_1 \ \not\delta\ V_2$. Do this for up to 10 strongly connected sets of Voronoï polygons.

5° Give a new implementation of Algorithm 16 that finds multiple sets of strongly connected polygons that are strongly near. In your implementation, highlight in yellow sets of Voronoï polygons V_1, V_2 that are strongly connected and V_1, V_2 are *not* separated (V_1, V_2 near enough to each other so that there is a polygon in V_1 that has an edge in common with a polygon in V_2), i.e., $V_1 \ \overset{\wedge\!\!\wedge}{\delta}\ V_2$. For sample multiple sets of strongly connected polygons that are strongly near, see Example 10.12. Do this for up to 10 strongly connected sets of Voronoï polygons.

6° Repeat the preceding steps for three different digital images of your own choosing. **N.B.**: Only use digital images taken by you, not found on the web. ■

Fig. 10.9 1870Punch image

Example 10.14 **Strongly Connected Sets of Polygons.**
Here are the steps to construct connected sets polygons in a Voronoï mesh on the dancing boy image in Fig. 10.9:
1° Select an image i.
2° Find the corners on i.
3° Generate a corner-based Voronoï mesh on i.
4° Produce the result as an image.
The result of carrying out these steps is shown in Fig. 10.10.1 and in Fig. 10.10.2. ■

Assume that the space X covered by a mesh $\mathscr{M}$ is topological space endowed with a proximal relator $\left\{\delta, \overset{\wedge\!\wedge}{\delta}\right\}$, where δ is the Wallman proximity. In general, every pair of polygons in the mesh $\mathscr{M}$ is not strongly connected nor is every pair of polygons connected. In general, the space covered by the Voronoï mesh is not connected, since all pairs of the Voronoï regions do not have a common edge.

Theorem 10.15 *A Vonoroï mesh in a Euclidean space E^2 containing only two polygons is connected.*

Proof Immediate from the geometry of a Voronoï mesh and Willard's Theorem 10.4. □

Theorem 10.16 *A Vonoroï mesh in a Euclidean space E^d containing only 2 polygons is connected.*

10.10.1: Image Corners 10.10.2: Voronoï Mesh

Fig. 10.10 Connected Voronoï regions

Proof Immediate from the geometry of a Voronoï mesh and Willard's Theorem 10.4. □

A natural extension of connectedness stems from our earlier work on strongly near sets.

Example 10.17 **Corner-Based Voronoï Mesh with Strongly Connected Sets.** Here are the steps to construct strongly connected sets polygons in a Voronoï mesh on the dancing boy image in Fig. 10.9:

1° Select an image i.
2° Find the corners on i.
3° Generate a corner-based Voronoï mesh on i.
4° Identify the Voronoï region with the greatest $\overset{\wedge\!\!\!\wedge}{\delta}$-connectivity, i.e., the greatest number of pairs of polygons so that the same polygon is common to each pair of polygons.
5° Produce the result as an image.

The result of carrying out these steps is shown in Fig. 10.8. The highlighted polygon in Fig. 10.8 is the most strongly connected polygon in the mesh. ■

Theorem 10.18 *Let X be a topological space, $\overset{\wedge\!\!\!\wedge}{\delta}$ a strong proximity on X, $\mathscr{M}$ a Voronoï mesh on X, $X = \bigcup_{A \in \mathscr{M}} A$. Strongly near polygons $A, B \in \mathscr{M}$ are $\overset{\wedge\!\!\!\wedge}{\delta}$-connnected.*

Proof Let $A, B \in \mathscr{M}$ be strongly near polygons. Consequently, A and B have more than one point in common. Hence, A and B are $\overset{\wedge\!\!\!\wedge}{\delta}$-connnected, i.e., $A \overset{\wedge\!\!\!\wedge}{conn} B$. □

Conjecture 10.19 *In a Voronoï diagram that covers a subset X in Euclidean space E^d, the subset X with $d+1$ subsets is connected, if and only if each pair of subsets in X has nonempty intersection.* ■

Example 10.20 **Strongly connected set of Voronoï polygons.**
In Fig. 10.11, let

$$\mathrm{conn}X = \{A_1, A_2, A_3\}, \text{ such that}$$

$$\mathrm{conn}X = \bigcup_{i=1}^{3} A_i, \text{ and}$$

$$A_1 \overset{\wedge\!\!\!\wedge}{\delta} A_2, A_1 \overset{\wedge\!\!\!\wedge}{\delta} A_3, A_2 \overset{\wedge\!\!\!\wedge}{\delta} A_3.$$ ■

Conjecture 10.21 *In a Voronoï diagram that covers a subset X in Euclidean space E^d, X with $d+1$ subsets is strongly connected, if and only if each pair of adjacent subsets in $A, B \subset X$ is strongly connected.* ■

Problem 10.22 Try finding a counterexample that contradicts Conjecture 10.19. ■

Fig. 10.11 $\text{conn}X = \{A_1, A_2, A_3\}$

Problem 10.23 ☕ Try finding a counterexample that contradicts Conjecture 10.21. ■

References

1. Peters, J., Guadagni, C.: Strongly hit and far miss hypertopology & hit and strongly far miss hypertopology. arXiv[Math.GN] **1503**(02587), 1–8 (2015)
2. Peters, J., Guadagni, C.: Strongly proximal continuity & strong connectedness. arXiv **1504**(02740), 1–11 (2015)
3. Naimpally, S., Peters, J.: Topology with Applications. Topological Spaces via Near and Far. World Scientific, Singapore (2013). Xv + 277 pp., Am. Math. Soc. MR3075111
4. Willard, S.: General Topology. Dover Pub. Inc, Mineola (1970). Xii + 369 pp, ISBN: 0-486-43479-6 54-02, MR0264581

Chapter 11
Helly's Theorem and Strongly Proximal Helly Theorem

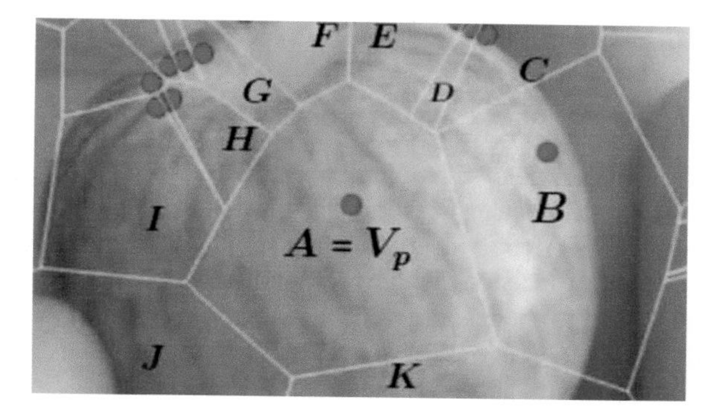

Fig. 11.1 Helly family of convex bodies in an apple mesh nerve

This chapter introduces a strongly proximal version of Helly's theorem for convex sets and convex bodies. The collection of labelled Voronoï regions in the mesh nerve on the apple surface in Fig. 11.1 is an example of a Helly family of convex bodies. Recall that a nonempty set in an n-dimensional Euclidean space is a *convex set*, provided every straight line segment between any two points in the set is also contained in the set. For example, let A be a nonempty set of points in the Euclidean plane. The set A is a *convex set*, provided

$$(1-\lambda)x + \lambda y \in A, \text{ for all } x, y \in A, 0 \leq \lambda \leq 1 \text{ (\textbf{Convexity property})}.$$

A nonempty set A is **strictly convex**, provided A is closed and

$$(1-\lambda)x + \lambda y \in A, \text{ for all } x, y \in A, x \neq y, 0 < \lambda < 1 \text{ (\textbf{Strict Convexity property})}.$$

J.F. Peters, *Computational Proximity*, Intelligent Systems Reference Library 102, DOI 10.1007/978-3-319-30262-1_11

Example 11.1 **Strictly Convex set.**
In Fig. 11.1, the Voronoï region $A = V_p$ is a strictly convex set, since A is closed and it has the Strict Convexity property. ■

Theorem 11.2 *Every Voronoï region is a strictly convex set.*

Proof Immediate from the definition of a strictly convex set. □

To reach the notion of a convex body, we need two other structures besides the convexity of nonempty sets, namely, cover of a set and compactness of set. A **cover** of a nonempty set A in a space X is a collection of nonempty subsets in X whose union is A. Put another way, let $\mathscr{C}(A)$ be a cover of A. Then

$$\mathscr{C}(A) = \bigcup_{B \subset X} B \text{ (Cover of the nonempty set } A).$$

Also recall that a nonempty subset A in a topological space is a **compact set**, provided every open cover of A has a finite subcover. A **convex body** is a compact convex set [1, Sect. 3.1, p.41].

The earliest appearance of convexity appears in definitions of a convex surface in about 250 B.C. by Archimedes in his Sphere and Cylinder. On Paraboloids, Hyperboloids and Ellipsoids [2]. Archimedes' definitions of a convex surface are given by P.M. Gruber [1, Sect. 3.1]. A good introduction to Archimedes in given by T.L. Heath in [2]. For a complete introduction to convex sets, see P. Mani-Levitska [3]. And for an introduction to convex geometry, see P. M. Gruber and J.M. Wills [4]. For an introduction to convex groupoids, see J.F. Peters, M.A. Öztürk and M. Uçkun [5]. For a simple algebraic approximation of convex bodies, see W.J. Firey [6]. Firey gives a proof of P.C. Hammer's theorem on the approximation of convex surfaces by algebraic surfaces [7]. For a study of the maximum number of mutually overlapping translates of a convex body (i.e., translative kissing number of convex bodies), see I. Talata [8]. ■

Example 11.3 **Compact Picture Sets.**
The interior of every nonempty set of picture points A in the Euclidean plane is compact, since int$A$ has a covering $\mathscr{C}(\text{int}A) = \bigcup_{X \in 2^{IntA}} X$. And $\mathscr{C}$(intA) always contains a finite collection that also covers int$A$, i.e., a finite subcover that is an open cover of int$A$. ■

Example 11.4 **Compact Region in a Voronoï Tessellation on a Picture.**
Let S be a set of sites (generating points) in a picture A in the Euclidean plane, $\mathscr{V}(S)$ the Voronoï diagram on A, Voronoï region $V \in \mathscr{V}(S)$. The interior of a Voronoï

region is an open set. Let $V \in \mathscr{V}(S)$ be a Voronoï region. int$V$ is compact, since int$V$ has a covering $\mathscr{C}(\text{int}V) = \bigcup_{X \in 2^{\text{int}V}} X$, i.e., $V \subseteq \mathscr{C}$. And $\mathscr{C}$ always contains a finite collection that also covers int$V$. For example, let $S' \subset V$ be a set of sites in the Voronoï diagram $D = \mathscr{V}'(S')$ and let $\mathscr{C}(D) = \bigcup_{X \in 2^D} X$. Then $V \subseteq \mathscr{C}(D) \subseteq \mathscr{C}$. ■

A compact convex set is a **convex body** [1, Sect. 3.1, p. 41]. A **proper convex body** has a nonempty interior. Otherwise, a convex body is *improper*.

For properties of convex bodies, see Q. Guo and S. Kaijser [9].

Let A be a convex body in the Euclidean plane. Let $M_A(x)$ be a measure of symmetry of A for x in the plane and let $\|x\|$ be the Euclidean norm of x. A closed **unit ball** of x (denoted $N_1(x)$) is a neighbourhood of x with radius 1, i.e.,

$$N_1(x) = \{y \in X : \|x - y\| \leq 1\} \text{ (\textbf{Closed unit ball})}.$$

If A is centrally symmetric with center at the origin with A as a unit ball, then Q. Guo and S. Kaijser prove that

$$M_A(x) = max\left\{1, \frac{2\,\|x\|}{\|x\| + 1}\right\}.$$

Theorem 11.5 *A Voronoï region in the Euclidean plane is a proper convex body.*

Proof Let V be a Voronoï region in a Voronoï diagram $\mathscr{V}(S)$ constructed from a set of sites S in the Euclidean plane. The interior of V is nonempty and the collection of subsets $\mathscr{F}$ of V is a cover of V. Choose S' to be a set of sites in V and let $\mathscr{V}'(S')$ be the Voronoï diagram on V, which is a covering of V. By definition, $\mathscr{F}$ contains the collection $\mathscr{V}'(S')$. Consequently, $V \subseteq \mathscr{V}'(S') \subseteq \mathscr{F}$. Hence, V is a proper convex body. □

Theorem 11.6 *A Voronoï diagram in the Euclidean plane is a space of proper convex bodies.*

Problem 11.7 Prove Theorem 11.6. ■

Example 11.8 **Delaunay Triangle in a Triangulation on a Picture.**
Let S be a set of sites (generating points) in a picture A in the Euclidean plane, $\mathscr{D}(S)$ the Delaunay Triangulation on A, Delaunay triangle $D \in \mathscr{D}(S)$. The triangle D is an improper convex body, since the interior of D is empty. The interior of a intcl(D) is an open set. Using a line of reasoning similar to the one in Example 11.4, it can be shown that intcl(D) is an improper convex body. ■

Let A be a nonempty subset in the Euclidean plane E. The intersection of all convex sets in E that contain A is called the **convex hull** of A (denoted by conv A).

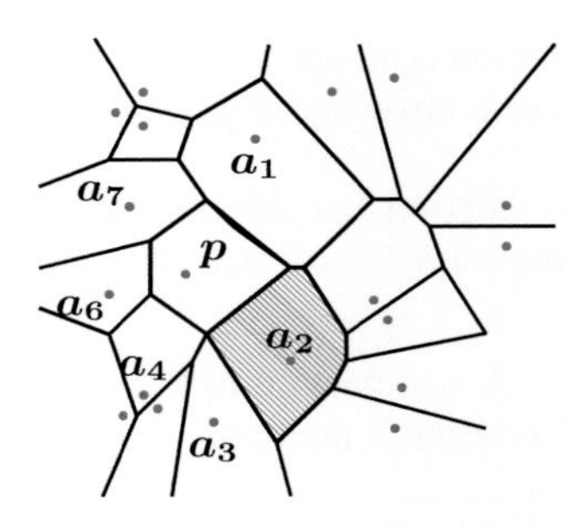

Fig. 11.2 $\mathcal{F} = \{V_p, V_{a_1}, V_{a_2}, V_{a_3}, V_{a_4}, V_{a_6}, V_{a_7}\}$

11.1 Helly's Theorem

There are a number of variations of Helly's theorem. Let $\mathbb{R}^d$ denote the d-dimensional Euclidean space.

Theorem 11.9 Helly's Theorem [1, Sect. 3.2, p. 47] *Let $\mathcal{F}$ be a family of convex bodies in $\mathbb{R}^d$. If any $d+1$ convex bodies have nonempty intersection, then the intersection of all convex bodies in $\mathcal{F}$ is nonempty.*

Proof The proof is limited to the Euclidean plane with $d = 2$.
$\Rightarrow$: Let $A_i \subset E, i \in \{1, 2\}$ be a pair of convex bodies in $\mathcal{F}$ with $\bigcap A_i \neq \emptyset$. Let $B \subset E, B \in \mathcal{F}\ \ \{A_i\}$ and assume $B \cap (\bigcap A_i) \neq \emptyset$. Since B is arbitrary, the conclusion follows. □

H. Edelsbrunner and J.L. Harer [10, Sect. III.2,p.58] point out that convexity is an unnecessary strong requirement in Helly's Theorem. It is not needed in the proof of the theorem. Alternatively, a necessary and sufficient condition for Helly's Theorem is given next.

Theorem 11.10 [10, Sect. III.2, p. 57] *Let $\mathcal{F}$ be a finite collection of closed convex sets in $\mathbb{R}^d$. Every $d+1$ of the convex sets have nonempty intersection, if and only if they all have a nonempty common intersection.*

Proof The proof is limited to the Euclidean plane with $d = 2$.
$\Rightarrow$: See the proof for Theorem 11.9.
$\Leftarrow$: Assume that the intersection of all convex sets in $\mathcal{F}$ is nonempty. Consequently, $\bigcap_{i \in \{1,2\}} A_i \neq \emptyset$. Choose $B \in \mathcal{F}\ \ \{A_i\}$. By assumption, $B \cap \bigcap A_i \neq \emptyset$. Hence, since B is arbitrary, every $d+1$ of the convex sets in $\mathcal{F}$ have nonempty intersection. □

Example 11.11 **Helly Family of Convex Bodies**.
Let the family of $\mathcal{F} = \{V_p, V_{a_1}, V_{a_2}, V_{a_3}, V_{a_4}, V_{a_5}, V_{a_6}, V_{a_7}\}$ be represented in the partial Voronoï diagram in Fig. 11.2. V_p and each $V_{a_i}, i \in \{p, 1, .., 7\}$ is a Voronoï region. From Theorem 11.6, each member of $\mathcal{F}$ is a convex body. Starting with $V_p \cap V_{a_2} \neq \emptyset$, each V_{a_i} has nonempty intersection with $V_p \cap V_{a_2}$. Hence, Helly's theorem holds true for the family of convex bodies in $\mathcal{F}$. ■

Problem 11.12 Prove Theorem 11.10 for $d = n$ in an n-dimensional Euclidean space $\mathbb{R}^n$. ■

Problem 11.13 Give an example of a family of proper convex bodies in the 3-dimensional Euclidean space $\mathbb{R}^3$ that satisfies Theorem 11.10, for $d = 3$. ■

Problem 11.14 Give an example of a family of improper convex bodies in the 2-dimensional Euclidean space $\mathbb{R}^2$. ■

Problem 11.15 Give an example of a family of improper convex bodies in the 3-dimensional Euclidean space $\mathbb{R}^3$. ■

For a family $\mathcal{F} \in 2^{2^X}$ of convex sets in a proximity space (X, δ), we have another version of Helly's theorem.

Theorem 11.16 *Let $\mathcal{F}$ be a finite collection of closed convex sets in $\mathbb{R}^d$. Every $d + 1$ of the convex sets are near, if and only if they all are near.*

Proof The proof is limited to the Euclidean plane with $d = 2$.
$\Rightarrow$: Let $A, B \subset E$ be a pair of convex bodies in $\mathcal{F}$ with $A\ \delta\ B$. Let $H \in \mathcal{F}\quad \{A, B\}$ and assume $H\ \delta\ A$ and $H\ \delta\ B$. Since B is arbitrary, the conclusion follows.
$\Leftarrow$: Assume that all convex sets in $\mathcal{F}$ are near. Consequently, $A\ \delta\ B$ for $A, B \subset E$. Choose $H \in \mathcal{F}\quad \{A, B\}$. By assumption, $H\ \delta\ A$ and $H\ \delta\ B$. Hence, since B is arbitrary, every $d + 1$ of the convex sets in $\mathcal{F}$ have nonempty intersection. □

Problem 11.17 Prove Theorem 11.16 for $d = 2$ in the 2-dimensional Euclidean space $\mathbb{R}^2$ and for $\mathcal{F}$ in the Lodato proximity space (X, δ_{LO}), i.e., $\mathcal{F}$ is a collection of convex sets in the power set of X and δ_{LO} is the Lodato proximity. ■

Problem 11.18 Prove Theorem 11.16 for $d = 2$ in the 2-dimensional Euclidean space $\mathbb{R}^2$ and for $\mathcal{F}$ in the Wallman proximity space (X, δ_0), i.e., $\mathcal{F}$ is a collection of convex sets in the power set of X and δ_0 is the Wallman proximity. ■

Problem 11.19 Prove Theorem 11.16 for $d = 2$ in the 2-dimensional Euclidean space $\mathbb{R}^2$ and for $\mathcal{F}$ in the Descriptive Lodato proximity space (X, δ_Φ), i.e., $\mathcal{F}$ is a collection of convex sets in the power set of X and δ_Φ is the Descriptive Lodato proximity. Give an example of a family $\mathcal{F}$ that satisfies the descriptive Lodato form of Theorem 11.16. ■

Problem 11.20 Prove Theorem 11.16 for $d = 2$ in the 2-dimensional Euclidean space $\mathbb{R}^2$ and for $\mathcal{F}$ in the Descriptive Wallman proximity space (X, δ_Φ), i.e., $\mathcal{F}$ is a collection of convex sets in the power set of X and δ_Φ is the Descriptive Wallman proximity. Give an example of a family $\mathcal{F}$ that satisfies the descriptive Wallman form of Theorem 11.16. ■

11.2 Strongly Near Version of Helly's Theorem

The extension of Theorem 11.16 is a natural outcome of considering the nonempty intersection of the interiors of convex bodies with a particular convex body.

Theorem 11.21 *Let $\mathcal{F}$ be a finite collection of closed convex sets in $\mathbb{R}^d$. Every $d+1$ of the convex sets are strongly near, if and only if they all are strongly near.*

Proof Symmetric with the proof of Theorem 11.16 for $d = 2$. □

Next, Theorem 11.21 is extended to the case where members in a family of convex sets are strongly near a particular nonempty set.

Theorem 11.22 *Let $\mathcal{F}$ be a finite collection of closed convex sets in $\mathbb{R}^d$. Every $d+1$ of the convex sets are strongly near a nonempty set A, if and only if they all are strongly near A.*

Example 11.23 **Helly Family of Convex Bodies in a Mesh Nerve.**
Let the family of $\mathcal{F}$ in a mesh nerve in a Voronoï diagram $\mathscr{V}(S)$ and let $\mathcal{F}$ be a collection of convex sets in the powerset of X endowed with the strong proximity δ, $S \subset X$. From Theorem 11.6, each member of $\mathcal{F}$ is a convex body. Let $V_p \in \mathscr{V}(S)$.

$$\mathcal{F} := NCL_{V_p}\mathscr{M}. \text{ (convex sets in a mesh nerve)}$$

Recall that a mesh nerve $NCL_{V_p}\mathscr{M}$ is defined by

$$NCL_{V_p}\mathscr{M} = \left\{ X \in \mathscr{M} : \text{cl}V_p \ \delta \ X \right\}. \text{ (Mesh nerve)}$$

Assume that the cardinality of $NCL_{V_p}\mathscr{M} \geq 3$. Then, from the definition of a mesh nerve, every pair of convex bodies A, B in a mesh nerve are strongly near V_p, i.e., $A \ \delta \ V_p, B \ \delta \ V_p$. Let $H \in \mathcal{F} \quad \{A, B\}$. By definition, $H \ \delta \ V_p$. Consequently, $d+1$ convex bodies in a mesh nerve are strongly near V_p, if and only if all of the convex bodies in $\mathcal{F}$ are strongly near V_p. Hence, Theorem 11.22 holds true for the family of convex bodies in $\mathcal{F}$. ■

Example 11.24 **Helly Family of Convex Bodies in a Mesh Nerve on a Apple Surface.**
Let the family of $\mathcal{F}$ in a mesh nerve in a Voronoï diagram $\mathscr{V}(S)$ and let $\mathcal{F}$ be a collection of convex sets in the powerset of X endowed with the strong proximity δ, $S \subset X$. Let the mesh nerve $NCL_{V_p}\mathscr{M}$ be represented in Fig. 11.1. From Theorem 11.6, each member of $\mathcal{F}$ is a convex body. Let $V_p \in \mathscr{V}(S)$.

$$NCL_{V_p}\mathscr{M} = \{B, C, D, E, F, G, H, I, J, K\}. \text{ (Apple surface mesh nerve)}$$

and let

$$\mathcal{F} := NCL_{V_p}\mathscr{M}.$$

Clearly, the convex bodies in $\mathcal{F}$ satisfy Theorem 11.22. ■

Problem 11.25 Prove Theorem 11.21 for $d = n$ in an n-dimensional Euclidean space $\mathbb{R}^n$. ■

Problem 11.26 Give an example of a family of proper convex bodies in the 3-dimensional Euclidean space $\mathbb{R}^2$ that satisfies Theorem 11.21, for $d = 2$. ■

Problem 11.27 Give an example of a family of proper convex bodies in the 3-dimensional Euclidean space $\mathbb{R}^3$ that satisfies Theorem 11.21, for $d = 3$. ■

Problem 11.28 Prove Theorem 11.22 for $d = n$ in an n-dimensional Euclidean space $\mathbb{R}^n$. ■

Problem 11.29 Give an example of a family of proper convex bodies in the 3-dimensional Euclidean space $\mathbb{R}^2$ that satisfies Theorem 11.22, for $d = 2$. ■

Problem 11.30 Give an example of a family of proper convex bodies in the 3-dimensional Euclidean space $\mathbb{R}^3$ that satisfies Theorem 11.22, for $d = 3$. ■

11.3 Descriptive Helly's Theorem

Let $E^d = \mathbb{R}^d$ denote the d-dimensional Euclidean space. Let $\Phi(A), \Phi(B)$ denote the descriptions of the points in $A, B \subset E$, respectively. Recall that $A \underset{\Phi}{\cap} B \neq \varnothing$ is the descriptive intersection of A and B, i.e., $\Phi(A) \cap \Phi(B) \neq \varnothing$ (A and B have points in common with the same description).

Theorem 11.31 Descriptive Helly's Theorem *Let $\mathcal{F}$ be a family of convex bodies in E^d with $\mathcal{F}$ in the descriptive proximity space (X, δ_Φ). If any $d+1$ convex bodies in $\mathcal{F}$ have nonempty descriptive intersection $\underset{\Phi}{\cap}$, then the descriptive intersection of all convex bodies in $\mathcal{F}$ is nonempty.*

Proof The proof is limited to the Euclidean plane with $d = 2$.
$\Rightarrow$: Let A, B be a pair of convex bodies in $\mathcal{F}$ with $A \underset{\Phi}{\cap} B \neq \varnothing$. Let $H \subset E, H \in \mathcal{F}\ \ \{A, B\}$ and assume $H \underset{\Phi}{\cap} (A \underset{\Phi}{\cap} B) \neq \varnothing$. Since H is arbitrary, the conclusion follows. □

Theorem 11.32 *Let $\mathcal{F}$ be a finite collection of closed convex sets in $E^d = \mathbb{R}^d$ with $\mathcal{F}$ in the descriptive proximity space (X, δ_Φ). Every $d+1$ of the convex sets have nonempty descriptive intersection $\underset{\Phi}{\cap}$, if and only if they all have a nonempty common descriptive intersection.*

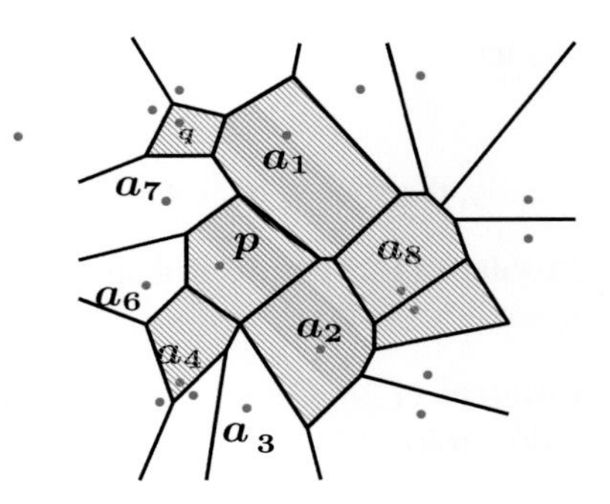

Fig. 11.3 $\mathcal{F} = \{V_p, V_q, V_{a_1}, V_{a_2}, V_{a_4}, V_{a_8}\}$

Proof The proof is limited to the Euclidean plane with $d = 2$.
$\Rightarrow$: See the proof for Theorem 11.31.
$\Leftarrow$: Assume that the descriptive intersection of all convex sets in $\mathcal{F}$ is nonempty. Let $A, B \subset E$. Consequently, $A \underset{\Phi}{\cap} B \neq \emptyset$. Choose $H \in \mathcal{F} \quad \{A, B\}$. By assumption, $H \underset{\Phi}{\cap} (A \underset{\Phi}{\cap} B) \neq \emptyset$, since all convex sets in $\mathcal{F}$ have nonempty descriptive intersection. Hence, since H is arbitrary, every $d + 1$ of the convex sets in $\mathcal{F}$ have nonempty descriptive intersection. □

Example 11.33 **Descriptive Helly's Theorem Instance**.
Let $\mathcal{F}$ be a finite collection of Voronoï regions in $E^d = \mathbb{R}^d$ with $\mathcal{F}$ in the descriptive proximity space (X, δ_Φ). From Theorem 11.5, each $V_p \in \mathcal{F}$ is a convex body. The space X is a nonempty set of picture points represented in Fig. 11.3 with red and green colours and let $\Phi : X \to \mathbb{R}$ be a description on X representing the colour of a picture point, where 0 stands for red (r), 1 for green (g) and 2 for blue (b). Suppose the range is endowed with the topology given by $\tau = \{\emptyset, \{r\}, \{g\}, \{r, g\}\}$. By choosing this topology, the focus is on red and green. In Fig. 11.3, let $\mathcal{F}$ be a family of Voronoï regions in E^d be represented by

$$\mathcal{F} = \left\{V_p, V_q, V_{a_1}, V_{a_2}, V_{a_4}, V_{a_8}\right\}.$$

Observe that $\Phi(V_p) = \Phi(V_{a_2}) = \{g\}$. Consequently, $V_p \; \delta_\Phi \; V_{a_2}$, i.e., $V_p \underset{\Phi}{\cap} V_{a_2} \neq \emptyset$. Likewise, $\Phi(V_p) = \Phi(V_q) = \{g\}$. Hence, $V_q \underset{\Phi}{\cap} \left(V_p \underset{\Phi}{\cap} V_{a_2}\right) \neq \emptyset$. This is true for all Voronoï regions in $\mathcal{F}$. For this reason, $\mathcal{F}$ illustrates Theorem 11.32. ■

Theorem 11.34 *Let $\mathcal{F}$ be a finite collection of Voronoï regions in $E^d = \mathbb{R}^d$ with $\mathcal{F}$ in the descriptive proximity space (X, δ_Φ). Every $d + 1$ of the convex sets are descriptively strongly near, if and only if they all are descriptively strongly near.*

Problem 11.35 Prove Theorem 11.34 for $d = 2$. ■

Next, Theorem 11.34 is extended to the case where members in a family of convex sets are descriptively strongly near a particular nonempty set. In this case, Theorem 11.34 is extended to mesh nerves in Theorem 11.36.

Theorem 11.36 *Let $\mathcal{F}$ be a finite collection of Voronoï regions in $E^d = \mathbb{R}^d$ with $\mathcal{F}$ in the descriptive proximity space (X, δ_Φ). Every $d+1$ of the convex sets are descriptively strongly near a nonempty set A, if and only if they all are descriptively strongly near A.*

Problem 11.37 Prove Theorem 11.36 for $d = 2$. ∎

Example 11.38 **Descriptive Helly's Theorem Applied to Mesh Nerves.**
Let $\mathcal{F}$ be a finite collection of Voronoï regions Example 11.33 and let $V_p \in \mathcal{F}$ as shown in Fig. 11.3. The mesh nerve $NCL_{V_p}\mathscr{F}$ is defined by

$$\mathcal{F} := NCL_{V_p}\mathscr{F} = \left\{X \in \mathscr{F} : \mathrm{cl}V_p\ \delta_\Phi\ X\right\}. \text{ (Mesh nerve)}$$

That is, we identify the collection of convex bodies $\mathcal{F}$ with a mesh nerve in the δ_Φ space represented by Fig. 11.3.

From the definition of the mesh nerve $NCL_{V_p}\mathscr{F}$, every pair of convex bodies A, B in the mesh nerve are descriptively strongly near V_p, i.e., $A\ \delta_\Phi\ V_p, B\ \delta_\Phi\ V_p$. Let $H \in \mathcal{F}\ \ \{A, B\}$. By definition, $H\ \delta_\Phi\ V_p$. Consequently, $d+1$ convex bodies in a mesh nerve are δ_Φ near V_p, if and only if all of the convex bodies in $\mathcal{F}$ are δ_Φ near V_p. Hence, Theorem 11.36 holds true for the family of convex bodies in $\mathcal{F}$. ∎

11.4 Polyforms

Polyforms in art were introduced by D.G. Markovic [11]. A *polyform* is a collection of finite, edge-connected set of cells on a square, triangular or hexagonal grid of any number of dimensions [12]. In the plane, a *polyform* is a collection of connected copies of a base shape such as a square. When a polyform contains a collection of connected copies of squares of equal size, it called a **polyomino**.

In two dimensions, a cell shape is a square (⟶ **polymino**), equilateral triangle (⟶ **polydiamond**, (see, e.g., Fig. 11.4) also called a **polyiamond** [13]) or a regular hexagon (⟶ polyhex). Examples of the base shapes are isosceles right triangle (**polyabolo**), cubes (**polycube**), lines from a square grid (**polystick**), and rhombus (**polyrhomb**). In general, a **polyform** is a plane figure or solid compound containing multiple copies of a given base shape.[1] In this section, only polyiamond and polyhex are considered in the Euclidean plane.

[1]E.W. Weisstein, polyform, Mathworld–A Wolfram web resourcem, http://mathworld.wolfram.com/Polyform.html.

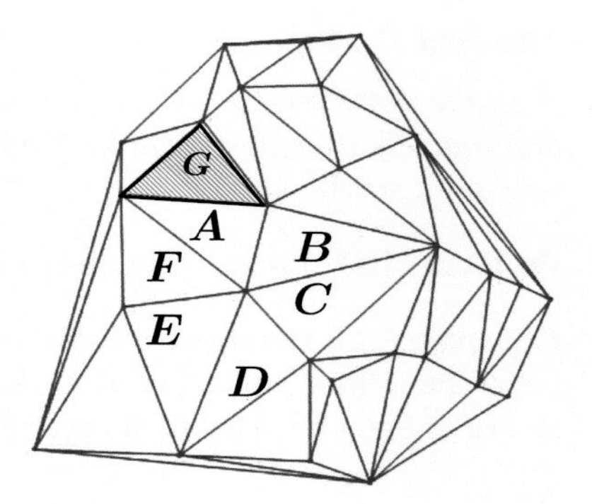

Fig. 11.4 3D Delaunay triangulation

Recall that a **polyiamond** is a collection of equal-sized equilateral triangles arranged with coincident sides.[2]

Theorem 11.39 *Let $\mathscr{D}(S)$ be a set of mesh sites that are corners of n squares of equal size with coincident sides and arranged in a 2-dimensional grid. A Delaunay triangulation of the sites in $\mathscr{D}(S)$ is a polyiamond.*

Proof Immediate from the definition of a polyiamond. □

Theorem 11.40 *If Voronoï diagram $\mathscr{V}(S)$ on a set of sites on a plane surface contains a repetition of a shape, then $\mathscr{V}(S)$ contains a polyform.*

Proof Immediate from the definition of a polyform. □

Example 11.41 **Delaunay Triangulation of a Rectangular Grid.**
From Theorem 11.39, a Delaunay triangulation on the corners in Fig. 11.5 is a polyiamond. ■

Example 11.42 **Mesh Nerve in a Delaunay Polydiamond.**
Consider Example 11.33 and let $V_p \in \mathcal{F}$ as shown in Fig. 11.3. The mesh nerve $NCL_{V_p}\mathscr{F}$ is defined by

$$\mathcal{F} := NCL_{V_p}\mathscr{F} = \left\{X \in \mathscr{F} : \text{cl}V_p\ \delta_\Phi\ X\right\}. \text{ (Mesh nerve)}$$

That is, we identify the collection of convex bodies $\mathcal{F}$ with a mesh nerve in the δ_Φ space represented by Fig. 11.3.

From the definition of the mesh nerve $NCL_{V_p}\mathscr{F}$, every pair of convex bodies A, B in the mesh nerve are descriptively strongly near V_p, i.e., $A\ \delta_\Phi\ V_p, B\ \delta_\Phi\ V_p$.

[2]For much more about this, see http://mathworld.wolfram.com/Polyiamond.html.

Let $H \in \mathcal{F} \quad \{A, B\}$. By definition, $H \ \delta_{\Phi} \ V_p$. Consequently, $d + 1$ convex bodies in a mesh nerve are δ_{Φ} near V_p, if and only if all of the convex bodies in $\mathcal{F}$ are δ_{Φ} near V_p. Hence, Theorem 11.36 holds true for the family of convex bodies in $\mathcal{F}$. ■

Theorem 11.43 *Let $\mathscr{P}$ be a collection of triangles in a polydiamond nerve in $\mathbb{R}^2$. Every 3 triangles surrounding a mesh nerve center are strongly near, if and only if all of the nerve triangles are near.*

Proof All triangles strongly near the central nerve triangle have a vertex in common. Hence, all such nerve triangles are near each other. □

Problem 11.44 Prove or disprove that the 3D Delaunay triangulation in Fig. 11.4 is a 3D polyiamond. ■

Problem 11.45 For polyform nerves that are collections of square-shaped cells, state and prove a theorem analogous to Theorem 11.43. ■

Problem 11.46 For polyform nerves that are collections of hexagons, state and prove a theorem analogous to Theorem 11.43. ■

Fig. 11.5 8×8 grid

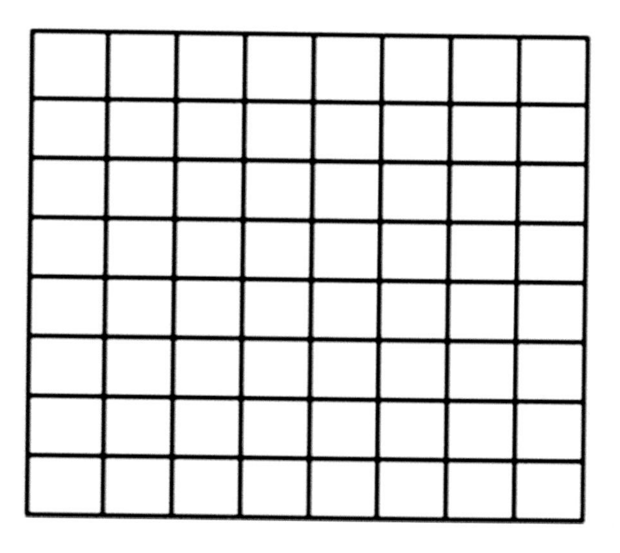

References

1. Gruber, P.: Convex and Discrete Geometry, Grundlehren der Mathematischen Wissenschaften, vol. 336, Xiv + 578 pp. Springer, Berlin (2007). MCS2000 52XX,11HXX, ISBN: 978-3-540-71132-2, MR2335496
2. Archimedes: Sphere and cylinder. On paraboloids, hyperboloids and ellipsoids, trans. and annot. by A. Czwalina-Allenstein. Cambridge University Press, UK (1897). Reprint of 1922, 1923, Geest & Portig, Leipzig 1987, The Works of Archimedes, Ed. by T.L. Heath, Cambridge University Press, Cambridge, 1897
3. Mani-Levitska, P.: Characterizations of convex sets. In: Gruber, P.M., Wills, J.M. (eds.) Handbook of Convex Geometry, vol. A, B, pp. 19–41. North-Holland, Amsterdam (1993). MR1242975
4. Gruber, P.M., J.M. Wills, E.: Handbook of Convex Geometry. North-Holland, Amsterdam (1993), vol. A: lxvi + 735 pp.; vol. B: ilxvi and 7371438 p. ISBN: 0-444-89598-1, MR1242973
5. Peters, J., Öztürk, M., Uçkun, M.: Klee-Phelps convex groupoids **0934**, 1–5 (2014). arXiv:1411.0934, Mathematica Slovaca (2015, accepted)
6. Firey, W.: Approximating convex bodies by algebraic ones. Archiv der Mathematik **25**(1), 424–425 (1974). MR0353146
7. Hammer, P.: Approximation of convex surfaces by algebraic surfaces. Mathematika **10**, 6471 (1963). MR0154184
8. Talata, I.: On extensive subsets of convex bodies. Periodica Mathematica Hungarica **38**(3), 231–246 (1999). MR1756241
9. Guo, Q., Kaijser, S.: On asymmetry of some convex bodies. Discrete Comput. Geom. **27**(2), 239–247 (2002). MR1880940
10. Edelsbrunner, H., Harer, J.: Computational Topology. An Introduction. American Mathematical Society, Providence (2010). Xii + 110 pp. MR2572029
11. Markovic, D.: Geometric Polyforms. 3M Makarije, Podgorica (2006)
12. Yang, W.: Adding layers to bumped-body polyforms with minimum perimeter preserves minimum perimeter. Electron. J. Comb. **13**(r6), 1–11 (2006)
13. Rhodes, G.: Planar tilings by polyominoes, polyhexes, and polyiamonds. J. Comput. Appl. Math. **174**(2), 329–353 (2005). MR 2005h:05042

Chapter 12
Nerves and Strongly Near Nerves

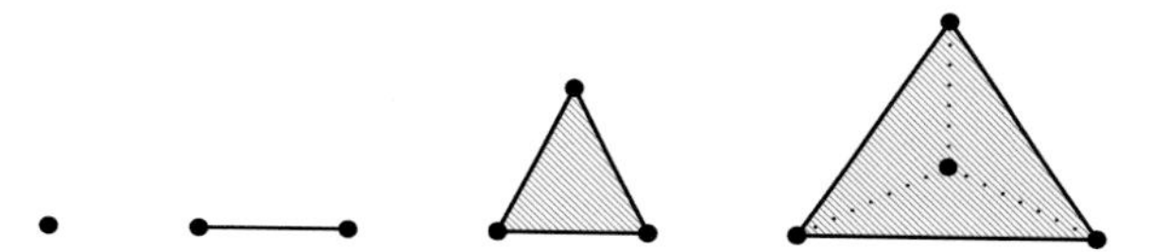

Fig. 12.1 0-, 1-, 2-, 3-simplical complexes

This chapter introduces various forms of geometric nerves, which are usually collections of what known as simplical complexes in a normed linear space. This chapter also introduces geometric nerves called polyform nerves that vary from the usual view of simplical complexes. In general, a nerve is a simplical complex [1].

A d**-simplex** (or **d-simplical complex**) has $v_0, v_1, \ldots, v_d$ vertices and $d - 1$ faces in $\mathbb{R}^d$. A d-simplex is the convex hull of $d + 1$ **linearly independent** points (vectors). Let $X \subset \mathbb{R}^d$. Recall that a **convex hull** of X (denoted conv X) is defined by

$$\text{conv } X = \bigcap \{C : C \text{ is a convex set that contains } X\}.$$

A set of vectors $v_0, v_1, \ldots, v_d$ in a d-dimensional space are linearly independent over a real field F, if and only if, for all scalars $\alpha_i \in F$,

$$\alpha_0 v_0 + \alpha_1 v_1, \ldots, \alpha_d v_d = 0 \text{ implies } \alpha_0 = \alpha_1 = \cdots = \alpha_d = 0.$$

0-simplex (point), 1-simplex (line segment), 2-simplex (solid triangle) and 3-simplex (3-sided polyhedron) are represented in Fig. 12.1. For more about simplicial complexes, see [2, Sect. 5.6, p.181ff] (see, also, [3, Sect. III.1], R. Street [4, 5]).

Here, the focus is on 2-simplices that are solid triangles in $\mathbb{R}^2$ (Euclidean plane). Let X be the set of face points plus the vertices in a 2-simplex, which equals the convex hull conv$X$ of the linearly independent points in X. The important thing to notice is that a 2-simplex is a solid triangle that has $d + 1 = 3$ vertices of a solid,

J.F. Peters, *Computational Proximity*, Intelligent Systems
Reference Library 102, DOI 10.1007/978-3-319-30262-1_12

triangular shaped face with no holes. A 2-simplex is the basic building block in what is known as a polyform nerve. A **polyform nerve** is a collection of the same shapes that have points in common with a single shape called the nerve center.

12.1 Polyform Nerves

Let $\blacktriangle(pqr)$ be a **solid triangle** (3 vertices on a triangular-shaped face) in the Euclidean plane E. A **face** is a 2D flat surface of a 3D solid. A **2-simplical complex** (briefly, **2sim**) is $\blacktriangle$ (pqr) is defined by

$$\blacktriangle(pqr) = \{x \in E : \|p - x\| \leq \|s - r\| \, \forall s \in \overline{pq} \setminus p\}.$$

Let $\mathscr{P}$ be a collection of 2-simplices (solid triangles with $2 + 1$ vertices and 2-1 face) in a poly2sim W_{pqr}. Let W_{pqr} denote a wedge with vertices p, q, r. Let $W' \subset \mathscr{P}$ be a poly2sim such that W' has an edge in common with W_{pqr}. Let $\mathscr{W}$ be a collection of 2-simplices in a poly2sim. A **poly2sim** nerve $NCL_{V_p}\mathscr{M}$ in the Euclidean plane is a convex set that is defined by

$$NCL_{W_{pqr}}\mathscr{W} = \left\{W' \in \mathscr{P} : \mathrm{cl}W_{pqr} \ \overset{\wedge\!\!\wedge}{\delta} \ W'\right\}. \text{ (Poly2sim nerve)}$$

In other words, a **poly2sim nerve** in the Euclidean plane is a collection of solid triangles that are strongly near a particular solid triangle.

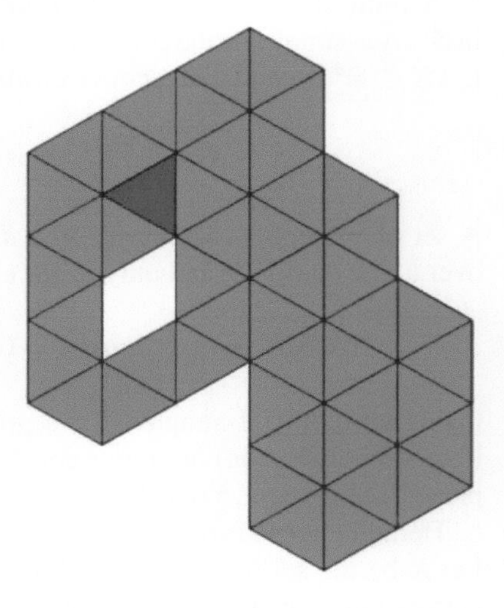

Fig. 12.2 Multiple polydiamond nerves

Fig. 12.3 Poly2sim nerve

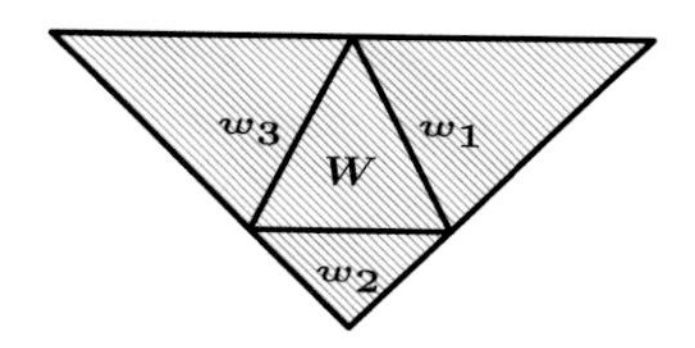

Example 12.1 The red ▲ in Fig. 12.2 is the center of a poly2sim nerve.[1] ■

Example 12.2 Let the poly2sim nerve $NCL_{W_{pqr}}\mathscr{W}$ be represented by the collection of $\overset{\wedge\wedge}{\delta}$-2sim in Fig. 12.3. In $NCL_{W_{pqr}}\mathscr{W}$, $w_i \overset{\wedge\wedge}{\delta} W, i \in \{1,2,3\}$. $NCL_{W_{pqr}}\mathscr{W}$ is defined by

$$NCL_{W_{pqr}}\mathscr{W} = W_{pqr} \cup \bigcup_{i\in\{1,2,3\}} w_i,\ w_i \overset{\wedge\wedge}{\delta} W_{pqr}.$$

Observe that there is no 2sim $A \in \mathscr{W}$ that has only a vertex in common with W_{pqr}. Hence, $NCL_{W_{pqr}}\mathscr{W}$ is itself a 2sim. For this reason, $NCL_{W_{pqr}}\mathscr{W}$ is a convex set (see Theorem 12.6). ■

Lemma 12.3 *Every 2-simplex is a convex set.*

Proof Immediate from the definition of a 2-simplex. □

A quadrilateral-shaped poly2sim nerve contains a pair of $\overset{\wedge\wedge}{\delta}$ 2-simplexes, i.e., the 2-simplexes in the nerve have a common edge.

Lemma 12.4 *A quadrilateral-shaped poly2sim nerve is a convex set.*

Proof Let $C :=$ ▲ (pqr) be the poly2sim nerve center and let $T :=$ ▲ (pqt) so that $C \overset{\wedge\wedge}{\delta} T$, i.e., the base of the non-center 2-simplex in the nerve is one of the edges of the nerve center. Let $\overline{pq}$ be the edge shared by the 2-simplices. Then the nerve is a quadrilateral $C \cup T$, which is convex, since a straight edge connected between any point on the edge $\overline{pq}$ to any other point in $C \cup T$ lies inside the nerve. □

Lemma 12.5 *A poly2sim nerve with more than one 2-simplex is a convex set, if only if the nerve has either a quadrilateral shape or a triangular shape with one of the edges of each triangle parallel with the nerve center.*

Proof Let $NCL_{W_{pqr}}\mathscr{W}$ be a poly2sim nerve with center 2-simplex W_{pqr} and let A be the only 2-simplex $NCL_{W_{pqr}}\mathscr{W}$. The 2-simplexes W_{pqr}, A have a common edge, forming a quadrilateral. From Lemma 12.4, the nerve is convex, if and only if $W_{pqr} \overset{\wedge\wedge}{\delta} A$. Similarly, a nerve containing a pair of edge-connected 2-simplexes is a quadrilateral that is convex, if and only if each of the 2-simplices are $\overset{\wedge\wedge}{\delta}$ the nerve center, forming a triangular shape. For the rest of the proof, see Problem 12.8. □

---

[1] For many more examples of polydiamond nerves, see http://demonstrations.wolfram.com/ChainsOfRegularPolygonsAndPolyhedra/.

Theorem 12.6 *A maximal poly2sim nerve in the Euclidean plane has 3 poly2sims $\overset{\wedge\!\!\wedge}{\delta}$ the nerve center, if and only if the nerve has a triangular shape.*

Proof Let $\mathscr{W}$ be a collection of 2-simplexes in a poly2sim. $NCL_{W_{pqr}}\mathscr{W}$ a poly2sim nerve in the Euclidean plane, $A \in NCL_{W_{pqr}}\mathscr{W}$. By definition, $A \overset{\wedge\!\!\wedge}{\delta} W_{pqr}$. This excludes 2-simplexes $B \in \mathscr{W}$ that are near $NCL_{W_{pqr}}\mathscr{W}$ but not $\overset{\wedge\!\!\wedge}{\delta} W_{pqr}$. In other words, a 2-simplex in $NCL_{W_{pqr}}\mathscr{W}$ must have more than one point in common with W_{pqr}. Hence,

$$NCL_{W_{pqr}}\mathscr{W} = \bigcup_{A \in \mathscr{W}} A, A \overset{\wedge\!\!\wedge}{\delta} W_{pqr}.$$

That is, a poly2sim nerve is the union of strongly near 2-simplexes (see, e.g., Fig. 12.3), which excludes 2-simplexes that have only a vertex in common with W_{pqr}. From Lemma 12.3, W_{pqr} is a convex set. From Lemma 12.5, $NCL_{W_{pqr}}\mathscr{W}$ is convex, since all 2-simplexes the poly2sim nerve are $\overset{\wedge\!\!\wedge}{\delta} W_{pqr}$, the nerve center. To complete the proof, see Problem 12.9. □

Next, consider a variation of Helly's theorem.

Theorem 12.7 *Let $\mathscr{N}$ be a polydsim nerve with a d-simplex center in $\mathbb{R}^d$. Every $d+1$ of the non-center d-simplexes are strongly near the $\mathscr{N}$-center, if and only if $\mathscr{N}$ is convex.*

Proof See Problem 12.10. □

Problem 12.8 Complete the proof of Lemma 12.5. That is, prove that the union of $\overset{\wedge\!\!\wedge}{\delta}$-poly2sim nerves is a convex set. **Hint**: If a third 2-simplex has an edge in common with a poly2sim nerve center, the nerve has a triangular shape, where one of the edges of each triangle is parallel with a nerve center edge. ■

Problem 12.9 Complete the proof of Theorem 12.6. That is, prove that the maximal union of $\overset{\wedge\!\!\wedge}{\delta}$-poly2sims has a triangular shape. ■

Problem 12.10 Prove Theorem 12.7. **Hint**: For $d > 2$, prove that every polydsim nerve has a surface in common with the polydsim nerve center and that the surface is parallel with a nerve center surface. ■

In addition to a 2-simplex nerve, many other polyform nerves are possible in the Euclidean plane. Here are some examples.

Polyform Nerves

RegularPolyhex nerve A polyhex nerve with a regular hexagon center (see, for example, the red ⬢ in Fig. 12.4, which is the center of a polyhex nerve).

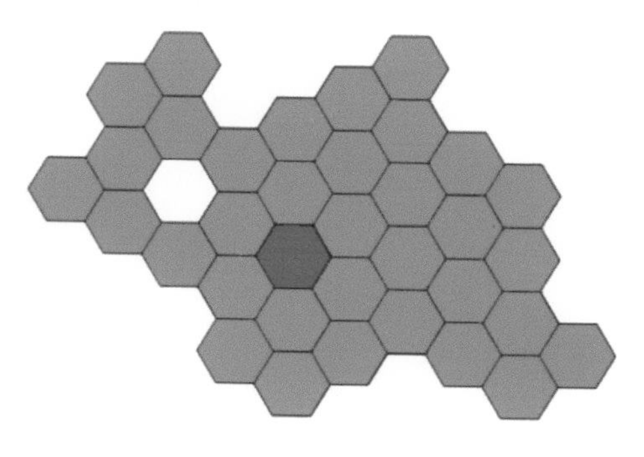

Fig. 12.4 Multiple polyhex nerves

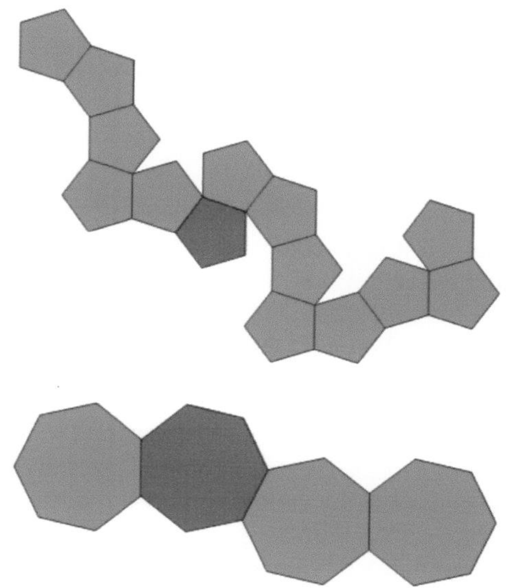

Fig. 12.5 Multiple polypentagon nerves

Fig. 12.6 Multiple polyheptagon nerves

IrregularPolyhex nerve A polyhex nerve containing non-regular hexagons. See Problem 12.11.

RegularPolypentagon nerve A polypentagon nerve with a regular pentagon center (see, for example, the red pentagon in Fig. 12.5, which is the center of a polypentagon nerve).

IrregularPolypentagon nerve A polypentagon nerve containing non-regular pentagons.

RegularPolyhex nerve A polyheptagon nerve with a regular heptagon center (see, for example, the red heptagon in Fig. 12.6, which is the center of a polypentagon nerve containing several heptagons).

IrregularPolyheptagon nerve A polyheptagon nerve containing non-regular heptagons.

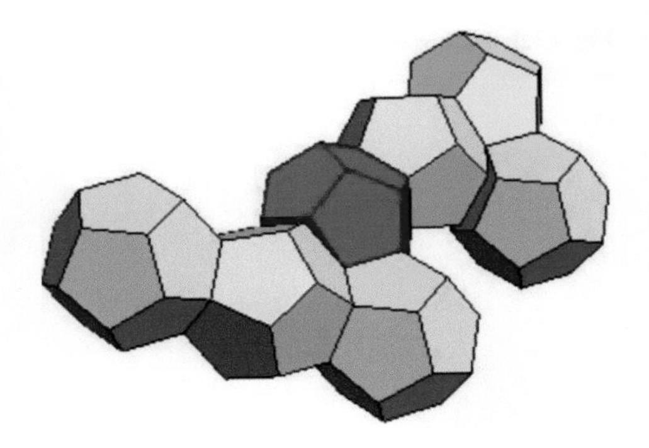

Fig. 12.7 Multiple polydecahedron nerves

Regular Polydodecahedron nerve A polydodecahedron nerve with a regular dodecahedron center (see, for example, the red heptagon in Fig. 12.7, which is the center of a dodecahedron nerve containing several dodecahedron[2]).

Problem 12.11 Write a Mathematica script that finds a polyhex nerve in a Voronoï mesh. Use false-colouring so that each polyhex nerve found has a red central hexagon and the yellow non-central hexagons. Give three examples of sites used to construct the Voronoï mesh. For each type of mesh, choose three digital images and superimpose a Voronoï mesh showing the location of the polyhex nerve on each image. ■

Problem 12.12 Write a Mathematica script that finds multiple polyhex nerves in a Voronoï mesh. Use false-colouring so that each polyhex nerve found has a red central hexagon and the yellow non-central hexagons. Give three examples of sites used to construct the Voronoï mesh. For each type of mesh, choose three digital images and superimpose a Voronoï mesh showing the location of the polyhex nerve on each image. ■

Problem 12.13 Write a Mathematica script that finds edge-connected multiple polyhex nerves in a Voronoï mesh. Use false-colouring so that each polyhex nerve found has a red central hexagon and the yellow non-central hexagons. If there separated, edge-connected multiple polyhex nerves in a Voronoï mesh, use false-colouring to display each of the collections of edge-connected polyhex nerves. Give three examples of sites used to construct the Voronoï mesh. For each type of mesh, choose three digital images and superimpose a Voronoï mesh showing the location of the edge-connected multiple polyhex nerves on each image. ■

A **polydodecahedron nerve** is a collection of dodecahedrons each of which has a surface in common with a dodecahedron center.

---

[2]For other examples of polydodecahedron nerves, see http://demonstrations.wolfram.com/ChainsOfRegularPolygonsAndPolyhedra/.

Problem 12.14 Write a Mathematica script that finds surface-connected one or more polydodecahedron nerves in Euclidean space $\mathbb{R}^3$. Use false-colouring so that each polydodecahedron nerve found has a red central polydodecahedron and the yellow non-central polydodecahedron. ■

References

1. Grünbaum, B.: Nerves of simplical complexes. Aequationes Mathematicae **2**(2–3), 400–401 (1969)
2. Basu, S., Pollack, R., Roy, M.F.: Algorithms in Real Algebraic Geometry, p. 658. Springer, Berlin (2006)
3. Edelsbrunner, H., Harer, J.: Computational Topology. An Introduction, Xii+110 pp. American Mathematical Society, Providence (2010). MR2572029
4. Street, R.: The algebra of oriented simplexes. J Pure Appl. Algebra **49**, 283–335 (1987)
5. Street, R.: Homotopy classification of filtered complexes, Ph.D. thesis, University of Sydney (1968). Supervisor: G.M. Kelly, 131 pp

Chapter 13
Connnectedness Patterns

Fig. 13.1 Voronoï mesh on archway image

Historical Note 1 *The picture in Fig. 13.1 contains afternoon archway shadows on a wall of the Rızvaniye Mosque in Şanlıurfa, south-eastern Turkey (see, also, Fig. 13.2). This image provides a good hunting ground for image patterns because there is a clearcut separation between the archway and its surroundings.* ■

This chapter introduces the study of connectedness patterns, which are collections of sets that exhibit a repetition of strongly connected polygons in a tiling of the plane.

J.F. Peters, *Computational Proximity*, Intelligent Systems Reference Library 102, DOI 10.1007/978-3-319-30262-1_13

Fig. 13.2 Shadows

A **plane tiling** is a countable collection of closed sets that cover the plane without gaps or overlaps [1, Sect. 1.1, p. 16]. Tilings and patterns have a history spanning thousands of years. Their mathematical treatment was begun by J. Kepler.[1]

An n-sided **polygon** is a geometric object containing n vertices and n sides that are straight edges between each pair of the vertices. In other words, a polygon is a closed broken line lying on a plane [2, p. 51]. The polygons in such a tiling can either be regular or non-regular. A **regular polygon** is an n-sided polygon in which all sides are the same length symmetrically arranged around a common center. Typically, a **connectedness pattern** contains **non-regular polygons**, which are polygons in which all sides do not have the same length and all of the polygons do not have the same number of sides.

In general, a **pattern** is an arrangement of the parts of a set. When the parts of a pattern are sets, then we have a **set pattern** [3–5]. Each connectedness pattern is an example of a set pattern called a proximal pattern. A **proximal pattern** is an arrangement of the parts of a set endowed with a proximity so that the parts have nonempty intersection. A **proximal ngon pattern** is an arrangement of ngons so two or more of the ngons have common borders. For example, a connectedness pattern is proximal, provided the pattern contains pairs of polygons whose union is the set having the pattern and whose members have nonempty intersection. Such patterns are geometric. Connectedness patterns are also visual, easily detected once we know what to look for. These geometric patterns are commonly found in tilings covering a space endowed with a proximity relation.

Typically, connectedness patterns will contain subsets containing shapes that are near each other. The nearness of the tiling subsets can either be spatial in the traditional sense (shapes that share points) or descriptive (polygons that have matching descriptions such as shape, area, convexity, colour) or both. Three examples of proximal patterns are a repetition of sets of strongly connected Voronoï regions or strongly connected sets of Delaunay triangles and a repetition of strongly connected mesh nerves, poly2sim nerves or polyform nerves or the charts of manifolds that are near each other, forming a chart nearness pattern. The polygonal Voronoï regions are convex sets. However, the Delaunay triangles are not convex sets, since each Delaunay triangle has an empty interior.

[1]H.S.M. Coxeter, review of the tilings and patterns book by B. Grünbaum and G.C. Shephard, MathSciNet review MR0857454.

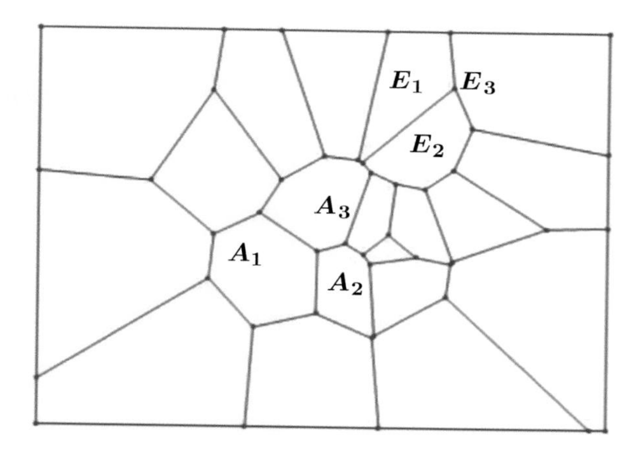

Fig. 13.3 Connectedness patterns on a Voronoï mesh

Although the polygons themselves in a set of strongly connected Voronoï regions are convex, the set of points in the union of connected Voronoï regions usually is not a convex set (Fig. 13.3).

Example 13.1 **Non-Convex Set of Connected Voronoï Regions.**

, .

From Fig. 13.1, let X be represented by the strongly near Voronoï regions A_1, A_2, A_3 in Fig. 13.4. We have

$$X = \bigcup_{A_i, i \in \{1,2,3\}} A_i (\text{Union of Voronoï regions}).$$

$$\bigcap_{A_i, i \in \{1,2,3\}} A_i \neq \emptyset,$$

$$A_i \overset{\wedge\!\!\wedge}{\delta} A_j, i, j \in \{1, 2, 3\}.$$

Hence, $\overset{\wedge\!\!\wedge}{\text{conn}}\, X$, i.e., X is a strongly connected space. The union of the Voronoï regions is not convex. To see this, notice, for example, the straight line segment $\overline{xy}$ between vertex $x \in A_1$ and vertex $y \in A_2$ is not contained in the $A_1 \cup A_2$. ■

Problem 13.2 ☕ If the edges of the Voronoï regions were curved instead of straight in a Voronoï diagram in Example 13.1, would the union of the curved Voronoï regions be convex? For examples of Voronoï regions with curved edges, see J.-D. Boissonnat, C. Wormser and M. Yvinec [6, Chap. 2], A. Efrat, F. Hoffmann and C. Knauer [7]. ■

Fig. 13.4 Strongly Connected Voronoï Regions whose Union is not Convex

For non-random selections of sites (mesh generating points), what we can observe is that sets of either connected Voronoï regions or connected Delaunay triangles tend to surround dominant objects in pictures. An example of this phenomenon is shown in the Voronoï mesh covering a picture in Fig. 13.1. The clustering of the mesh polygons are grouped around the shadow of the Rzvaniye Mosque archway shadow in Fig. 13.2. The patterns considered here bring to the forefront pattern-based delineation of dominant picture parts.

Conjecture 13.3 *Every set of strongly connected, non-rectangular or non-triangular Voronoï regions is not convex.* ■

Problem 13.4 ☕ Prove that Conjecture 13.3 is true or find a counterexample that disproves it. ■

13.1 Connectedness in Voronoï Meshes

Strongly connected sets of polygons are common in a Voronoï mesh. Hence, one can expect to find various connectedness patterns in such meshes. Such patterns become significant when the strongly connected sets of polygons have at least one common edge. This means that a polygon in one set of connected polygons shares an edge with a polygon in a different set of connected polygons. Let ngon$A$, ngon$B$ denote n-sided polygons in a Voronoï mesh $\mathscr{V}(S)$ on a set of sites S. The expression ngon$A$ $\overset{\wedge\wedge}{\text{conn}}$ ngon$B$ specifies that the polygons ngon$A$, ngon$B$ are strongly connected, i.e., have an edge in common. This leads to the notion of a strong intersection of the closures of subsets, denoted by $\overset{\wedge\!\!\!\wedge}{\bigcap} \mathrm{cl}A$ for subsets $\mathrm{cl}A$ in a strongly connected set.

Example 13.5 **Strong Connectedness Patterns.**
The Voronoï mesh $\mathscr{V}(S)$ in Fig. 13.3 on a set of image corners S in Fig. 13.2 was produced using MScript 53 in Appendix A.13. This corner-based mesh contains a number of $\overset{\wedge\!\!\!\wedge}{\mathrm{conn}}$ (strongly connected) sets of convex polygons. Two such sets of convex polygons are considered, here. For example, the strongly connected set A is defined by

$$\overset{\wedge\!\!\!\wedge}{\mathrm{conn}}\, A = \bigcup_{i=1}^{3} \mathrm{cl}A_i,$$

$$\overset{\wedge\!\!\!\wedge}{\bigcap} A_i \neq \emptyset, \text{ so that}$$

$$A_1 \overset{\wedge\!\!\!\wedge}{\delta} A_2, A_1 \overset{\wedge\!\!\!\wedge}{\delta} A_3, A_2 \overset{\wedge\!\!\!\wedge}{\delta} A_3,$$

The subsets in A are $\overset{\wedge\!\!\!\wedge}{\mathrm{conn}}$ (strongly connected), since each pair of subsets has a common edge. The second strongly connected set E in Fig. 13.3 is defined by

$$\overset{\wedge\!\!\!\wedge}{\mathrm{conn}}\, E = \bigcup_{i=1}^{3} \mathrm{cl}E_i,$$

$$\overset{\wedge\!\!\!\wedge}{\bigcap} E_i \neq \emptyset, \text{ so that}$$

$$E_1 \overset{\wedge\!\!\!\wedge}{\delta} E_2, E_1 \overset{\wedge\!\!\!\wedge}{\delta} E_3, E_2 \overset{\wedge\!\!\!\wedge}{\delta} E_3. \quad \blacksquare$$

13.2 Nearness of Collections of Strongly Connected Sets

In the case where there are collections of strongly connected sets A, E of Voronoï mesh polygons, there are instances where $A \overset{\wedge\!\!\!\wedge}{\delta} E$, i.e., the collections of strongly connected sets can themselves be strongly near. Recall that nonempty sets in a proximity space are *strongly near*, provided the sets have more than one point in common.

Definition 13.6 Strongly Proximal Connectedness.
Let X be a proximity space endowed with the proximity $\overset{\wedge\!\!\!\wedge}{\delta}$ and let $A, E \in 2^{2^X}$ be collections of strongly connected sets. **Strongly proximal connectedness** is defined in terms of collections of strongly connected sets. The collection of subsets collection

A is strongly proximal (strongly near) the collection E (denoted $A \overset{\wedge\wedge}{\delta} E$), if and only if

$$A_i \overset{\wedge\wedge}{\delta} E_j, \textit{ for some } A_i \in A, E_j \in E. \quad \blacksquare$$

Algorithm 17: Strong Connectedness Mesh Pattern on a Digital Image

Input : Read digital image img, $MaxFeatures$.

Output: Connnected set patttern pat$V$ of Voronoï regions V in mesh $\mathscr{V}(S)$ on img generated by a set of sites S containing selected img picture points.

1 $img \longmapsto cornerCoordinates$;
2 /* $MaxFeatures$ is an upper bound on the number of detected corners */ ;
3 $S \leftarrow cornerCoordinates$;
4 /* S contains corner coordinates used as mesh generating points (sites). */ ;
5 $S \longmapsto Vorono\ddot{i}\ Mesh\ \mathscr{V}(S)$;
6 $Vorono\ddot{i}\ Mesh\ \mathscr{V}(S) \longmapsto img$;
7 *Identify each set of connected regions* $connX_i, i \in \{1, 2, \dots\}$;
8 *Identify all sets of connected regions* $\mathscr{X} := \{connX_i, i \in \{1, 2, \dots\}\}$;
9 $patV := \mathscr{X}$;
10 /* The collection of sets $\mathscr{X}$ defines the mesh pattern pat$V$. */ ;

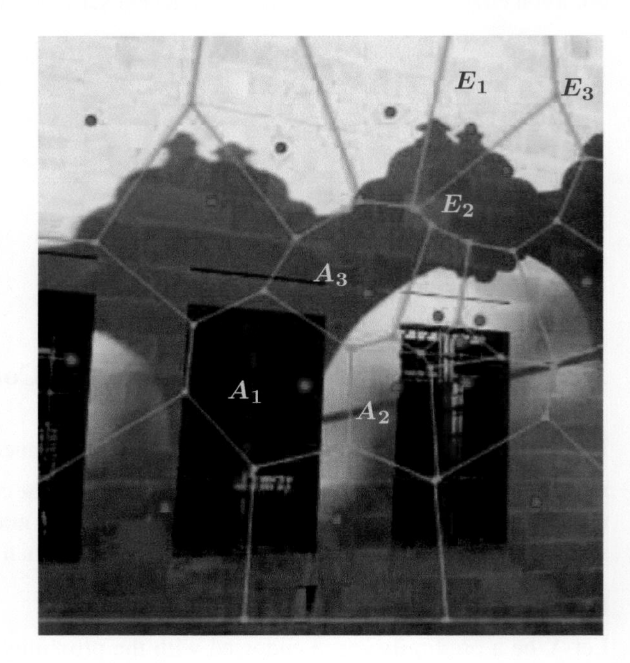

Fig. 13.5 Connectedness on a Voronoï mesh

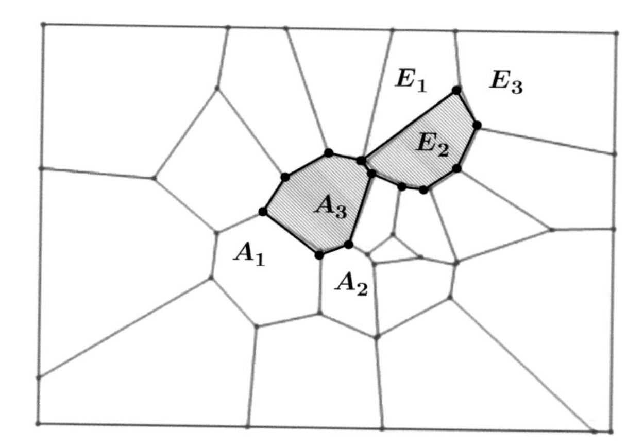

Fig. 13.6 Connectedness shape pattern on a Voronoï mesh

Example 13.7 **Nearness of Strongly Connected Sets.**
Let the strongly connected sets A, E ($A\ \overset{\wedge\!\!\wedge}{\text{conn}}\ E$) be those defined in Example 13.5. Since $A_3 \in A$ has an edge in common with $E_2 \in E$. Hence, $A\ \overset{\wedge\!\!\wedge}{\delta}\ E$. See Fig. 13.6. ■

13.3 Picture Mesh Patterns

Picture mesh patterns are introduced in this section. The focus here is on patterns in the tiling of digital images, also called pictures. A **mesh pattern** on a nonempty set X (denoted by pat$X$) is a collection of polygons with some form of repetitive arrangement in a mesh $\mathscr{M}$. The term **picture** is synonymous with *digital image* or *digital picture* or *scanned photograph*, a convention that was begun by A. Rosenfeld in the 1970s [8, Sect. 1]. A **mesh** on a picture is plane tiling of the picture.

A mesh pattern pat$X$ becomes proximal in the case where the pattern contains an arrangement of near sets in a proximity space. A **picture mesh pattern** is a collection of sets of polygons having a recognized arrangement in a mesh covering a picture Img.

Example 13.8 Let pat$X$ be a collection of sets of strongly connected Voronoï regions such that each set of regions contains an n-sided polygon with the same shape. For example, in Fig. 13.6 (see also Fig. 13.5), the 7-sided polygons A_3 and E_2 belong to sets of strongly connected sets of polygons, namely, A and E. In addition, $A_3\ \overset{\wedge\!\!\wedge}{\delta}\ E_2$. ■

Depending on a number of choices (e.g., type of mesh, type of generating points), a picture mesh pattern reveals hidden picture structures, sharply delineating the structures that are surrounded by neighbouring mesh polygons with some form of regularity. Sets of connected mesh polygons that share a common edge typify what one can expect to find in mesh picture patterns. ■

The basic steps to follow in detecting strong connectedness mesh patterns in a picture are given in Algorithm 17.

Theorem 13.9 *Every Voronoï mesh in the Euclidean space E^d with at least $d + 1$ regions contains a strongly connected collection of $d + 1$ regions.*

Proof The proof is given for $d = 2$. Let S be a set of generating points and let $\mathscr{A} = \{A_1, A_2\}$ be a set strongly connected regions in a Voronoï mesh $\mathscr{V}(S)$ covering the Euclidean plane E^d with $d + 1$ regions. If $\mathscr{V}(S)$ contains only $d + 1$ regions A_1, A_2, A_3, then the strongly connected set pattern pat$P$ is defined by

$$\mathscr{A} = \bigcup_{i=1}^{3} A_i,$$

$$\bigcap_{i=1}^{3} \text{cl}A_i \neq \varnothing,$$

$$\text{pat}P := \mathscr{A}.$$

By definition, $\mathscr{V}(S)$ is generated by a set of sites containing $d + 1$ points and each pair of regions have a common edge. Since $\mathscr{V}(S)$ covers the space, A_1 and A_2 have an edge in common with A_3. Hence,

$$A_1 \overset{\wedge\!\!\!\wedge}{\delta} A_2, A_1 \overset{\wedge\!\!\!\wedge}{\delta} A_3, \text{ and } A_2 \overset{\wedge\!\!\!\wedge}{\delta} A_3.$$

If $\mathscr{V}(S)$ contains more than $d + 1$ regions, then regions A_1 and A_2 will have an edge in common with one of the remaining regions C so that $\overset{\wedge\!\!\!\wedge}{\text{conn}}\, \mathscr{A} := \{A_1, A_2, C\}$, such that C has an edge in common with A_1 and an edge in common with A_2. That this will always happens needs to be proved (see Problem 13.10). □

Problem 13.10 ☕ Complete the proof of Theorem 13.9. That is, let $\mathscr{V}(S)$ be a Voronoï mesh on the Euclidean plane E^3, where $d = 2$. Assume that $\mathscr{V}(S)$ has more than three regions. Let $A_1, A_2 \in \mathscr{V}(S)$ such that $A_1 \overset{\wedge\!\!\!\wedge}{\text{conn}} A_2$, i.e., A_1 and A_2 have an edge in common. Let C be strongly connected to A_1 (i.e., $C \overset{\wedge\!\!\!\wedge}{\text{conn}} A_1$ so that C has an edge in common with A_1). Prove that $C \overset{\wedge\!\!\!\wedge}{\text{conn}} A_2$, i.e., C has an edge in common with A_2.

Hint: Consider the following cases.

Case.1 Every Voronoï region in $\mathscr{V}(S)$ is a rectangle.
Case.2 Every Voronoï region in $\mathscr{V}(S)$ is a quadrilateral.
Case.3 Every Voronoï region in $\mathscr{V}(S)$ is a polygon with more than 3 sides.
Case.4 Every Voronoï region in $\mathscr{V}(S)$ is a mixture of n-sided polygons.
Case.5 Every Voronoï region in $\mathscr{V}(S)$ is a mixture of n-sided convex regions with curved edges. ∎

13.4 Pattern Axioms and Object Signature Axioms

This section introduces pattern and object signature axioms useful in evaluating mesh patterns. Many different shape patterns can be found in a Voronoï mesh or in a Delaunay mesh. A **shape pattern** is a collection of polygons with some form of repetition. For example, a shape pattern is defined by repeated polygons with the same number of edges.

Example 13.11 **Mesh Shape Pattern.**
This is a continuation of Example 13.5, where A and E are strongly connected sets of polygons in a Voronoï mesh covering an image img. Polygons $A_3 \in A$, $E_2 \in E$ are represented in Fig. 13.6. A_3 and E_2 are irregular seven-sided polygons (also called *7gons* or **septagons**). These are the only 7gons in the mesh in Fig. 13.6. These 7gons are $\overset{\wedge\!\!\!\wedge}{\delta}$ (strongly near) each other, since they have a common edge. In addition to being strongly near, these polygons mostly cover the top part of the archway shadow, which is a dominant object in the archway picture in Fig. 13.6. ∎

For multiple reasons, the 7gons A_3 and E_2 define a mesh shape pattern that provides a signature of an underlying object in the picture in Fig. 13.6. First, A_3 and E_2 have the same shape. Second, A_3 and E_2 are strongly connected (these 7gons share an edge). Third, pat$A$ is strongly near pat$B$, since A_3 and E_2 are strongly near each other. A **signature** of an object is a distinctive characteristic by which an object can be identified. A **cluster of sets** is a collection of sets with a common characteristic such as connectedness.

Mesh patterns satisfy a number of axioms.

Pattern Axioms

(pat0) Each strongly connected set of mesh polygons defines a shape pattern.
(pat1) The higher the number of edges in the polygons in a shape pattern on a picture in the plane, the greater the proximity of the shape patterns to objects in a picture.
(pat2) Clusters of sets of strongly connected polygons serve as a signature for picture objects.

(pat3) A repetition of polygons with the same number of sides in a cluster of sets of strongly connected polygons covering a picture provides a strong signature for a picture object.

(pat4) If disconnected sets of mesh polygons belong to a picture shape pattern, then there are separated objects in the picture. ■

Example 13.12 **Axiomatic View of Mesh Shape Patterns**.
This is a continuation of Example 13.11, where A and E are strongly connected sets of polygons in a Voronoï mesh $\mathscr{V}(S)$ covering an image img. Polygons $A_3 \in A, E_2 \in E$ are represented in Fig. 13.6. Let $\mathscr{X} := \{A, E\}$.

(pat0) The mesh in Fig. 13.6 contains strongly connected sets of polygons contain 7gons that are near each other (the two 7gons share a common edge). Hence, from Axiom pat0, $\mathscr{X}$ defines a mesh pattern, i.e., pat(ngon$A$) := $\mathscr{X}$.

(pat1) Polygons A_3 and E_2 have the highest number of edges in $\mathscr{V}(S)$. Hence, from Axiom pat1 has high proximity to a picture object, namely, the archway shadow.

(pat2) The two collections of sets in pat(ngon$A$) form clusters of strongly connected polygons. From Axiom pat2, these clusters provide a signature of picture object.

(pat3) pat(ngon$A$) contains a repetitions of polygons with the same number of sides, namely, the 7gons $A_3 \in A, E_2 \in E$. From Axiom pat3, this repetition of 7gons provide a strong signature of picture object.

(pat4) There are disconnected sets of mesh polygons covering the picture in Fig 13.7.2. Hence, from Axiom pat4, there are separated objects in the picture. ■

To see pattern Axiom 5° at work, consider increasing the number of sites in MScript 53 in Appendix A.13.

Example 13.13 To see more than one mesh pattern containing strongly connected sets of polygons, try changing the number of sites in MScript 53 in Appendix A.13 from 30 to 100. The result is shown in the Voronoï mesh in Fig. 13.7.1, which is superimposed on the archway image in Fig. 13.7.2. Now observe that there separated mesh patterns, one for the archway shadow and others patterns that overlay the windows in the picture. See Problem 13.14. ■

Problem 13.14 🚲 Identify all separated mesh patterns containing strongly connected polygons in Fig. 13.7.2. ■

Let obj$E$ denote an object in a picture. The term *object* is left undefined. The nature of an object is clear from the context in which an object appears. Each object has its own signature characterized the mesh polygons covering the object. Let (X, δ) be a Lodato proximity space that satisfies the Wallman proximity axiom, obj$E$, obj$G \subset X$. And let $\mathscr{V}(S)$ be a Voronoï mesh that covers X,

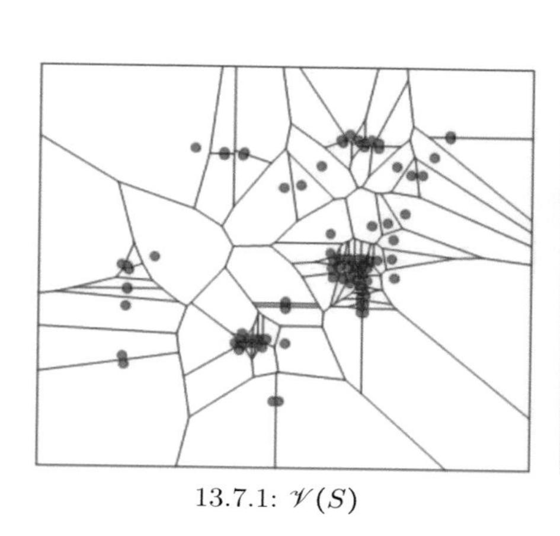

13.7.1: $\mathscr{V}(S)$

13.7.2: Picture

Fig. 13.7 Sample picture mesh patterns

ngon$A$, ngon$B \in \mathscr{V}(S)$. A **polygon pattern** pat(ngon$A$) is a collection of strongly connected polygons containing ngon$A$. The following axioms characterize an object signature.

Object Signature Axioms

(obj0) obj$E \subset X \Leftrightarrow$ obj$E \overset{\wedge\!\wedge}{\delta}$ pat(ngon$A$) for some ngon$A \in \mathscr{V}(S)$.
(obj1) (ngon$A$ $\overset{\wedge\!\wedge}{\text{conn}}$ ngon$B$) δ obj$E \Leftrightarrow$ (cl$A \cap$ cl$B$) δ obj$E$.
(obj2) ngon$A$ $\overset{\wedge\!\wedge}{\delta}$ ngon$B \Leftrightarrow$ (cl$B \cap$ cl$A$) $\overset{\wedge\!\wedge}{\delta}$ obj$E$. ■

Further patterns pat(ngon$A$), pat(ngon$B$) with obj$E$ δ pat(ngon$A$), obj$G$ δ pat (ngon$B$) are *separated*, provided

(obj3) pat(ngon$A$) $\overset{\wedge\!\wedge}{\not\delta}$ pat(ngon$B$) $\Leftrightarrow$ obj$E \neq$ obj$G$. (**Object Separation**) ■

Problem 13.15 🚲 Identify separated mesh patterns in Fig. 13.7.2 that satisfy Object Signature Axiom (obj3). In each case, identify the corresponding separated objects. ■

Algorithm 18: Voronoï Mesh Shape Pattern on a Digital Image

Input : Read digital image img, MaxFeatures, n = number of edges in n-sided polygons to find in img.

Output: Shape-based connnected set pattern pat$V$ of Voronoï regions V in mesh $\mathscr{V}(S)$ on img generated by a set of sites S containing selected img picture points.

1 $img \longmapsto cornerCoordinates$;
2 $S \leftarrow cornerCoordinates$;
3 /* S contains corner coordinates used as mesh generating points (sites). */ ;
4 $S \longmapsto Vorono\ddot{i}\ Mesh\ \mathscr{V}(S)$;
5 $Vorono\ddot{i}\ Mesh\ \mathscr{V}(S) \longmapsto img$;
6 $ngonA \in \mathscr{V}(S)$;
7 $\mathscr{V}(S) := \mathscr{V}(S) \setminus ngonA$;
8 $continue := true$;
9 **while** *(continue)* **do**
10   **if** $\exists ngonB \in \mathscr{V}(S)$ **then**
11   pat$V$:= pat$V \cup B$;
12     $\mathscr{V}(S) := \mathscr{V}(S) \setminus$ ngon$B$;
13   /* ngon$B$ belongs to shape-based pattern pat$V$ */
14   **else**
15     $continue := false$;

Problem 13.16 Do the following:

1° Use Mathematica to implement Algorithm 18. Give three sample runs of your Mathematica script using images of your own choosing. **N.B.**: Use only digital images (pictures) captured by you, not images found on the web.

2° Use Matlab to implement Algorithm 18. Give three sample runs of your Matlab script using images of your own choosing. **N.B.**: Use only digital images (pictures) captured by you, not images found on the web. ■

Theorem 13.17 *Let pat(ngonA) be a picture mesh pattern containing ngonA. $objE \overset{\wedge\!\!\wedge}{\delta} pat(ngonA)$ for some $ngonA \in \mathscr{V}(S)$ with $ngonA\ \delta\ ngonB$ for some $ngonB \in \mathscr{V}(S)$ implies $objE\ \delta\ ngonB$.*

Proof Immediate from Axioms (obj0) and (obj1). □

Theorem 13.18 *Different objects are near different patterns.*

Proof Immediate from Axioms (pat5) and (obj3). □

Let ngon$E$ denote a polygon with n sides. Algorithm 18 gives the steps for finding an ngon-based shape pattern in a Voronoï mesh that covers a picture. For the sake of clarity, the set of sites S contains picture corners. This is one among many different choices for the points in S.

By experimenting with the upper bound on the number corners to find in a picture, it is possible to find an optimum number of sites useful in detecting objects and patterns.

13.8.1: Corner Sites ≤ 100

13.8.2: Corner Sites ≤ 30

Fig. 13.8 Increasing number of corner sites

Example 13.19 **Experimenting with the Number of Corner Sites.**
To see more than one mesh pattern containing strongly connected sets of polygons, try changing the number of sites in MScript 53 in Appendix A.13 from 30 to 100. The result is shown in Fig. 13.8, which is superimposed on the archway image in Fig. 13.7.2. Increasing the number of corner sites leads to a more complete delineation of the dominant object in the picture, namely, the archway shadow. See Problem 13.20. ■

Problem 13.20 ☕ Derive a formula that computes the optimal number of corner sites in the generation of a Voronoï mesh on a picture. The number of sites is optimum, provided it the number of sites leads to the construction of a mesh that provides good separation between picture objects. ■

13.5 Keypoints Mesh Sites

So far, mesh patterns have been defined relative to corner-based Voronoï meshes. Keypoints provide a promising alternative to corners as a source of sites in constructing a Voronoï mesh. This is the case, since sites that are keypoints tend to follow more closely the contour of dominant picture objects.

A **dominant object** has border (usually edge) points with intensities that sharply contrast with neighbouring points. Dominant objects are commonly found in pictures, musical scores, gardens and countryside plants (flora), animal populations (fauna), animal behaviour, electrical signals, stellar formations, fossil deposits, weather, ocean waves, communication networks and social networks. The focus here is on object recognition and the recognition of patterns in the fabric of pictures. A **picture** can be a drawing, painting, photograph, or digital image.

> Pattern recognition often reduces to the discovery of repeated dominant objects revealed by strongly connected mesh polygons. The trick is to select sites that best delineate dominant objects. ■

Example 13.21 **Experimenting with the Number of Keypoint Sites.**
To see more than one mesh pattern containing strongly connected sets of polygons, try changing the number of keypoints sites in MScript 54 in Appendix A.13 from 30 to 100. The result is shown in Fig. 13.9. The contrast between the corners in Fig. 13.8 and the keypoints in Fig. 13.9 suggests that keypoints provide a better source of mesh sites. Increasing the number of keypoint sites leads to a more complete delineation of the dominant object in the picture, namely, the archway shadow. See Problem 13.20. ■

Example 13.22 **Snowdrop Keypoint Sites.**
Try changing the number of keypoints sites in MScript 54 in Appendix A.13 from 30 to 100 with different images. The result is shown in Fig. 13.10. Increasing the number of keypoint sites leads to a more complete delineation of the dominant object in the picture, e.g., contours of a pair of snowdrops. See Problem 13.20. ■

13.9.1: Keypoints Sites ≤ 100

13.9.2: Keypoints Sites ≤ 30

Fig. 13.9 Increasing number of keypoint sites

13.10.1: Keypoints Sites ≤ 100 13.10.2: Keypoints Sites ≤ 30

Fig. 13.10 Increasing number of keypoint sites on snowdrops image

Problem 13.23 ☕ Derive a formula that computes the optimal number of keypoint sites in the generation of a Voronoï mesh on a picture. The number of keypoints sites is optimum, provided it the number of keypoints leads to the construction of a mesh that provides good separation between picture objects. ■

13.6 Keypoints Mesh Patterns

This section is a continuation of the approach in Sect. 13.1. Here the focus is finding strongly connected sets of polygons in a keypoints-based Voronoï mesh.

Example 13.24 **Snowdrop Keypoints-based Voronoï mesh.**
MScript 54 in Appendix A.13 is used to construct a keypoints-based Voronoï mesh on . For simplicity, the mesh is constructed with up to 30 keypoint sites. The result is shown in Fig. 13.11. ■

In Fig. 13.11, the following strongly connected sets of snowdrop polygons are shown. Let $a_i, e_i, i \in \{1, 2, 3\}$ be keypoint sites. The strongly connected set $\overset{\wedge\!\!\wedge}{\text{conn}}\, A$ is defined by

$$\overset{\wedge\!\!\wedge}{\text{conn}}\, A = \bigcup_{i=1}^{3} \text{cl}\, V(a_i). \text{ union of Voronoï regions}$$

$$\overset{\wedge\!\!\wedge}{\bigcap} V(a_i) \neq \emptyset, \text{ so that}$$

$$V(a_1) \overset{\wedge\!\!\wedge}{\delta} V(a_2), V(a_1) \overset{\wedge\!\!\wedge}{\delta} V(a_3), V(a_2) \overset{\wedge\!\!\wedge}{\delta} V(a_3).$$

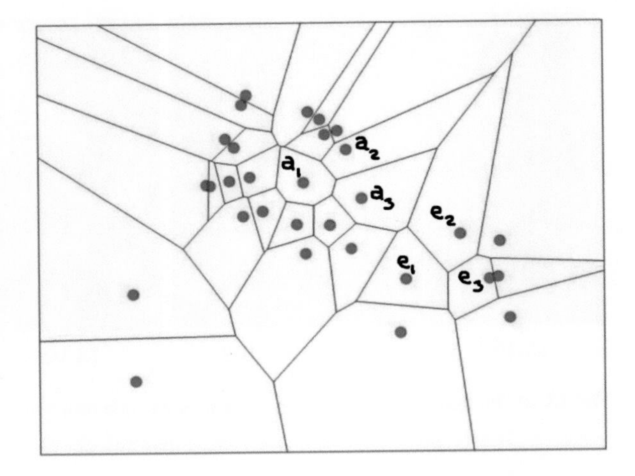

Fig. 13.11 Connectedness on a snowdrops Voronoï mesh

Similarly, the strongly connected set of Voronoï regions $\overset{\wedge\!\!\!\wedge}{\text{conn}}\, E$ is defined by

$$\overset{\wedge\!\!\!\wedge}{\text{conn}}\, E = \bigcup_{i=1}^{3} \text{cl}V(e_i). \text{ union of Voronoï regions}$$

$$\overset{\wedge\!\!\!\wedge}{\bigcap} V(e_i) \neq \varnothing, \text{ so that}$$

$$V(e_1) \overset{\wedge\!\!\!\wedge}{\delta} V(e_2), V(e_1) \overset{\wedge\!\!\!\wedge}{\delta} V(e_3), V(e_2) \overset{\wedge\!\!\!\wedge}{\delta} V(e_3).$$

13.12.1: Keypoints Mesh ≤ 100 13.12.2: Keypoints Mesh ≤ 30

Fig. 13.12 Keypoints Voronoï meshes on snowdrops image

The strongly connected collections $\overset{\wedge\wedge}{\text{conn}}\, A$ and $\overset{\wedge\wedge}{\text{conn}}\, E$ contain 5gons $V(a_2)$,

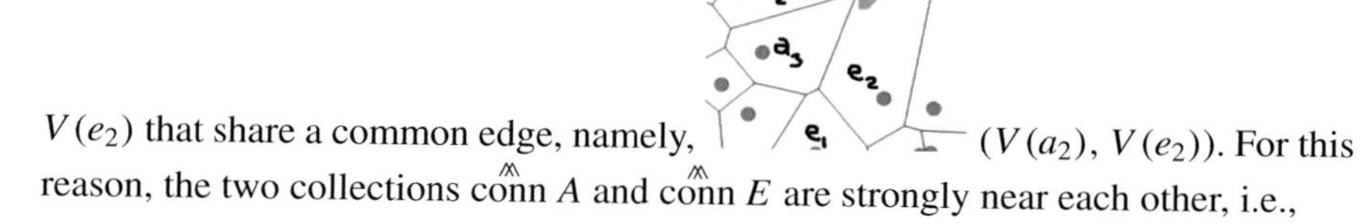

$V(e_2)$ that share a common edge, namely, $(V(a_2), V(e_2))$. For this reason, the two collections $\overset{\wedge\wedge}{\text{conn}}\, A$ and $\overset{\wedge\wedge}{\text{conn}}\, E$ are strongly near each other, i.e.,

$$\overset{\wedge\wedge}{\text{conn}}\, A \; \overset{\wedge\wedge}{\delta} \; \overset{\wedge\wedge}{\text{conn}}\, E, \text{ since } V(a_2) \; \overset{\wedge\wedge}{\delta} \; V(e_2).$$

Example 13.25 **Patterns in Snowdrop Keypoints-Based Voronoï Mesh**. MScript 54 in Appendix A.13 is used to construct a pair of keypoints-based Voronoï meshes in Fig. 13.12. The strongly connected sets $\overset{\wedge\wedge}{\text{conn}}\, A$ and $\overset{\wedge\wedge}{\text{conn}}\, E$ in a Voronoï mesh with up to 30 keypoint sites appear in Fig. 13.12.2. The situation changes dramatically in the mesh with up to 100 keypoint sites in Fig. 13.12.1. See Problem 13.26. ■

Problem 13.26 For three different images of your own choosing, do the following using Mathematica to construct and identify connected set patterns in a Voronoï mesh similar to the one in Fig. 13.12.1.

1° Identify, label and highlight with yellow the repeated 5gons.

2° Identify, label and highlight with orange the repeated 5gons with approximately the same area. Let areaA, areaE denote the areas of 5gons A and E. Choose a small value of ε (upper bound on the difference between polygon areas). The areas of 5gons A and E are approximately the same, provided

$$|areaA - areaE| < \varepsilon.$$

3° Identify, label the sites in strongly connected sets of polygons containing at least one 5gon.

4° In a pattern containing a repetition of strongly connected sets of polygons, each containing at least one 5gon, identify sets of polygons that are strongly near, i.e., share a edge.

5° Identify, label and highlight with yellow the repeated 6gons.

6° Identify, label and highlight with orange the repeated 6gons with approximately the same area.

7° Identify, label the sites in strongly connected sets of polygons containing at least one 6gon.

8° In a pattern containing a repetition of strongly connected sets of polygons, each containing at least one 6gon, identify sets of polygons that are strongly near, i.e., share a edge. ■

References

1. Grünbaum, B., Shephard, G.: Tilings and Patterns. W.H. Freeman and Co., New York (1987). Xii+700 pp., MR0857454
2. Coxeter, H., Greitzer, S.: Geometry revisited. New Mathematical Library, 19. Random House Inc, NY (1967). Xiv+193 pp., MR3155265
3. Peters, J., Ramanna, S.: Treasure trove at banacha. set patterns in descriptive proximity spaces. Fund. Inform. **127**(1—4), 357–367 (2013). MR3134960
4. Peters, J.: Topology of Digital Images - Visual Pattern Discovery in Proximity Spaces, Intelligent Systems Reference Library, vol. 63. Springer, : Xv + 411 pp. Zentralblatt MATH Zbl **1295**, 68010 (2014)
5. Naimpally, S., Peters, J.: Topology with Applications. Topological Spaces via Near and Far. World Scientific, Singapore (2013). Xv + 277 pp, Amer. Math. Soc. MR3075111
6. Boissonnat, J.D., Teillaud, M.: Effective computational geometry for curves and surfaces. Springer, Berlin (2007). Xii+343 pp. ISBN: 978-3-540-33258-9; 3-540-33258-8, MR2387755
7. Efrat, A., Hoffmann, F., Knauer, C.: Covering with ellipses. Algorithmica **38**, 45–160 (2004)
8. Rosenfeld, A.: Digital Picture Analysis, p. Xi + 351. Springer, Berlin (1976)

Chapter 14
Nerve Patterns

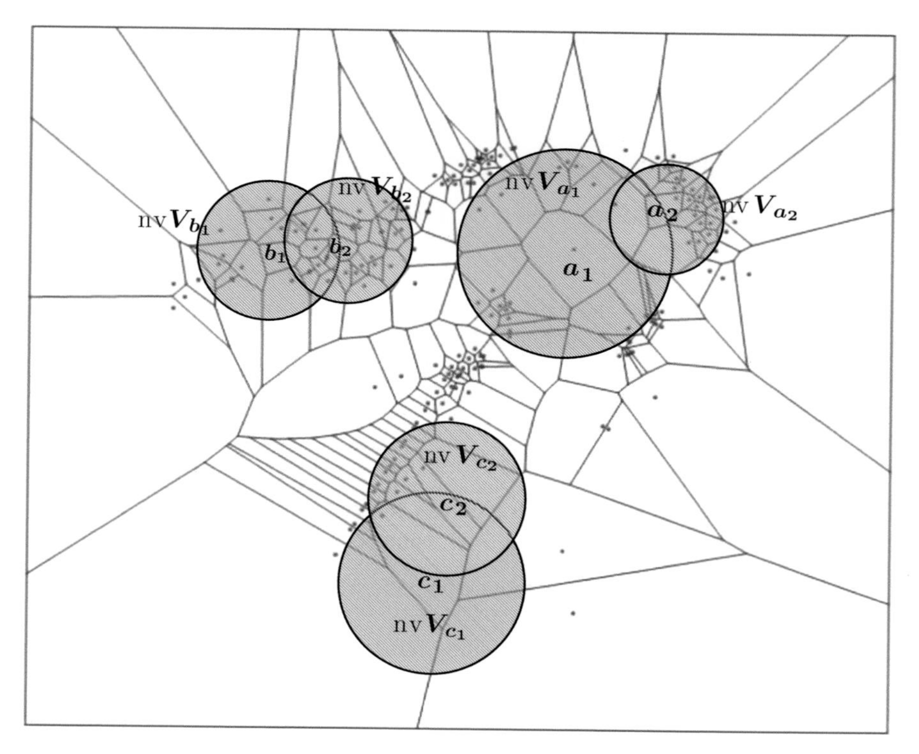

Fig. 14.1 Nucleus nerve patterns on a Voronoï mesh

This chapter introduces mesh nerve patterns. A **nucleus nerve pattern** is an arrangement of mesh nerves based on a partial ordering of the nuclei of the nerves. A **nerve nucleus** in a Voronoï diagram is an ngon that is strongly connected to polygons along

J.F. Peters, *Computational Proximity*, Intelligent Systems Reference Library 102, DOI 10.1007/978-3-319-30262-1_14

Fig. 14.2 Source of nucleus nerve patterns in Fig. 14.1

the border of the nucleus. The partial ordering relation $\leq$ on either the number of polygon sides or the areas of the nuclei leads to the arrangement of the nerves in a mesh nerve pattern. There can be more than one nerve pattern in the same picture. For example, a Voronoï mesh on the plant image in Fig. 14.2 is shown in Fig. 14.1. Let $\varphi(P) = $ area of P.

Example 14.1 **Plant Picture Mesh Nucleus Nerve Patterns**
Three nerve patterns are shown in Fig. 14.1, namely,

1° pat nv$\mathcal{N}_a = \left\{\text{nv}V_{a_1}, \text{nv}V_{a_2}\right\}$ on plant with pattern induced by the partial ordering of the nuclei areas: $\varphi(V_{a_2}) \leq \varphi(V_{a_1})$.
2° pat nv$\mathcal{N}_b = \left\{\text{nv}V_{b_1}, \text{nv}V_{b_2}\right\}$ on building structure with pattern induced by the partial ordering of the nuclei areas: $\varphi(V_{a_2}) \leq \varphi(V_{b_1})$.
3° pat nv$\mathcal{N}_c = \left\{\text{nv}V_{c_1}, \text{nv}V_{c_2}\right\}$ on plant shadow with rainbow with pattern induced by the partial ordering of the nuclei areas: $\varphi(V_{c_2}) \leq \varphi(V_{c_1})$.
■

In general, a **nerve pattern** is an arrangement in a collection of nerves $\mathcal{N}$. Usually, more than one nerve pattern is present in $\mathcal{N}$. Recall that an Edelsbrunner–Harer nerve [1, Sect. III.2, p.50] (denoted nv $\mathcal{N}$) is defined by

$$\text{nv}\mathcal{N} = \left\{X \in \mathcal{N} : \bigcap X \neq \varnothing\right\}. \text{ (model for a nerve)}$$

This basic model of a nerve is extended in a number of ways. For example, closure nerves (denoted nv_{cl}) are definable in a topological space X. A **closure nerve** is defined by

$$\text{nv}_{\text{cl}}\mathcal{N} = \left\{X \in \mathcal{N} : \bigcap \text{cl}X \neq \varnothing\right\}. \text{ (closure nerve)}$$

Another useful extension of the Edelsbrunner–Harer nerve is for those nerves that reside in a proximity space. For example, in a Lodato proximity space (X, δ), an **LO nerve** $\text{nv}_{LO}\ \mathcal{N}$ in the family of subsets of X is defined by

$$\text{nv}_{LO}\mathcal{N}(A) = \{B \in \mathcal{N} : A\ \delta\ B\}. \text{ (model for an LO nerve)}$$

That is, a Lodato nerve is the set of all subsets in $\mathcal{N}$ that are close to a particular subset A. The set A is called the **nucleus** of the nerve $\text{nv}_{LO}\mathcal{N}(A)$.

Example 14.2 **Lodato Nerve**.
In Fig. 14.4, let $\mathcal{N} := \{A, E_1, E_2, E_3\}$ be in a space endowned with the Lodato proximity. The interior of A has points in common with the interior each of the other subsets in $\mathcal{N}$, i.e., $A\ \delta\ E_i, i \in \{1, 2, 3\}$. Hence, $\mathcal{N}$ is a Lodato nerve. The nucleus A is a connected subset of $\mathcal{N}$, since $A\ \overset{\wedge\!\!\wedge}{\delta}\ E_i, i \in \{1, 2, 3\}$. ■

14.1 Voronoï Mesh Nerves

The focus in this chapter is on Voronoï mesh nerves. Let $\mathcal{M}$ denote a Voronoï mesh, $\mathcal{N} \subset \mathcal{M}$, $A, B \in \mathcal{N}$ (A and B are Voronoï regions of sites). A **Voronoï mesh nerve** (denoted by $\overset{\wedge\!\!\wedge}{\text{nv}}\ \mathcal{N}$)in a space endowed with the $\overset{\wedge\!\!\wedge}{\delta}$ (strongly near) proximity is defined by

$$\overset{\wedge\!\!\wedge}{\text{nv}}\,\mathcal{N} = \bigcup_{A \in \mathcal{N}} A,$$

$$A\ \overset{\wedge\!\!\wedge}{\delta}\ B \text{ for all } A, B \in \mathcal{N}. \ (\overset{\wedge\!\!\wedge}{\text{nv}} \text{ mesh nerve})$$

That is, each pair of Voronoï regions in a mesh nerve have a common edge and all of the Voronoï regions in a mesh nerve have a common vertex.

Example 14.3 **Voronoï Mesh Nerves** In Fig. 14.3, let the members of a mesh $\mathcal{M}$ denote Voronoï regions of sites (not shown), $\mathcal{N} \subset \mathcal{M}$, i.e.,

$$\mathscr{M} = \left\{\text{cl}V_p, X_1, X_2, X_3, X_4, X_5, X_6, X_7, X_8, X_9, X_{10}\right\}.$$

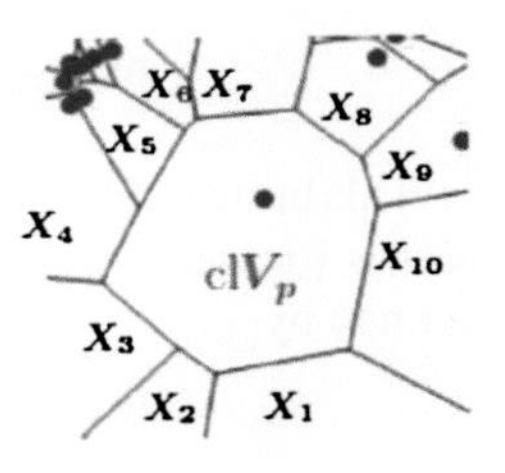

Fig. 14.3 Mesh nerve

Then a number of mesh nerves can be observed. Here are several mesh nerves.

$$\overset{\wedge\wedge}{\mathrm{nv}}\,\mathcal{N}_1 = \left\{\mathrm{cl}V_p, \mathrm{cl}X_1, \mathrm{cl}X_2\right\}.$$
$$\overset{\wedge\wedge}{\mathrm{nv}}\,\mathcal{N}_2 = \left\{\mathrm{cl}V_p, \mathrm{cl}X_2, \mathrm{cl}X_3\right\}.$$
$$\overset{\wedge\wedge}{\mathrm{nv}}\,\mathcal{N}_3 = \left\{\mathrm{cl}V_p, \mathrm{cl}X_3, \mathrm{cl}X_4\right\}. \quad \blacksquare$$

14.2 Nerves with a Nucleus

A nerve can be endowed with a nucleus. A **nerve nucleus** is a subset of a nerve that is strongly connected to every other subset of the nerve in an $\overset{\wedge\wedge}{\delta}$ proximity space X. A nerve with a nucleus A is called a **nucleus nerve** (denoted $\overset{\wedge\wedge}{\mathrm{nv}}\,\mathcal{N}(A)$), defined by

$$\overset{\wedge\wedge}{\mathrm{nv}}\,\mathcal{N}(A) = \{E \subset X : \mathrm{cl}A \cap \mathrm{int}E \neq \emptyset\}. \text{ (Nucleus nerve)}$$

Nucleus Nerve Property

Every subset of a nucleus nerve is strongly near the nucleus of the nerve. ■

That is, given a nonempty subset A in an $\overset{\wedge\wedge}{\delta}$ proximity space X, a collection of subsets in the space that satisfies the Nucleus Nerve Property is a nucleus nerve.

Example 14.4 **Nucleus Nerves** In Fig. 14.4, the set A behaves like a nucleus, since each subset $E_i, i \in \{1, 2, 3\}$ is strongly near A. Consequently,

$$\overset{\wedge\wedge}{\mathrm{nv}}\,\mathcal{N}(A) = \{A, E_1, E_2, E_3\}$$

satisfies the Nucleus Nerve Property. Hence, $\overset{\wedge\wedge}{\mathrm{nv}}\,\mathcal{N}(A)$ is a nucleus nerve. ■

The notion of a nucleus nerve carries over in looking for patterns in either a Voronoï or a Delaunay mesh. Let V_p be a Vororonï region of site p in a Vororonï mesh. For consistency with the more general notation for a nucleus nerve, $\overset{\wedge\wedge}{\mathrm{nv}}\,\mathcal{N}(V_p)$

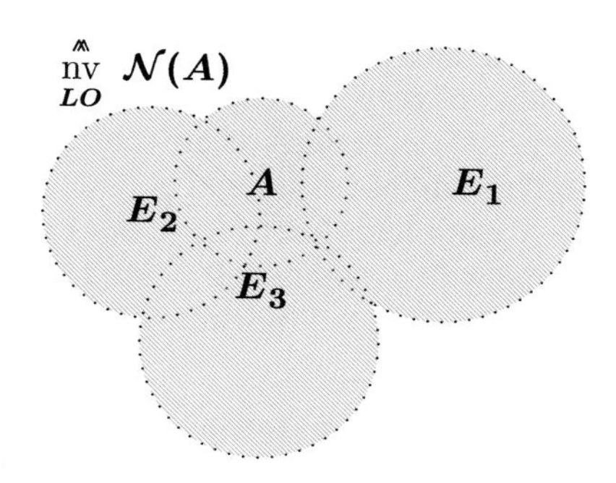

Fig. 14.4 Lodato nerve with a nucleus A

is synonymous with $NCL_{V_p}\mathscr{M}$ (from Sect. 1.10). A *mesh nucleus nerve* is defined by

$$\overset{\wedge\!\!\!\wedge}{\text{nv}}\,\mathcal{N}(V_p) = \left\{E \in \mathscr{N} : \text{cl}V_p \cap \text{int}E \neq \varnothing\right\}. \text{ (Mesh nucleus nerve)}$$

Clearly, if $\overset{\wedge\!\!\!\wedge}{\text{nv}}\,\mathcal{N}(V_p)$ is a mesh nucleus nerve, then

$$\overset{\wedge\!\!\!\wedge}{\text{nv}}\,\mathcal{N}(V_p) = \left\{E \in \mathscr{N} : \text{cl}V_p \ \overset{\wedge\!\!\!\wedge}{\delta}\ \text{cl}(\text{int}E)\right\}.$$

Example 14.5 **Mesh Nucleus Nerves** In Fig. 14.3, let the members of a mesh $\mathcal{M}$ denote Voronoï regions of sites (not shown), $\mathcal{N} \subset \mathcal{M}$, i.e.,

$$\mathscr{N} = \left\{\text{cl}V_p, X_1, X_2, X_3, X_4, X_5, X_6, X_7, X_8, X_9, X_{10}\right\}.$$

Then the nucleus nerve $\overset{\wedge\!\!\!\wedge}{\text{nv}}\,\mathcal{N}(V_p)$ is defined by

$$\overset{\wedge\!\!\!\wedge}{\text{nv}}\,\mathcal{N}(V_p) = \left\{E \subset \mathcal{N} : \text{cl}V_p \cap \text{cl}(\text{int}E) \neq \varnothing\right\}.$$

For example, for $E := X_1$, the closure of the interior of the Voronoï region E shares an edge with the nucleus V_p. Hence, $V_p \ \overset{\wedge\!\!\!\wedge}{\delta}\ \text{cl}(\text{int}E)$. ■

14.3 Partial Ordering of Nuclei

Nuclei can be partially ordered in a number of ways.

Recall that a nonempty set P equipped with a relation $\leq$ that is reflexive, anti-symmetric and transitive is called a **partially ordered set** (or **poset**). Let G be a geometric structure such as a polygon, which is a bounded set of points in the plane. Let $\mathcal{P}$ be a set of geometric structures. A bounded geometric structure has features such as area, number of sides, convexity. Recall that a **probe function** on a nonempty set X or on a powerset 2^X is a mapping φ on X or 2^X to $\mathbb{R}$ (reals). For example, let P be a set of polygons and define the probe function $\varphi : 2^P \longrightarrow \mathbb{R} \times \cdots \times \mathbb{R}$ defined by $(\varphi(A))$ = feature vector containing a single feature value of A such as the area and the number of sides of A.

A **complete poset** $(L, \leq, \mathbf{0}, \mathbf{1})$ is a poset L equipped with an ordering relation $\leq$ and with a minimum $\mathbf{0} = \inf(L)$ and maximum $\mathbf{1} = \sup(L)$ [2, Sect. 2.1.3, p. 15] (see, also, [3–5]). From this, we can derive various posets useful in pattern and object recognition.

Example 14.6 **Mesh Nuclei Complete Poset.**
Let $\mathscr{N}$ be a set of Voronoï mesh nerve nuclei defined by

$$\mathscr{N} = \{P : P \text{ is an ngon}\}.$$

Let $P_1, P_2, \ldots, P_n \in \mathscr{N}$, $\varphi(P_i)$ = number of sides on A_i, $1 \leq i \leq n \in \mathbb{N}^+$. Assume that $\varphi(P_1) \leq \varphi(P_2) \leq \cdots \leq \varphi(P_n)$. Then $\left(\{\varphi(P_i)\}_{i=1}^{n}, \leq\right)$ is a poset. Every $\mathscr{N}$ has a polygon with the smallest number of sides and a polygon with the largest number of sides. Then $(\varphi(P_i), \leq, \mathbf{0}, \mathbf{1})$ is a complete poset. Let $\Phi(P)$ be the set of areas of the polygons in $\mathscr{N}$ with a smallest polygon area and largest polygon area. Hence, $(\Phi(P), \leq, \mathbf{0}, \mathbf{1})$ is another complete poset. ■

Complete posets of mesh nerve nuclei provide a basis for pattern recognition on either a Voronoï or Delaunay mesh. The basic approach is start with an ngon viewed as the nucleus of a nerve. The choice of the initial nerve nucleus (called **seed nucleus**) will depend on the form of mesh. The simplest form of mesh contains a cluster of polygons with small areas bounded by polygons with large areas. In that case, the seed nucleus will be an ngon with small area (by comparison with the boundary polygons). The choice of a seed nucleus leads to a partially ordered set of nuclei. Each pair of nuclei define a pair of connected subsets that are mesh nerves (see, e.g., Fig. 14.5).

Definition 14.7 A collection of connected mesh nerves in a complete poset of nuclei defines a **nerve pattern**. ■

Example 14.8 **Partially Ordered Voronoï Mesh Nuclei.**
The Voronoï mesh in Fig. 14.6 has been derived from an edge-keypoint-based mesh on the image of the Manitoba chipmunk in Fig. 14.7. Let $\mathscr{N}$ be a set of Voronoï mesh nerve nuclei defined by

$$\mathscr{N} = \left\{P_a, P_e, P_g\right\}.$$

Let $\varphi(P_x)$ = number of sides on $P_x \in \mathscr{N}$, where

$$\varphi(P_g) \leq \varphi(P_e) \leq \varphi(P_a).$$

Then $\left(\left\{\varphi(P_x)_{x\in\{a,e,g\}}\right\}, \leq\right)$ is a poset. Every $\mathscr{N}$ has a polygon with the smallest number of sides $\mathbf{0}$ (namely, 4) and a polygon with the largest number of sides $\mathbf{1}$ (namely, 5). Then $(\varphi(P_x)_{x\in\{a,e,g\}}, \leq, \mathbf{0}, \mathbf{1})$ is a complete poset. ■

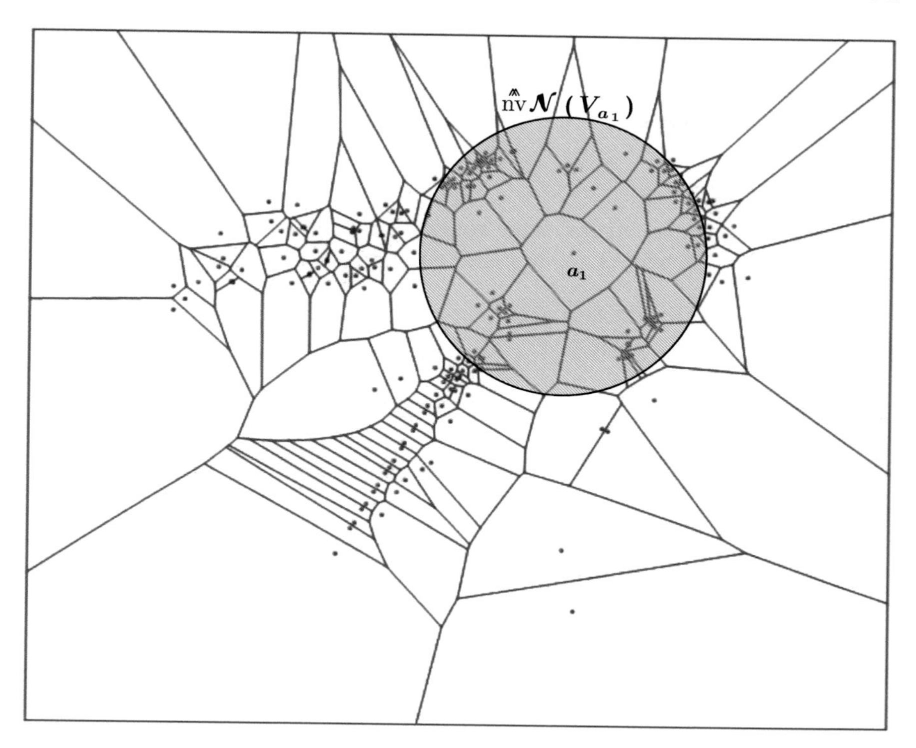

Fig. 14.5 Nucleus nerve in a Voronoï mesh

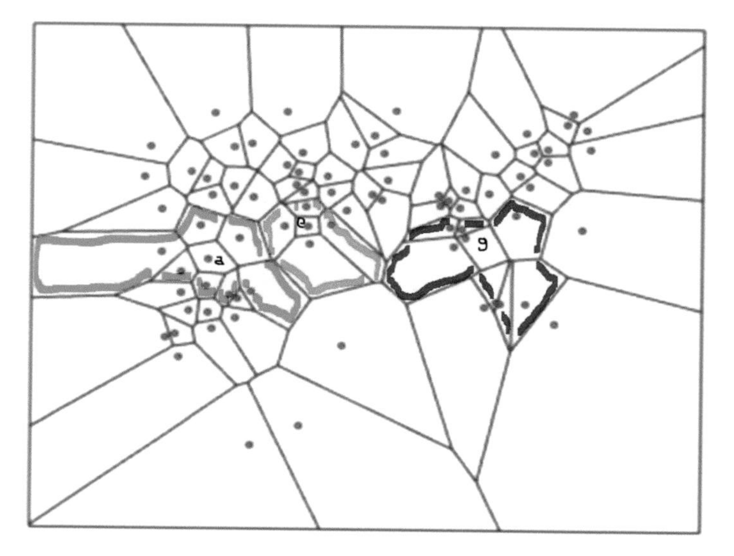

Fig. 14.6 Nuclei complete poset on Voronoï mesh

Fig. 14.7 Chipmunk

The Voronoï region polygons on the boundary of a mesh nerve nucleus are called **satellites**. Each mesh nerve satellite shares an edge with the nerve nucleus.

Example 14.9 V_p is the nucleus of the nerve shown in Fig. 14.3. This nerve has 10 satellites of nucleus V_p, namely, $X_1, \ldots, X_{10}$. ■

Lemma 14.10 *Every satellite of a Voronoï mesh nerve is* $\overset{\wedge\!\!\wedge}{\delta}$*-near the nucleus of the nerve.*

Proof Immediate from the definition of a nerve satellite. □

Theorem 14.11 *A Voronoï mesh nerve is a collection of strongly connected sets.*

Proof Let $\overset{\wedge\!\!\wedge}{\text{nv}}\mathcal{N}$ be a Voronoï mesh nerve. Let $E \in \mathcal{N}$ be the nerve nucleus and let $\mathscr{S}$ the set of satellite polygons of $\mathcal{N}$. Let $X = A \cup E$. From Lemma 14.10, A is strongly near E, i.e., $\text{cl}A \cap \text{cl}E \neq \emptyset$. So X is strongly connected. Hence, $\overset{\wedge\!\!\wedge}{\text{nv}}\mathcal{N}$ is a collection of strongly connected sets X. □

Each pair of nerves in a Voronoï nerve pattern shares one or more satellite edges. In other words, a **Voronoï nerve pattern** is a collection of pairs of strongly near mesh nerves.

Example 14.12 **Strongly Connected Voronoï Mesh Nerves.**
V_a, V_e are nuclei of mesh nerves shown in Fig. 14.6:

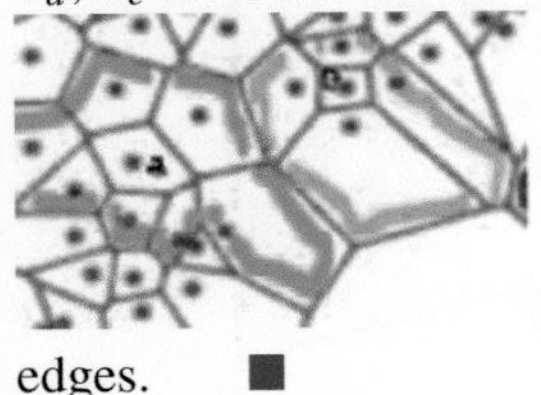

This pair of nerves has two pairs of satellites with common edges. ■

Theorem 14.13 *The nuclei in a complete poset on a Voronoï mesh belong to a nerve pattern.*

Proof See Problem 14.15. □

Problem 14.14 Prove Theorem 14.13. ■

The mesh in Fig. 14.6 is edge-keypoint-based, extracted from the chipmunk image in Fig. 14.7 using MScript 55 in Appendix A.14.

The first step in the construction of an edge-keypoint-based Voronoï mesh is to detect the edges in an image (see Fig. 14.8.1). After that, the keypoints are found among the image edges (see Fig. 14.8.2). Both steps are the result of making reasonable choices for the edge density and the maximum number of keypoints to detect among the edges.

Next, a Voronoï mesh is found using the edge keypoints (see Fig. 14.9.1). This mesh contrasts sharply with a Voronoï mesh constructed using only image keypoints (see Fig. 14.9.2). The edge-keypoints approach forces the Voronoï polygons to be more tightly grouped around image objects than the keypoints-only approach to mesh construction. For this reason, the edge-keypoints approach is considered a more reliable source of mesh nerve patterns.

A comparison of the edge-keypoints mesh and keypoints-only mesh on the chipmunk image is shown in Fig. 14.10. The tight grouping of the edges-keypoints-based

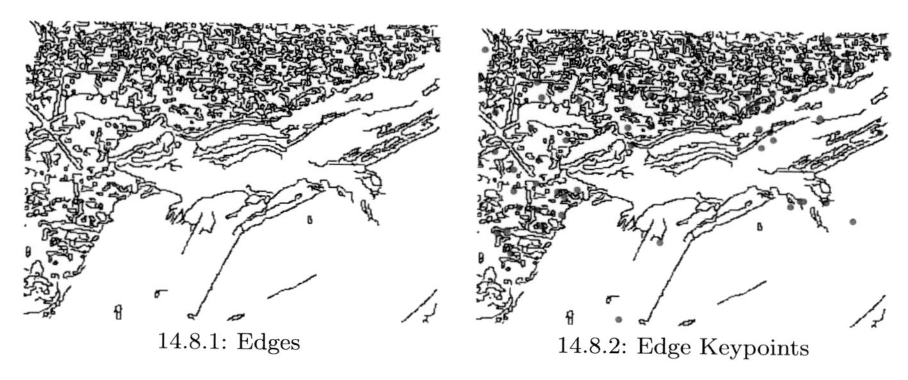

14.8.1: Edges 14.8.2: Edge Keypoints

Fig. 14.8 Sample keypoints on picture edges

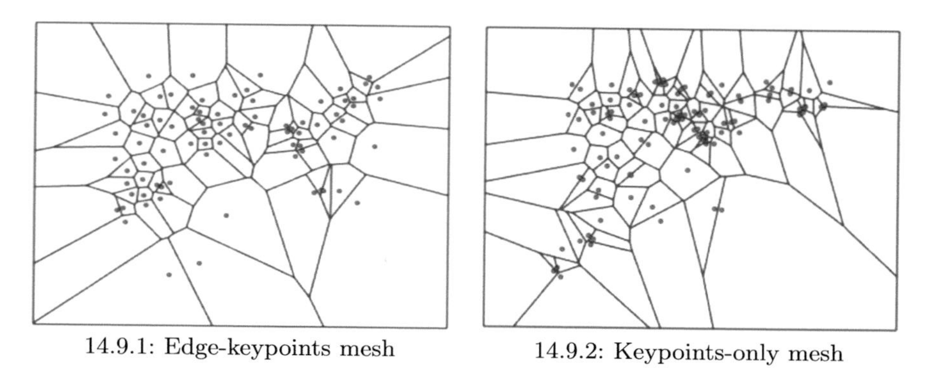

14.9.1: Edge-keypoints mesh 14.9.2: Keypoints-only mesh

Fig. 14.9 Chipmunk edge-keypoints versus keypoints mesh

14.10.1: Edge-keypoints image mesh 14.10.2: Keypoints-only image mesh

Fig. 14.10 Chipmunk edge-keypoints versus keypoints-only picture mesh

Voronoï mesh on the chipmunk image in Fig. 14.10.1 leads to a complete poset containing connected mesh nerves that surround the main object in the picture, namely, the chipmunk.

14.4 Voronoï Mesh Nerve Patterns

Voronoï Mesh Nerve Patterns are found by identifying complete posets of mesh nerve nuclei. This is not an easy problem to solve. The steps in solving this problem are given in Algorithm 19.

Algorithm 19: Complete Posets of Voronoï Mesh Nuclei

Input : Read Voronoï mesh $\mathscr{M}$, $\boldsymbol{glb}, \boldsymbol{lub}$ (estimated largest and smallest number of edges in n-sided polygons in $\mathscr{M}$.

Output: Complete poset containing mesh nerve pattern pat$\mathcal{N}$ in mesh $\mathscr{M}$.

1 $\varphi(P) = n$, *number of edges in polygon* $P \in \mathscr{M}$;
2 $\mathbf{0} \in \mathscr{M} : \varphi(\mathbf{0}) \geq \boldsymbol{glb}$;
3 $\mathbf{1} \in \mathscr{M} : \varphi(\mathbf{1}) \leq \boldsymbol{lub}$;
4 $pat\mathcal{N} := \{\mathbf{0}, \mathbf{1}\}$;
5 $\mathscr{M} := \mathscr{M} \setminus \{\mathbf{0}, \mathbf{1}\}$;
6 /* pat$\mathcal{N}$ contains initial complete poset. */ ;
7 $continue$:= $true$;
8 **while** *(continue)* **do**
9   **if** $\exists ngonE \in \mathscr{M} : \varphi(\mathbf{0}) \leq \varphi(ngonE) \leq \varphi(\mathbf{1})$ **then**
10   pat$\mathcal{N}$:= pat$\mathcal{N} \cup$ ngon$E$;
11         $\mathscr{M} := \mathscr{M} \setminus$ ngon$E$;
12   /* ngon$E$ belongs to complete poset pat$\mathcal{N}$ */
13   **else**
14         $continue$:= $false$;

Problem 14.15 ☕ Give a method to determine the minimum and maximum number of polygon sides used to choose **0**, **1** needed to initialize the complete poset in Algorithm 19. ■

References

1. Edelsbrunner, H., Harer, J.: Computational Topology. An Introduction. American Mathematical Society, Providence (2010). Xii+110 pp., MR2572029
2. Shukla, A.: Theory of Generalized Measures and Applications, Ph.D. thesis. University of Allahabad (2015). Supervisor: M. Khare, 99 pp
3. Khare, M., Gupta, S.: Extensions of nonadditive measures on locally complete σ-continuous lattices. Nova Sad J. Math. **38**(2), 15–23 (2008)
4. Kaplansky, I.: Any orthocomplemented complete modular lattice is a continuous geometry. Ann. Math. **61**, 524–541 (1955)
5. Kaplansky, I.: Orthomodular Lattices. Academic Press, London (1983)

Appendix A
Mathematica and Matlab Scripts

This Appendix contains Mathematica and Matlab scripts referenced in the chapters. Wolfram Mathematica® 10 is used to write the MScripts (Math scripts) and Matlab® 7.10.0 (R2010a) is also used to write the Matlab scripts.

A.1 Scripts from Chap. 1

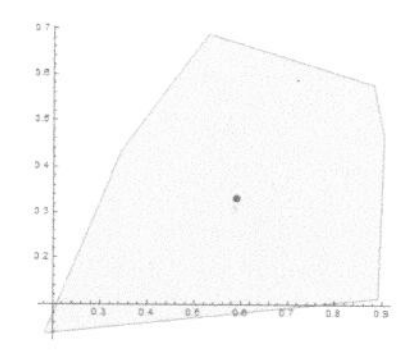

Mscript 16 Convex 2D region centroid.

```
(* 2D Region centroid *)
R = ConvexHullMesh[RandomReal[1, {8, 2}]];
c = RegionCentroid[R]
Show[HighlightMesh[R, Style[{2, Black}, Opacity[0.2]]],
Graphics[{Red, PointSize[0.02], Point[RegionCentroid[R]]}]]
```
■

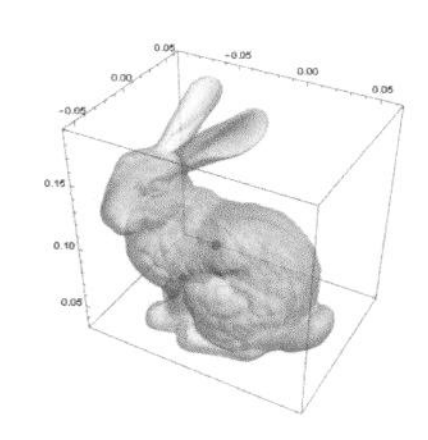

J.F. Peters, *Computational Proximity*, Intelligent Systems Reference Library 102, DOI 10.1007/978-3-319-30262-1

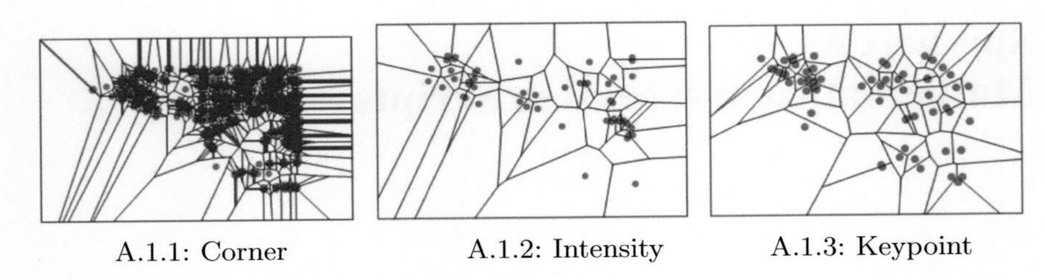

A.1.1: Corner A.1.2: Intensity A.1.3: Keypoint

Fig. A.1 Three Dirichlet tessellations

Mscript 17 StanfordBunny centroid.

```
(* 3D Region centroids *)
gr = ExampleData[{"Geometry3D", "StanfordBunny"}];
gm = DiscretizeGraphics[gr];
c2 = RegionCentroid[gm]
Show[Graphics3D[Prepend[First[gr], Opacity[0.5]]],
Graphics3D[{Red, PointSize[0.03], Point@c2}]]
```

■

Mscript 18 UNISA coin centroid.

```
(* Digital Image Region Centroid *)
img =;
c = ComponentMeasurements[img, "Centroid"][[All, 2]][[1]];
Show[img, Graphics[{Black, PointSize[0.02], Point[c]}]]
```

■

See, e.g., (Fig. A.1) for three types tessellations obtained using Mscript 19.

Mscript 19 Tessellations from Three Different Types of Seed Points.

```
(*Corner, intensity, keypoint – based Dirichlet Tessellations*)
img =;
ImageDimensions[img];
corncoords = ImageCorners[img, MaxFeatures → 50];
HighlightImage[img, corncoords]
ListPlot[corncoords, PlotStyle → {Red, PointSize[.01]},
PlotLegends → {"corner"}, AxesLabel → {wpixels, ht(pixels)},
LabelStyle → Directive[Blue, Bold]]
(* Give the Voronoï mesh and display on the image *)
```

```
vm = VoronoiMesh[corncoords];
t12 = Graphics[{Red, {PointSize[0.02], Point[corncoords]}, Blue,
MeshPrimitives[vm, 1], {Opacity[0.1], EdgeForm[{Thick, Yellow}],
FaceForm[Yellow]}}]
Show[img, t12]
(*p = ImageValue[img, {300, 200}]*)
pts = ImageValuePositions[img, White, 0.45];
vm2 = VoronoiMesh[pts];
t13 = Graphics[{Red, {PointSize[0.02], Point[pts]}, Blue,
MeshPrimitives[vm2, 1], {Opacity[0.1], EdgeForm[{Thick, Yellow}],
FaceForm[Yellow]}}]
Show[img, t13]
corncoords2 = ImageKeypoints[img, MaxFeatures → 60];
vm3 = VoronoiMesh[corncoords2];
t15 = Graphics[{Red, {PointSize[Large], Point[corncoords2]}, Blue,
MeshPrimitives[vm3, 1], {Opacity[0.1], EdgeForm[{Thick, Yellow}],
FaceForm[Yellow]}}]
Show[img, t15]   ■
```

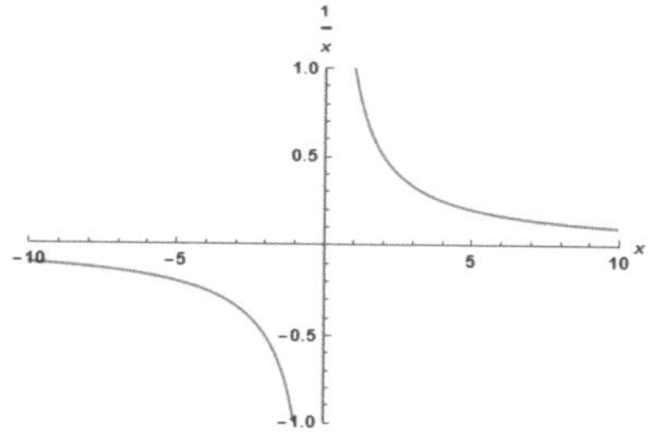

Mscript 20 Conventional near sets.

```
(*Conventional near sets : Cech, Efremovich, Lodato, ...*)   ■

Plot[1/x, {x, −10, 10}, AxesLabel → {x, 1/x},
PlotRange → {{−10, 10}, {−1, 1}},
LabelStyle → Directive[Blue, Bold]]   ■
```

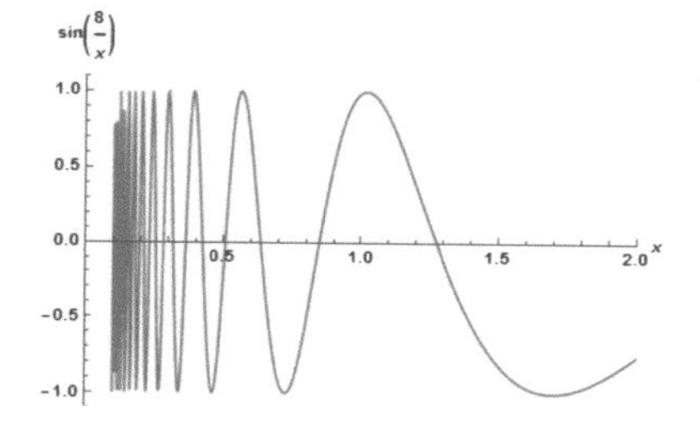

Mscript 21 Strongly near sets.

```
(*Strongly near sets : intersection of A, B > 1*)

Plot[Sin[8/x], {x, 0.1, 10}, AxesLabel → {x, Sin[8/x]},
PlotRange → {{0, 2}, {−1.1, 1.1}},
LabelStyle → Directive[Blue, Bold]] ■
```

Mscript 22 Mesh on a Hooke Needle Point Image.

```
(* Mesh Containing Multiple Nearves *)

img = ;
corncoords = ImageCorners[img, MaxFeatures → 60];
HighlightImage[img, corncoords]

(* Give the Voronoï mesh and display on the image *)
vm = VoronoiMesh[corncoords];
t11 = Graphics[{Blue, {PointSize[Large], Point[corncoords]}, Black,
MeshPrimitives[vm, 1]}]
t12 = Graphics[{Red, {PointSize[Large], Point[corncoords]}, Green,
MeshPrimitives[vm, 1], {Opacity[0.1], EdgeForm[{Thick, Yellow}], FaceForm[Yellow]}}]
Show[img, t12] ■
```

Mscript 23 Image Raster Mesh.

```
(* Nearness of image pixel keypoint mesh parts *)
```

```
img = ;
Rasterize[img, RasterSize → 32](*32pixelwiderasterization.*)
x1 = ImageKeypoints[img];
HighlightImage[img, x1, "HighlightColor" → White];
P1 = ImageKeypoints[img, {"Position", "Scale", "Orientation", "ContrastSign"},
MaxFeatures → 100];
Show[img,
Graphics[
Table[{{If[p[[4]] == 1, Red, Green], Circle[p[[1]], p[[2]] * 3.5],
Line[{p[[1]], p[[1]] + p[[2]] * 1.5 * {Cos[p[[3]]], Sin[p[[3]]]}}]}}, {p, P1}]]] ■
```

Mscript 24 Image Keypoint Mesh Nerves.

```
(* Mesh Containing Multiple Keypoint Nearves *)
```

```
img = ;
```

```
corncoords = ImageKeypoints[img, MaxFeatures → 60];
HighlightImage[img, corncoords]

(* Give the Voronoï mesh and display on the image *)
vm = VoronoiMesh[corncoords];
tl1 = Graphics[{Blue, {PointSize[Large], Point[corncoords]}, Black,
MeshPrimitives[vm, 1]}]
tl2 = Graphics[{Red, {PointSize[Large], Point[corncoords]}, Green,
MeshPrimitives[vm, 1], {Opacity[0.1], EdgeForm[{Thick, Yellow}], FaceForm[Yellow]}}]
Show[img, tl2]
```
■

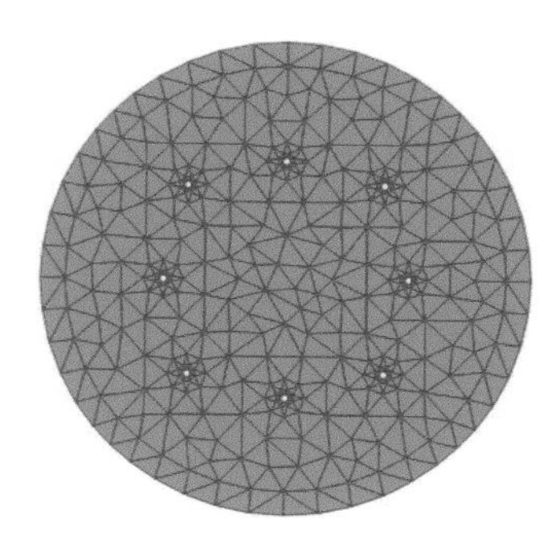

Mscript 25 Mesh on an Annulus.

```
(*Set up a boundedness that is amesh circular region. Each
inner circular region is aproximal nerve.*)

annulus[x_, y_]:=(9/10)^2 ≤ x^2 + y^2 ≤ 1^2
holes[{x0_, y0_}, r_]:=((x + x0)^2 + (y + y0)^2 ≤ (r)^2)
crds = {{−1/2, 0}, {1/2, 0}, {0, −1/2}, {0, 1/2}, {2/5, 2/5}, {−2/5, −2/5},
{2/5, −2/5}, {−2/5, 2/5}};
sd = Or@@(holes[#, 1/64]&/@crds);
Ω2 = ImplicitRegion[Or[annulus[x, y], sd], {x, y}];
ToElementMesh[Ω2, "BoundaryMeshGenerator" → {"Continuation"}, "RegionHoles" → crds][
"Wireframe"["MeshElementStyle" → {Directive[FaceForm[Green], EdgeForm[{Red}]]}]]
```
■

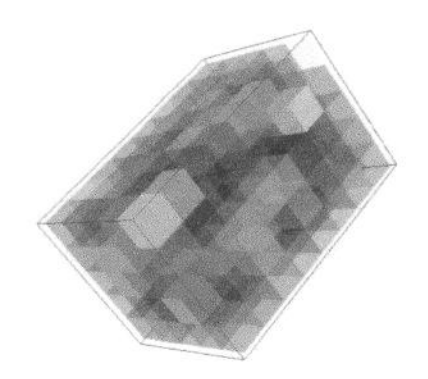

Mscript 26 Four-Dimensional Array of Cubes.

```
(*Four − dimensional array of cubes.*)
Graphics3D[Raster3D[RandomReal[1, {8, 5, 5, 3}],
ColorFunction →
(RGBColor[#[[1]], #[[2]], #[[3]], 0.299#[[1]] + 0.587#[[2]] + 0.114#[[3]]]&)]]
```
■

A.2 Scripts from Chap. 2

Mscript 27 Delaunay and Voronoï Mesh on Hand Image CornerDelaunay.nb.

```
(*Corner – based Delaunay triangulation*)
```

```
img = [image];
ImageDimensions[img]
corncoords = ImageCorners[img, MaxFeatures → 50]
HighlightImage[img, corncoords]
ListPlot[corncoords, PlotStyle → {Black, PointSize[.01]},
PlotLegends → {"corner"}, AxesLabel → {wpixels, ht(pixels)},
LabelStyle → Directive[Blue, Bold]]
(* Give the Delaunay mesh and display on the image *)
vm = DelaunayMesh[corncoords];
dm = VoronoiMesh[corncoords];
t11 = Graphics[{Blue, {PointSize[0.02], Point[corncoords]}, Black,
MeshPrimitives[vm, 1]}]
t12 = Graphics[{Red, {PointSize[0.02], Point[corncoords]}, Blue, MeshPrimitives[vm, 1],
{Opacity[0.1], EdgeForm[{Thick, Yellow}], FaceForm[Yellow]}}]
t13 = Graphics[{Red, {PointSize[0.02], Point[corncoords]}, Yellow,
MeshPrimitives[dm, 1], {Opacity[0.1], EdgeForm[{Thick, Yellow}], FaceForm[Yellow]}}]
t14 = Graphics[{Red, {PointSize[0.02], Point[corncoords]}, Blue, MeshPrimitives[dm, 1],
{Opacity[0.1], EdgeForm[{Thick, Yellow}], FaceForm[Yellow]}}]
Show[img, t12]
Show[img, t13]
Show[img, t12, t13]
```
■

```
% centroid-based image Delaunay and Voronoi mesh
clc, clear all, close all
im = imread('redcarFrame.jpg');
% if size(im,3)==3
%      g=rgb2gray(im);
% end
[m,n]=size(im);
bw = im2bw(im,0.5); % threshold at 50%
bw = bwareaopen(bw,2); % remove objects less 2 than pixels
stats = regionprops(bw,'Centroid'); % centroid coordinates
centroids = cat(1,stats.Centroid);
fc=[1 1;n 1; 1 m; n m]; % identify image corners
centroids=[centroids;fc];
imshow(im),hold on
plot(centroids(:,1),centroids(:,2),'y+')
hold on;
X=centroids(:,1);
Y=centroids(:,2);
%find delaunay triangulation
TRI = delaunay(X,Y);
triplot(TRI,X,Y,'y');
%find Voronoi tessellation
[vx,vy] = voronoi(X,Y);
plot(vx,vy,'g-');
```

Listing A.1 Matlab script in `findCentroidalMesh.m` to construct a centroidal-based mesh.

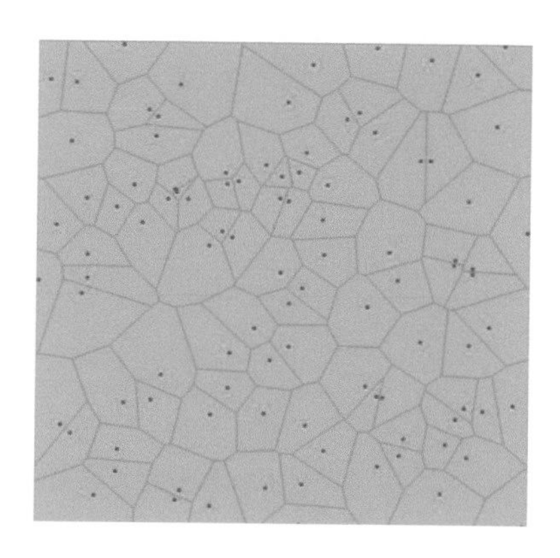

Mscript 28 Centroid-based Voronoï mesh: ch2centroidalMesh.nb.

```
(*Mesh experiment with centroids.*)
pts = RandomReal[1, {100, 2}];
ℛ = VoronoiMesh[pts, {{0, 1}, {0, 1}}];
```

```
vcf[pts_]:=
RegionCentroid/@MeshPrimitives[VoronoiMesh[pts, {{0, 1}, {0, 1}}], "Faces"];
centroids = NestList[vcf, pts, 3];
vm = VoronoiMesh[#, {{0, 1}, {0, 1}}]&/@centroids
MapThread[Show[#1, Graphics[{Black, Point[#2], Red, Point[vcf[#2]]}]]&,
{vm, centroids}]
Show[R, Graphics[{Red, Point[pts]}]]   ■
```

Mscript 29 Centroid-based Voronoï mesh: centroidCornerVoronoiMesh.nb.

```
(*Centroids N.B. position of PointSize[0.01]*)
Im1 = ;
im1corners = ImageCorners[Im1, 3, 0.01, 5];
im1vm = VoronoiMesh[im1corners];
im1centroids = RegionCentroid/@MeshPrimitives[im1vm, "Faces"];
centroids = Graphics[{PointSize[0.01], Red, Point[im1centroids]}];
corners = Graphics[{PointSize[0.01], Yellow, Point[im1corners]}];
ptsCorners = MeshCells[im1vm, 1];
mesh = VoronoiMesh[im1centroids];
mesh2 = VoronoiMesh[centroids];
pts = MeshCoordinates[mesh];
lines = MeshCells[mesh, 1];
vm = MeshRegion[pts, lines]
mesh3 = HighlightMesh[vm, Style[1, Green]]
Show[Im1, vm]
Show[Im1, centroids]
Show[Im1, vm, centroids]
Show[Im1, mesh3, centroids]
Show[Im1, mesh3, centroids, corners]
   ■
```

A.3 Scripts from Chap. 3

Mscript 30 Unbounded descriptive Nbd: UnboundedDescriptiveNbd.nb.

(* Unbounded Descriptive Neighbourhood of a Point *)

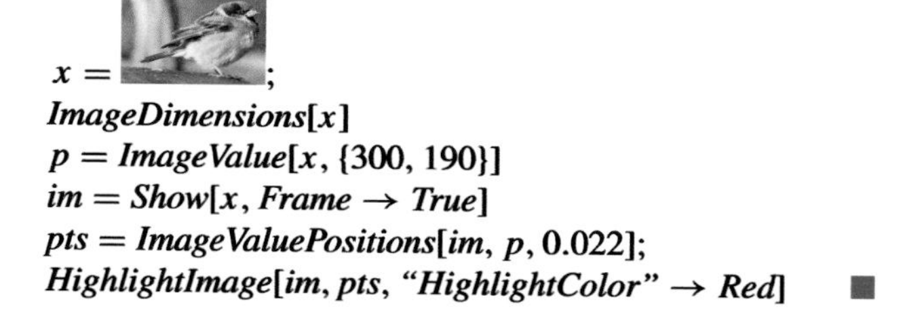

```
x = [image];
ImageDimensions[x]
p = ImageValue[x, {300, 190}]
im = Show[x, Frame → True]
pts = ImageValuePositions[im, p, 0.022];
HighlightImage[im, pts, "HighlightColor" → Red]
```

■

Note The following picture neighbourhood GUI is a minor revision of a script written by Binglin Li. A sample bounded descriptive nhdb using the neigbourhood GUI is shown in Fig. A.2.

```
function varargout = Nbds(varargin)
% NBDS MATLAB code for Nbds.fig
%       NBDS, by itself, creates a new NBDS or raises the existing
%       singleton*.
%
%       H = NBDS returns the handle to a new NBDS or the handle to
```

Fig. A.2 Bounded descriptive Nbd at (347,320), $r = 100$, $\varepsilon = 50$

```
%       the existing singleton*.
%
%       NBDS('CALLBACK',hObject,eventData,handles,...) calls the local
%       function named CALLBACK in NBDS.M with the given input arguments.
%
%       NBDS('Property','Value',...) creates a new NBDS or raises the
%       existing singleton*.  Starting from the left, property value pairs are
%       applied to the GUI before Nbds_OpeningFcn gets called.  An
%       unrecognized property name or invalid value makes property application
%       stop.  All inputs are passed to Nbds_OpeningFcn via varargin.
%
%       *See GUI Options on GUIDE's Tools menu.  Choose "GUI allows only one
%       instance to run (singleton)".
%
% See also: GUIDE, GUIDATA, GUIHANDLES

% Edit the above text to modify the response to help Nbds

% Last Modified by GUIDE v2.5 05-Feb-2015 15:58:09

% Begin initialization code - DO NOT EDIT
gui_Singleton = 1;
gui_State = struct('gui_Name',       mfilename, ...
                   'gui_Singleton',  gui_Singleton, ...
                   'gui_OpeningFcn', @Nbds_OpeningFcn, ...
                   'gui_OutputFcn',  @Nbds_OutputFcn, ...
                   'gui_LayoutFcn',  [] , ...
                   'gui_Callback',   []);
if nargin && ischar(varargin{1})
    gui_State.gui_Callback = str2func(varargin{1});
end

if nargout
    [varargout{1:nargout}] = gui_mainfcn(gui_State, varargin{:});
else
    gui_mainfcn(gui_State, varargin{:});
end
% End initialization code - DO NOT EDIT


% --- Executes just before Nbds is made visible.
function Nbds_OpeningFcn(hObject, eventdata, handles, varargin)
% This function has no output args, see OutputFcn.
% hObject    handle to figure
% eventdata  reserved - to be defined in a future version of MATLAB
% handles    structure with handles and user data (see GUIDATA)
% varargin   command line arguments to Nbds (see VARARGIN)

% Choose default command line output for Nbds
handles.output = hObject;

% Update handles structure
guidata(hObject, handles);

% UIWAIT makes Nbds wait for user response (see UIRESUME)
% uiwait(handles.figure1);


% --- Outputs from this function are returned to the command line.
function varargout = Nbds_OutputFcn(hObject, eventdata, handles)
% varargout  cell array for returning output args (see VARARGOUT);
% hObject    handle to figure
% eventdata  reserved - to be defined in a future version of MATLAB
% handles    structure with handles and user data (see GUIDATA)

% Get default command line output from handles structure
varargout{1} = handles.output;
```

```
function edit1_Callback(hObject, eventdata, handles)
% hObject    handle to edit1 (see GCBO)
% eventdata  reserved - to be defined in a future version of MATLAB
% handles    structure with handles and user data (see GUIDATA)

% Hints: get(hObject,'String') returns contents of edit1 as text
%        str2double(get(hObject,'String')) returns contents of edit1 as a double


% --- Executes during object creation, after setting all properties.
function edit1_CreateFcn(hObject, eventdata, handles)
% hObject    handle to edit1 (see GCBO)
% eventdata  reserved - to be defined in a future version of MATLAB
% handles    empty - handles not created until after all CreateFcns called

% Hint: edit controls usually have a white background on Windows.
%       See ISPC and COMPUTER.
if ispc && isequal(get(hObject,'BackgroundColor'), get(0,'defaultUicontrolBackgroundColor'))
    set(hObject,'BackgroundColor','white');
end



function edit2_Callback(hObject, eventdata, handles)
% hObject    handle to edit2 (see GCBO)
% eventdata  reserved - to be defined in a future version of MATLAB
% handles    structure with handles and user data (see GUIDATA)

% Hints: get(hObject,'String') returns contents of edit2 as text
%        str2double(get(hObject,'String')) returns contents of edit2 as a double


% --- Executes during object creation, after setting all properties.
function edit2_CreateFcn(hObject, eventdata, handles)
% hObject    handle to edit2 (see GCBO)
% eventdata  reserved - to be defined in a future version of MATLAB
% handles    empty - handles not created until after all CreateFcns called

% Hint: edit controls usually have a white background on Windows.
%       See ISPC and COMPUTER.
if ispc && isequal(get(hObject,'BackgroundColor'), get(0,'defaultUicontrolBackgroundColor'))
    set(hObject,'BackgroundColor','white');
end


% --- Executes on button press in pushbutton1.
function pushbutton1_Callback(hObject, eventdata, handles)
% hObject    handle to pushbutton1 (see GCBO)
% eventdata  reserved - to be defined in a future version of MATLAB
% handles    structure with handles and user data (see GUIDATA)


% --- Executes on button press in pushbutton2.
function pushbutton2_Callback(hObject, eventdata, handles)
% hObject    handle to pushbutton2 (see GCBO)
% eventdata  reserved - to be defined in a future version of MATLAB
% handles    structure with handles and user data (see GUIDATA)


% --- Executes on button press in pushbutton3.
function pushbutton3_Callback(hObject, eventdata, handles)
% hObject    handle to pushbutton3 (see GCBO)
% eventdata  reserved - to be defined in a future version of MATLAB
% handles    structure with handles and user data (see GUIDATA)
```

```
%—— Executes on button press in pushbutton4.
function pushbutton4_Callback(hObject, eventdata, handles)
% hObject    handle to pushbutton4 (see GCBO)
% eventdata  reserved – to be defined in a future version of MATLAB
% handles    structure with handles and user data (see GUIDATA)


%————————————————————————————————————
function Untitled_1_Callback(hObject, eventdata, handles)
% hObject    handle to Untitled_1 (see GCBO)
% eventdata  reserved – to be defined in a future version of MATLAB
% handles    structure with handles and user data (see GUIDATA)


%————————————————————————————————————
function Open_Callback(hObject, eventdata, handles)
% hObject    handle to Open (see GCBO)
% eventdata  reserved – to be defined in a future version of MATLAB
% handles    structure with handles and user data (see GUIDATA)
[FileName,PathName,FilterIndex] = uigetfile('*.*');
axes(handles.axes1);
I=imread(strcat(PathName, FileName));
handles.rgbimage=[];
%%%%% comment four lines below if we want to find Nbd in color images.
if size(I,3)==3
    handles.rgbimage=I;
    I=rgb2gray(I);
end;
%%%%%%%%%%
imshow(I,[]);
handles.I=I;
guidata(hObject, handles);


%—— Executes on mouse press over axes background.
function axes1_ButtonDownFcn(hObject, eventdata, handles)
% hObject    handle to axes1 (see GCBO)
% eventdata  reserved – to be defined in a future version of MATLAB
% handles    structure with handles and user data (see GUIDATA)
%figure;imshow(handles.I);

%—— Executes on mouse press over axes background.
function figure1_WindowButtonDownFcn(hObject, eventdata, handles)
% hObject    handle to axes1 (see GCBO)
% eventdata  reserved – to be defined in a future version of MATLAB
% handles    structure with handles and user data (see GUIDATA)

radius = str2double(get(handles.edit1,'String')); %read the radius value
epsilon = str2double(get(handles.edit2,'String'));

point = get(gca,'CurrentPoint'); %get the point click on the figure
y=point(1,1);
x=point(1,2);
if (get(handles.radiobutton1,'Value')==get(handles.radiobutton1,'Max'))
    disp('Bounded Descriptive Nbd selected');
    BDNbd(handles.I,x,y,epsilon,radius);
    disp(strcat('X(xpos,ypos):(',num2str(round(x)),',',num2str(round(y)),')'));
end

if (get(handles.radiobutton2,'Value')==get(handles.radiobutton2,'Max'))
    disp('Unbounded Descriptive Nbd selected');
     UBDNbd(handles.I,x,y,epsilon);
    disp(strcat('X(xpos,ypos):(',num2str(round(x)),',',num2str(round(y)),')'));
end

if (get(handles.radiobutton3,'Value')==get(handles.radiobutton3,'Max'))
    disp('Spherical Nbd selected');
```

```
    sphNbd(handles.I,x,y,radius);
    disp(strcat('X(xpos,ypos):(',num2str(round(x)),',',num2str(round(y)),')'));
end

if (get(handles.radiobutton4,'Value')==get(handles.radiobutton4,'Max'))
    disp('Bounded Indistinguable Descriptive Nbd selected');
    BDIndDNbd(handles.I,x,y,radius);
    disp(strcat('X(xpos,ypos):(',num2str(round(x)),',',num2str(round(y)),')'));
end


function BDNbd(I,x,y,epsilon,radius)
%this is a description based neighbourhood works on color
color=double(I(round(x),round(y),:));
% measure the descriptive distance
if (numel(color))==3 %means that I is a color image
    I=double(I);
  [row,col,v]=find(sqrt((I(:,:,1)-color(:,:,1)).^2+(I(:,:,2)-color(:,:,2)).^2+(I(:,:,3)-color(:,:,3))
        .^2)<epsilon);
end;

if (numel(color)==1)%==gray level image
    I=double(I);
    [row,col]=find(color-epsilon<I&I<color+epsilon);
    I3=zeros(size(I,1),size(I,2),3);
    I3(:,:,1)=I;I3(:,:,2)=I;I3(:,:,3)=I;
    I=I3;
end;
% caculate the spatial distance
  P=repmat([round(x),round(y)],[length(row),1]);
  Q=[row,col];
  V=Q-P;
  r=find(sqrt(V(:,1).^2+V(:,2).^2)<radius);
  l=length(r);
  row1=row(r(1:l));
  col1=col(r(1:l));
 I3=zeros(size(I,1),size(I,2),3);
% make and show the Nbd
    I3(:,:,1)=I(:,:,1);
    I3(:,:,2)=I(:,:,2);
    I3(:,:,3)=I(:,:,3);
    for i=1:l
    I3(row1(i),col1(i),2)=255;
    end

figure;imshow(I3./255);
hold on;
rectangle('position',[y-radius,x-radius,2*radius,2*radius],'Curvature',[1,1],'EdgeColor','b','
      LineWidth',2);
plot(y,x,'.g');
title(['Bounded Descriptive Nbd around point (',num2str(y),',',num2str(x),')']);

function UBDNbd(I,x,y,epsilon)
%this is a description based neighbourhood works on color
color=double(I(round(x),round(y),:));
% measure the descriptive distance
if (numel(color))==3 %means that I is a color image
    I=double(I);
  j=find(sqrt((I(:,:,1)-color(:,:,1)).^2+(I(:,:,2)-color(:,:,2)).^2+(I(:,:,3)-color(:,:,3)).^2)<
        epsilon);
end;
if (numel(color)==1)%==gray level image
    I=double(I);
    j=find(color-epsilon<I&I<color+epsilon);
    I3=zeros(size(I,1),size(I,2),3);
    I3(:,:,1)=I;I3(:,:,2)=I;I3(:,:,3)=I;
    I=I3;
```

```
end;
 I3=zeros(size(I,1),size(I,2),3);
% make and show the Nbd
     I1=I(:,:,2);
     I1(j)=255;
     I3(:,:,1)=I1;
     I3(:,:,2)=I(:,:,2);
     I3(:,:,3)=I(:,:,3);

 figure;imshow(I3./255);
 hold on;
 plot(y,x,'.g');
 title(['Unbounded Descriptive Nbd around point (',num2str(y),',',num2str(x),')'])

 function sphNbd(I,x,y,radius)
%this is a description based neighbourhood works on color
% caculate the spatial distance
[m,n,k]=size(I);
P=repmat([round(x),round(y)],[m*n,1]);
[X,Y]=ind2sub([m,n],1:m*n);
Q=[X;Y]';
V=Q-P;
D=sqrt(V(:,1).^2+V(:,2).^2);
j=D<radius;
if (numel(I(1,1,:))==1)%graylevel image
        I=double(I);
        I3=zeros(size(I,1),size(I,2),3);
        I3(:,:,1)=I;I3(:,:,2)=I;I3(:,:,3)=I;
        I=I3;
 end;
 I3=zeros(size(I,1),size(I,2),3);
% make and show the Nbd
     I1=I(:,:,2);
     I1(j)=255;
     I3(:,:,1)=I1;
     I3(:,:,2)=I(:,:,2);
     I3(:,:,3)=I(:,:,3);

 figure;imshow(I3./255);
 hold on;
 plot(y,x,'.g');
 title(['Spherical Nbd around point (',num2str(y),',',num2str(x),')'])

 function BDIndDNbd(I,x,y,radius)
%this is a description based neighbourhood works on color
color=double(I(round(x),round(y),:));
% epsilon equals zero
if (numel(color))==3 %means that I is a color image
     I=double(I);
   [row,col,v]=find(sqrt((I(:,:,1)-color(:,:,1)).^2+(I(:,:,2)-color(:,:,2)).^2+(I(:,:,3)-color(:,:,3))
         .^2)==0);
end;
if (numel(color)==1)%==gray level image
     I=double(I);
     [row,col,v]=find(I==color);
     I3=zeros(size(I,1),size(I,2),3);
     I3(:,:,1)=I;I3(:,:,2)=I;I3(:,:,3)=I;
     I=I3;
end;
% caculate the spatial distance
   P=repmat([round(x),round(y)],[length(row),1]);
   Q=[row,col];
   V=Q-P;
   r=find(sqrt(V(:,1).^2+V(:,2).^2)<radius);
   l=length(r);
   row1=row(r(1:l));
   col1=col(r(1:l));
```

```
  I3=zeros(size(I,1),size(I,2),3);
% make and show the Nbd
    I3(:,:,1)=I(:,:,1);
    I3(:,:,2)=I(:,:,2);
    I3(:,:,3)=I(:,:,3);
    for i=1:1
    I3(row1(i),col1(i),1)=255;
    end

 figure;imshow(I3./255);
 hold on;
 rectangle('position',[y-radius,x-radius,2*radius,2*radius],'Curvature',[1,1],'EdgeColor','b','
      LineWidth',2);
 plot(y,x,'.g');
 title(['Bounded Indistinguable Descriptive Nbd around point (',num2str(y),',',num2str(x),')'])
```

Listing A.2 Matlab script in `Nbds.m` to construct bounded and unbounded neighbourhoods of picture points.

Mscript 31 Prob. 3.18: Highest Adjacency Voronoï Polygon.

```
(*Problem 3.18, based on solution by Binglin Li*)
(* Find the Voronoï region with the highest number of adjacent regions *)
img = ;
corncoords = ImageCorners[img, MaxFeatures → 80];
HighlightImage[img, corncoords]
vm = VoronoiMesh[corncoords];
t11 = Graphics[{Green, {PointSize[Large], Point[corncoords]}, Black, MeshPrimitives[vm, 1]}];
Show[img, t11]
(*FindthelistofpolygonsfromtheVoronoïmesh.Thepolygonsarerepresentedbythevertices
intheorderoftheascendingdimensionoftheverticesofeachpolygon.Thenumberofpolygons
isthenumberofsites, thatisdv = 50*)
pv = MeshPrimitives[vm, 2];
dv = Length[pv];
(* We find the dimension of vertices of each polygon *)
numv = Length/@Level[pv, {2}];
(*Findthemaximumvalueinthedimensionarray.AsforaVoronoïregion,
thenumberofstronglynearregionsaroundaparticularregionequalstothenumberofitsedges
ornumberofitsvertices*)
sv = Max[numv]
(* Count the number of Voronoï regions that have the maximum adjacent regions *)
nv = Count[numv, sv]
(* Get the Position of the Voronoï regions that have the maximum adjacent regions*)
index1 = Level[Position[numv, sv], {2}];
(*Voronoïregionsincludetheboundaryandinterior,
sohereEdgeFormandFaceFormarebothusedtodisplaythecolorofregions*)
t12 = Graphics[{Green, {PointSize[Large], Point[corncoords]}, Black, MeshPrimitives[vm, 1],
{EdgeForm[{Thick, Yellow}], FaceForm[Yellow], pv[[index1]]}}];
Show[img, t12]
```
■

A.4 Scripts from Chap. 4

Mscript 32 Keypoint-Based Dirichlet Tessellation.

```
(*Keypoint − based Dirichlet Tesselation*)
img =;
  □
ImageDimensions[img]
corncoords2 = ImageKeypoints[img, MaxFeatures → 60];
vm3 = VoronoiMesh[corncoords2];
t15 = Graphics[{Red, {PointSize[Large], Point[corncoords2]}, Blue,
MeshPrimitives[vm3, 1], {Opacity[0.1], EdgeForm[{Thick, Yellow}],
FaceForm[Yellow]}}]
Show[img, t15]   ■
```

Mscript 33 EdgeCorner-Based Mesh.

```
(*Geometry of an Image Scene from train station in Salerno, Italy*)

im0 = Pruning[]
corncoords = ImageCorners[im0, MaxFeatures → 80];
vm = VoronoiMesh[corncoords];
t13 = Graphics[{Red, {PointSize[Large], Point[corncoords]}, Green,
MeshPrimitives[vm, 1], {Opacity[0.1], EdgeForm[{Thick, Yellow}],
```

```
FaceForm[Yellow]}}];
dm = DelaunayMesh[corncoords];
t15 = Graphics[{Red, {PointSize[Large], Point[corncoords]}, Orange,
MeshPrimitives[dm, 1], {Opacity[0.1], EdgeForm[{Thick, Orange}],
FaceForm[Yellow]}}];
Show[im0, t13]
Show[im0, t15]
Show[im0, t13, t15]
```
■

A.5 Scripts from Chap. 5

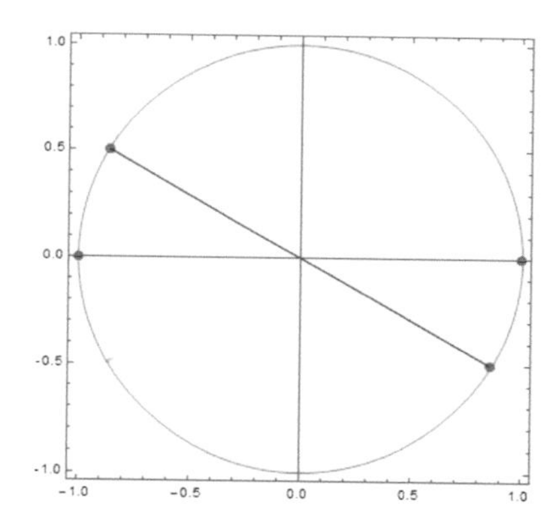

Mscript 34 Classical Borsuk-Ulam Theorem.

```
(*Classical Borsuk − Ulam Theorem for S2 *)
S2 = Circle[{0, 0}, 1.0];
p1 = {−0.86, 0.5}; p2 = {0.86, −0.5};
p5 = {−1.0, 0.0}; p8 = {1.0, 0.0};
Graphics[{{Gray, S2}, {PointSize[Large], Red, Point[p1], Point[p2],
Point[p5], Point[p8]}},
Frame → True, Axes → True]
```
■

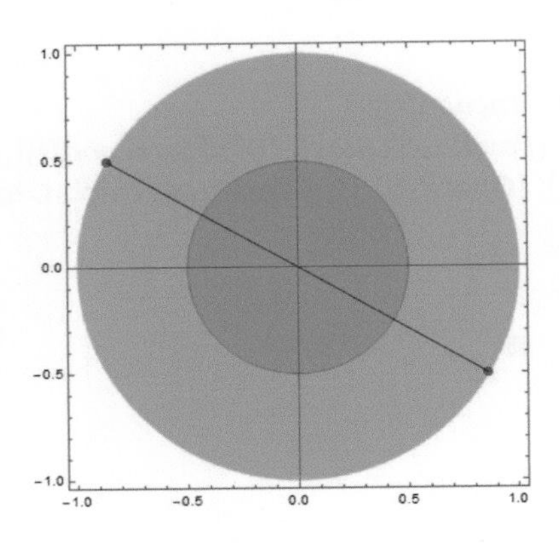

Mscript 35 Proximal Borsuk-Ulam Theorem.

```
(*Proximal Borsuk − Ulam Theorem for S2 *)
S3 = Disk[{0, 0}, 1.0];
S4 = Disk[{0, 0}, 0.5];
S5 = RegionBoundary[S4];
p3 = {−0.86, 0.5}; p4 = {0.86, −0.5};
Graphics[{{Hue[1/3, 1, 1, .5], S3, Green, S4, {Blue, S5}},
{PointSize[Large], Red, Point[p3], Point[p4]}},
Frame → True, Axes → True]
```
■

Mscript 36 Homotopic Mapping from ? to J with a dot on top.

```
(*This sample homotopy visualization is an application of the script given in
http : //mathematica.stackexchange.com/questions/59463/homotopy − visualization
*)
img = ? j;
Show[img]
boundaries = Thinning@EdgeDetect[img, 1]
comp = MorphologicalComponents@boundaries;
pdata = Position[comp, #].{{0, −1}, {1, 0}}&/@{1, 2, 3, 4};
curves = FindCurvePath/@pdata;
cIFNs =
MapThread[
```

```
Interpolation[Transpose@{Rescale@Range@Length@First@#2, #1[[First@#2]]},
PeriodicInterpolation → True]&, {pdata, curves}];
(*offset between middle of ? and middle of J with a dot on top*)
offset =
First@Differences[Mean@Through[{Min, Max}[#]]&/@pdata[[{3, 4}, All, 1]]]
Manipulate[
ParametricPlot[{(1 − p)cIFNs[[1]][t] + p(cIFNs[[2]][t] + {−offset, 0}),
(1 − p)cIFNs[[3]][t] + p(cIFNs[[4]][t] + {−offset, 0})}, {t, 0, 1},
Axes → False, Frame → True,
PlotRange → {{0, Total[Through[{Min, Max}[pdata[[3, All, 1]]]]]},
{−Last@ImageDimensions[img], 0}}], {p, 0, 1}
]
```
■

A.6 Scripts from Chap. 6

```
function varargout = TA_Solution13(varargin)
% TA_SOLUTION13 M-file for TA_Solution13.fig
%       Solution by H. Fashandi, 2013
%       Revised by J.F. Peters, 2015
%       TA_SOLUTION13, by itself, creates a new TA_SOLUTION13 or raises the existing
%       singleton*.
%
%       H = TA_SOLUTION13 returns the handle to a new TA_SOLUTION13 or the handle to
%       the existing singleton*.
%
%       TA_SOLUTION13('CALLBACK',hObject,eventData,handles,...) calls the local
%       function named CALLBACK in TA_SOLUTION13.M with the given input arguments.
%
%       TA_SOLUTION13('Property','Value',...) creates a new TA_SOLUTION13 or raises the
%       existing singleton*. Starting from the left, property value pairs are
%       applied to the GUI before TA_Solution13_OpeningFcn gets called. An
%       unrecognized property name or invalid value makes property application
%       stop. All inputs are passed to TA_Solution13_OpeningFcn via varargin.
%
%       *See GUI Options on GUIDE's Tools menu. Choose "GUI allows only one
```

```
%       instance to run (singleton)".
%
% See also: GUIDE, GUIDATA, GUIHANDLES

% Edit the above text to modify the response to help TA_Solution13

% Last Modified by GUIDE v2.5 10-Aug-2012 17:35:11

% Begin initialization code - DO NOT EDIT
gui_Singleton = 1;
gui_State = struct('gui_Name',       mfilename, ...
                   'gui_Singleton',  gui_Singleton, ...
                   'gui_OpeningFcn', @TA_Solution13_OpeningFcn, ...
                   'gui_OutputFcn',  @TA_Solution13_OutputFcn, ...
                   'gui_LayoutFcn',  [] , ...
                   'gui_Callback',   []);
if nargin && ischar(varargin{1})
    gui_State.gui_Callback = str2func(varargin{1});
end

if nargout
    [varargout{1:nargout}] = gui_mainfcn(gui_State, varargin{:});
else
    gui_mainfcn(gui_State, varargin{:});
end
% End initialization code - DO NOT EDIT


% --- Executes just before TA_Solution13 is made visible.
function TA_Solution13_OpeningFcn(hObject, eventdata, handles, varargin)
% This function has no output args, see OutputFcn.
% hObject    handle to figure
% eventdata  reserved - to be defined in a future version of MATLAB
% handles    structure with handles and user data (see GUIDATA)
% varargin   command line arguments to TA_Solution13 (see VARARGIN)

% Choose default command line output for TA_Solution13
handles.output = hObject;

% Update handles structure
guidata(hObject, handles);

% UIWAIT makes TA_Solution13 wait for user response (see UIRESUME)
% uiwait(handles.figure1);


% --- Outputs from this function are returned to the command line.
function varargout = TA_Solution13_OutputFcn(hObject, eventdata, handles)
% varargout  cell array for returning output args (see VARARGOUT);
% hObject    handle to figure
% eventdata  reserved - to be defined in a future version of MATLAB
% handles    structure with handles and user data (see GUIDATA)

% Get default command line output from handles structure
varargout{1} = handles.output;



function edit1_Callback(hObject, eventdata, handles)
% hObject    handle to edit1 (see GCBO)
% eventdata  reserved - to be defined in a future version of MATLAB
% handles    structure with handles and user data (see GUIDATA)

% Hints: get(hObject,'String') returns contents of edit1 as text
%        str2double(get(hObject,'String')) returns contents of edit1 as a double
```

```
%—— Executes during object creation, after setting all properties.
function edit1_CreateFcn(hObject, eventdata, handles)
% hObject     handle to edit1 (see GCBO)
% eventdata   reserved – to be defined in a future version of MATLAB
% handles     empty – handles not created until after all CreateFcns called

% Hint: edit controls usually have a white background on Windows.
%       See ISPC and COMPUTER.
if ispc && isequal(get(hObject,'BackgroundColor'), get(0,'defaultUicontrolBackgroundColor'))
    set(hObject,'BackgroundColor','white');
end


%—— Executes on button press in pushbutton1.
function pushbutton1_Callback(hObject, eventdata, handles)
% hObject     handle to pushbutton1 (see GCBO)
% eventdata   reserved – to be defined in a future version of MATLAB
% handles     structure with handles and user data (see GUIDATA)


%————————————————————————————————————————
function File_Callback(hObject, eventdata, handles)
% hObject     handle to File (see GCBO)
% eventdata   reserved – to be defined in a future version of MATLAB
% handles     structure with handles and user data (see GUIDATA)


%————————————————————————————————————————
function Untitled_1_Callback(hObject, eventdata, handles)
% hObject     handle to Untitled_1 (see GCBO)
% eventdata   reserved – to be defined in a future version of MATLAB
% handles     structure with handles and user data (see GUIDATA)


%————————————————————————————————————————
function Open_Callback(hObject, eventdata, handles)
% hObject     handle to Open (see GCBO)
% eventdata   reserved – to be defined in a future version of MATLAB
% handles     structure with handles and user data (see GUIDATA)

[FileName,PathName,FilterIndex] = uigetfile('*.*');
axes(handles.axes1);
I=imread(strcat(PathName, FileName));
handles.rgbimage=[];
% if size(I,3)==3
%     handles.rgbimage=I;
%     I=rgb2gray(I);
% end;
imshow(I,[]);
handles.I=I;
guidata(hObject, handles);


%—— Executes on mouse press over axes background.
function axes1_ButtonDownFcn(hObject, eventdata, handles)
% hObject     handle to axes1 (see GCBO)
% eventdata   reserved – to be defined in a future version of MATLAB
% handles     structure with handles and user data (see GUIDATA)
%figure;imshow(handles.I);


%—— Executes on mouse press over figure background, over a disabled or
%—— inactive control, or over an axes background.
function figure1_WindowButtonDownFcn(hObject, eventdata, handles)
% hObject     handle to figure1 (see GCBO)
% eventdata   reserved – to be defined in a future version of MATLAB
% handles     structure with handles and user data (see GUIDATA)
```

```
epsilon = str2double(get(handles.edit1,'String'));

point = get(gca,'CurrentPoint');
y=point(1,1);
x=point(1,2);
if (get(handles.radiobutton2,'Value')==get(handles.radiobutton2,'Max'))
    disp('2 selected');
    neighbourhood = illuminate_neighbourhood(handles.I,x,y,epsilon,1);
    setAllPixels = 1:size(handles.I,1)*size(handles.I,2);
    outNeighbourhood = setdiff(setAllPixels, neighbourhood);
    z=randi(numel(outNeighbourhood));%==randeom selection of a point outside the neighbourhood
    z=outNeighbourhood(z);
    [z_x,z_y]= ind2sub([size(handles.I,1),size(handles.I,2)],z);
    neighbourhood2=illuminate_neighbourhood(handles.I,z_x,z_y,epsilon,2);
    illuminate_both(handles.I,neighbourhood,neighbourhood2);
    disp('Summary: ');
    disp(strcat('X(xpos,ypos):(',num2str(round(x)),',',num2str(round(y)),')'));
    disp(strcat('Z(xpos,ypos):(',num2str(round(z_x)),',',num2str(round(z_y)),')'));
else
    disp('1 selected');
    neighbourhood = illuminate_neighbourhood_spherical(handles.I,x,y,epsilon,1);
    setAllPixels = 1:size(handles.I,1)*size(handles.I,2);
    outNeighbourhood = setdiff(setAllPixels, neighbourhood);
    z=randi(numel(outNeighbourhood));%==randeom selection of a point outside the neighbourhood
    z=outNeighbourhood(z);
    [z_x,z_y]= ind2sub([size(handles.I,1),size(handles.I,2)],z);
    neighbourhood2=illuminate_neighbourhood_spherical(handles.I,z_x,z_y,epsilon,2);
     illuminate_both(handles.I,neighbourhood,neighbourhood2);
    disp('Summary: ');
    disp(strcat('X(xpos,ypos):(',num2str(round(x)),',',num2str(round(y)),')'));
    disp(strcat('Z(xpos,ypos):(',num2str(round(z_x)),',',num2str(round(z_y)),')'));
end;
```

```
function j=illuminate_neighbourhood(I,x,y,epsilon,sw)
%this description-based neighbourhood works on color images
color=double(I(round(x),round(y),:));
if (numel(color))==3 %means that I is a color image
    I=double(I);
    Z=zeros(size(I,1),size(I,2));
    j=find(sqrt((I(:,:,1)-color(:,:,1)).^2+(I(:,:,2)-color(:,:,2)).^2+(I(:,:,3)-color(:,:,3)).^2)<
          epsilon);


end;
if (numel(color)==1)%==gray level image
    I=double(I);
    j=find(color-epsilon<I&I<color+epsilon);
    I3=zeros(size(I,1),size(I,2),3);
    I3(:,:,1)=I;I3(:,:,2)=I;I3(:,:,3)=I;
    I=I3;
end;
 I3=zeros(size(I,1),size(I,2),3);
 if (sw==1)
    I1=I(:,:,2);
    I1(j)=255;
    I3(:,:,1)=I(:,:,1);
    I3(:,:,2)=I1;
    I3(:,:,3)=I(:,:,3);
 else
     I1=I(:,:,1);
    I1(j)=255;
    I3(:,:,1)=I1;
    I3(:,:,2)=I(:,:,2);
```

```
    I3(:,:,3)=I(:,:,3);
 end;
 figure;imshow(I3./255);
%==========================================================
 function j=illuminate_both(I,j,j1)
    if (numel(I(1,1,:))==1)%graylevel image
        I=double(I);
        I3=zeros(size(I,1),size(I,2),3);
        I3(:,:,1)=I;I3(:,:,2)=I;I3(:,:,3)=I;
        I=I3;
    end;
    I3=zeros(size(I,1),size(I,2),3);
    I1=I(:,:,2);
    I1(j)=255;
    I3(:,:,1)=I(:,:,1);
    I3(:,:,2)=I1;
    I3(:,:,3)=I(:,:,3);

     I1=I3(:,:,1);
    I3(j1)=255;
    I3(:,:,1)=I3(:,:,1);
    I3(:,:,2)=I3(:,:,2);
    I3(:,:,3)=I3(:,:,3);

 figure;imshow(I3./255);


function j=illuminate_neighbourhood_spherical(I,x,y,epsilon,sw)

%this spherical neighbourhood works on color images
[m,n,k]=size(I);
P=repmat([round(x),round(y)],[m*n,1]);
[X,Y]=ind2sub([m,n],1:m*n);
Q=[X;Y]';
V=Q-P;
D=sqrt(V(:,1).^2+V(:,2).^2);
j=D<epsilon;
if (numel(I(1,1,:))==1)%graylevel image
        I=double(I);
        I3=zeros(size(I,1),size(I,2),3);
        I3(:,:,1)=I;I3(:,:,2)=I;I3(:,:,3)=I;
        I=I3;
 end;
 I3=zeros(size(I,1),size(I,2),3);
 if (sw==1)
    I1=I(:,:,2);
    I1(j)=255;
    I3(:,:,1)=I(:,:,1);
    I3(:,:,2)=I1;
    I3(:,:,3)=I(:,:,3);
 else
     I1=I(:,:,1);
    I1(j)=255;
    I3(:,:,1)=I1;
    I3(:,:,2)=I(:,:,2);
    I3(:,:,3)=I(:,:,3);
 end;
 figure;imshow(I3./255);
```

Listing A.3 Matlab script in `TASolution13.m` to construct spatially close descriptive neighbourhoods.

A.7 Scripts from Chap. 7

See (Figs. A.3 and A.4)

Mscript 37 Image Tessellation Cells and Segmentation.

*(*Tessellation cells + image segments *)*

```
img = ;
sites = ImageKeypoints[img, MaxFeatures → 60];
HighlightImage[img, sites]
vm = VoronoiMesh[sites];
t15 = Graphics[{Red, {PointSize[Large], Point[sites]}, Blue,
MeshPrimitives[vm, 1], {Opacity[0.1], EdgeForm[{Thick, Yellow}],
FaceForm[Yellow]}}]
tessellatedImage = Show[img, t15]
ImageForestingComponents[, sites]//Colorize
```
■

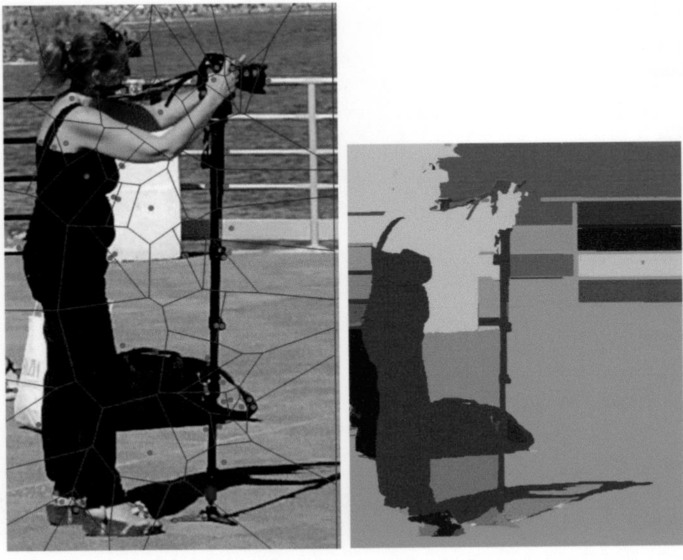

A.3.1: Tessellation A.3.2: Segmentation

Fig. A.3 Image keypoint-based Dirichlet tessellation cells versus segmentation

A.4.1: Tessellation

A.4.2: Segmentation

Fig. A.4 Default ImageForestingComponents segmentation

Mscript 38 Default ImageForestingComponents Segmentation.

```
(* Sample Image Segmentation *)
```

```
ImageForestingComponents[ ]//Colorize
```

■

Mscript 39 EdgeCorner-Based Mesh.

```
(*Edge – Corner – Based Voronoï Mesh. Solution by Binglin Li, winter2015.*)
im0 = ;
edge = EdgeDetect[im0, Method → "ShenCastan"];
im2 = ImageValuePositions[edge, White];
imgBounds = Transpose[{{0, 0}, ImageDimensions[im0]}];
voronoi = HighlightMesh[VoronoiMesh[im2, imgBounds],
Style[2, Opacity[0.1], Yellow]]
(*Display edge – based mesh on im0.*)
Show[im0, voronoi]
corners = ImageCorners[im0, MaxFeatures → 500];
im3 = Array[0&, {500, 1}];
(*Findcornersamongedges.*)
For[i = 1, i ≤ 500, i++, im3[[i]] = Count[im2, corners[[i]]]];
index = Position[im3, 1];
im4 = corners[[#]]&/@index;
m = VoronoiMesh[Level[im4, {2}], imgBounds]
t12 = Graphics[{Green, {PointSize[Large], Point[Level[im4, {2}]]},
Blue, MeshPrimitives[m, 1]}];
(*Display edge – corner – based Voronoï mesh on im0.*)
Show[im0, t12]
```

■

Mscript 40 EdgeKeypoints-Based Mesh.

```
(*Problem5.12 : High Common Vertices in a Delaunay mesh.
Based on Solution by Binglin Li.*)
im0 = ;
(* Find the image edges and assign edge point coordinates to im2 *)
im2 = EdgeDetect[im0, Method → "ShenCastan"]
edge = ImageValuePositions[edge1, White];
im5 = ImageKeypoints[im2, MaxFeatures → 60];
HighlightImage[im0, im5]
imgBounds = Transpose[{{0, 0}, ImageDimensions[im0]}];
voronoi = HighlightMesh[DelaunayMesh[im5, imgBounds],
Style[2, Opacity[0.1], Yellow]]
t11 = Graphics[{Green, {PointSize[Large], Point[im5]},
Black, MeshPrimitives[DelaunayMesh[im5, imgBounds], 1]}];
Show[im0, t11]
pd = MeshPrimitives[DelaunayMesh[im5, imgBounds], 2];
dd = Length[pd]; (* number of triangles *)
vd = Level[pd, {3}](* triangle vertices *)
sd1 = Array[0&, {dd, 1}];
sd2 = Array[0&, {dd, 1}];
sd3 = Array[0&, {dd, 1}];
Stri = Array[6&, dd](*Max.adjacenttrianglesforeachtriangle*)
For[i = 1, i<=dd, i++, vd1 = Level[pd[[i]], {2}]; sd1[[i]] = Count[vd, vd1[[1]]];
sd2[[i]] = Count[vd, vd1[[2]]]; sd3[[i]] = Count[vd, vd1[[3]]]];
sd = sd1 + sd2 + sd3 − Stri;
index = Level[Position[sd, n_/;n ≥ 15], {2}];
nn = Length[index];
t12 = Graphics[{Green, {PointSize[Large], Point[im5]}, Blue,
MeshPrimitives[DelaunayMesh[im5, imgBounds], 1],
{EdgeForm[{Thick, Yellow}], FaceForm[{Opacity[0]}], pd[[index]]}}];
Show[im0, t12]
index = Ordering[sd, −3];
index1 = index[[1]];
index2 = index[[2]];
index3 = index[[3]];
img0 = ImageCompose[im0, , RegionCentroid[pd[[index1]]]];
img1 = ImageCompose[img0, , RegionCentroid[pd[[index2]]]];
img2 = ImageCompose[img1, , RegionCentroid[pd[[index3]]]];
Show[img2, t12]
```
■

Mscript 41 Edge-Keypoint-Based Object Detection.

```
(*Object detection using image keypoints based on solution
by S. Younas*)
```

```
img2 = ;
ImageDimensions[img2]
EdgeDetect[img2]
kpImage2 = ImageKeypoints[EdgeDetect[img2], MaxFeatures → 250];
HighlightImage[img2, kpImage2]
voronoi2 = HighlightMesh[VoronoiMesh[kpImage2],
{Style[1, Black], Style[2, Opacity[0]]}]
delaunay2 = HighlightMesh[DelaunayMesh[kpImage2],
{Style[1, Black], Style[2, Opacity[0]]}]
cl = FindClusters[kpImage2, 10];
ListPlot[cl]
t2 = Graphics[{Green, {PointSize[Large], Point[kpImage2]}, Yellow,
MeshPrimitives[VoronoiMesh[kpImage2], 1]}];
Show[img2, t2]
t2 = Graphics[{Red, {PointSize[Large], Point[kpImage2]}, Orange,
MeshPrimitives[DelaunayMesh[kpImage2], 1]}];
Show[img2, t2]
```
■

Mscript 42 Branch Points.

```
MorphologicalBranchPoints[Binarize[]]//Colorize
```
■

A.8 Scripts from Chap. 8

Remark A.1 Many thanks to Piotr Artiemjew for contributing the macro image of the dragonfly in Fig. A.5, which is either the the female of Cordulegaster boltonii or Onychogomphus forcipatus, captured during the summer of 2015 in Poland. ■

Mscript 43 Colour image manifold.

```
(* image manifold *)
img = ;
Show[img]
ColorConvert[img, "Grayscale"]
```
■

A.5.1: Colour Image Manifold

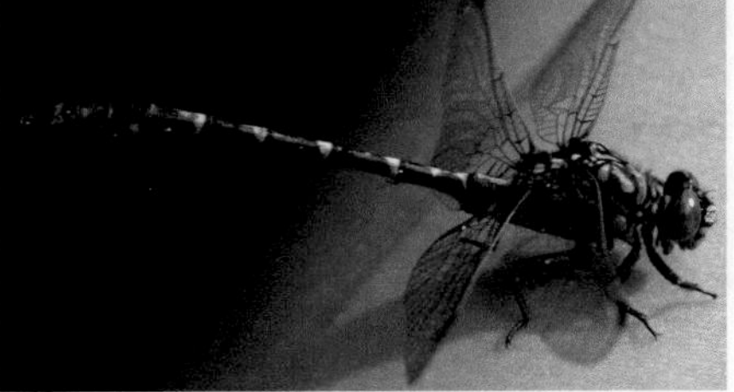

A.5.2: Greyscale Image

Fig. A.5 Colour image manifold $\mathbb{R}^5$ mapped to greyscale image $\mathbb{R}^3$

A.6.1: Hand Image

A.6.2: Segmentation

Fig. A.6 Watershed segments

A.9 Scripts from Chap. 9

See (Figs. A.6 and A.7).

Mscript 44 Watershed Segmented Hand Image.

```
(*MMexperiments, ignoringcornerneighbours*)
img = [image];
WatershedComponents[img]//MatrixForm
WatershedComponents[img]//Colorize
WatershedComponents[img, CornerNeighbors → False]//MatrixForm
WatershedComponents[img, CornerNeighbors → False]//Colorize
img2 = WatershedComponents[img, Method → {"MinimumSaliency", .35}]//Colorize
```

■

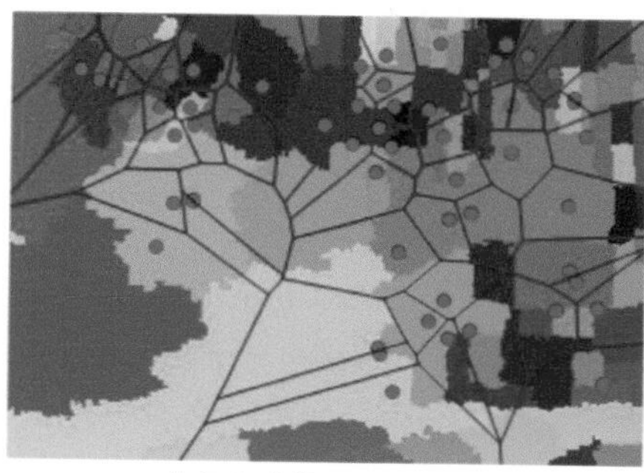

A.7.1: MM Segments

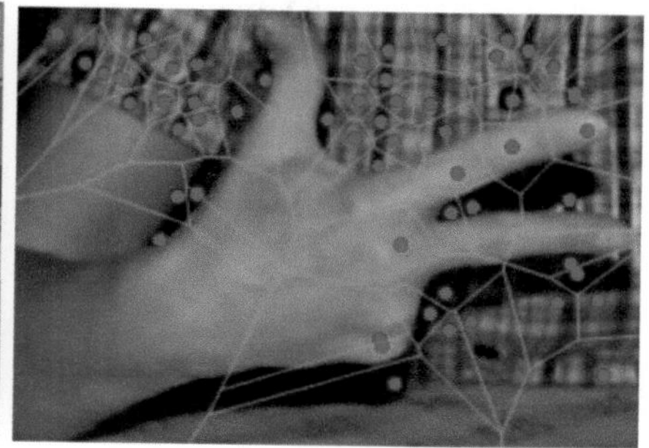

A.7.2: See-thru Segments

Fig. A.7 See-through Watershed segments on image

Mscript 45 See-through MM segments on original image.

```
(*See − through image segments with centroid − based Dirichlet
tessellation on morphological components*)
img = ▪;
WatershedComponents[img, CornerNeighbors → False]//Colorize;
img2 = WatershedComponents[img, Method → {"MinimumSaliency", .3}]//Colorize
binImg = Dilation[Binarize[img2, 2/3], 1];
centroidData = ComponentMeasurements[binImg, {"Centroid"}];
coords = centroidData[[All, 2, 1]];
HighlightImage[img2, coords]
vm3 = VoronoiMesh[coords];
t12 = Graphics[{Red, {PointSize[Large], Point[coords]}, Blue,
MeshPrimitives[vm3, 1], {Opacity[0.1], EdgeForm[{Thick, Yellow}],
FaceForm[Yellow]}}]
t15 = Graphics[{Red, {PointSize[Large], Point[coords]}, Orange,
MeshPrimitives[vm3, 1], {Opacity[0.1], EdgeForm[{Thick, Yellow}],
FaceForm[Yellow]}}];
Show[img2, t12]
Show[img, t15]
coords2 = ImageKeypoints[img, MaxFeatures → 60];
HighlightImage[img2, coords2]
vm5 = VoronoiMesh[coords2];
t13 = Graphics[{Red, {PointSize[Large], Point[coords2]}, Blue,
MeshPrimitives[vm5, 1], {Opacity[0.1], EdgeForm[{Thick, Yellow}],
FaceForm[Yellow]}}]
t18 = Graphics[{Red, {PointSize[Large], Point[coords2]}, Orange,
MeshPrimitives[vm5, 1], {Opacity[0.1], EdgeForm[{Thick, Yellow}],
FaceForm[Yellow]}}];
Show[img2, t13]
Show[img, t18]
```
■

A.10 Scripts from Chap. 10

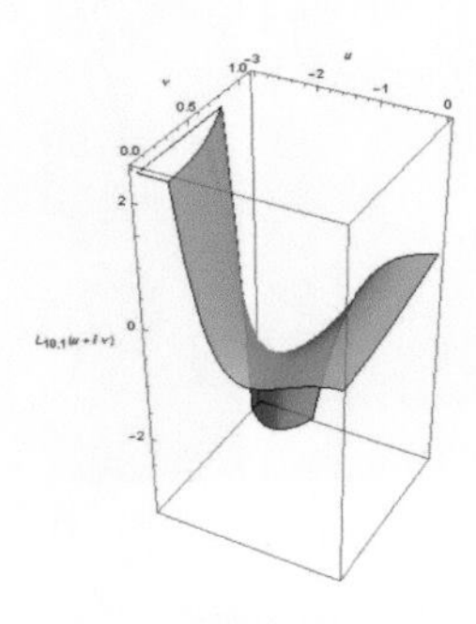

Mscript 46 3D View of a Subset Defined by the Dirichlet L-Function.

```
(*Set A := points in Dirichlet L function plot*)
Plot3D[Re@DirichletL[10, 1, u + iv], {u, 0, −3}, {v, 0, 1.1}, Mesh → None,
AxesLabel → {u, v, DirichletL[10, 1, u + iv]},
LabelStyle → Directive[Blue, Bold],
PlotStyle → Directive[Opacity[0.7], Green, Specularity[10]], BoxRatios → {1, 1, 2}]
```
■

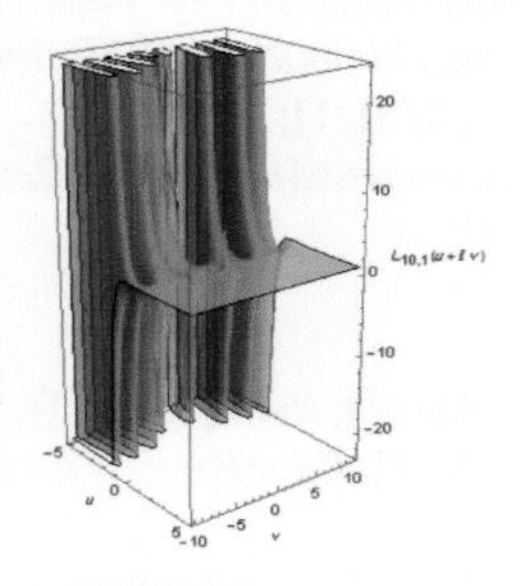

Mscript 47 3D View of Connected Subsets Defined by the Dirichlet L-Function.

```
(* Connected subset in Euclidean 3D space *)

Plot3D[Re@DirichletL[10, 1, u + iv], {u, −5, 5}, {v, −10, 10}, Mesh → None,
PlotStyle → Directive[Opacity[0.5], Green, Specularity[10]], BoxRatios → {1, 1, 2}]
```
■

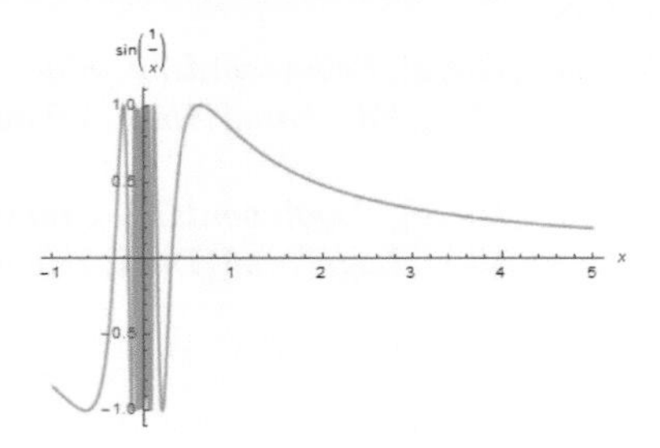

Mscript 48 topologist's sine curve.

```
(*Topologist's sine curve*)

Plot[Re@Sin[1/x], {x, −1, 5},
AxesLabel → {x, Sin[1/x]},
PlotStyle → Directive[Opacity[0.8], Orange]]

Plot[Re@Sin[1/x], {x, −1, 5},
AxesLabel → {x, Sin[1/x]},
PlotStyle → Directive[Opacity[0.8], Green]]

Plot[Re@Sin[1/x], {x, −1, 5},
AxesLabel → {x, Sin[1/x]},
PlotStyle → Directive[Opacity[0.8], Magenta]]
```

■

Mscript 49 Visualised Gradients with Mathematica9.

i =  ;

```
orientation = GradientOrientationFilter[i, 1]//ImageAdjust;
magnitude = GradientFilter[i, 1]//ImageAdjust;
ImageReflect[ColorCombine[{orientation, magnitude, magnitude}, "HSB"],
Left]
```

■

Mscript 50 Connected Cellular Components.

Colorize[*MorphologicalComponents*[*CellularAutomaton*[30, {{1}, 0}, 5^3], *CornerNeighbors* → *False*]]//*Magnify* ■

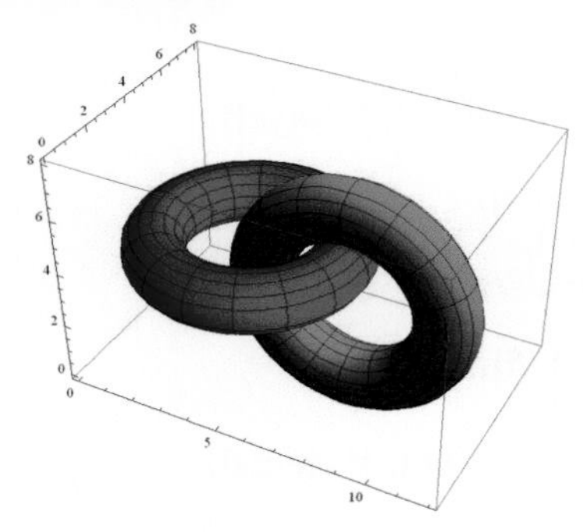

Mscript 51 Mathematica9 Script produces interlocking 3D Rings.

$u = 0$;
For[$i = 0, i < 200, u = i/100; i{+}{+}$,
ParametricPlot3D[{x, y, $2(x * y * Abs[x - y] + 0.2 * x * y) + Abs[x - y] * Sin[5 * u]$},
{x, 0, 1}, {y, 0, 1}, *ViewPoint* → {$1 - u$, $2 - u$, $1.5 + u * (1 - u)$},
Axes → *False*, *PlotRange* → {−1.2, 1.4}, *Boxed* → *False*]] ■

A.11 Scripts from Chap. 11

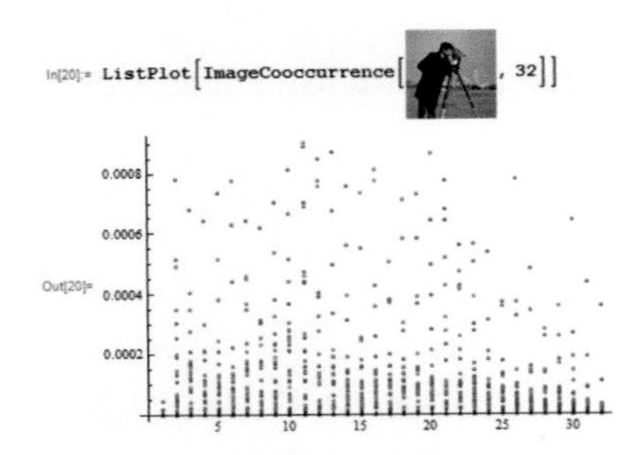

Mscript 52 Co-occurrence matrix.

ListPlot3D[*ImageCooccurrence*[, 40], *Mesh* → *Automatic*, *MeshFunctions* → {#3&}, *Mesh* → *False*, *Boxed* → *False*, *Axes* → *None*,

ColorFunction → "DarkBands"]
Show[%1, Boxed → True, Axes → True] ■

A.12 Scripts from Chap. 12

A.13 Scripts from Chap. 13

Mscript 53 cornerMeshPattern.

*(*Corner – basedMeshPatterns*)*

img = ;
ImageDimensions[img]
cornerCoords = ImageCorners[img, MaxFeatures → 30]
HighlightImage[img, cornerCoords]
ListPlot[cornerCoords, PlotStyle → {Black, PointSize[.01]},
PlotLegends → {"corner"}, AxesLabel → {wpixels, ht(pixels)},
LabelStyle → Directive[Blue, Bold]]
(Give the Delaunay mesh and display on the image *)*
vm = DelaunayMesh[cornerCoords];
dm = VoronoiMesh[cornerCoords];
t11 = Graphics[{Blue, {PointSize[0.02], Point[cornerCoords]}, Black,
MeshPrimitives[vm, 1]}]
t12 = Graphics[{Red, {PointSize[0.02], Point[cornerCoords]}, Blue,
MeshPrimitives[vm, 1], {Opacity[0.1], EdgeForm[{Thick, Yellow}], FaceForm[Yellow]}}]
t13 = Graphics[{Red, {PointSize[0.02], Point[cornerCoords]}, Magenta,
MeshPrimitives[dm, 1], {Opacity[0.1], EdgeForm[{Thick, Magenta}],
FaceForm[Yellow]}}]
t14 = Graphics[{Red, {PointSize[0.02], Point[cornerCoords]}, Blue,
MeshPrimitives[dm, 1], {Opacity[0.1], EdgeForm[{Thick, Yellow}], FaceForm[Yellow]}}]
Show[img, t12]
Show[img, t13]
Show[img, t12, t13] ■

Mscript 54 keypointsMeshPattern.

```
(*Keypoint – based Mesh Patterns*)
img = [image];
ImageDimensions[img]
keypointsCoords = ImageKeypoints[img, MaxFeatures → 100]
HighlightImage[img, keypointsCoords]
ListPlot[keypointsCoords, PlotStyle → {Black, PointSize[.01]},
PlotLegends → {"keypoints"}, AxesLabel → {wpixels, ht(pixels)},
LabelStyle → Directive[Blue, Bold]]
(* Give the Delaunay mesh and display on the image *)
vm = DelaunayMesh[keypointsCoords];
dm = VoronoiMesh[keypointsCoords];
t11 = Graphics[{Blue, {PointSize[0.02], Point[keypointsCoords]}, Black,
MeshPrimitives[vm, 1]}]
t12 = Graphics[{Red, {PointSize[0.02], Point[keypointsCoords]}, Blue,
MeshPrimitives[vm, 1], {Opacity[0.1], EdgeForm[{Thick, Yellow}], FaceForm[Yellow]}}]
t13 = Graphics[{Red, {PointSize[0.02], Point[keypointsCoords]}, Magenta,
MeshPrimitives[dm, 1], {Opacity[0.1], EdgeForm[{Thick, Magenta}],
FaceForm[Yellow]}}]
t14 = Graphics[{Red, {PointSize[0.02], Point[keypointsCoords]}, Blue,
MeshPrimitives[dm, 1], {Opacity[0.1], EdgeForm[{Thick, Yellow}], FaceForm[Yellow]}}]
Show[img, t12]
Show[img, t13]
Show[img, t12, t13] ■
```

A.14 Scripts from Chap. 14

Mscript 55 ChipmunkKeyptEdgesVoronoïMeshOnImage.

*(*Keypoints on Edges Pattern*)*

im0 = ;

```
Show[im0]
s = ImageDimensions[im0]
edgeCoords = ImageValuePositions[
EdgeDetect[im0, 5, "0.08", Method → "ShenCastan"], 1];
t11 = Graphics[{Black, {PointSize[0.0005], Point[edgeCoords]}, Black}]
keyPts = ImageKeypoints[t11, MaxFeatures → 80];
t110 = Image[Graphics[{Black, {PointSize[0.0005], Point[edgeCoords]}, Black}]];
t12 = Image[Graphics[{Red, {PointSize[0.015], Point[keyPts]}, Black}]];
ImageMultiply[t110, t12]
vm = VoronoiMesh[keyPts];
t14 = Graphics[{Red, {PointSize[0.01], Point[keyPts]}, Blue,
MeshPrimitives[vm, 1], {Opacity[0.1], EdgeForm[{Thick, Yellow}],
FaceForm[Yellow]}}]
t15 =
Image[Graphics[{Red, {PointSize[0.01], Point[keyPts]}, Blue,
MeshPrimitives[vm, 1], {Opacity[0.1], EdgeForm[{Thick, Yellow}],
FaceForm[Yellow]}}]]
vorIm = ImageResize[t15, ImageDimensions[]]
ImageCompose[im0, vorIm]
```
■

Appendix B
Kuratowski Closure Axioms

The closure axioms were introduced by Kuratowski in [1, Sect. 4.III] [2, Chap. X, Sect. 1]. Let X be a topological space, $A, B \subset X$. The closure mapping $\mathrm{cl} : 2^X \longrightarrow 2^X$ satisfies the following axioms.

***cl*.1** $\mathrm{cl}A \cup \mathrm{cl}B = \mathrm{cl}(A \cup B)$.
***cl*.2** $A \subset \mathrm{cl}A$.
***cl*.3** $\mathrm{cl}\varnothing = \varnothing$.
***cl*.4** $\mathrm{cl}(\mathrm{cl}A) = \mathrm{cl}A$.

Theorem B.1 [2, Chap. X, Sect. 5, Theorem1]
$cl(A \cup B) = clA \cup clB$.

Theorem B.2 [2, Chap. X, Sect. 5, Theorem2]
$cl(A \cap B) = clA \cap clB$.

References

1. Kuratowski, C.: Topologie I. Panstwowe Wydawnictwo Naukowe, Warsaw (1958). XIII + 494pp
2. Kuratowski, K.: Introduction to Set Theory and Topology, 2nd edn. Pergamon Press, Oxford (1962, 1972). 349 pp

J.F. Peters, *Computational Proximity*, Intelligent Systems Reference Library 102, DOI 10.1007/978-3-319-30262-1

Appendix C
Sets. A Topological Perspective

$$\begin{aligned}
\boldsymbol{X} &: \text{nonempty set, usually a topological space.}\\
\mathbf{2}^{X} &: \text{collection of all subsets of } X.\\
\mathbf{2}^{2^{X}} &: \text{all collections of subsets of } 2^{X}.\\
\boldsymbol{x} \in \boldsymbol{X} &: \text{member (element) of } X.\\
\boldsymbol{A}, \boldsymbol{B} \subset \boldsymbol{X} &: \text{proper subsets of } X.\\
\boldsymbol{A} \subseteq \boldsymbol{X} &: A \text{ is a subset of } X (A \text{possibly the same as } X).\\
|\boldsymbol{A}| &: \text{cardinality (number of members) of } A.\\
\boldsymbol{A} \cap \boldsymbol{B} &= \{x \in A : x \in B\} \text{ (intersection).}\\
\boldsymbol{A} \text{ meets } \boldsymbol{B} &\Leftrightarrow A \cap B \neq \emptyset \text{ (nearness).}\\
\boldsymbol{A} \cup \boldsymbol{B} &= \{x \in X : x \in A \vee x \in B\} \text{ (union).}\\
\mathrm{cl}\boldsymbol{A} = \bar{\boldsymbol{A}} &= \{x \in X : x\ \delta\ A\} \text{ (closure of } A).\\
\mathrm{cl}_{\boldsymbol{\Phi}}\boldsymbol{A} &= \{x \in X : x\ \delta_{\Phi}\ A\} \text{ (descriptive closure of } A).\\
\mathbf{int}\boldsymbol{A} = \mathring{\boldsymbol{A}} &: \text{interior of } A.\\
\mathrm{bdy}\boldsymbol{A} &: \text{boundary (frontier) of } A.\\
\emptyset &: \text{empty set.}\\
\boldsymbol{X} - \boldsymbol{A} = \boldsymbol{X} \setminus \boldsymbol{A} = \boldsymbol{A}^{c} &= \{x \in X : x \notin A\} \text{ (complement of } A).\\
\mathscr{B} &: \text{boundedness.}\\
\text{meet of } \boldsymbol{B} &: \text{inf (greatest lower bound) of } A.\\
\wedge\boldsymbol{B} &: \text{inf } A.\\
\text{join of } \boldsymbol{B} &: \text{sup (least upper bound) of } A.\\
\vee\boldsymbol{B} &: \text{sup } A.\\
\leq &: \text{partial order.}
\end{aligned}$$

J.F. Peters, *Computational Proximity*, Intelligent Systems Reference Library 102, DOI 10.1007/978-3-319-30262-1

Appendix D
Basics of Proximities

X (nonempty set, usually a topological space).

$A, B \subset X$ (proper subsets of X).

δ (close to, near).

$\not\delta$ (not close to).

δ_C (Čech proximity).

$A \cap B \neq \emptyset \Rightarrow A\ \delta\ B$ (Čech Axiom).

δ_E (Efremovič proximity).

$A\ \not\delta\ B \Rightarrow \exists C, D \subset X, C \cup D = X : A\ \not\delta\ C$ and $B\ \not\delta\ D$ (EF Axiom).

δ_L (Lodato proximity),

$A\ \delta\ B$ and $\{b\}\ \delta\ C\ \forall b \in B \Rightarrow A\ \delta\ C$ (LO Axiom).

δ_0 (Wallman proximity).

$A\ \delta\ B \Leftrightarrow \mathrm{cl}A \cap \mathrm{cl}B \neq \emptyset$ (Wallman Axiom).

δ_Φ (Peters proximity).

$A\ \delta_\Phi\ B$ and $\{b\}\ \delta_\Phi\ C\ \forall b \in B \Rightarrow A\ \delta_\Phi\ C$ (Descriptive LO Axiom [1, Sect. 4.15.2]).

$\overset{\wedge\!\!\wedge}{\delta}$ (Peters-Guadagni proximity).

$A\ \overset{\wedge\!\!\wedge}{\delta}\ B \Rightarrow A \cap B \neq \emptyset$ (Peters-Guadagni Axiom [2, Sect. 2]).

J.F. Peters, *Computational Proximity*, Intelligent Systems Reference Library 102, DOI 10.1007/978-3-319-30262-1

References

1. Peters, J.: Topology of Digital Images—Visual Pattern Discovery in Proximity Spaces, Intelligent Systems Reference Library, vol. 63. Springer, Zentralblatt MATH Zbl 1295 68010 (2014) Xv + 411 pp
2. Peters, J., Guadagni, C.: Strongly near proximity and hyperspace topology, 1–6 (2015). arXiv:1502.05913

Appendix E
Set Theory Axioms, Operations and Symbols

This section gives a brief overview of set theory axioms, symbols, operations from an ordinary perspective and from a computational proximity perspective. A ***set*** is either a finite or infinite collection of objects in which order has no significance.[1]

The first set of axioms for a general set theory was by given by E. Zermelo in 1908 [1]. The axioms were later developed by A.A. Fraenkel and are known as the Zermelo-Fraenkel (ZF) axioms [2, 3]. The following formulation of the ZF axioms comes from P.J. Cohen [4] and A. Abian [5].

E.1 Zermelo-Fraenkel Set Theory Axioms

This section briefly introduces the Zermelo-Fraenkel (ZF) axioms.

Extensionality. A set is determined by its members. A. Abian [5] puts the Extensionality axiom in another way: *equal sets are elements of the same sets*.

Null (Empty) Set. For all Y, there exists X, such that no Y is a member of X. The set X defined by this axiom is the empty set or null set, denoted by $\emptyset$.

Unordered Pairs. Given two sets X, Y, there is a set that contains precisely X and Y as its members and that set is called the unordered pair of X and Y.

Sum Set or Union. Given two sets X, Y, there is a set Z such that $Z = X \cup Y$, i.e., the ***sum set*** of a set X is a set whose elements are exactly the elements of the elements of X [5]. If X has a sum set and the axiom of Extensionality is valid, then the sum set of X is denoted by $\bigcup X$.

Power Set. Each set X has a power set (denoted 2^X), i.e., the ***power set*** of a set X is a set whose elements are exactly the subsets of X [5].

[1]For a good overview of the notion of a set, see http://mathworld.wolfram.com/Set.html.

J.F. Peters, *Computational Proximity*, Intelligent Systems Reference Library 102, DOI 10.1007/978-3-319-30262-1

Infinity. If x is an integer, then x is defined by $x \cup \{x\}$. Put another way, given x such that $\varnothing \in x$ and for all y, we have $y \in x$, which implies that $y \cup \{y\} \in x$. This axiom guarantees the existence of a set that contains all the integers and hence is infinite.

Axiom of Choice. A set is ***disjointed set***, provided no two distinct elements of the set have an element in common [5]. If a set A has a unique element in common with each non-empty set of X and if A has no other elements, then A is called a ***selection set*** of X. The Axiom of Choice says that each disjointed set has a selection set.

■

Remark E.1 P.J. Cohen [4] includes the ZF Axiom of Replacement and the Axiom of Regularity, which are not included, here. ■

E.2 Set Theory Symbols, Sets, and Operations

Let X be a nonempty set, A, B subsets in X, x, y in X.

Expression	Explanation
X	Space
2^X	Collection (family) of all subsets of X
2^{2^X}	Collections of subsets in the family of all subsets of X
$x \in A$	Element (member) x in subset A
$x \notin A$	x not in A
$A \subset X$	A proper subset in X
$A \subseteq X$	A proper subset in X or $A = X$
$A \not\subset 2^X$	A not a subset in 2^X
$A \not\subset X$	A not a subset in X
$\mathscr{A} \in 2^{2^X}$	Collection of subsets $\mathscr{A}$ in 2^{2^X}
$A \supset B$	A is superset of (contains) B
$A \supseteq B$	A is superset of (contains) B or $A = B$

E.3 More Set Symbols

Expression	Explanation
$X \setminus A$	$= \{x \in X : x \notin A\}$
$A \setminus B$	$= \{a \in A : a \notin B \text{ and } A \subset X\}$
$A \cup B$	A union B combining the members of A and B such that $= \{x \in X : x \in A \text{ or } x \in B\}$
$\bigcup\limits_{A_i \subset X, i \in \{0,\ldots,n\}}$	Union of n subsets
$A \cap B$	Intersection of A and B
$\bigcap\limits_{A_i \subset X, i \in \{0,\ldots,n\}}$	Intersection of n subsets
$A \underset{\Phi}{\cap} B$	Descriptive intersection of A and B
$\lvert x \rvert$	Absolute value of point x
$\lvert A \rvert$	Cardinality of set A

E.4 Proximal Sets Symbols

Expression	Explanation
φ	Probe function
$\varphi(x)$	Feature value of x
Φ	A set of probe functions
$\Phi(A)$	Set of descriptions of members of A
$A\ \delta\ B$	A near (close to) B
$A\ \not\delta\ B$	A far from (not close to) B
$A\ \delta_{\Phi}\ B$	A descriptively near (descriptively close to) B
$A\ \not\delta_{\Phi}\ B$	A descriptively far from (not descriptively close to) B
$A\ \overset{\wedge\!\!\wedge}{\delta}\ B$	A strongly near (contacts) B
$A\ \overset{\wedge\!\!\wedge}{\delta_{\Phi}}\ B$	A strongly descriptively near (descriptively contacts) B
$A\ \overset{\not\delta}{\vee\!\!\vee}\ B$	A strongly far from B
$A\ \overset{\wedge\!\!\wedge}{\not\delta}\ B$	A not strongly near B
$A\ \overset{\not\delta}{\vee\!\!\vee_{\Phi}}\ B$	A strongly descriptively far from (descriptively does not contact) B

E.5 Topology

Expression	Explanation
cl$A$	Closure of A
$\text{cl}_{\Phi} A$	Descriptive closure of A
int$A$	Interior of A
bdy$A$	Boundary of A
nhbd$A(x)$	Neighbourhood A of x
$\overset{\wedge\!\!\wedge}{\text{conn}}\ A$	Strongly connected subset A
$\overset{\wedge\!\!\wedge}{\bigcap}\text{cl}A$	Strong intersection of subsets of cl$A$
CLX	Hyperspace X (collection of closed sets)

E.6 Other Symbols

Expression	Explanation
pat$\mathscr{A}$	Pattern $\mathscr{A}$
obj$X$	Object X
ngon$E$	n-sided polygon E
nv$\mathcal{F}$	Nerve $\mathcal{F}$
ct(M, φ)	Chart (M, φ)
at$\mathcal{A}$	Atlas $\mathcal{A}$
conv$E$	Convex hull E
$x := k$	Assign value of k to x

References

1. Zermelo, E.: Untersuchungen über die grundlagen der mengenlehre. Math. Ann. **65**, 261–281 (1908)
2. Fraenkel, A.: Abstract set theory. Studies in Logic and the Foundations of Mathematics, North-Holland Publishing Co., Amsterdam (1953). Xii + 479 pp., MR0056660
3. Fraenkel, A., Bar-Hillel, Y.: Foundations of set theory. Studies in Logic and the Foundations of Mathematics, North-Holland Publishing Co., Amsterdam (1958). X + 415 pp., MR0101841
4. Cohen, P.: Set theory and the continuum hypothesis. W. A. Benjamin Inc., NewYork (1966). Vi + 154 pp., MR0232676
5. Abian, A.: On the independence of set theoretical axioms. Am. Math. Mon. **76**(7), 787–790 (1969). MR0258618

Appendix F
Topology of Digital Images

This appendix briefly introduces the topology of digital images from a computational proximity perspective.

F.1 Open Sets

Open sets arise naturally in considering different types of neighbourhoods of a point. Notice, however, that the notion of the openness of a set is defined in terms of the *sufficient nearness* of points in the set that is open.

> **Open Set.**
> Let X denote a nonempty open set. A set A in X is an ***open set***, if and only if, for each $x \in A$, all points in X *sufficiently near* x belong to A [1, Sect. 2, p. 19]. ■

Example F.1 **Open Sets in a Digital Image.**
Recall that a set A is ***open***, provided, for each point $x \in A$, all points sufficiently near x belong to A. Let a colour image img be represented by the set of picture elements in Fig. F.1. The picture elements (pixels) are represented by tiny squares. A pixel can be viewed as fat point, i.e., a physical point that has area, which contrasts with a point in the Euclidean plane. An example of an open set is the interior of img:

$$X = \{■, ■, ■, ■, ■, ■\} \text{ (Open set).}$$

In this case, a pixel $\{p\}$ (written simply as p) is an open set that belongs to X, provided p has a hue intensity value sufficiently close to one of the hue intensities in X. If the hue of p is blue, then $p = ■$ is a border pixel of X. The interior intimg

J.F. Peters, *Computational Proximity*, Intelligent Systems Reference Library 102, DOI 10.1007/978-3-319-30262-1

Fig. F.1 Open set
$X =$ 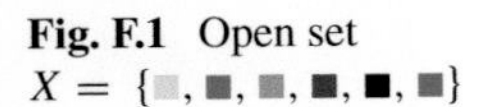

equals X and the blue pixels in img do not belong to X. The set X is an example of an open set in digital geometry. The set X is an example of a ***digital open set***. ■

F.2 Topology

This section briefly introduces topological spaces. Open sets are central in the definition of a topological space.

Definition F.2 Topology [2, Sect. 12, p. 76], [3, Sect. 1.2, p. 1], [4, Sect. 1.6, p. 11]. A collection of open sets $\mathcal{T}$ on a nonempty open set X is a *topology* on X, provided

1° The empty set $\emptyset$ is open and $\emptyset$ is in $\mathcal{T}$.
2° The set X is open and X is in $\mathcal{T}$.
3° If $\mathcal{A}$ is a sub-collection of open sets in $\mathcal{T}$, then

$$\bigcup_{B \in \mathcal{A}} B \text{ is a open set in } \mathcal{T}.$$

In other words, the union of open sets in $\mathcal{T}$ is another open set in $\mathcal{T}$.
4° If $\mathcal{A}$ is a sub-collection open sets in $\mathcal{T}$, then

$$\bigcap_{B \in \mathcal{A}} B \text{ is a open set in } \mathcal{T}.$$

In other words, the intersection of open sets in $\mathcal{T}$ is another open set in $\mathcal{T}$. ■

A nonempty set X with a topology $\mathcal{T}$ on it, is a ***topological space***. The pair $(X, \mathcal{T})$ is called a ***topological space***.

Brief History of Topology.
For a detailed history of topology in four stages (eras), see J. Milnor [5]. Part 1 of this history traces the origins of topology back to 1736 and the first topological statement with L. Euler's solution to the seven bridges of Königsberg problem. Part 2 covers the introduction of 2-dimensional manifolds in the 19th century, starting with the work by S. L'Huilier, 1812-1823, on the surface of a polyhedron in Euclidean 3-space that is drilled through with n holes. Part 3 of this history covers the study of 3-dimensional manifolds, starting with the 1898 work by P. Heegard on the decomposition of closed orientable 3-manifolds as the union of two handle-bodies of the same genus and which intersect only along their boundaries. Part 4 covers 4-dimensional manifolds, starting with the work by A.A. Markov, 1958, and the work by J.H.C. Whitehead in 1949. ■

Example F.3 The set of open intervals in the set of real numbers $\mathbb{R}$ is a basis for the usual topology. ■

F.3 Digital Topology

This section briefly introduces a digital topology on a nonempty open set of physical (non-abstract) points. The hallmark of a digital topology is a collection of what are known as digital open sets.

Digital Open Set.
A set A of digitized points on a Euclidean metric space X is a ***digital open set***, if and only if, for each $x \in A$, all points *sufficiently close* $x \in X$ belong to A. The closeness of x to each y in A defined by $\|x - y\| < \varepsilon$ for a small real number $\varepsilon > 0$. ■

Example F.4 **Digital Open set**. Every singleton set $\{p\}$ containing an individual picture element (pixel) p in a digital image is a digital open set, since $\{p\}$ does not include its boundary pixels. Each of the pixels represented by the tiny squares in Fig. F.1 and the set of all pixels img defined by

$$img = \{\blacksquare, \blacksquare, \blacksquare, \blacksquare, \blacksquare, \blacksquare, \blacksquare\} \text{ (Image Digital Open set).}$$

is an example of a digital open set, which contains the digital open subset X defined by

$$X = \{\blacksquare, \blacksquare, \blacksquare, \blacksquare, \blacksquare, \blacksquare\} \subset img \text{ (Image Digital Open subset).}$$

■

With digital open sets, we can consider collections of digital open sets that have the required properties for a digital topology.

Definition F.5 Digital Topology [6, Sect. 1.24.1, p. 64], [4, Sect. 2.10.1, p. 65].
A collection of digital open sets τ is a *digital topology* on a nonempty digital open set X, provided

1° The empty set $\emptyset$ is a digital open set and $\emptyset$ is in τ.

2° The set X is a digital open and X is in τ.

3° The union of members of any subcollection of τ are in τ. If $\mathcal{A}$ is a sub-collection of digital open sets in τ, then

$$\bigcup_{B \in \mathcal{A}} B \text{ is a open set in } \tau.$$

In other words, the union of digital open sets in τ is another digital open set in τ.

4° If $\mathcal{A}$ is a sub-collection of digital open sets in τ, then

$$\bigcap_{B \in \mathcal{A}} B \text{ is a digital open set in } \tau.$$

In other words, the intersection of digital open sets in τ is another digital open set in τ. ■

A nonempty set X of physical members with a digital topology τ on it, is a ***digital topological space***. The pair (X, τ) is called a ***digital topological space***.

F.4 Standard Basis and Digital Basis

We can construct a topology on a set by finding the unions of all open subsets in the basis.

Definition F.6 Basis [3 Sect. 2.1, p. 47].
Let X be a nonempty open set. A ***basis*** $\mathscr{B}$ for a topology τ is a collection of open subsets in the family of subsets 2^X such that the collection of all unions of subsets in $\mathscr{B}$ equals τ, i.e.,

$$\tau = \{A \cup B : A, B \in \mathscr{B}\} \text{ (**Topology construction property**)}.$$

In other words, a topology τ can be recovered by taking all possible unions of subcollections from the basis $\mathscr{B}$ [7, Sect. 5.1]. ■

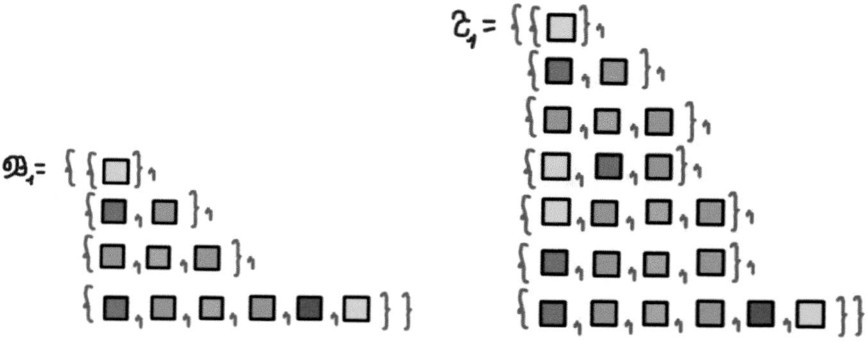

F.2.1: basis

F.2.2: topology

Fig. F.2 $\mathscr{B}$ = basis, τ = topology

Example F.7 **Basis for a Coloured-Picture Elements Topology.**
Let

$X =$ {■, ■, ■, ■, ■, ■} (Set of coloured picture of open singleton sets)

The coloured picture elements (pixels) in X are represented by tiny boxes, which is a common convention in raster images. Further, let the basis $\mathscr{B}_1$ be represented by Fig. F.2.1 and the topology τ_1 is represented in Fig. F.2.2. ■

$\mathscr{B}_1$ in Example F.7 is a digital basis, since this basis is a collection of open sets of pixels, which are physical objects accessible in a computer.

Definition F.8 Digital Basis.
Let X be a nonempty open set of physical objects. A ***digital basis*** $\mathscr{B}$ for a topology τ is a collection of digital open subsets in the digital open set X such that the collection all unions of open subsets in $\mathscr{B}$ equals τ, i.e.,

$$\tau = \{A \cup B : A, B \in \mathscr{B}\} \text{ (\textbf{Digital Topology construction property})}.$$

A topology constructed from a digital basis is called a *digital topology*. In other words, a digital topology τ can be recovered by taking all possible unions of subcollections from the digital basis $\mathscr{B}$. ■

Example F.9 **Basis for Second Coloured-Picture Elements Topology.**
Let

$X =$ {■, ■, ■, ■, ■, ■} (Set of coloured picture elements)

F.3.1: basis

F.3.2: topology

Fig. F.3 $\mathscr{B}_2$ = basis, τ_2 = topology

The coloured picture elements (pixels) in X are represented by tiny boxes, which is a common convention in raster images. Further, let the basis $\mathscr{B}_2$ be represented by Fig. F.3.1 and the topology τ_2 is represented in Fig. F.3.2. ■

From Examples F.7 and F.9, we have an example of two topologies τ_1, τ_2 defined on the same nonempty set X. These examples provide what is known as a digital bitopological space (denoted by (X, τ_1, τ_2)).

Definition F.10 Bitopological space.
Let X be a nonempty set with a pair of topologies τ_1, τ_2 on it. Then (X, τ_1, τ_2) is a *bitopological space*. Notice that a tri-topological space results from introducing three different topologies on X. An ***n**-topological space* is a nonempty set with n different topologies defined on it. ■

For more about bitopological spaces, see I. Dochviri [8], I. Dochviri and T. Noiri [9].

Definition F.11 Digital Bitopological space.
Let X be a nonempty set of physical objects with a pair of topologies τ_1, τ_2 on it. Then (X, τ_1, τ_2) is a *digital bitopological space*. Notice that a *tri-digital topological* space results from introducing three different topologies on X. An ***n**-digital topological space* is a nonempty set of physical objects with n different topologies defined on it. ■

Example F.12 **Digital bitopological space.**
Let $X =$ (Set of coloured picture elements) Let τ_1, τ_2 be topologies on X from Examples F.7 and F.9, respectively. From Definition F.11, (X, τ_1, τ_2) is a digital bitopological space. ■

Problem F.13 Give an example of a digital tri-topological space with three different topologies on a 2D digital image. ■

Examples F.7 and F.9 each illustrate what is known as a topology on a digital image in the Euclidean plane.

Definition F.14 Topology on a Digital Image.
Let X be a nonempty open set of picture elements (pixels, picture point or, briefly, points) in a digital image in the Euclidean plane, namely, $\mathbb{R}^2$. In addition, let $\mathscr{B}$ be a digital basis set which is a collection of sets of vectors in X, i.e.,

$$\mathscr{B} = \left\{A : A \in 2^X\right\} \text{ (picture basis).}$$

Then $\tau \in 2^X$ is a topology on X, provided

$$\tau = \left\{\bigcup_{A \in \mathscr{C}} A : \mathscr{C} \subset \mathscr{B}\right\}.$$

The pair (X, τ) is a ***topology on a digital image*** (briefly, ***digital topology***). ■

By equipping a digital topology with a proximity relation, we obtain a proximal digital topology.

Definition F.15 Proximal Digital Topology.
Let X be a nonempty digital open set such as pixels in a digital image or nodes in a social network. In addition, let $\mathscr{B}$ be a digital basis set which is a collection of digital open sets in X, i.e.,

$$\mathscr{B} = \left\{A : \text{ open set } A \in 2^X\right\} \text{ (picture basis).}$$

and the $\tau \in 2^X$ is a topology on X derived from the digital basis, i.e.,

$$\tau = \left\{\bigcup_{A \in \mathscr{C}} A : \mathscr{C} \subset \mathscr{B}\right\}.$$

In addition to the topology τ on X, the topological space X is equipped with the descriptive proximity δ_Φ. The triple (X, τ, δ_Φ) is a ***proximal digital topology***. ■

Remark F.16 **Proximal Topology of Digital Images.**
Let X be an open set that is a digital image endowed with a proximity δ and let $\mathscr{B}$ be a digital basis that is a collection of digital open sets that are subimages in X. The topology τ with digital basis $\mathscr{B}$ is a ***topology of digital images***. If X is endowed with a proximity δ, then τ is a ***proximal topology of digital images***. ■

For more about about the topology of digital images, see [6].

F.5 Rosenfeld Digital Topology Versus Proximal Topology of Digital Images

The term **digital topology** was introduced by A. Rosenfeld in 1979 [10] and has many variations [11–15] and is also called topology of digital images [16].

This book presents a near set-based digital topology, which is not the same as Rosenfeld digital topology. The basic building block for the proposed form of digital topology is a digital open set (see Definition F.5). A proximal form of digital topology is obtained by endowing a set of picture points in a digital image with either a spatial or a descriptive proximity relation. A digital topology on a digital open set equipped with a proximity relation leads naturally to a proximal digital topology (see Definition F.15).

In other words, in the proposed approach, one starts by proximitising a digital open set in a digital image that is a subset of either Euclidean space $\mathbb{R}^2$ or $\mathbb{R}^3$, instead of considering the geometry associated with digital points in a digital image that is a subset in $\mathbb{Z}^2$. In the digital geometry approach to topologizing a picture, a picture is viewed as a collection of square grids modelled as finite subsets of $\mathbb{Z}^2$ [15, 17]. A picture grid consists of digital points. An **Eckhardt-Latechi digital point** is a point with integer coordinates [15].

Eckhardt-Latechi digital points are distinguished from picture points in either Euclidean 2-space or in Euclidean 3-space, since digital points have location but do not have measurable content such as greylevel intensity or colour channel brightness. By contrast, a ***picture point*** is a small region in a digital image. Viewed abstractly, an ***abstract picture point*** (also called a ***Euclidean picture point***) is a small region in either Euclidean 2-space $\mathbb{R}^2$ or in Euclidean 3-space $\mathbb{R}^3$. The centroid of a Euclidean picture point is its location. Its measurable features are, for example, diameter and area. Viewed non-abstractly, a Euclidean picture point has a 1-1 correspondence with a physical picture point (a small region in a digital image) that is endowed with features such as diameter, area, density, colour and intensity.

F.6 Stepping Stones to a Proximal Topology of Digital Images

Metric spaces provide stepping stones to many topological structures that are derived from metric spaces and that are useful in a topology of digital images. In fact, any metric space generates a topological space. A proximal metric topology of digital images results from equipping a metric topological space X with one or more of the proximities introduced in this book. That availability of a metric on X makes it possible to define each proximity in terms of the Hausdorff distance between sets in the space. For more about proximal metric topologies, see [4, Sects. 1.24.2 and 1.24.3] and [6, Sects. 2.10.2 and 2.10.3]. In addition, metric spaces are also related to analysis and geometry in intuitively appealing ways. A good introduction to metric and pseudometric spaces is given by [18, pp. 119–124], [19, Sect. 5], [20, Sect. 1–4, p. 11] and [21, Chaps. 5–6]. One of the most complete, formal presentations of metric spaces is given by C. Kuratowski [22, vol. 1, Chap. 2].

Another place to look for stepping stones to a topology of digital images is recent work on descriptive EF-proximity spaces [4, 23–25]. A topological space X endowed with a descriptive EF-proximity δ_{Φ} is called a **descriptive EF-proximity space**. Recall that a **descriptive EF-proximity relation** on a set is an extension of the usual EF-proximity relation, where the descriptive closeness of sets in the space is defined relative to probe functions that represent features of the non-abstract points in the space. The abstract points in an EF-proximity space are replaced by non-abstract points, where each non-abstract point has both a location and measurable content.

In a topology of digital images, the focus is on families of picture points that are descriptive topologies. A descriptive topological space is the natural outcome of endowing a set of picture points with a descriptive proximity relation. And what Ju. M. Smirnov [26] and S. Leader [27] observed about EF-proximity spaces holds true for descriptive EF-spaces. That is, a uniform topology in an EF-space is the natural outcome of finding all sets near each given set in the space. Recall that sets are near in a Wallman EF-space, provided the closures of the sets have common elements. In effect, a uniform topology derived from a Wallman EF-proximity space is spatial in character. By contrast, a **descriptive uniform topology** results from finding all sets that are descriptively near each given set of picture points in the descriptive proximity space.

As befits an interest in classifying structures in digital images, nonempty sets A, B of picture points are descriptively near, provided the sets have picture points $a \in A, b \in B$ with matching descriptions. Notice the sets A and B can be disjoint (no points in common) and can still be descriptively close to each other. The analogue of this typical situation can be found in the experience of a visitor to a museum of fine arts containing paintings and statuary. Typically, one compares a painting with its neighbours. One painting is stylistically close to another painting, provided structures in the painting being observed resemble structures in another painting (e.g., matching colors, shapes, texture of parts of the pictures). From this line of reasoning, a descriptive uniform topology of digital images naturally arises by endowing a picture-based topological space with a descriptive proximity relation.

1. Bourbaki, N.: Elements of Mathematics. General Topology, Part 1. Hermann and Addison-Wesley, Reading (1966). I-vii, 437 pp
2. Munkres, J.: Topology, 2nd edn. Prentice-Hall, Englewood Cliffs (2000). Xvi + 537 pp., 1st edn., in 1975, MR0464128
3. Krantz, S.: A Guide to Topology. The Mathematical Association of America, Washington (2009). Ix + 107pp
4. Naimpally, S., Peters, J.: Topology with Applications. Topological Spaces via Near and Far. World Scientific Publishing, Singapore (2013). Xv + 277 pp., Am. Math. Soc., MR3075111
5. Milnor, J.: Topology through the centuries: Low dimensional manifolds. Bull. (New Ser.) Amer. Math. Soc. **52**(4), 545–584 (2015)

6. Fraenkel, A., Bar-Hillel, Y.: Foundations of set theory. Studies in Logic and the Foundations of Mathematics, North-Holland Publishing Co., Amsterdam (1958). X + 415 pp., MR0101841
7. Willard, S.: General Topology. Dover Publishing Inc., Mineola (1970). Xii + 369 pp., ISBN: 0-486-43479-6 54-02, MR0264581
8. Dochviri, I.: On submaximality of bitopological saces. Kochi J. Math. **5**, 121–128 (2010). MR2656713
9. Dochviri, I., Noiri, T.: On some properties of stable bitopological spaces. Topol. Proc. **45**, 111–119 (2015). MR3218261
10. Rosenfeld, A.: Digital topology. Am. Math. Mon. **86**(8), 621–630 (1979). Am. Math. Soc., MR0546174
11. Kovalevsky, V.: Finite topology as applied to image analysis. Comput. Vis. Gr. Image Process. **46**, 141–161 (1989)
12. Latecki, L.: Topological connectedness and 8-connectedness in digital pictures. Comput. Vis. Gra. Image Process. **57**, 261–262 (1993)
13. Kong, T., Rosenfeld, A.: Topological Algorithms for Digital Image Processing. North-Holland, Amsterdam (1996)
14. Pták, P., Kofter, H., Kropatsch, W.G.: Digital topologies revisited: an approach based on the topological point-neighbourhood. In: Proceedings of the 7th International Workshop on Discrete Geometry for Computer Imagery, pp. 151–159 (1997)
15. Eckhardt, U., Latecki, L.J.: Topologies for the digital spaces $\mathbb{Z}^2$ and $\mathbb{Z}^3$. Comput. Vis. Image Underst. **90**(3), 295–312 (2003)
16. Kronheimer, E.: The topology of digital images. Special issue on digital topology. Topol. Appl. **46**(3), 279–303 (1992). MR1198735
17. Latecki, L., Conrad, C., Gross, A.: Preserving topology by a digitization process. J. Math. Imaging Vis. **8**, 131–159 (1998)
18. Kelley, J.: General Topology. Springer, Berlin (1955). xiv + 298 pp
19. Steen, L., Seebach, J.: Counterexamples in Topology. Springer, New York (1970). Dover edition, 1995
20. Hocking, J., Young, G.: Topology. Dover, NewYork (1988)
21. Sutherland, W.: Introduction to Metric and Topological Spaces, 2nd edn. Oxford University Press, Oxford (1974, 2009); 2008
22. Kuratowski, C.: Topologie I. Panstwowe Wydawnictwo Naukowe, Warsaw (1958). XIII + 494pp
23. Peters, J., Naimpally, S.: Applications of near sets. Not. Am. Math. Soc. **59**(4), 536–542 (2012). http://dx.doi.org/10.1090/noti817, MR2951956
24. Peters, J.: Local near sets: pattern discovery in proximity spaces. Math. Comput. Sci. **7**(1), 87–106 (2013). doi:10.1007/s11786-013-0143-z, MR3043920
25. Peters, J.: Near sets: an introduction. Math. Comput. Sci. **7**(1), 3–9 (2013). doi:10.1007/s11786-013-0149-6, MR3043914
26. Smirnov, J.M.: On proximity spaces. Math. Sb. (N.S.) **31**(73), 543–574 (1952). English translation: Am. Math. Soc. Trans. Ser. 2, 38, 1964, 5–35
27. Leader, S.: On clusters in proximity spaces. Fundam. Math. **47**, 205–213 (1959)

Author Index

A

A-iyeh, E., x, 84, 86, 87
A-iyeh, N., xi
Abian, A., 397
Adams, G., 12
Aichholzer, O., 142
Alexandroff, P., 183, 184
Alliez, P., 34
Anton, F., 34
Archimedes, 306
Arkowitz, M., 4
Artiemjew, P., xi, 381
Aurenhammer, F., 1, 34
Aygün, H., 268

B

Báñez, J.M., 142
Bank, R.E., 84
Bar-Hillel, Y., 397
Basu, S., 318
Beer, G., xi, 1, 2, 64, 196, 243
Bennett, B., 102
Berger, M., 272
Beyer, W., 171
Bhatia, R.P., 84, 86
Boissonnat, J.-D., 11, 34, 328
Boothby, W., 171
Boots, B., 34
Borkowski, M., xi, 103, 292
Borsuk, K., ix, 143, 145, 147, 272
Bourbaki, N., 42, 401
Bovik, A.C., 87
Brennan, J., 268
Burak, G., 145

C

Caraballo, L.E., 142
Cattaneo, G., x
Cecatto, J.R., 103
Čech, E., 14, 64, 66, 70
Çetkin, V., 268
Chattopadhyay, K.C., 268
Chitcharoen, D., x, xi, 103
Chitraprasad, D., 34
Chiu , S.N., 34
Cocurjolly, D., 87
Cohen, M.M., 147
Cohen, P.J., 397, 398
Collins, G.P., 147
Conrad, C., 120
Coxeter, H.S.M., 326
Crabb, M.C., 145
Cross, B., x, xi, 64
Czyżewski, A., x

D

Düntsch, I., 102
Davis, A.S., 179
De Floriani, L., 34
Deritei, D., 5
Deza, E., 98, 99
Deza, M.D., 98
Deza, M.M., 99
Di Concilio, A., x, xi, 1–3, 9, 10, 16, 20, 23, 42, 64, 103, 164, 196, 197, 243
Di Maio, G., vii, xi, 1, 2, 42, 64, 120, 243
Dimov, G., 102
Dochviri, I., xi, 1, 70, 275, 406
Dodson, C.T.J., 144, 145
Dori, D., 87
Du, Q., 11, 120

J.F. Peters, *Computational Proximity*, Intelligent Systems Reference Library 102, DOI 10.1007/978-3-319-30262-1

Dumonceaux, R., xi

E
Eckhardt, U., 120, 408
Edelsbrunner, H., 1, 4, 11, 30, 32, 248, 308, 318, 345
Efrat, A., 328
Efremovič, V.A., ix, 14, 64, 70, 179
Efremovich, V.A., 11
Ehresmann, C., 171
Engelking, R., 187
Ercsey-Ravasz, M., 5
Euclid, 4, 6

F
Faber, V., 11, 120
Fabila-Monroy, R., 142
Fashandi, H., x, xi, 268
Favorskaya, M.N., 3
Fell, J., 243
Field, D.A., 84
Firey, W.J., 306
Fréchet, M., 98, 185
Fraenkel, A.A., 397
Frank, N., 33
Frege, G., 3
Friesen, N., xi
Friston, K., 171

G
Gardner, M., 120
Gaudreau, C., x, xi
Gavrilova, M.L., 120
Gerla, G., xi, 10, 20, 64, 103, 164, 196, 197
Ghosh, S., 175
Gmurczyk, A., 143
Gold, C., 34
Gomolińska, A., x
Grzymała-Busse, J., xi
Grzymała-Busse, J.W., x
Grünbaum, B., 11, 120, 326
Grünbuam, B., 317
Greitzer, S.L., 326
Gross, A., 120
Gruber, P.A., 307, 308
Gruber, P.M., 306
Guadagni, C., vii, x, xi, 1, 4, 19, 20, 23, 28, 46, 197, 240, 242–244, 247, 292
Gunzburger, M., 11, 120
Guo, Q., 307
Gupta, S., 348

H
Hadamard, J., 98
Hairer, E., 170
Hammer, P.C., 306
Hankley, W., xi
Harer, J., 318
Harer, J.L., 1, 4, 308, 345
Hart, S., 33
Hashimoto, K., 153
Hata, M., 253
Hatcher, A., 4
Hausdorff, F., 98, 156, 178, 187
Hazra, H., 268
Heath, T.L., 306
Henry, C.J., x, xi, 103
Hettiarachchi, R., x, xi, 1, 10, 255
Hilhorst, H.J., 34
Hocking, J.G., 409
Hoffmann, F., 328
Holá, L., 120
Holý, D., 120
Hooke, R., 3, 37
Hopf, H., 183, 184

I
İnan, E., viii, xi, 1
Install, M., 43

J
Járai-Szabó, F., 5
Jain, L.C., 3
Jaworowski, J., 145
Jost, J., 171

K
Kaijser, S., 307
Kaimal, M., 34
Kak, A., 120
Kalmbach, G., 348
Kaplansky, I., 348
Karaca, I., 145
Kay, D.C., 24
Kelley, J.L., 409
Kepler, J., 326
Khare, M., 348
Kiermeier, K.D., 104
Kinsner, W., xi
Klette, R., 11, 120
Knauer, C., 328
Kong, T., 120

Kong, T.Y., 408
Kostek, B., x
Kovalevsky, V., 11
Kovalevsky, V.A., 408
Krantz, S., 4, 146, 237, 238
Krantz, S.G., 42, 402
Kronheimer, E., 120
Kronheimer, E.H., 408
Kropatsch, P., 2
Kropatsch, W., 180
Kuba, A., 87
Kuratowski, C., 15, 66, 409

L
Lázár, Z.I., 5
Lai, R., 120
Latecki, J.L., 408
Lateki, L., 120
Lawrence, K.L., 84, 86
Leader, S., ix, 42, 51, 55, 179, 275
Li, B., x, xi, 69
Liebling, T.M., 33
Liu, H., x, 34
Lange, J., xi
Liu, Y.-M., 268
Liziér, M., 106
Llapa, E., 103
Lodata, M., 1
Lodato, C., ix, 23
Lodato, M., 128
Lodato, M.A., 14, 15
Lubich, C., 170
Luo, M.-K., 268
Lusternik, L.A., 145

M
Malgouyres, R., 148
Mandelbrot, S., 98
Manetti, M., 147–149
Mani-Levitska, P., 306
Mao, X., 253
Markovic, D.G., 314
Martin, E., 268
Matheron, G., 259
McCoy, R., 120
Meghdadi, A.H., x, xi, 103, 268
Milnor, J., 238, 403
Mioc, D., 34
Mozzochi, C.J., 42
Munkres, J.R., 42, 143, 147, 402
Murrell, H., 153

N
Naimpally, S., vii, x, xi, 2
Naimpally, S.A., 1, 2, 4, 11, 15, 23, 42, 64, 65, 102, 104, 120, 178, 187, 243, 294, 326, 402, 407, 409
Nicoletti, M., x
Nicoletti, M.C., x, 103
Nigsch, P., 142
Noiri, T., 406
Nyúl, L.G., 87

O
Ochoq, C., 142
Okabe, A., 33
Orłowska, E., x, xi
O'Rourke, J., 175
Öztürk, M.A., viii, xi, 1, 4, 306

P
Pal, S.K., x, xi
Palágyi, K., 87
Papp, I., 5
Pareek, C., 64
Pareek, C.M., 2
Parker, P.E., 144, 145
Pavel, M., 103, 195
Pawlak, M., xi
Pawlak, Z., x, xi, 103
Pedrycz, W., xi
Penny, W., 171
Peters, J.F., vii, viii, x, 1, 2, 4, 10, 11, 14–16, 19, 20, 23, 27, 28, 34, 42, 46, 64, 65, 84, 87, 103, 104, 120, 123, 128, 145, 178, 187, 197, 240, 242–244, 247, 255, 275, 292, 294, 306, 326, 402, 407, 409
Plato, 3
Platonism, 3
Poli, G., 103
Polkowski, L., x, xi
Pollack, R., 318
Pourin, L., 33
Preparata, F.P., 120
Prince, S.J.D., 3
Pták, W.G., 2
Pták, P., 180, 408
Puzio, L., x, xi

Q
Qiu, W., x

R
Ramanna, S., x, xi, 1, 34, 64, 103, 326
Regan, E.R., 5
Rhoads, G.C., 314
Riesz, F., 11, 64, 185
Rocchi, N., 27
Romesburg, H.C., 12
Ronse, C., 4
Roscoe, A., 120
Rosenfeld, A., 11, 120, 331, 408
Roy, M.-F., 318

S
Saito, J.H., 103
Sangoz, C., xi
Schilmoeler, L., xi
Schmidt, D.A., xi
Schnirelmann, L., 145
Seebach, J.A., 183, 409
Sengupta, B., 171
Serra, J., 259
Shafar, S., xi
Shamos, M.I., 120
Sheikh, H.R., 87
Shephard, G., 120
Shephard, G.C., 11, 326
Shewchuk, J.R., 86
Shukla, A., 348
Simoncelli, .P., 87
Skowron, A., x, xi, 103
Slowiński, R., x
Smirnov, Ju.M., ix, 179, 267
Snyman, J.A., 171
Solan, V.P., 24
Šostak, A.P., 268
Spagnuolo, M., 34
Steen, L.A., 183, 409
Stepaniuk, J., x, xi, 103
Stover, C., 6, 34
Street, R., 318
Su, F.E., 145
Sugihara , K., 34
Sumi, R., 5
Suraj, Z., x, xi
Surendran, S., 34
Sussman, I., xi
Sutherland, W., 409
Száz, Á., 128
Száz, A., 123

T
Taimanov, I.A., xi
Takatori, D., 153
Talata, I., 306
Teillaud, E., 34
Teillaud, M., 328
Thron, W., 178
Tiwari, S., x, xi, 103
Tougne, L., 87
Toussaint, G., 120
Toyoura, M., 253
Tozzi, A., xi, 145, 170
Tuz, V.V., 24

U
Uçkun, M., viii, xi, 1, 4, 306
Uchime, C., x, xi
Umdu, Ö., xi

V
Vacavant, A., 87
Vakarelov, D., 102
Varga, L., 5
Vietoris, L., 243
Villar, D., x, xi
Voronoï, G., 120

W
Wallman, H., ix, 14, 15, 70
Wang, L., x
Wang, Z., 87
Wanner, G., 170
Warrack, B., 64
Warrack, B.D., x, 102
Wasilewski, P., x, xi, 103
Weihrauch, K., 185
Weil, A., 98
Weisstein, E., 43
Weisstein, E.W., 6, 33, 34, 142–144, 216, 314
Wenyin, L., 87
Whitehead, A.N., 3
Whitehead, J.H.C., 4, 164, 166
Willard, S., vii, 43, 46, 146, 216, 238, 295
Wills, J.M., 306
Wolski, M., x, xi, 11, 64, 103, 409
Womble, E.W., 24
Wormser, C., 34, 328

X
Xu, J., 84

Y

Yao, Y., x
Younas, S., x, xi
Yvinec, M., 34, 328

Z

Zadeh, L.A., xi, 268
Zardecki, A., 171
Zelins'kyi, Y.B., 24
Zermelo, E., 397

Subject Index

Symbols

$(X, \delta_{\Phi}, \mu_{\delta_{\Phi}})$, 279
1_X, 148
$D(A, B)$, 102
D_{Φ}, 104
$NCL_{W_{pqr}}\mathscr{W}$, 318
$N_{\Phi(x)}$, 184
R^2, 150
R^n, 150
R_0, 179
  descriptive, 179
R_0 (symmetric), 179
R_0^{Φ}, 183
R_0^{Φ} (descriptive), 183
S^1, S^2, S^3, 144
S^2, 143
S^3, 143
S^4, 144
S^n, 143, 144
T_0, 184
T_0^{Φ}, 184, 185
T_1, 185
T_1^{Φ}, 185, 186
T_2, 178, 187
T_2 (Hausdorff), 187
T_2^{Φ}, 187, 194
T_3 (regular), 204
T_3^{Φ} (regular), 204
T_4 (normal), 204
T_4^{Φ} (normal), 205
int, 196
Φ, 197
$\Phi(A)$, 9, 15, 18, 27
$\Phi(X)$, 158
$\Phi(x)$, 15, 27, 152, 153
$\Phi^{-1}(A)$, 9
0, 348
1, 348
$\cap$, 66
$\circ_{\Phi}$, 194, 203
$\cup$, 66
$\underset{\Phi}{\cap}$, 15, 27, 73, 103, 130, 312
δ, 14, 15, 18, 51, 66, 102, 128, 129, 157, 161, 242, 244
$\delta_{\varepsilon_{\Phi}, \upsilon_{\overset{\wedge\wedge}{\delta}_{\Phi}}}$, 281
$\not\delta_{\Phi}$, 27, 128, 130, 190, 192
δ_{Φ}, 15, 16, 18, 27, 128, 130, 150, 200, 279, 407, 409
$\overset{\wedge\wedge,\Phi}{\text{conn}}$, 197, 292
$\overset{\wedge\wedge,\Phi}{\text{conn}}(A)$, 196
$\overset{\wedge\wedge}{\text{conn}}$, 196, 197, 292, 329
$\varnothing$, 66
$\not\delta$, 65, 128, 129
æ$B30D$·æ $B30D$, 102, 156
æ$B30Dx - y$æ $B30D$, 99
æ$B30Dx - y$æ $B30D_2$, 99
æ$B30Dx$æ $B30D$, 99
$\mathbb{D}_{\Phi}$, 104
$\mathbb{R}$, 99, 152
$\mathbb{R}^1$, 99
$\mathbb{R}^2$, 99, 156–158, 272, 318, 407
$\mathbb{R}^3$, 99, 154, 156, 272
$\mathbb{R}^4$, 154
$\mathbb{R}^7$, 155
$\mathbb{R}^n$, 15, 18, 27, 99, 152, 153, 238, 248
$\overset{\wedge\wedge}{\delta}_{\Phi}$, 17, 29, 128, 156, 157, 161, 162, 197, 246, 312
$\overset{\wedge\wedge}{\delta}$, 16, 19, 20, 24, 26, 28, 44, 52, 128, 156, 157, 160, 197, 243, 246, 329
  naming, 20

J.F. Peters, *Computational Proximity*, Intelligent Systems Reference Library 102, DOI 10.1007/978-3-319-30262-1

$\overset{\wedge\!\!\wedge}{\delta}$-connectedness, 295
  Delaunay triangles, 296
$\overset{\wedge\!\!\wedge}{\bigcap}$, 329
$\overset{\wedge\!\!\wedge}{\not\delta}_{\Phi}$, 128
$\overset{\wedge\!\!\wedge}{\not\delta}$, 19
$\overset{\wedge\!\!\wedge}{\mathrm{nv}}$, 345
$\overset{\wedge\!\!\wedge}{\delta}$, 219
$\overset{\wedge\!\!\wedge}{\not\delta}_{\Phi}$, 190, 191
$\overset{\wedge\!\!\wedge}{\mathrm{conn}}$, 43
$\overset{\not\delta}{\mathbb{w}}$, 128, 192, 22, 23
$\overset{\not\delta}{\mathbb{w}}_{\Phi}$, 17
$\mu_{\delta_{\Phi}}$, 279
μ_{δ}, 268, 271
ϕ, 196
$\phi(x)$, 27, 152
$\not\delta$, 15
CP, 1
Re.s.p.c., 163
Re, 152, 168
cl$A$, 71
cl, 15, 66, 102, 156, 157, 196
cl_{Φ}, 16, 18, 157
conn, 43, 293
conv, 318
ngon, 329, 337
nhbd, 194, 245
$\|\cdot\|$, 102, 155
$\|x - y\|$, 99
$\|x - y\|_2$, 99
$\|x\|$, 99
obj, 335
pat, 332
s.p.c., 160, 161
$\neg A$, 163
$\neg x$, 162
$\searrow$, 147
$\varphi(Re)$, 152
d_{taxi}, 101
id_X, 148
n-sphere, 143, 144, 238
  S^n, 143
n-topological space, 406
  definition, 406
$near$, 55
$\overset{\wedge\!\!\wedge}{\delta}_{\mathcal{A}}$, 242
pat, 329
$\blacktriangle(pqr)$, 318
Čech, 98
  distance, 98

A
Algebra, 4, 194, 195, 198, 201, 203
  convex groupoids, 4
  Delaunay groupoid, 198
  feature space groupoid, 199
  near partial descriptive groupoids, 195
  partial descriptive groupoid, 194
  partial descriptive semigroup, 195
  proximal Delaunay groupoid, 198
  Voronoï feature space groupoid, 203
  Voronoï Groupoid, 201
  Voronoï-Hausdorff Convex Groupoid, 201
Algorithm, 35
  bounded descriptive Nbd, 111
  bounded indistinguishable descriptive Nbd, 113
  centroid-based regions, 261
  centroidal Voronoi mesh, 89
  CentroidDirichletTessellation, 261
  complete poset pattern, 353
  deformation mapping, 165
  Delaunay $\overset{\wedge\!\!\wedge}{\delta}$-set, 296
  Descriptively Sorted Near Sets, 277
  DirichletTessellation, 216
  homotopy, 165
  image geometry, 121
  image mesh nerves, 38
  keypoint-based regions, 216, 261
  keypoint-marker segments, 216
  KeypointsDirichletTessellation, 261
  mesh nerves construction, 35
  morphology segments, 261
  Spatially Sorted Near Sets, 275
  unbounded descriptive Nbd, 108
  Voronoï ngon shape pattern, 335
  Voronoï connectedness, 298
Antipodal points, 143, 170
  Borsuk–Ulam, 143
  concave surface, 170
  flat surface, 170
Antipodal polygon, 142
  5gon, 142
  on a circle, 142
Antipode of a point, 142
  mirror, 142
  on a circle, 142
Atlas, 241, 242
  property, 242
  smoothly compatible, 241
Axiom, 14, 103, 178, 242, 268, 279
  R_0, 179
  R_0^{Φ}, 183

δ_{Φ}, 27
δ_{Φ}-Lodato, 15, 130
δ-Lodato, 129
$\overset{\wedge\!\!\wedge}{\delta}_{\Phi}$, 17, 29
$\overset{\wedge\!\!\wedge}{\delta}$, 16, 19
Čech proximity, 14
descriptive EF proximity, 27
descriptive Lodato, 73
descriptive Lodato proximity, 16, 73
descriptive nearness, 103
descriptive Wallman proximity, 16
dL(4), 183
Efremovič proximity, 15
Etd Descriptive Lodato, 130
fuzzy descriptive Lodato proximity, 279
fuzzy Lodato proximity, 268
Kuratowski, 15
Kuratowski closure, 66
Lodato point-set proximity, 67
Lodato proximity, 15, 72
LodatoWallman Object Signature, 335
ltd Lodato, 129
manifold strong proximity, 242
Mesh pattern, 334
point process, 34
separation, 178, 179, 183
strong contact, 16
strong proximity, 16
Trennungsaxiome, 183
Wallman–Lodato, 271
Wallman proximity, 15
Weak Descriptive Lodato, 130

B
Ball, 307
neighbourhood of x, 307
unit, 307
Basis, 404
definition, 404
digital image, 239
Bitopological space, 406
definition, 406
Bornology, ix, 49
nerve, ix, 50
Borsuk–Ulam Theorem, 143, 145, 170
S^n-Free s.p.c.BUT, 162
$S^n \mapsto \mathbb{R}^n$, 145
antipodal points, 143, 145
brain activity, 145
BUT, 162
digital, 145
direct proof, 145
Euclidean n-sphere, 143
hyperbolic surfaces, 170
quantum entanglement, 145
s.p.c.BUT, 162
Boundary, 156
set, 156
Boundedness, ix, 49, 51
bornology, ix
$\overset{\wedge\!\!\wedge}{\delta}$-boundedness nerve, 52
mesh nerve, 49
proximal, ix
proximal nerves, ix
proximal structure, 51
strong proximal, ix, 52

C
Cell, 33, 158, 216
interior, 216
keypoint-based, 216
quality, 158
Voronoï, 379
Voronoï polygon, 33
Voronoï region, 33
Centroid, 6, 7
2D region, 6
3D region, 6
digital image, 7
weighted, 7
Chart, 241, 242
$\overset{\wedge\!\!\wedge}{\delta}_{\Phi}$, 246
smoothly compatible, 241
strongly near, 242
Voronoï, 245
Closed set, 66, 67
Delaunay triangle, 67
Voronoï region, 67
Closure, 15, 16, 66, 71, 157
cl$A$, 157
cl$_{\Phi}A$, 157
axioms, 15
descriptive, 16, 157
mapping, 15
spatial, 102
union property, 71
usual, 157
Colour space, 154, 240
HSB, 154
painter's palette, 240
RGB, 155, 240
Compact, 306
convex body, 307
picture, 306

Voronoï region, 307
Compact set, 306
definition, 306
Compactness, 307
compact set, 307
picture set, 307
Voronoï picture set, 307
Complete poset, 348
nerve nuclei, 348
Complex, 4, 164
nucleus, 4, 164
set of sets, 164
symbolic, 4
Complexes, 4
strongly near, 4
Computational proximity, vii, 1, 11
applications, viii
algorithmic approach, 1
boundedness, 3
classifying images, 4
closeness, 3
closeness of sets, 4
components, 1
computational geometry, 4
computer vision, 3
connectedness, 3
CP, vii
digital image, 42
framework, 1, 5
geometric representation, 3
geometry, 11
geometry of objects, 4
image patterns, 4
local proximity spaces, 3
mesh polygons, 4
nearness, 3
object recognition, 39
point-free geometry, 6
practical outcomes, 4
proximity spaces, 1
region-approach, viii
separation, 3
setting, 42
study of structures, 1
topology of digital images, 42
utility, 5
Computer vision, 3
object recognition, 3
Concrete object, 27
pixel, 27
social network node, 27
Conj, 302
Conjecture, 169
100 theorems, 272
Borsuk, 272, 280
Delaunay groupoid semigroup, 201
digital Borsuk, 272
Etd Descriptive Lodato, 130
feature-based deformable shapes, 169
high quality image, 86
Ltd Lodato, 129
mesh nerves, 41
non-conexity, 328
regions with strong resemblance, 169
Voronoï E^d connectedness, 302
Voronoï E^d $\overset{\wedge\wedge}{\delta}$-connectedness, 302
Weak Descriptive Lodato, 130
Connected, 292
$\mathbb{R}^2$ subset, 293
$\mathbb{R}^3$ subset, 294
$\overset{\wedge\wedge}{\delta}$-connected, 246
$\overset{\wedge\wedge}{\delta}$-connectedness, 295
$\overset{\wedge\wedge}{\delta}$-convexity structure, 44
convexity structure, 44
Delaunay connectedness, 298
descriptive strong, 292
DirichletL, 294
space, 293
strong, 292
subset, 293
topologist's sine, 294
Voronoï E^2 3-connectedness, 302
Voronoï connectedness, 298–300
Wallman proximity, 293
Willard's theorem, 295
Connected regions, 10
common boundary point, 10
overlap, 10
Connected space, 43
geometric, 43
Connectedness, ix, 43, 196, 197, 292, 326
$\overset{\wedge\wedge}{\delta}$-connected, 46, 196, 197, 246
$\overset{\wedge\wedge}{\delta}$-connected closures, 46
$\overset{\wedge\wedge}{\delta}$-connected interiors, 46
$\overset{\wedge\wedge,\Phi}{\text{conn}}$-connected, 197
$\overset{\wedge\wedge,\Phi}{\text{conn}}$-relation, 196
$\overset{\wedge\wedge,\Phi}{\text{conn}}$-space, 196
connected set, 43
connected space, 43
descriptive strong, 197, 292
in proximity space, 43
pattern, 326
portrait, 292

  portrait space, 292
  probe, 196
  strong, 46, 197, 292, 329
  strongly connected, 43
  strongly near, 43
  strong proximity, 246, 329
Connection, 171
  Ehresmann, 171
  Levi-Civita, 171
Connection relation, 10
  relation between regions, 10
Continuity, 120
  Borsuk–Ulam Theorem, 120
  homotopy theory, 120
  preservation, 120
Continuous, 238
Continuous map, 143, 146, 163, 167
  $\overset{\wedge\wedge}{\delta}$, 146
  2Re.s.p.c., 167
  Borsuk–Ulam Theorem, 143, 146
  homotopy, 146
  Re.s.p.c., 163
  region $\overset{\wedge\wedge}{\delta}$, 163
  region-to-region $\overset{\wedge\wedge}{\delta}$, 167
  s.p.c., 160–162
Convex body, 306, 307
  definition, 306
  descriptive Helly's theorem, 311
  Helly's theorem, 308
  improper, 307
  proper, 307
  translative kissing number, 306
  Voronoï region, 307
Convex geometry, 306
  introduction, 306
Convex groupoid, 306
  introduction, 306
Convex hull, 318
Convexity, 24, 307
  axiomatic, 24
  convex body, 307
  convex set, 307
  probe, 196
  region, 196
  strictly convex set, 307
  strong contact, 24
  Zelins'kyi-Kay-Womble, 24
Convexity space, 24, 26
  definition, 24
  proximal Zelins'kyi-Kay-Womble, 26
  Zelins'kyi-Kay-Womble, 24
Convexity structure, ix, 24
  $\overset{\wedge\wedge}{\delta}$-connected, 44
  connected, 44
  intersection, ix
  Zelins'kyi-Kay-Womble, 24
Convex set, 306
  Convexity property, 306
  definition, 306
  introduction, 306
  Strict Convexity property, 306
  strictly convex, 306
Convex surface, 306
  Archimedes, 306
  definitions, 306
Cover, 306
  covering, 306
  definition, 306
CP, ix, 6
  algebra, xi
  algorithms, xi
  applications, xi
  bornology, ix
  Borsuk-Ulam Theorem, ix
  boundedness, 3
  computational geometry, 11
  connectedness, ix, 3
  convexity structures, ix
  digital image manifolds, ix
  homotopy theory, ix
  local proximity spaces, ix, 3
  Mathematica® scripts, xi
  Matlab® scripts, xi
  mesh bornological nerves, ix
  mesh nerves, ix
  primitives, 6
  problem classification, xii
  proximal Boundedness, ix
  proximities, ix
  proximities, ix
  proximity space theory, ix
  topological sorts, xi
  topology, ix
CP components, 1
  algebra, 1
  axioms, 1
  basis, 1
  connectedness, 6
  finite topology, 1
  geometry, 1
  methods, 1
  point-free geometry, 6
  points, 6
  probes, 1
  properties, 1

proximities, 1
pseudometrics, 1
regions, 6
relator, 1
rules, 1
theorems, 1

D

Deformation retract, 149
disks, 150
star shape center, 149
Delaunay, 123, 198
descriptively near, 200
feature vector, 199, 203
groupoid, 198
proximity, 200
spatially near, 199
topological space, 200
triangulation, 123, 199, 203
Delaunay feature vector, 199
definition, 199
Delaunay mesh, 176
strong invisibility, 177
visibility, 176
Delaunay triangle, 326
empty interior, 326
Description, 10, 197
feature vector, 10, 197
picture point, 10
set-based, 197
Descriptive intersection, 15, 27
$A \underset{\Phi}{\cap} B$, 27
$\underset{\Phi}{\cap}$, 15
Descriptive nearness, 10, 27
$A \ \delta_{\Phi} \ B$, 27
analogy, 10
Descriptive remoteness, 27
$A \ \not\delta_{\Phi} \ B$, 27
Descriptive similarity distance, 104
definition, 104
Descriptive strong proximity, 17
$\overset{\wedge\!\!\!\wedge}{\delta}_{\Phi}$, 17
axiom, 17
Digital, 270
Borsuk conjecture, 272
fuzzy Lodato proximity space, 270
fuzzy Wallman–Lodato proximity, 274
Digital n-topological space, 406
definition, 406
Digital basis, 405
definition, 405
on a picture, 405
physical objects, 405
pixels, 405
Digital bitopological space, 406
definition, 406
Digital image, 2, 216, 272
2D, 272
connectedness, 2
geometric view, 42
picture points, 272
pixel, 68
segmentation, 216
subset in Euclidean space, 272
tessellation, 216
Digital open set
definition, 404
Digital proximal topology
definition, 407
Digital topology, 120, 404, 407, 408
definition, 404, 407
digital geometry, 120
on digital open set, 404
proximal, 407
Rosenfeld, 120
Digital tri-topological space, 406
definition, 406
Dirichlet, 202
tessellation, 202
Dirichlet tessellation, 4
tiling, 4
Disconnectedness, 43
disconnected set, 43
Distance, 99, 156, 195, 272
between Euclidean plane points, 272
between pixels, 101, 272
Euclidean, 99, 156
Hausdorff, 156
Manhattan, 101, 195
point-to-set, 156
similarity, 195
taxicab, 101

E

Edge, 165
boundary, 165
EdgeDetect, 165, 373
thinning, 165, 373
Euclid, 4
elements, 4
Euclidean plane, 272
2-tuples of real numbers, 272
definition, 272

gaps between picture points, 272
points, 272
Euclidean space, 99, 158, 248, 272
$\Phi(X)$, 158
3-space, 99
distance, 99
feature space, 158
locally Euclidean, 248
manifold, 248
norm, 99
n-space, 99, 248
object space, 158
plane, 99, 272
real line, 99

F

Feature, 6, 152
brightness, 154
colour, 162
corner pixel, 152
Cornerness, 152
gradient orientation, 162
hue, 154
object characteristic, 6
object location, 6
pixel gradient magnitude, 154
pixel gradient orientation, 154
point-based probe, 152
region-based probe, 152
saturation, 154
types, 6
Feature space, 142, 157, 168
$\mathcal{O} \mapsto \mathbb{R}^n$, 142
collection of descriptions, 168
descriptions, 150
Euclidean, 150
Euclidean space R^n, 150
feature vectors, 150, 157
greyscale, 157
normed linear, 150
point-based, 150
region-based, 150
regions, 142
Feature vector, 27, 152, 154, 199, 202, 203
$\Phi(x) \in \mathbb{R}^n$, 152
$x \in \mathbb{R}^n$, 152
$y \in \mathbb{R}^n$, 152
Delaunay, 199, 203
description, 27
in $\mathbb{R}^3$, 154
in $\mathbb{R}^n$, 27
point in $\mathbb{R}^n$, 152
point-based, 152, 154
region-based, 152
Voronoï, 202
Filter, 51
Fréchet, 98
functional calculus, 98
Function, 292, 294
DirichletL, 292
topologist's sine, 294
Fuzzy proximity, 268, 271
Lodato, 268
Lodato axioms, 268
Wallman–Lodato, 271
Fuzzy proximity space, 279
$\delta_{\varepsilon_\Phi, \upsilon_{\overset{\wedge\!\!\wedge}{\delta}_\Phi}}$, 281
descriptive Lodato, 279
descriptive strong, 281

G

Geometric structure, 348
features, 348
polygon, 348
Geometry, 6, 11, 87, 120, 128, 164, 170, 272, 306, 318
$\overset{\wedge\!\!\wedge}{\delta}_\Phi$ nearness, 128
$\overset{\wedge\!\!\wedge}{\delta}$ nearness, 128
$\overset{\wedge\!\!\wedge}{\not\delta}_\Phi$ farness, 128
adjacent polygons, 128
computational, 11, 120, 175, 195
convex, 306
Di Concilio-Gerla, 164
digital, 10, 120, 402
Euclidean, 6, 10
face, 318
hyperbolic, 170
image, 87
numeric integration, 170
plane, 272
point-free, 10, 164, 195
polyform, 314
polygon, 348
region-based, 10
solid triangle, 318
strongly connected, 329
Groupoid, 194, 198, 201, 306
behaviour, 198
convex, 306
Delaunay, 198
descriptive, 194
descriptive Delaunay, 200
descriptive neighbourhood, 194

descriptively near, 200
Hausdorff Space, 194
near groupoids, 195
partial, 194
proximal Delaunay, 198
Groupoid Voronoï-Hausdorff Convex, 201

H

Hadamard, 98
geodesics, 98
Ham Sandwich Theorem, 171
Implied by BUT, 171
Ulam, 171
Hausdorff, 98
distance, 98
set theory, 98
Hausdorfness, 190
descriptive, 190
separated, 190
Homeomorphic, 248
Homeomorphism, 248
continuous 1-1,onto mapping, 248
Homology theory, 4
Homotopic, 147
maps, 147
Homotopy, 4, 147, 164, 166, 167
$\mathbf{j} \mapsto$?, 166
deformation, 147, 166
deformation mapping, 164
equivalence, 148
homotopic maps, 148
maps, 147
Mathematica MapThread, 373
Mathematica widget, 166
Re.s.p.c., 167
region-based, 167
region-to-region, 167
shape, 147
shapes with holes, 147
theory, 164
type, 148
Homotopy theory, ix
homotopic maps, ix
shapes, ix
Hooke, 37, 38
needle, 37
needle mesh, 37
needle mesh nerve, 39
needle point tip, 38, 39
Hyperspace, 243
Hypersphere, 143, 144
S^2, 143
S^3, 143
S^4, 144
S^n, 143
$S^n, n \geq 4$, 144
Hypertopology, 243
Vietoris, 243
Voronoï regions, 244

I

Ideal point, 10
small region approximation, 10
Image, 7, 37, 87, 104, 120
background, 37
bumps, 41
cluster of nerves, 38
corners, 37, 41
dragonfly, 381
foreground, 37
geometric reconstruction, 87
geometry, 41, 87, 104, 120
Hooke needle corners, 37
Manitoba chipmonk, 347
matching problems, 104
mesh, 37, 38
microscope, 39–41
needle point tip, 40
nerve, 38
Nikon, 40
object, 37
object structure, 87
quality, 87
similarity, 104
tolerance relations, 104
Voronoï mesh, 38, 39, 41
weighted centroid, 7
Zeiss, 40, 41
Image analysis, 104
content-based retrieval, 104
geometry, 104
matching problems, 104
spatial nearness, 104
Indeterminate, 198
set-based probe, 198
Interior, 156
set, 156

L

Leader, 38
uniform topology, 38
Linearly independent, 318

M

Manifold, 170, 238, 239, 248, 250–255
  $\overset{\wedge\!\!\wedge}{\delta}_{\Phi}$, 246
  $\mathbb{R}^2$, 252
  $\mathbb{R}^3$, 253
  atlas, 241
  chart, 241
  colour image, 253
  concave Riemannian, 171
  convex, 170
  curves, 241
  digital image, 239
  greyscale image, 252
  hyperbolic, 170
  mean shift, 252
  properties, 238
  proximal, 255
  saliency, 253
  salient region, 253
  selfie Voronoï, 254
  smooth, 238
  smooth boundary, 248
  topological, 238
  Voronoï, 245
Manifold strong proximity, 242
  $\overset{\wedge\!\!\wedge}{\delta}_{\mathcal{A}}$, 242
  axioms, 242
  definition, 242
Map, 146
  continuous, 146
  function, 146
  homotopic, 147
  identity, 148
  mapping, 146
  Re.s.p.c., 163
  s.p.c., 160
  transition, 241
Mapping, 143, 171, 238, 248, 250
  $\overset{\wedge\!\!\wedge}{\delta}$-continuous, 160
  1-1, 248
  bijection, 248
  bijective, 238
  continuous, 238, 250
  deformation, 149
  homeomophic, 250
  homeomorphic, 238, 248
  homeomorphic mapping, 250
  homeomorphism, 250
  homotopic, 143
  logarithmic, 171
  mean shift, 251
  Object-to-Feature, 153
  onto, 248
  open, 160
  proximally continuous, 250
  Re.s.p.c., 163
  retraction, 143
  Riemannian exponential, 171
  s.p.c., 160, 161
Matching, 103
  coarse, 103
  dense, 103
  feature, 103
  point, 103
Mathematica, 249, 253
  ColorNegate, 287
  Dilation, 287
  FillingTransform, 287
  For, 386, 387
  GradientFilter, 287
  HighlightImage, 287
  homotopy widget, 164
  Image, 287
  ImageKeypoints, 388
  ImageSaliencyFilter, 253
  MaxIterations, 249
  MeanShiftFilter, 249, 252
  MorphologicalBinarize, 282
  ParametricPlot3D, 386, 387
  Threshold, 287
  WatershedComponents, 287
Mathematical morphology, 281, 282
  cachement area, 287
  dilation, 281, 282
  MM, 282
  reflection, 282
  translation, 282
  watershed components, 287
  watershed segment, 287
  watershed segmentation, 287
Mathematics, 4
  closeness of lines, 4
  cover, 4
  equilateral triangles, 4
  interest in nearness, 4
  parallel lines, 4
Matlab
  neighourhood GUI, 363
Membership function, 268
  characteristic, 268
  fuzzy Lodato–Smirnov proximity, 268
  Zadeh, 268
Mesh, ix, 32, 120, 123, 254, 331
  bornological nerve, ix
  closed, 32

3D, viii
Delaunay, 123
edge-keypoint, 350
generating point, 32, 37, 38
generators, 120
geometry, 39
nerves construction, 35
nervous system, 35
plane tiling, 331
proximal Delaunay, 123
torus surface, 32
Voronoï, 350
Voronoï mesh, 38
Voronoï, 254
Metric, 98, 156, 409
distance function, 98
Euclidean, 156
Hausdorff, 156
norm, 156
pseudometrics, 98
topology, 409
Metric space, 98
Metric space conditions
identity of indiscernibles, 98
non-negativity, 98
symmetry, 98
triangle inequality, 98
Micrographia, 3
little pictures, 3
MM, 259, 282
mathematical morphology, 259
See mathematical morphology, 282
Möbius diagram, 34
distance, 34
extendable, 34
region, 34
MScript, 356
annulus i mesh, 356
centroid Voronoï mesh, 363
centroid Voronoï mesh on image, 363
corner Delaunay mesh, 356
cornerVoronoï mesh, 329, 356
DirichletL function, 295, 386
DirichletL set, 386
edgeCorner Voronoï, 381
edgeKeyPoints object detection, 381
edgeKeyPoints Voronoï, 381
highestAdjacency Voronoï, 370
image manifold, 382
MMwatershed Segments, 383
See-thru MM segments, 384
topologist's sine curve, 294, 384

N
Near, 102, 103, 199, 202, 203
Delaunay feature vectors, 199, 203
descriptively, 103
descriptively near, 103
matching, 103
spatially, 102, 190
Voronoï feature vectors, 202
Near set-based topology, 408
Near sets, viii, 102
point-based, viii
descriptive, 103
distance, 102
region-based, viii
spatial, 102
Nearness, 2, 12, 64, 102, 103
sn, 16
connected sets, 329
connection relation, 102
descriptive, 103
descriptive nearness, 103
digital images, 2
groupoids, 4
mathematics, 4
musical sounds, 2
nextness, 12
overlapping nextness, 12
Riesz, 64
space, 103
spatial, 102
strong, 16
tolerance, 103
Nearness of sets
connection relation, 102
descriptively near, 102
spatially near, 102
Needle, 37, 39
Hooke, 37
Singer, 39
Neighbourhood, 178
nhbd, 245
bounded descriptive, 184
bounded descriptive neighbourhood, 188
descriptive, 178
descriptive open, 184
disjoint, 178
Nerve, ix, 30, 50, 317, 345
$\overset{\wedge\!\wedge}{nv}$, 345
adjacent nerve, ix
adjacent polygons, 35
bornological, 50
bornological nerve, ix
closure, 31

computational proximity, 43
connectedness, 43
Edelsbrunner-Harer, 345
extension, 31
family, 31
geometric, 317
Lodato, 345
mesh, 32
mesh nerves, ix
mesh nucleus, 347
nucleus, 344, 345
nucleus property, 346
pattern, 345, 348
poly2sim, 318
polydodecahedron, 322
polyform, 318
polyheptagon, 322
polyhex, 322
polypentagon, 322
proximal bornological, ix, 50
satellite, 350
strengthening, 43
strongly connected, ix
torus, 31
Voronoï, 345, 346
Voronoï nucleus, 347
Voronoï nerve pattern, 350
Norm, 156
Euclidean, 156
Nuclei, 350
poset, 350
Nucleus, 4, 164, 348
seed, 348
Whiteheadean, 164

O

Object, 6, 103, 120, 152, 156, 157, 335
abstract, 103
antipodal, 145
brain signal, 156
characteristic, 6
concrete, 103
dominant, 337
edges, 120
location, 6, 152
physical, 103
picture element, 156
picture point, 6, 103
pixel, 156, 157
signature, 333
signature axioms, 335
skeletonization, 120
social network node, 156
voxel, 156
Object recognition, 3, 145
antipodal regions, 145
geometric models, 3
separating objects, 145
shape matching, 3
Object space, 142, 150, 156, 157, 168
Axiom, 151
brain activity, 156
collection of regions, 168
digital image, 156
Euclidean topological, 156
frame of reference, 156
hypersphere, 156
picture, 157
picture elements, 157
planar, 142
planar shapes, 150
points, 150
proximal, 156
regions, 150
social network, 156
strong proximal, 156
Open set, 42, 401
definition, 401
digital, 402, 404
sufficiently near, 401, 404
Overlap, 10, 13, 194
theorem, 194
tightly twisted, 13

P

Painting, 68
spot, 68
Parallel transport, 170
2nd-order differential equation, 170
conjugate gradient-descent, 171
Ehresmann connection, 171
first-order approximation, 170
Laplace-Beltrami operator, 171
Levi-Civita connection, 171
options, 171
Pattern, 38, 326, 329, 332
arrangement of parts, 326
axioms, 334
chart nearness, 326
cluster of nerves, 38
connectedness, 326, 329
geometric, 326
mesh, 331
mesh nerve, 326, 347

nearness of clusters, 38
nerve, 345, 348
nucleus nerve, 344
picture, 331
polygon, 335
proximal, 326
proximal connectedness, 329
proximal ngon, 326
shape, 333
spatial, 326
Physical object, 67
boundary, 67
perception, 67
Picture, 68, 331, 337
digital image, 331
digital picture, 331
dominant object, 337
pixel, 68
scanned photograph, 331
Picture point, 68
adjacent, 68
closure, 68
descriptively remote, 68
spatially remote, 68
Pixel, 4, 157
cellular pixel, 4
greyscale, 157
Plane figure, 11
closed set, 11
Plane tiling, 11
Plato
allegory of the cave, 3
Forms, 3
mathematical objects, 3
Republic, 3
Platonism
abstract mathematical objects, 3
Philosophy of Mathematics, 3
Point, 6, 27, 99, 143, 152, 178, 409
abstract, 6
abstract picture, 408
antipodal, 143, 144, 371
concrete, 6
descriptively distinct, 184
distinct, 178
Eckhardt-Latechi digital point, 408
Euclidean picture, 408
feature space, 152
feature vector, 150, 152
ideal, 6
in $\mathbb{R}^n$, 27
non-abstract, 409
physical, 408
picture, 408
region, 408
spatially distinct, 184
vector, 99
Point process, 34
axioms, 34
Poisson, 34
Point-free geometry, 10, 145
antipodal regions, 145
connection relation, 10
primitives, 10
region, 10
Poly2sim nerve, 318
quadrilateral-shaped, 319
Polyform, 314
polyabolo, 314
polycube, 314
polydiamond, 314
polyheptagon, 322
polyhex, 314, 322
polyiamond, 314
polymino, 314
polypentagon, 322
polyrhomb, 314
polystick, 314
Polygon, 326, 333
ngon, 337
7gon, 333
non-regular, 326
regular, 326
Poset, 348
complete, 348, 350
nuclei, 348
Primitives, 6
points, 6
regions, 6
Probe, 10, 27, 103, 152, 196, 203, 409
$\overset{\Join,\Phi}{conn}$, 196
choice, 10
colour brightness, 10
connectedness, 196
gradient orientation, 10
point-based, 103, 152, 203
real-valued function, 27
region, 195
region-based, 103, 152, 197
region-part, 10
set-based, 10, 195, 196, 203
Probe function, 10, 348
values, 10
Problem
matching, 103
Proximal, 26

Zelins'kyi-Kay-Womble convexity, 26
Proximal convexity space, 26
axioms, 26
definition, 26
strong contact, 26
Proximal digital topology
definition, 407
Proximal topology of digital images, 407
definition, 407
Proximity, ix, 10, 12–20, 22, 63, 242, 256, 268, 409
δ Wallman, 161
δ_{Φ} Wallman, 16
$\overset{\wedge\!\!\wedge}{\delta}_{\Phi}$ strong descriptive, 161
Čech, 14
Čech, 64, 65
sn, 16
applications, 64
closeness of sets, 14
common, 14
descriptive, ix, 13–15, 27
descriptive EF, 409
descriptive Efremovich, 10
descriptive Lodato, 16, 64, 73, 128, 130
descriptive Lodato axioms, 16
descriptive L-proximity, 181
descriptive strong Lodato, 17
descriptively near, 10
Efremovič, ix, 14, 15
Efremovič, 64
extension, 23
far, 65
fuzzy Lodato–Smirnov, 268
inherent geometry, vii
L-proximity, 179
landscape, 63
Lodato, ix, 14, 15, 72, 128, 129, 179, 244
Ltd Lodato, 129
manifold strong, 242
nearness relation, 14
nextness, 12
remote, 65
separated, 65
space, vii, 14, 42
spatial, 13
strong, 16, 128
strong contact, 16
strong descriptive, 128, 161
strong descriptive Lodato, 64
strong descriptive remoteness, 128
strong form, 13
strong Lodato proximity, 19
strong proximity, 16, 20, 28
strongly connected, 329
strongly contacted, 19
strongly far, 22, 23
strongly near, 19, 256, 329
until 1970, 63
Wallman, ix, 14, 15, 18, 44, 71, 161, 256
Wallman descriptive, 18
Weak Descriptive Lodato, 130
Proximity measure, 267
descriptive Smirnov, 274
set inclusion, 267
Smirnov, 267
spatial Smirnov, 268
strong Smirnov, 274
Proximity space, vii, 42, 55, 279
annulus local, 55
close, 66
descriptive Lodato, 279
finite, 65
fuzzy Lodato, 268
fuzzy Lodato axioms, 268
fuzzy Lodato–Smirnov, 268
local, ix, 55
near, 66
roots in antiquity, 3
Pullback, 9

Q

Quality, 87
fragmentation, 87
SSIM, 87

R

Region, 6, 10, 120, 152, 161, 164, 168, 170, 195, 196
Re, 152
n-feature vector, 10
antipodal, 145, 161, 373
Borsuk–Ulam Theorem, 164
boundedness, 10
centroid, 152
characteristic feature, 6
complex, 164
concave shape, 170
concrete, 10
description, 168
diameter, 10
features, 10, 170
feature vector, 150, 152
geometric centroid, 6
ideal, 10
location, 152

Möbius, 34
nucleus, 164
overlap, 10
probe, 10
Re.s.p.c., 160, 163
simplest, 6
single set, 6
spatial, 10
surface shape, 6
Voronoï, 120
Region description, 168
resemblance, 168
strong resemblance, 168
Region feature, 142
area, 142
colour, 142
density, 142
Relator, 123
proximal, 123
Száz, 123
Remote, 190
descriptive, 192
descriptively, 190
spatial, 192
Retraction, 143
Ronse, 4
cellular representation, 4

S
Saliency map, 253
Gabor orientation detection, 253
lateral inhibition mechanism, 253
Scene, 120, 121
analysis, 120
mesh cover, 121
Salerno Poste Auto, 121
Seed nucleus, 348
ngon, 348
Seed point, 11
centroid, 11
corner, 11
critical, 11
edge, 11
generator, 11
key, 11
pixel intensity, 11
salient, 11
site, 11
types, 12
Segmentation
default, 379
ImageForestingComponents, 379
pixel clusters, 281
site-based, 379
Semigroup, 195
partial descriptive, 195
Separated sets, 65
descriptively different, 65
remote, 65
Separation, 175, 178, 190
axioms, 178
descriptive, 190
disjoint sets, 175
natural, 190
neighbourhoods, 183
spatial, 190, 192
strong descriptive, 191
strong spatial, 191
Set, 10, 156, 306
boundary, 156
closed, 66
closure, 156
convex, 306
descriptively near sets, 181
disjoint, 175
distant, 65
empty, 66
far, 65
interior, 156
power, 66
regular closed, 67
regular open, 67
remote, 65
separated, 65
spatial region, 10
spatially near sets, 179
spatially visible, 176
strictly convex, 306
strong inclusion, 23
structured, 150
Set theory, 397
axioms, 397
Choice Axiom, 398
Extensionality Axiom, 397
Infinity Axiom, 398
Null Set Axiom, 397
Power Set Axiom, 397
proximal sets symbols, 399
Sum Set or Union Axiom, 397
symbols, 398
Unordered Pairs Axiom, 397
Zermelo-Fraenkel, 397
Shape, 13, 86, 120, 147, 148
analysis, 120
approximation, 86

  - Borsuk, 147
  - class, 148
  - deformation, 86
  - descriptively near, 150
  - equivalent, 148
  - geometry, 147
  - homotopy type, 148
  - planar, 150
  - planer sets, 147
  - skeletonization, 120
  - tightly twisted, 13
  - with holes, 147
  - without holes, 148
- Signature, 333
  - clusters of sets, 334
  - object, 335
  - separated objects, 334
  - shape pattern, 333
  - strong, 334
- Similarity, 87, 103, 104
  - distance, 104
  - SSIM, 87
  - structural, 87
- Simplex, 4
  - n-simplex, 4
  - empty, 4
- Simplical complex, 318
  - 2-simplex, 318
- Sites, 32, 33, 84, 86, 119, 120, 337
  - $\barwedge$Centroids, 86
  - ⌐Centroids, 86
  - ⚽, 86
  - centroid, 120
  - choice, 84
  - connected, 86
  - connected-sites shape, 86
  - corners, 337
  - generating point, 32, 35
  - keypoint, 119, 120, 337, 371
  - non-random, 328
  - optimal number, 337
  - path, 86
  - pathwise-connected, 86
  - Voronoï region, 33
- Sort, 275, 278
  - descriptively near, 278
  - spatially near, 275
- Space, 32, 144, 178, 268
  - T_1^{Φ}, 186
  - n-sphere, 144
  - descriptive separation, 178
  - descriptively symmetric, 183
  - feature, 144
  - fuzzy Lodato–Smirnov proximity, 268
  - Hausdorff, 178
  - Hausdorffness, 190
  - Naimpally Family T_2^{Φ}, 212
  - normed linear, 32
  - object, 144
  - symmetric, 179
  - Tychonoff, 183
- Strictly convex, 306
  - definition, 306
- Strong, 12
  - Tightly twisted, 12
- Strong contact, 16, 19
  - $\overset{\wedge\wedge}{\delta}$, 16, 19
  - axiom, 16
  - contactedness, 20
  - convexity, 24
  - definition, 19
  - overlap, 20
  - relations, 19
- Strong contactedness, 20
  - strong closeness, 20
- Strong proximity, 12, 16, 19
  - $\overset{\wedge\wedge}{\delta}_{\Phi}$, 17
  - $\overset{\wedge\wedge}{\delta}$, 16
  - axiom, 16, 19
  - definition, 19
  - name, 20
  - relations, 19
  - strong contact, 16
  - strong contactedness, 20
  - strongly near, 16, 19
  - tightly twisted overlapping nextness, 12
- Strongly connected, 329
- Strongly far, 22
  - $\underset{\vee\vee}{\not\delta}$, 22
- Strongly near, 19, 329
  - $\overset{\wedge\wedge}{\delta}$, 19
  - strong contact, 19
- Structure
  - descriptive L-proximity space, 181
  - L-proximity space, 179
  - motif, 210
  - separation space, 206
  - set pattern, 206
  - set pattern generator, 211
  - spatially near sets, 102
- subject, 387
- Symbols, 397

T

Tessellation, ix, 11, 86, 88, 119, 216
- cell, 216
- Delaunay, 11
- digital image, ix
- Dirichlet, 11, 34, 119, 371
- keypoint-based, 216
- Möbius, 34
- overlap, 88
- quality, 86
- Voronoï, 119, 371

Theorem, 250, 254, 256
- S^n-Free s.p.c.BUT, 162
- T_1^{Φ} segmented image, 186
- T_1^{Φ} space, 186
- $\overset{\wedge\wedge}{\delta}$ Voronoï regions, 243
- $\overset{\wedge\wedge}{\delta}$ proximity, 20
- $\overset{\wedge\wedge}{\delta}$-connected, 46
- $\overset{\wedge\wedge}{\delta}$-connectedness, 295
- $\overset{\wedge\wedge}{\delta}_{\Phi}$-Helly, 312
- about pat(ngon$A$), 336
- Borsuk–Ulam, 142–144
- closure nerve, 31, 36
- Delaunay triangles, 200
- descriptive Delaunay, 200
- descriptive Edelsbrunner-Harer-Helly, 312
- descriptive Gruber-Helly, 311
- descriptively disjoint, 192
- Edelsbrunner-Harer-Helly, 308
- fuzzy descriptive Lodato–Smirnov proximity, 279
- fuzzy Lodato–Smirnov proximity, 269
- Gruber-Helly, 308
- Helly dNerve, 313
- image manifold, 254, 256
- invisibility, 192
- manifold strong proximity, 242
- Manifold Wallman proximity, 256
- mesh nerve, 36
- nerve family, 31
- open mapping, 160
- overlapping, 194
- polydiamond, 314
- polydiamond nerve, 315
- polyform, 314
- proper convex body, 307
- Proximal Borsuk–Ulam, 160
- proximal Helly, 309
- proximity, 242
- s.p.c.BUT, 162
- selfie manifold, 250
- separated objects, 336
- strongly near, 256
- strongly near charts, 242
- strongly near Helly, 310
- Voronoï nerve connectedness, 350
- Voronoï nerve satellite, 350
- Voronoï E^2 connectedness, 301
- Voronoï E^d connectedness, 302
- Voronoï strong connectedness, 302
- Voronoï diagram, 307
- Voronoï region, 35, 307
- Willard connected, 46
- Willard connectedness, 295

Tightly twisted, 12
- overlapping nextness, 12

Tiling, 216, 314, 326
- plane, 216, 326, 345
- polyiamond, 314

Tolerance, 103
- space, 103

Topological manifold, 238
- Milnor, 238

Topological space, vii, 42, 238, 250, 402, 404
- R_0 (symmetric), 179
- R_0^{Φ}, 183
- T_0 (Kolmogorov), 184
- T_0^{Φ}, 184
- T_0^{Φ}(descriptive), 184
- T_1(spatial), 185
- T_1^{Φ}, 186
- T_1^{Φ}(descriptive), 186
- T_2, 178, 187
- T_2 (Hausdorff), 187
- T_2^{Φ} (descriptive Hausdorff), 187
- T_2^{Φ}, 187
- T_3 (regular), 204
- T_3^{Φ} (descriptively regular), 204
- T_4 (normal), 204
- T_4^{Φ} (normal), 205
- base, 250
- basis, 238
- descriptive Hausdorff, 187
- descriptively symmetric, 183
- digital, 404
- Hausdorff, 178, 187, 238, 250
- manifold, 238
- properties, 42
- second countable, 238, 250

Topology, ix, 42, 187, 238, 250, 402, 409
- base, 238
- basis, 238, 250
- definition, 402

digital, 402, 407
digital image, 42, 65, 239
history, 403
Leader uniform, ix
load of gravel, 187
metric, 65, 409
near set-based, 408
of digital images, 164, 409
on a digital image, 407
on open set, 402
open set, 402
proximal, 409
Topology of digital images, 402, 408
definition, 407
geometry-based, 408
open set, 402
proximity-based, 408
Torus, 18
chain of near sets, 18
closed mesh, 32
connected set, 45
descriptive closeness, 18
strongly near, 23
surface mesh, 18
surface polygons, 18
Tri-topological space, 406
definition, 406
Triangulation, 119, 123
Delaunay, 119, 123

V

Visibility, 175, 190
art gallery theorems, 175
computational geometry, 175
Delaunay, 176
descriptive, 193
spatial, 176
strongly near, 176
Vision, 119
visual field, 119
visual scene, 119
Vononoï
mesh, 37
Voronoï, 32, 38, 120, 160, 201, 350
archway mesh, 325
edge, 33
feature vector, 202
groupoid, 201
mesh, 38, 39, 120, 350
polygon, 33
region, 32, 160
region of a site, 33
tessellation, 33, 202
vertex, 33
Voronoï diagram, 33, 34, 327
application, 34
computational, 33
curved, 34, 328
infinite, 34
origins, 33
Poisson, 34
sites, 328
social network, 34
surface 3D, 34
Voronoï feature vector, 203
definition, 203
Voronoï groupoid, 201
definition, 201
Voronoï region, 243, 307
proper convex body, 307
strongly near, 243

Printed in the United States
By Bookmasters